VOLUME FOUR HUNDRED AND FIFTY-SEVEN

METHODS IN
ENZYMOLOGY

Mitochondrial Function, Part B: Mitochondrial Protein Kinases, Protein Phosphatases and Mitochondrial Diseases

METHODS IN ENZYMOLOGY

Editors-in-Chief

JOHN N. ABELSON AND MELVIN I. SIMON

Division of Biology
California Institute of Technology
Pasadena, California, USA

Founding Editors

SIDNEY P. COLOWICK AND NATHAN O. KAPLAN

VOLUME FOUR HUNDRED AND FIFTY-SEVEN

Methods in
ENZYMOLOGY

Mitochondrial Function, Part B: Mitochondrial Protein Kinases, Protein Phosphatases and Mitochondrial Diseases

EDITED BY

WILLIAM S. ALLISON
Department of Chemistry and Biochemistry
University of California, San Diego
La Jolla, CA, USA

ANNE N. MURPHY
Department of Pharmacology, School of Medicine
University of California, San Diego
La Jolla, CA, USA

AMSTERDAM • BOSTON • HEIDELBERG • LONDON
NEW YORK • OXFORD • PARIS • SAN DIEGO
SAN FRANCISCO • SINGAPORE • SYDNEY • TOKYO
Academic Press is an imprint of Elsevier

Academic Press is an imprint of Elsevier
525 B Street, Suite 1900, San Diego, CA 92101-4495, USA
30 Corporate Drive, Suite 400, Burlington, MA 01803, USA
32 Jamestown Road, London NW1 7BY, UK

First edition 2009

For information on all Academic Press publications
visit our website at elsevierdirect.com

ISBN: 978-0-12-374622-1
ISSN: 0076-6879

Printed and bound in United States of America
09 10 11 12   10 9 8 7 6 5 4 3 2 1

CONTENTS

Contributors   xv
Preface   xxv
Volumes in Series   xxvii

**Section I. Defining the Mitochondrial Proteome, Including
Phosphorylated, Acetylated, and Palmitoylated Proteins   1**

1. The Mitochondrial Proteome Database: MitoP2   3

M. Elstner, C. Andreoli, T. Klopstock, T. Meitinger, and H. Prokisch

1. Introduction   4
2. Yeast, Human, and Mouse Proteome Information   6
3. Mitochondrial Reference Set, Candidates, and SVM Prediction Score   8
4. Integrated Genome-Wide Approaches   9
5. Other Search Options: Restrictions to Functional Categories,
    Gene Locus and Disease-Causing Proteins   12
6. Output Lists and Practical Examples   14
    References   17

**2. Predicting Proteomes of Mitochondria and Related Organelles
from Genomic and Expressed Sequence Tag Data   21**

Daniel Gaston, Anastasios D. Tsaousis, and Andrew J. Roger

1. Introduction   22
2. Data Preparation   28
3. Screening Large Datasets for Mitochondrial Proteins   29
4. Detailed Analysis   33
5. Comparison of Methods   41
    Acknowledgments   44
    References   44

**3. Proteome Characterization of Mouse Brain Mitochondria
Using Electrospray Ionization Tandem Mass Spectrometry   49**

X. Fang and Cheng S. Lee

1. Introduction   50

2. Analysis of Mitochondrial Proteome by Two-Dimensional PAGE
    and LC–MS Techniques   51
3. Selective Proteome Enrichment by Capillary Isotachophoresis
    (CITP)-Based Separations   51
4. Analysis of Mouse Brain Mitochondria Using the CITP-Based
    Proteome Platform   52
5. Proteome Characterization of Mouse Brain Mitochondria   55
References   60

4. ^{32}P Labeling of Protein Phosphorylation and Metabolite Association in the Mitochondria Matrix   63

Angel M. Aponte, Darci Phillips, Robert A. Harris, Ksenia Blinova,
Stephanie French, D. Thor Johnson, and Robert. S. Balaban

1. Introduction   64
2. Methods   65
3. Results and Discussion   72
4. Summary   78
References   79

5. Selective Enrichment in Phosphopeptides for the Identification of Phosphorylated Mitochondrial Proteins   81

Gabriella Pocsfalvi

1. Introduction   82
2. Off-Line Phosphopeptide Enrichment Methods   84
3. On-Line 2D-LC Phosphopeptide Enrichment Methods   90
4. Mass Spectrometry-Based Methods   92
5. Quantitative Phosphoproteomics   94
References   94

6. Post-translational Modifications of Mitochondrial Outer Membrane Proteins   97

Anne M. Distler, Janos Kerner, Kwangwon Lee, and Charles L. Hoppel

1. Introduction   98
2. Isolation of Mitochondria and Mitochondrial Membranes   99
3. Detection of Post-translational Modifications Using One-dimensional
    Gel Electrophoresis   101
4. Focused Proteomics Using Immunological Approaches   105
5. Shotgun Methods for Identification of Post-translational Modifications 109
6. Conclusion   111
Acknowledgments   111
References   112

7. Analysis of Tyrosine-Phosphorylated Proteins in Rat Brain Mitochondria   117

Urs Lewandrowski, Elena Tibaldi, Luca Cesaro, Anna M. Brunati, Antonio Toninello, Albert Sickmann, and Mauro Salvi

1. Introduction	118
2. Identification of Src Tyrosine Kinase Substrates in Rat Brain Mitochondria	119
3. Mass Spectrometric Analysis of Tyrosine-Phosphorylated Peptides in Mitochondria	125
4. Bioinformatic Tools to Analyze the Potential Role of Tyrosine Phosphorylation	131
Acknowledgment	134
References	134

8. Acetylation of Mitochondrial Proteins   137

Matthew D. Hirschey, Tadahiro Shimazu, Jing-Yi Huang, and Eric Verdin

1. Introduction	138
2. Purification of Enzymatically Active SIRT3	138
3. SIRT3 Enzymatic Deacetylation Assay	141
4. Detection of Acetylated Proteins in Mitochondria	144
5. Conclusion	146
References	147

9. Non-radioactive Detection of Palmitoylated Mitochondrial Proteins Using an Azido-Palmitate Analogue   149

Morris A. Kostiuk, Bernd O. Keller, and Luc G. Berthiaume

1. Introduction	150
2. Synthesis of 12-Azidododecanoate and 14-Azidotetradecanoate	153
3. Synthesis of Tagged Phosphine Probes	155
4. Preparation of Azido-Fatty Acids and Tagged Phosphine Stock Solutions	155
5. Production of 14-Azidotetradecanoyl-CoA Derivatives	156
6. *In Vitro* Labeling of Proteins with 14-Azidotetradecanoyl-CoA and Tagged Phosphines	156
7. Isolation of Intact Mitochondria from Livers of Sprague–Dawley Rats	157
8. Chromatography of Soluble Mitochondrial Proteins	158
9. Labeling of Anion Exchange Eluates with 14-Azidotetradecanoyl-CoA and Phosphine–Biotin	158
10. Identification of Labeled Proteins by Mass Spectrometry	159

11. Demonstration that Acylation of Cysteine Residues Occurs via a
    Thioester Bond   160
12. Concluding Remarks   162
References   163

Section II. Characterization of Mitochondrial Kinases, Phosphatases, Dynamics, and Respiratory Function   167

10. Detection of a Mitochondrial Kinase Complex that Mediates PKA–MEK–ERK-Dependent Phosphorylation of Mitochondrial Proteins Involved in the Regulation of Steroid Biosynthesis   169

Cristina Paz, Cecilia Poderoso, Paula Maloberti, Fabiana Cornejo Maciel,
Carlos Mendez, Juan J. Poderoso, and Ernesto J. Podestá

1. Introduction   170
2. Steroidogenic Cells   171
3. Analysis of a Mitochondrial Kinase Complex   172
4. Analysis of a Mitochondrial Acyl-CoA Thioesterase, Acot2
   as a Phosphoprotein   186
References   190

11. Isolation of Regulatory-Competent, Phosphorylated Cytochrome c Oxidase   193

Icksoo Lee, Arthur R. Salomon, Kebing Yu, Lobelia Samavati, Petr Pecina,
Alena Pecinova, and Maik Hüttemann

1. Introduction   194
2. Purification of Mitochondria Maintaining Protein Phosphorylation   195
3. Isolation of Cytochrome c Oxidase from Mitochondria   199
4. Analysis of Cytochrome c Oxidase Phosphorylation   204
5. Concluding Remarks   208
References   209

12. Using Functional Genomics to Study PINK1 and Metabolic Physiology   211

Camilla Scheele, Ola Larsson, and James A. Timmons

1. Introduction   212
2. Studying Mitochondrial Modulation in Humans In Vivo   214
3. Cellular Mitochondrial Models   216
4. Considerations When Using Short Interfering RNAs   218
5. Ex Vivo Gene Expression Analysis   222
References   226

13. Functional Characterization of Phosphorylation Sites in Dynamin-Related Protein 1   231

J. Thomas Cribbs and Stefan Strack

1. Introduction   232
2. Replacement of Endogenous with Phosphorylation-Site Mutant Drp1   234
3. Analysis of Drp1 Phosphorylation   237
4. Production and Assays of Recombinant Drp1   240
5. Cell Death Assays   243
6. Mitochondrial Shape Analysis with ImageJ   245
7. Appendix: Morphometry Macro   250
References   251

14. Functional Characterization of a Mitochondrial Ser/Thr Protein Phosphatase in Cell Death Regulation   255

Gang Lu, Haipeng Sun, Paavo Korge, Carla M. Koehler,
James N. Weiss, and Yibin Wang

1. Introduction   256
2. Identification of Protein Phosphatases in Mitochondria   258
3. Characterization of Mitochondrial Localization of PP2Cm in
   Mammalian Cells   260
4. Functional Characterization of PP2Cm in Cultured Cells   264
5. Functional Characterization of PP2Cm in Cell Death and
   Mitochondrial Regulation   267
6. Mitochondrial Phosphatase in Cell Death Regulation   270
Acknowledgments   271
References   271

15. Distinguishing Mitochondrial Inner Membrane Orientation of Dual Specific Phosphatase 18 and 21   275

Matthew J. Rardin, Gregory S. Taylor, and Jack E. Dixon

1. Introduction   276
2. Analysis of Phosphatase Activity of DSP18   277
3. Isolation of Highly Purified Rat Kidney Mitochondria
   and Subfractionation   279
4. Mitochondrial Inner Membrane Association of DSP18 and DSP21   283
Acknowledgments   286
References   287

16. **Monitoring Mitochondrial Dynamics with Photoactivateable Green Fluorescent Protein**   **289**

Anthony J. A. Molina and Orian S. Shirihai

1. Mitochondrial Dynamics   290
2. PAGFPmt   292
3. Tracking Individual Fusion and Fission Events   293
4. Quantifying Networking Activity in Whole Cells: Whole
   Cell Mitochondrial Dynamics Assay   296
5. Potential Artifacts and Important Controls   301
6. Comparison with an Alternative Method   302
Acknowledgments   303
References   303

17. **Determination of Yeast Mitochondrial KHE Activity, Osmotic Swelling and Mitophagy**   **305**

Karin Nowikovsky, Rodney J. Devenish, Elisabeth Froschauer, and Rudolf J. Schweyen

1. Introduction   306
2. Disturbance of Mitochondrial K^+ Homeostasis by dox-Regulated
   Mdm38 Expression   307
3. Morphological Changes in Mitochondria in Mdm38-Depleted Cells:
   Confocal Microscopy and Electron Microscopy   309
4. Measurement of KHE Activity and Other
   Bioenergetic Parameters   310
5. Following Mitophagy Using a Novel Biosensor and
   Fluorescence Microscopy   313
6. Growth and Preparation of Cells for Fluorescence Microscopy
   Observations of Mitophagy   314
7. Fluorescence Microscopy   315
Acknowledgments   316
References   317

18. **Imaging Axonal Transport of Mitochondria**   **319**

Xinnan Wang and Thomas L. Schwarz

1. Introduction   320
2. Protocols to Image Axonal Transport of Mitochondria in
   Rat Hippocampal Neurons   324
3. Protocols to Image Axonal Transport of Mitochondria in
   Drosophila Larval Neurons   328
4. Conclusion   331
Acknowledgments   332
References   332

19. Generation of mtDNA-Exchanged Cybrids for Determination of the Effects of mtDNA Mutations on Tumor Phenotypes   335

Kaori Ishikawa and Jun-Ichi Hayashi

1. Introduction   336
2. Establishment of ρ^0 Cells of Mice   337
3. Preparation and Enucleation of mtDNA-Donor Cells   340
4. Cell Fusion between ρ^0 Cells and Cytoplasts   341
5. Concluding Remarks   343
References   344

Section III. Mitochondrial Function in Pancreatic Beta Cells and Insulin-Responsive Tissues   347

20. Functional Assessment of Isolated Mitochondria *In Vitro*   349

Ian R. Lanza and K. Sreekumaran Nair

1. Introduction   350
2. Mitochondrial Isolation Procedures   354
3. Mitochondrial ATP Production   360
4. Mitochondrial Respiration   365
5. Summary   369
References   369

21. Assessment of *In Vivo* Mitochondrial Metabolism by Magnetic Resonance Spectroscopy   373

Douglas E. Befroy, Kitt Falk Petersen, Douglas L. Rothman, and Gerald I. Shulman

1. Introduction   374
2. *In Vivo* Magnetic Resonance Spectroscopy   375
3. Insulin Resistance and Resting Mitochondrial Metabolism   386
4. Conclusions   389
References   389

22. Methods for Assessing and Modulating UCP2 Expression and Function   395

Lício A. Velloso, Giovanna R. Degasperi, Aníbal E. Vercesi, and Mário A. Saad

1. Introduction   396
2. Analysis of UCP2 Expression by Immunoblot   397
3. Analysis of UCP2 Expression by Real-Time PCR   399

4. Mitochondria Respiration Assay   400
5. Modulating UCP2 Expression by Cold Exposure   400
6. Inhibiting UCP2 Expression by Antisense Oligonucleotide   401
7. Inhibiting PGC-1α as an Indirect Means of Controlling
   UCP2 Expression   402
References   403

23. **Measuring Mitochondrial Bioenergetics in INS-1E
    Insulinoma Cells**   **405**

Charles Affourtit and Martin D. Brand

1. Introduction   406
2. Quantification of Cellular Bioenergetics—Theoretical Aspects   407
3. INS-1E Cells—A Valuable Pancreatic Beta Cell Model   411
4. Measurement of Coupling Efficiency in Trypsinized INS-1E Cells   416
5. Noninvasive Measurement of Cellular Bioenergetics   418
Acknowledgment   422
References   423

24. **Investigating the Roles of Mitochondrial and Cytosolic
    Malic Enzyme in Insulin Secretion**   **425**

Rebecca L. Pongratz, Richard G. Kibbey, and Gary W. Cline

1. Introduction   426
2. Malic Enzyme mRNA Expression in Rat Insulinoma INS-1 832/13 Cells,
   Rat Islets, and Mouse Islets   428
3. siRNA Knock-Down of ME1 and ME2 in INS-1 832/13 β-Cells   432
4. Enzymatic Assays to Determine Activity of Cytosolic and
   Mitochondrial Malic Enzymes   435
5. Calculating Relative Rates of Anaplerotic Pathways from ^{13}C-Glutamate
   Isotopomer Distribution   442
6. Discussion   448
References   449

25. **Insulin Secretion from β-Cells is Affected by Deletion
    of Nicotinamide Nucleotide Transhydrogenase**   **451**

Kenju Shimomura, Juris Galvanovskis, Michelle Goldsworthy, Alison Hugill,
Stephan Kaizak, Angela Lee, Nicholas Meadows, Mohamed Mohideen
Quwailid, Jan Rydström, Lydia Teboul, Fran Ashcroft, and Roger D. Cox

1. Introduction   452
2. Gene-Driven ENU Screens   453
3. Generating BAC Transgenics   457

4. Isolation of Islets of Langerhans from Mouse Pancreas   460
5. RNA Interference (RNAi) Methodologies on Insulinoma Cell Lines   464
6. Measuring Intracellular Calcium   467
7. Measuring Hydrogen Peroxide in Mitochondria   468
8. Imaging of Mitochondrial Membrane Potential Changes by
   Confocal Microscopy   471
9. Measuring NNT Activity   475
10. Conclusions   477
References   478

Author Index   481
Subject Index   509

CONTRIBUTORS

Charles Affourtit
MRC Dunn Human Nutrition Unit, Cambridge, United Kingdom

C. Andreoli
Institute of Human Genetics, Technical University Munich, Munich, Germany

Angel M. Aponte
Proteomics Core Facility, National Heart Lung and Blood Institute, DHHS, Bethesda, Maryland, USA

Fran Ashcroft
Henry Wellcome Center for Gene Function, Department of Physiology, Anatomy, and Genetics, University of Oxford, Oxford, United Kingdom

Robert. S. Balaban
Laboratory of Cardiac Energetics, National Heart Lung and Blood Institute, DHHS, Bethesda, Maryland, USA

Douglas E. Befroy
Departments of Diagnostic Radiology and Internal Medicine, Yale University School of Medicine, New Haven, Connecticut, USA

Luc G. Berthiaume
Department of Cell Biology, Faculty of Medicine and Dentistry, University of Alberta, Edmonton, Alberta, Canada

Ksenia Blinova
Laboratory of Cardiac Energetics, National Heart Lung and Blood Institute, DHHS, Bethesda, Maryland, USA

Martin D. Brand
MRC Dunn Human Nutrition Unit, Cambridge, United Kingdom, and Buck Institute for Age Research, Novato, California, USA

Anna M. Brunati
Department of Biological Chemistry, University of Padova, Padova, Italy

Luca Cesaro
Department of Biological Chemistry, University of Padova, Padova, Italy

Gary W. Cline
Department of Internal Medicine, Yale University School of Medicine, New Haven, Connecticut, USA

Roger D. Cox
MRC Harwell, Diabetes Group, Harwell Science and Innovation Campus, Oxfordshire, United Kingdom

J. Thomas Cribbs
Department of Pharmacology, University of Iowa Carver College of Medicine, Iowa City, Iowa, USA

Giovanna R. Degasperi
Departments of Internal Medicine and Clinical Pathology, University of Campinas, Campinas, SP, Brazil

Rodney J. Devenish
Department of Biochemistry and Molecular Biology, and ARC Centre of Excellence in Structural and Functional Microbial Genomics, Monash University, Clayton campus, Melbourne, Victoria, Australia

Anne M. Distler
Department of Pharmacology, and Center for Mitochondrial Disease, Case Western Reserve University, Cleveland, Ohio, USA

Jack E. Dixon
Departments of Pharmacology, Cellular and Molecular Medicine, and Chemistry and Biochemistry, and The Howard Hughes Medical Institute, University of California, San Diego, La Jolla, California, USA

M. Elstner
Institute of Human Genetics, Helmholtz Zentrum Munich—German Research Center for Environmental Health, Neuherberg, Germany, and Department of Neurology with Friedrich-Baur-Institute, Ludwig-Maximilians University, Munich, Germany

X. Fang
Department of Chemistry and Biochemistry, University of Maryland, College Park, Maryland, USA

Stephanie French
Proteomics Core Facility, National Heart Lung and Blood Institute, DHHS, Bethesda, Maryland, USA

Elisabeth Froschauer
Max F. Perutz Laboratories, Department of Genetics, University of Vienna, Campus Vienna Biocenter, Wien, Austria

Juris Galvanovskis
The Oxford Centre for Diabetes, Endocrinology, and Metabolism, Churchill Hospital, Oxford, United Kingdom

Daniel Gaston
Centre for Comparative Genomics and Evolutionary Bioinformatics, Department of Biochemistry and Molecular Biology, Dalhousie University, Halifax, Nova Scotia, Canada

Michelle Goldsworthy
MRC Harwell, Diabetes Group, Harwell Science and Innovation Campus, Oxfordshire, United Kingdom

Maik Hüttemann
Center for Molecular Medicine and Genetics, Wayne State University School of Medicine, Detroit, Michigan, USA

Robert A. Harris
Department of Biochemistry and Molecular Biology, Indiana University School of Medicine, Indianapolis, Indiana, USA

Jun-Ichi Hayashi
Graduate School of Life and Environmental Sciences, University of Tsukuba, Tennodai, Tsukuba, Ibaraki, Japan

Matthew D. Hirschey
Gladstone Institute of Virology and Immunology, University of California, San Francisco, California, USA

Charles L. Hoppel
Department of Pharmacology, Department of Medicine, and Center for Mitochondrial Disease, Case Western Reserve University, Cleveland, Ohio, USA

Jing-Yi Huang
Gladstone Institute of Virology and Immunology, University of California, San Francisco, California, USA

Alison Hugill
MRC Harwell, Diabetes Group, Harwell Science and Innovation Campus, Oxfordshire, United Kingdom

Kaori Ishikawa
Graduate School of Life and Environmental Sciences, University of Tsukuba, Tennodai, Tsukuba, Ibaraki, Japan, and Japan Society for the Promotion of Science (JSPS), Chiyoda-ku, Tokyo, Japan, and Pharmaceutical Research Division, Takeda Pharmaceutical Company Limited, Tsukuba, Ibaraki, Japan

D. Thor Johnson
Proteomics Core Facility, National Heart Lung and Blood Institute, DHHS, Bethesda, Maryland, USA, and Department of Biochemistry and Molecular Biology, Indiana University School of Medicine, Indianapolis, Indiana, USA

Stephan Kaizak
Henry Wellcome Center for Gene Function, Department of Physiology, Anatomy, and Genetics, University of Oxford, Oxford, United Kingdom

Bernd O. Keller
Department of Pathology and Laboratory Medicine, Child & Family Research Institute, University of British Columbia, Vancouver, British Columbia, Canada

Janos Kerner
Department of Pharmacology, and Center for Mitochondrial Disease, Case Western Reserve University, Cleveland, Ohio, USA

Richard G. Kibbey
Department of Internal Medicine, Yale University School of Medicine, New Haven, Connecticut, USA

T. Klopstock
Department of Neurology with Friedrich-Baur-Institute, Ludwig-Maximilians University, Munich, Germany

Carla M. Koehler
Department of Biological Chemistry, Division of Molecular Medicine, UCLA, Los Angeles, California, USA

Paavo Korge
Department of Medicine, Division of Molecular Medicine, UCLA, Los Angeles, California, USA

Morris A. Kostiuk
Department of Cell Biology, Faculty of Medicine and Dentistry, University of Alberta, Edmonton, Alberta, Canada

Ian R. Lanza
Division of Endocrinology, Endocrinology Research Unit, Mayo Clinic College of Medicine, Rochester, Minnesota, USA

Ola Larsson
Department of Biochemistry, McGill University, Montreal, Canada

Kwangwon Lee
Department of Pharmacology, and Center for Mitochondrial Disease, Case Western Reserve University, Cleveland, Ohio, USA

Angela Lee
MRC Harwell, Diabetes Group, Harwell Science and Innovation Campus, Oxfordshire, United Kingdom

Cheng S. Lee
Department of Chemistry and Biochemistry, University of Maryland, College Park, Maryland, USA

Icksoo Lee
Center for Molecular Medicine and Genetics, Wayne State University School of Medicine, Detroit, Michigan, USA

Urs Lewandrowski
ISAS–Institute for Analytical Sciences, Dortmund, Germany

Gang Lu
Department of Anesthesiology, Division of Molecular Medicine, UCLA, Los Angeles, California, USA

Fabiana Cornejo Maciel
IIMHNO, Department of Biochemistry, School of Medicine, University of Buenos Aires, Paraguay, Buenos Aires, Argentina

Paula Maloberti
IIMHNO, Department of Biochemistry, School of Medicine, University of Buenos Aires, Paraguay, Buenos Aires, Argentina

Nicholas Meadows
MRC Harwell, Diabetes Group, Harwell Science and Innovation Campus, Oxfordshire, United Kingdom

T. Meitinger
Institute of Human Genetics, Helmholtz Zentrum Munich—German Research Center for Environmental Health, Neuherberg, Germany, and Institute of Human Genetics, Technical University Munich, Munich, Germany

Carlos Mendez
IIMHNO, Department of Biochemistry, School of Medicine, University of Buenos Aires, Paraguay, Buenos Aires, Argentina

Anthony J. A. Molina
Department of Medicine, Boston University, Massachusetts, USA

K. Sreekumaran Nair
Division of Endocrinology, Endocrinology Research Unit, Mayo Clinic College of Medicine, Rochester, Minnesota, USA

Karin Nowikovsky
Max F. Perutz Laboratories, Department of Genetics, University of Vienna, Campus Vienna Biocenter, Wien, Austria

Cristina Paz
IIMHNO, Department of Biochemistry, School of Medicine, University of Buenos Aires, Paraguay, Buenos Aires, Argentina

Petr Pecina
Center for Molecular Medicine and Genetics, Wayne State University School of Medicine, Detroit, Michigan, USA

Alena Pecinova
Center for Molecular Medicine and Genetics, Wayne State University School of Medicine, Detroit, Michigan, USA

Kitt Falk Petersen
Department of Internal Medicine, Yale University School of Medicine, New Haven, Connecticut, USA

Darci Phillips
Laboratory of Cardiac Energetics, National Heart Lung and Blood Institute, DHHS, Bethesda, Maryland, USA

Gabriella Pocsfalvi
Istituto di Biochimica delle Proteine-Consiglio Nazionale delle Ricerche, Naples, Italy

Cecilia Poderoso
IIMHNO, Department of Biochemistry, School of Medicine, University of Buenos Aires, Paraguay, Buenos Aires, Argentina

Juan J. Poderoso
Laboratory of Oxygen Metabolism, University Hospital, University of Buenos Aires, Buenos Aires, Argentina

Ernesto J. Podestá
IIMHNO, Department of Biochemistry, School of Medicine, University of Buenos Aires, Paraguay, Buenos Aires, Argentina

Rebecca L. Pongratz
Department of Internal Medicine, Yale University School of Medicine, New Haven, Connecticut, USA

H. Prokisch
Institute of Human Genetics, Helmholtz Zentrum Munich—German Research Center for Environmental Health, Neuherberg, Germany, and Institute of Human Genetics, Technical University Munich, Munich, Germany

Mohamed Mohideen Quwailid
MRC Harwell, Diabetes Group, Harwell Science and Innovation Campus, Oxfordshire, United Kingdom

Matthew J. Rardin
Departments of Pharmacology, Cellular and Molecular Medicine, and Chemistry and Biochemistry, and Biomedical Sciences Graduate Program, University of California, San Diego, La Jolla, California, USA

Andrew J. Roger
Centre for Comparative Genomics and Evolutionary Bioinformatics, Department of Biochemistry and Molecular Biology, Dalhousie University, Halifax, Nova Scotia, Canada

Douglas L. Rothman
Departments of Diagnostic Radiology and Biomedical Engineering, Yale University School of Medicine, New Haven, Connecticut, USA

Jan Rydström
Biochemistry and Biophysics, Department of Chemistry, Lundberg Laboratory, Göteborg University, Göteborg, Sweden

Mário A. Saad
Departments of Internal Medicine and Clinical Pathology, University of Campinas, Campinas, SP, Brazil

Arthur R. Salomon
Department of Molecular Biology, Cell Biology, and Biochemistry, Brown University, Providence, Rhode Island, USA

Mauro Salvi
Department of Biological Chemistry, University of Padova, Padova, Italy

Lobelia Samavati
Department of Medicine, Division of Pulmonary/Critical Care and Sleep Medicine, Wayne State University School of Medicine, Detroit, Michigan, USA

Camilla Scheele
The Centre of Inflammation and Metabolism, Department of Infectious Diseases and CMRC, Rigshospitalet, The Faculty of Health Sciences, University of Copenhagen, Denmark

Thomas L. Schwarz
F. M. Kirby Neurobiology Center, Children's Hospital Boston, and Department of Neurobiology, Harvard Medical School, Boston, Massachusetts, USA

Rudolf J. Schweyen
Max F. Perutz Laboratories, Department of Genetics, University of Vienna, Campus Vienna Biocenter, Wien, Austria

Tadahiro Shimazu
Gladstone Institute of Virology and Immunology, University of California, San Francisco, California, USA

Kenju Shimomura
Henry Wellcome Center for Gene Function, Department of Physiology, Anatomy, and Genetics, University of Oxford, Oxford, United Kingdom

Orian S. Shirihai
Department of Medicine, Boston University, Massachusetts, USA

Gerald I. Shulman
Departments of Internal Medicine and Cellular and Molecular Physiology, Howard Hughes Medical Institute, Yale University School of Medicine, New Haven, Connecticut, USA

Albert Sickmann
ISAS–Institute for Analytical Sciences, Dortmund, Germany and Medizinisches Proteom-Center (MPC), Ruhr-Universitaet Bochum, Bochum, Germany

Stefan Strack
Department of Pharmacology, University of Iowa Carver College of Medicine, Iowa City, Iowa, USA

Haipeng Sun
Department of Anesthesiology, Division of Molecular Medicine, UCLA, Los Angeles, California, USA

Gregory S. Taylor
Department of Biochemistry and Molecular Biology, University of Nebraska, Nebraska Medical Center, Omaha, Nebraska, USA

Lydia Teboul
MRC Harwell, Diabetes Group, Harwell Science and Innovation Campus, Oxfordshire, United Kingdom

Elena Tibaldi
Department of Biological Chemistry, University of Padova, Padova, Italy

James A. Timmons
The Wenner-Gren Institute, Arrhenius Laboratories, Stockholm University, Stockholm, Sweden

Antonio Toninello
Department of Biological Chemistry, University of Padova, Padova, Italy

Anastasios D. Tsaousis
Centre for Comparative Genomics and Evolutionary Bioinformatics, Department of Biochemistry and Molecular Biology, Dalhousie University, Halifax, Nova Scotia, Canada

Lício A. Velloso
Departments of Internal Medicine and Clinical Pathology, University of Campinas, Campinas, SP, Brazil

Aníbal E. Vercesi
Departments of Internal Medicine and Clinical Pathology, University of Campinas, Campinas, SP, Brazil

Eric Verdin
Gladstone Institute of Virology and Immunology, University of California, San Francisco, California, USA

Xinnan Wang
F. M. Kirby Neurobiology Center, Children's Hospital Boston, and Department of Neurobiology, Harvard Medical School, Boston, Massachusetts, USA

Yibin Wang
Department of Anesthesiology, Division of Molecular Medicine, Department of Medicine, Division of Molecular Medicine, and David Geffen School of Medicine, Cardiovascular Research Laboratories, UCLA, Los Angeles, California, USA

James N. Weiss
Department of Medicine, Division of Molecular Medicine, and David Geffen School of Medicine, Cardiovascular Research Laboratories, UCLA, Los Angeles, California, USA

Kebing Yu
Department of Molecular Biology, Cell Biology, and Biochemistry, Brown University, Providence, Rhode Island, USA

PREFACE

The field of mitochondrial research broadened significantly in the 1990s with the discovery of the critical role that mitochondria play in cell death signaling and apoptosis. Participation in the field has continued to expand at a rapid pace as interest in mitochondrial research expands from other disciplines. A diverse array of topics has come to the forefront of mitochondrial research, including signaling via reversible phosphorylation, organellar trafficking, involvement in secretory events, and the pathogenesis of prevalent diseases including chronic forms of neurodegeneration and diabetes, each of which is addressed in this volume of *Methods in Enzymology*. Here we highlight some of the advances in methodology that have allowed further identification of components and posttranslational modifications of the mitochondrial proteome, characterization of the dynamic changes in structure and movement of the organelles, as well as assessment of the functional changes of mitochondria both in health and disease.

Specifically, advancements in bioinformatics and mass spectrometry that have provided more accurate prediction and identification of components of the mitochondrial proteome are described in Chapters 1–3. It is now recognized that the functions of certain proteins located in the inner membrane, the outer membrane, and the mitochondrial matrix are controlled by posttranslational modifications. Methods to detect protein phosphorylations are the focus of Chapters 4–7 and the detection of protein acylations are the focus of Chapters 8 and 9.

Experimental approaches arising from advances in proteomics include the characterization of functional changes of key mitochondrial enzymes induced by signaling proteins and posttranslational modifications. These approaches are addressed by techniques described in the second section of this volume. Methods for the study of mitochondrial kinases and phosphatases and their effects on mitochondrial function, morphology, and the control of cell death are the focus of Chapters 10–14. The important topic of determining submitochondrial localization of specific enzymes is addressed in Chapter 15, which is critical to identify relevant substrates for these signaling enzymes.

Other significant advances in mitochondrial research have risen from techniques that allow visualization of individual mitochondria within intact cells, revealing the highly dynamic nature of mitochondrial morphology, localization, and turnover. Imaging methods, including the use of photo-activateable GFP as well as other approaches, are described to monitor mitochondrial fusion and fission (Chapters 13 and 16), mitochondrial autophagy in yeast (Chapter 17), and mitochondrial movement in neuronal processes (Chapter 18). A method to establish a hybrid cell line containing exchanged mitochondrial-DNA is described in Chapter 19.

The third section of this volume describes methods developed to examine ATP production in human tissues and the participation of mitochondria in pancreatic insulin secretion and insulin responsiveness of various tissues. Chapter 20 describes a bioluminescence approach to measuring rates of ATP production in biopsies of human muscle and Chapter 21 describes the use of non-invasive magnetic resonance spectroscopy to examine modulation of ATP production in human muscle. Recent developments in mitochondrial research have revealed the importance of uncoupling proteins in both insulin secretion and insulin action. Methods to assess uncoupling protein 2 (UCP2) expression in pancreatic islets and the hypothalamus (Chapter 22) and the effects of UCP2 expression on respiratory function in monolayers of INS1 cells (Chapter 23). The importance of mitochondrial anaplerosis and the redox state of pyridine nucleotide coenzymes for insulin secretion have also been demonstrated recently. Consequently, Chapter 24 provides methods for the study of the mitochondrial isoforms of malic enzyme in insulin secretion, and Chapter 25 describes methods to examine the effects of deletion of the nicotinamide nucleotide transhydrogenase on insulin secretion.

The chapters in this volume of *Methods in Enzymology* describe useful approaches in mitochondrial research that can be used directly or modified to investigate emerging areas of mitochondrial research. We wish to thank all the contributors to this volume and hope that the methods described will be useful in facilitating future studies in this rapidly evolving and exciting field.

ANNE N. MURPHY AND WILLIAM S. ALLISON

METHODS IN ENZYMOLOGY

VOLUME I. Preparation and Assay of Enzymes
Edited by SIDNEY P. COLOWICK AND NATHAN O. KAPLAN

VOLUME II. Preparation and Assay of Enzymes
Edited by SIDNEY P. COLOWICK AND NATHAN O. KAPLAN

VOLUME III. Preparation and Assay of Substrates
Edited by SIDNEY P. COLOWICK AND NATHAN O. KAPLAN

VOLUME IV. Special Techniques for the Enzymologist
Edited by SIDNEY P. COLOWICK AND NATHAN O. KAPLAN

VOLUME V. Preparation and Assay of Enzymes
Edited by SIDNEY P. COLOWICK AND NATHAN O. KAPLAN

VOLUME VI. Preparation and Assay of Enzymes *(Continued)*
Preparation and Assay of Substrates
Special Techniques
Edited by SIDNEY P. COLOWICK AND NATHAN O. KAPLAN

VOLUME VII. Cumulative Subject Index
Edited by SIDNEY P. COLOWICK AND NATHAN O. KAPLAN

VOLUME VIII. Complex Carbohydrates
Edited by ELIZABETH F. NEUFELD AND VICTOR GINSBURG

VOLUME IX. Carbohydrate Metabolism
Edited by WILLIS A. WOOD

VOLUME X. Oxidation and Phosphorylation
Edited by RONALD W. ESTABROOK AND MAYNARD E. PULLMAN

VOLUME XI. Enzyme Structure
Edited by C. H. W. HIRS

VOLUME XII. Nucleic Acids (Parts A and B)
Edited by LAWRENCE GROSSMAN AND KIVIE MOLDAVE

VOLUME XIII. Citric Acid Cycle
Edited by J. M. LOWENSTEIN

VOLUME XIV. Lipids
Edited by J. M. LOWENSTEIN

VOLUME XV. Steroids and Terpenoids
Edited by RAYMOND B. CLAYTON

VOLUME XVI. Fast Reactions
Edited by KENNETH KUSTIN

VOLUME XVII. Metabolism of Amino Acids and Amines (Parts A and B)
Edited by HERBERT TABOR AND CELIA WHITE TABOR

VOLUME XVIII. Vitamins and Coenzymes (Parts A, B, and C)
Edited by DONALD B. MCCORMICK AND LEMUEL D. WRIGHT

VOLUME XIX. Proteolytic Enzymes
Edited by GERTRUDE E. PERLMANN AND LASZLO LORAND

VOLUME XX. Nucleic Acids and Protein Synthesis (Part C)
Edited by KIVIE MOLDAVE AND LAWRENCE GROSSMAN

VOLUME XXI. Nucleic Acids (Part D)
Edited by LAWRENCE GROSSMAN AND KIVIE MOLDAVE

VOLUME XXII. Enzyme Purification and Related Techniques
Edited by WILLIAM B. JAKOBY

VOLUME XXIII. Photosynthesis (Part A)
Edited by ANTHONY SAN PIETRO

VOLUME XXIV. Photosynthesis and Nitrogen Fixation (Part B)
Edited by ANTHONY SAN PIETRO

VOLUME XXV. Enzyme Structure (Part B)
Edited by C. H. W. HIRS AND SERGE N. TIMASHEFF

VOLUME XXVI. Enzyme Structure (Part C)
Edited by C. H. W. HIRS AND SERGE N. TIMASHEFF

VOLUME XXVII. Enzyme Structure (Part D)
Edited by C. H. W. HIRS AND SERGE N. TIMASHEFF

VOLUME XXVIII. Complex Carbohydrates (Part B)
Edited by VICTOR GINSBURG

VOLUME XXIX. Nucleic Acids and Protein Synthesis (Part E)
Edited by LAWRENCE GROSSMAN AND KIVIE MOLDAVE

VOLUME XXX. Nucleic Acids and Protein Synthesis (Part F)
Edited by KIVIE MOLDAVE AND LAWRENCE GROSSMAN

VOLUME XXXI. Biomembranes (Part A)
Edited by SIDNEY FLEISCHER AND LESTER PACKER

VOLUME XXXII. Biomembranes (Part B)
Edited by SIDNEY FLEISCHER AND LESTER PACKER

VOLUME XXXIII. Cumulative Subject Index Volumes I–XXX
Edited by MARTHA G. DENNIS AND EDWARD A. DENNIS

VOLUME XXXIV. Affinity Techniques (Enzyme Purification: Part B)
Edited by WILLIAM B. JAKOBY AND MEIR WILCHEK

VOLUME XXXV. Lipids (Part B)
Edited by JOHN M. LOWENSTEIN

VOLUME XXXVI. Hormone Action (Part A: Steroid Hormones)
Edited by BERT W. O'MALLEY AND JOEL G. HARDMAN

VOLUME XXXVII. Hormone Action (Part B: Peptide Hormones)
Edited by BERT W. O'MALLEY AND JOEL G. HARDMAN

VOLUME XXXVIII. Hormone Action (Part C: Cyclic Nucleotides)
Edited by JOEL G. HARDMAN AND BERT W. O'MALLEY

VOLUME XXXIX. Hormone Action (Part D: Isolated Cells, Tissues,
and Organ Systems)
Edited by JOEL G. HARDMAN AND BERT W. O'MALLEY

VOLUME XL. Hormone Action (Part E: Nuclear Structure and Function)
Edited by BERT W. O'MALLEY AND JOEL G. HARDMAN

VOLUME XLI. Carbohydrate Metabolism (Part B)
Edited by W. A. WOOD

VOLUME XLII. Carbohydrate Metabolism (Part C)
Edited by W. A. WOOD

VOLUME XLIII. Antibiotics
Edited by JOHN H. HASH

VOLUME XLIV. Immobilized Enzymes
Edited by KLAUS MOSBACH

VOLUME XLV. Proteolytic Enzymes (Part B)
Edited by LASZLO LORAND

VOLUME XLVI. Affinity Labeling
Edited by WILLIAM B. JAKOBY AND MEIR WILCHEK

VOLUME XLVII. Enzyme Structure (Part E)
Edited by C. H. W. HIRS AND SERGE N. TIMASHEFF

VOLUME XLVIII. Enzyme Structure (Part F)
Edited by C. H. W. HIRS AND SERGE N. TIMASHEFF

VOLUME XLIX. Enzyme Structure (Part G)
Edited by C. H. W. HIRS AND SERGE N. TIMASHEFF

VOLUME L. Complex Carbohydrates (Part C)
Edited by VICTOR GINSBURG

VOLUME LI. Purine and Pyrimidine Nucleotide Metabolism
Edited by PATRICIA A. HOFFEE AND MARY ELLEN JONES

VOLUME LII. Biomembranes (Part C: Biological Oxidations)
Edited by SIDNEY FLEISCHER AND LESTER PACKER

VOLUME LIII. Biomembranes (Part D: Biological Oxidations)
Edited by SIDNEY FLEISCHER AND LESTER PACKER

VOLUME LIV. Biomembranes (Part E: Biological Oxidations)
Edited by SIDNEY FLEISCHER AND LESTER PACKER

VOLUME LV. Biomembranes (Part F: Bioenergetics)
Edited by SIDNEY FLEISCHER AND LESTER PACKER

VOLUME LVI. Biomembranes (Part G: Bioenergetics)
Edited by SIDNEY FLEISCHER AND LESTER PACKER

VOLUME LVII. Bioluminescence and Chemiluminescence
Edited by MARLENE A. DELUCA

VOLUME LVIII. Cell Culture
Edited by WILLIAM B. JAKOBY AND IRA PASTAN

VOLUME LIX. Nucleic Acids and Protein Synthesis (Part G)
Edited by KIVIE MOLDAVE AND LAWRENCE GROSSMAN

VOLUME LX. Nucleic Acids and Protein Synthesis (Part H)
Edited by KIVIE MOLDAVE AND LAWRENCE GROSSMAN

VOLUME 61. Enzyme Structure (Part H)
Edited by C. H. W. HIRS AND SERGE N. TIMASHEFF

VOLUME 62. Vitamins and Coenzymes (Part D)
Edited by DONALD B. MCCORMICK AND LEMUEL D. WRIGHT

VOLUME 63. Enzyme Kinetics and Mechanism (Part A: Initial Rate and
Inhibitor Methods)
Edited by DANIEL L. PURICH

VOLUME 64. Enzyme Kinetics and Mechanism
(Part B: Isotopic Probes and Complex Enzyme Systems)
Edited by DANIEL L. PURICH

VOLUME 65. Nucleic Acids (Part I)
Edited by LAWRENCE GROSSMAN AND KIVIE MOLDAVE

VOLUME 66. Vitamins and Coenzymes (Part E)
Edited by DONALD B. MCCORMICK AND LEMUEL D. WRIGHT

VOLUME 67. Vitamins and Coenzymes (Part F)
Edited by DONALD B. MCCORMICK AND LEMUEL D. WRIGHT

VOLUME 68. Recombinant DNA
Edited by RAY WU

VOLUME 69. Photosynthesis and Nitrogen Fixation (Part C)
Edited by ANTHONY SAN PIETRO

VOLUME 70. Immunochemical Techniques (Part A)
Edited by HELEN VAN VUNAKIS AND JOHN J. LANGONE

VOLUME 71. Lipids (Part C)
Edited by JOHN M. LOWENSTEIN

VOLUME 72. Lipids (Part D)
Edited by JOHN M. LOWENSTEIN

VOLUME 73. Immunochemical Techniques (Part B)
Edited by JOHN J. LANGONE AND HELEN VAN VUNAKIS

VOLUME 74. Immunochemical Techniques (Part C)
Edited by JOHN J. LANGONE AND HELEN VAN VUNAKIS

VOLUME 75. Cumulative Subject Index Volumes XXXI, XXXII, XXXIV–LX
Edited by EDWARD A. DENNIS AND MARTHA G. DENNIS

VOLUME 76. Hemoglobins
Edited by ERALDO ANTONINI, LUIGI ROSSI-BERNARDI, AND EMILIA CHIANCONE

VOLUME 77. Detoxication and Drug Metabolism
Edited by WILLIAM B. JAKOBY

VOLUME 78. Interferons (Part A)
Edited by SIDNEY PESTKA

VOLUME 79. Interferons (Part B)
Edited by SIDNEY PESTKA

VOLUME 80. Proteolytic Enzymes (Part C)
Edited by LASZLO LORAND

VOLUME 81. Biomembranes (Part H: Visual Pigments and Purple Membranes, I)
Edited by LESTER PACKER

VOLUME 82. Structural and Contractile Proteins (Part A: Extracellular Matrix)
Edited by LEON W. CUNNINGHAM AND DIXIE W. FREDERIKSEN

VOLUME 83. Complex Carbohydrates (Part D)
Edited by VICTOR GINSBURG

VOLUME 84. Immunochemical Techniques (Part D: Selected Immunoassays)
Edited by JOHN J. LANGONE AND HELEN VAN VUNAKIS

VOLUME 85. Structural and Contractile Proteins (Part B: The Contractile Apparatus and the Cytoskeleton)
Edited by DIXIE W. FREDERIKSEN AND LEON W. CUNNINGHAM

VOLUME 86. Prostaglandins and Arachidonate Metabolites
Edited by WILLIAM E. M. LANDS AND WILLIAM L. SMITH

VOLUME 87. Enzyme Kinetics and Mechanism (Part C: Intermediates, Stereo-chemistry, and Rate Studies)
Edited by DANIEL L. PURICH

VOLUME 88. Biomembranes (Part I: Visual Pigments and Purple Membranes, II)
Edited by LESTER PACKER

VOLUME 89. Carbohydrate Metabolism (Part D)
Edited by WILLIS A. WOOD

VOLUME 90. Carbohydrate Metabolism (Part E)
Edited by WILLIS A. WOOD

VOLUME 91. Enzyme Structure (Part I)
Edited by C. H. W. HIRS AND SERGE N. TIMASHEFF

VOLUME 92. Immunochemical Techniques (Part E: Monoclonal Antibodies and
General Immunoassay Methods)
Edited by JOHN J. LANGONE AND HELEN VAN VUNAKIS

VOLUME 93. Immunochemical Techniques (Part F: Conventional Antibodies, Fc
Receptors, and Cytotoxicity)
Edited by JOHN J. LANGONE AND HELEN VAN VUNAKIS

VOLUME 94. Polyamines
Edited by HERBERT TABOR AND CELIA WHITE TABOR

VOLUME 95. Cumulative Subject Index Volumes 61–74, 76–80
Edited by EDWARD A. DENNIS AND MARTHA G. DENNIS

VOLUME 96. Biomembranes [Part J: Membrane Biogenesis: Assembly and
Targeting (General Methods; Eukaryotes)]
Edited by SIDNEY FLEISCHER AND BECCA FLEISCHER

VOLUME 97. Biomembranes [Part K: Membrane Biogenesis: Assembly and
Targeting (Prokaryotes, Mitochondria, and Chloroplasts)]
Edited by SIDNEY FLEISCHER AND BECCA FLEISCHER

VOLUME 98. Biomembranes (Part L: Membrane Biogenesis: Processing
and Recycling)
Edited by SIDNEY FLEISCHER AND BECCA FLEISCHER

VOLUME 99. Hormone Action (Part F: Protein Kinases)
Edited by JACKIE D. CORBIN AND JOEL G. HARDMAN

VOLUME 100. Recombinant DNA (Part B)
Edited by RAY WU, LAWRENCE GROSSMAN, AND KIVIE MOLDAVE

VOLUME 101. Recombinant DNA (Part C)
Edited by RAY WU, LAWRENCE GROSSMAN, AND KIVIE MOLDAVE

VOLUME 102. Hormone Action (Part G: Calmodulin and
Calcium-Binding Proteins)
Edited by ANTHONY R. MEANS AND BERT W. O'MALLEY

VOLUME 103. Hormone Action (Part H: Neuroendocrine Peptides)
Edited by P. MICHAEL CONN

VOLUME 104. Enzyme Purification and Related Techniques (Part C)
Edited by WILLIAM B. JAKOBY

VOLUME 105. Oxygen Radicals in Biological Systems
Edited by LESTER PACKER

VOLUME 106. Posttranslational Modifications (Part A)
Edited by FINN WOLD AND KIVIE MOLDAVE

VOLUME 107. Posttranslational Modifications (Part B)
Edited by FINN WOLD AND KIVIE MOLDAVE

VOLUME 108. Immunochemical Techniques (Part G: Separation and Characterization of Lymphoid Cells)
Edited by GIOVANNI DI SABATO, JOHN J. LANGONE, AND HELEN VAN VUNAKIS

VOLUME 109. Hormone Action (Part I: Peptide Hormones)
Edited by LUTZ BIRNBAUMER AND BERT W. O'MALLEY

VOLUME 110. Steroids and Isoprenoids (Part A)
Edited by JOHN H. LAW AND HANS C. RILLING

VOLUME 111. Steroids and Isoprenoids (Part B)
Edited by JOHN H. LAW AND HANS C. RILLING

VOLUME 112. Drug and Enzyme Targeting (Part A)
Edited by KENNETH J. WIDDER AND RALPH GREEN

VOLUME 113. Glutamate, Glutamine, Glutathione, and Related Compounds
Edited by ALTON MEISTER

VOLUME 114. Diffraction Methods for Biological Macromolecules (Part A)
Edited by HAROLD W. WYCKOFF, C. H. W. HIRS, AND SERGE N. TIMASHEFF

VOLUME 115. Diffraction Methods for Biological Macromolecules (Part B)
Edited by HAROLD W. WYCKOFF, C. H. W. HIRS, AND SERGE N. TIMASHEFF

VOLUME 116. Immunochemical Techniques
(Part H: Effectors and Mediators of Lymphoid Cell Functions)
Edited by GIOVANNI DI SABATO, JOHN J. LANGONE, AND HELEN VAN VUNAKIS

VOLUME 117. Enzyme Structure (Part J)
Edited by C. H. W. HIRS AND SERGE N. TIMASHEFF

VOLUME 118. Plant Molecular Biology
Edited by ARTHUR WEISSBACH AND HERBERT WEISSBACH

VOLUME 119. Interferons (Part C)
Edited by SIDNEY PESTKA

VOLUME 120. Cumulative Subject Index Volumes 81–94, 96–101

VOLUME 121. Immunochemical Techniques (Part I: Hybridoma Technology and Monoclonal Antibodies)
Edited by JOHN J. LANGONE AND HELEN VAN VUNAKIS

VOLUME 122. Vitamins and Coenzymes (Part G)
Edited by FRANK CHYTIL AND DONALD B. MCCORMICK

VOLUME 123. Vitamins and Coenzymes (Part H)
Edited by FRANK CHYTIL AND DONALD B. MCCORMICK

VOLUME 124. Hormone Action (Part J: Neuroendocrine Peptides)
Edited by P. MICHAEL CONN

VOLUME 125. Biomembranes (Part M: Transport in Bacteria, Mitochondria, and
Chloroplasts: General Approaches and Transport Systems)
Edited by SIDNEY FLEISCHER AND BECCA FLEISCHER

VOLUME 126. Biomembranes (Part N: Transport in Bacteria, Mitochondria, and
Chloroplasts: Protonmotive Force)
Edited by SIDNEY FLEISCHER AND BECCA FLEISCHER

VOLUME 127. Biomembranes (Part O: Protons and Water: Structure
and Translocation)
Edited by LESTER PACKER

VOLUME 128. Plasma Lipoproteins (Part A: Preparation, Structure, and
Molecular Biology)
Edited by JERE P. SEGREST AND JOHN J. ALBERS

VOLUME 129. Plasma Lipoproteins (Part B: Characterization, Cell Biology,
and Metabolism)
Edited by JOHN J. ALBERS AND JERE P. SEGREST

VOLUME 130. Enzyme Structure (Part K)
Edited by C. H. W. HIRS AND SERGE N. TIMASHEFF

VOLUME 131. Enzyme Structure (Part L)
Edited by C. H. W. HIRS AND SERGE N. TIMASHEFF

VOLUME 132. Immunochemical Techniques (Part J: Phagocytosis and
Cell-Mediated Cytotoxicity)
Edited by GIOVANNI DI SABATO AND JOHANNES EVERSE

VOLUME 133. Bioluminescence and Chemiluminescence (Part B)
Edited by MARLENE DELUCA AND WILLIAM D. MCELROY

VOLUME 134. Structural and Contractile Proteins (Part C: The Contractile
Apparatus and the Cytoskeleton)
Edited by RICHARD B. VALLEE

VOLUME 135. Immobilized Enzymes and Cells (Part B)
Edited by KLAUS MOSBACH

VOLUME 136. Immobilized Enzymes and Cells (Part C)
Edited by KLAUS MOSBACH

VOLUME 137. Immobilized Enzymes and Cells (Part D)
Edited by KLAUS MOSBACH

VOLUME 138. Complex Carbohydrates (Part E)
Edited by VICTOR GINSBURG

VOLUME 139. Cellular Regulators (Part A: Calcium- and
Calmodulin-Binding Proteins)
Edited by ANTHONY R. MEANS AND P. MICHAEL CONN

VOLUME 140. Cumulative Subject Index Volumes 102–119, 121–134

VOLUME 141. Cellular Regulators (Part B: Calcium and Lipids)
Edited by P. MICHAEL CONN AND ANTHONY R. MEANS

VOLUME 142. Metabolism of Aromatic Amino Acids and Amines
Edited by SEYMOUR KAUFMAN

VOLUME 143. Sulfur and Sulfur Amino Acids
Edited by WILLIAM B. JAKOBY AND OWEN GRIFFITH

VOLUME 144. Structural and Contractile Proteins (Part D: Extracellular Matrix)
Edited by LEON W. CUNNINGHAM

VOLUME 145. Structural and Contractile Proteins (Part E: Extracellular Matrix)
Edited by LEON W. CUNNINGHAM

VOLUME 146. Peptide Growth Factors (Part A)
Edited by DAVID BARNES AND DAVID A. SIRBASKU

VOLUME 147. Peptide Growth Factors (Part B)
Edited by DAVID BARNES AND DAVID A. SIRBASKU

VOLUME 148. Plant Cell Membranes
Edited by LESTER PACKER AND ROLAND DOUCE

VOLUME 149. Drug and Enzyme Targeting (Part B)
Edited by RALPH GREEN AND KENNETH J. WIDDER

VOLUME 150. Immunochemical Techniques (Part K: *In Vitro* Models of B and
T Cell Functions and Lymphoid Cell Receptors)
Edited by GIOVANNI DI SABATO

VOLUME 151. Molecular Genetics of Mammalian Cells
Edited by MICHAEL M. GOTTESMAN

VOLUME 152. Guide to Molecular Cloning Techniques
Edited by SHELBY L. BERGER AND ALAN R. KIMMEL

VOLUME 153. Recombinant DNA (Part D)
Edited by RAY WU AND LAWRENCE GROSSMAN

VOLUME 154. Recombinant DNA (Part E)
Edited by RAY WU AND LAWRENCE GROSSMAN

VOLUME 155. Recombinant DNA (Part F)
Edited by RAY WU

VOLUME 156. Biomembranes (Part P: ATP-Driven Pumps and Related Transport:
The Na, K-Pump)
Edited by SIDNEY FLEISCHER AND BECCA FLEISCHER

VOLUME 157. Biomembranes (Part Q: ATP-Driven Pumps and Related Transport: Calcium, Proton, and Potassium Pumps)
Edited by SIDNEY FLEISCHER AND BECCA FLEISCHER

VOLUME 158. Metalloproteins (Part A)
Edited by JAMES F. RIORDAN AND BERT L. VALLEE

VOLUME 159. Initiation and Termination of Cyclic Nucleotide Action
Edited by JACKIE D. CORBIN AND ROGER A. JOHNSON

VOLUME 160. Biomass (Part A: Cellulose and Hemicellulose)
Edited by WILLIS A. WOOD AND SCOTT T. KELLOGG

VOLUME 161. Biomass (Part B: Lignin, Pectin, and Chitin)
Edited by WILLIS A. WOOD AND SCOTT T. KELLOGG

VOLUME 162. Immunochemical Techniques (Part L: Chemotaxis and Inflammation)
Edited by GIOVANNI DI SABATO

VOLUME 163. Immunochemical Techniques (Part M: Chemotaxis and Inflammation)
Edited by GIOVANNI DI SABATO

VOLUME 164. Ribosomes
Edited by HARRY F. NOLLER, JR., AND KIVIE MOLDAVE

VOLUME 165. Microbial Toxins: Tools for Enzymology
Edited by SIDNEY HARSHMAN

VOLUME 166. Branched-Chain Amino Acids
Edited by ROBERT HARRIS AND JOHN R. SOKATCH

VOLUME 167. Cyanobacteria
Edited by LESTER PACKER AND ALEXANDER N. GLAZER

VOLUME 168. Hormone Action (Part K: Neuroendocrine Peptides)
Edited by P. MICHAEL CONN

VOLUME 169. Platelets: Receptors, Adhesion, Secretion (Part A)
Edited by JACEK HAWIGER

VOLUME 170. Nucleosomes
Edited by PAUL M. WASSARMAN AND ROGER D. KORNBERG

VOLUME 171. Biomembranes (Part R: Transport Theory: Cells and Model Membranes)
Edited by SIDNEY FLEISCHER AND BECCA FLEISCHER

VOLUME 172. Biomembranes (Part S: Transport: Membrane Isolation and Characterization)
Edited by SIDNEY FLEISCHER AND BECCA FLEISCHER

VOLUME 173. Biomembranes [Part T: Cellular and Subcellular Transport: Eukaryotic (Nonepithelial) Cells]
Edited by SIDNEY FLEISCHER AND BECCA FLEISCHER

VOLUME 174. Biomembranes [Part U: Cellular and Subcellular Transport: Eukaryotic (Nonepithelial) Cells]
Edited by SIDNEY FLEISCHER AND BECCA FLEISCHER

VOLUME 175. Cumulative Subject Index Volumes 135–139, 141–167

VOLUME 176. Nuclear Magnetic Resonance (Part A: Spectral Techniques and Dynamics)
Edited by NORMAN J. OPPENHEIMER AND THOMAS L. JAMES

VOLUME 177. Nuclear Magnetic Resonance (Part B: Structure and Mechanism)
Edited by NORMAN J. OPPENHEIMER AND THOMAS L. JAMES

VOLUME 178. Antibodies, Antigens, and Molecular Mimicry
Edited by JOHN J. LANGONE

VOLUME 179. Complex Carbohydrates (Part F)
Edited by VICTOR GINSBURG

VOLUME 180. RNA Processing (Part A: General Methods)
Edited by JAMES E. DAHLBERG AND JOHN N. ABELSON

VOLUME 181. RNA Processing (Part B: Specific Methods)
Edited by JAMES E. DAHLBERG AND JOHN N. ABELSON

VOLUME 182. Guide to Protein Purification
Edited by MURRAY P. DEUTSCHER

VOLUME 183. Molecular Evolution: Computer Analysis of Protein and Nucleic Acid Sequences
Edited by RUSSELL F. DOOLITTLE

VOLUME 184. Avidin-Biotin Technology
Edited by MEIR WILCHEK AND EDWARD A. BAYER

VOLUME 185. Gene Expression Technology
Edited by DAVID V. GOEDDEL

VOLUME 186. Oxygen Radicals in Biological Systems (Part B: Oxygen Radicals and Antioxidants)
Edited by LESTER PACKER AND ALEXANDER N. GLAZER

VOLUME 187. Arachidonate Related Lipid Mediators
Edited by ROBERT C. MURPHY AND FRANK A. FITZPATRICK

VOLUME 188. Hydrocarbons and Methylotrophy
Edited by MARY E. LIDSTROM

VOLUME 189. Retinoids (Part A: Molecular and Metabolic Aspects)
Edited by LESTER PACKER

VOLUME 190. Retinoids (Part B: Cell Differentiation and Clinical Applications)
Edited by LESTER PACKER

VOLUME 191. Biomembranes (Part V: Cellular and Subcellular Transport:
Epithelial Cells)
Edited by SIDNEY FLEISCHER AND BECCA FLEISCHER

VOLUME 192. Biomembranes (Part W: Cellular and Subcellular Transport:
Epithelial Cells)
Edited by SIDNEY FLEISCHER AND BECCA FLEISCHER

VOLUME 193. Mass Spectrometry
Edited by JAMES A. McCLOSKEY

VOLUME 194. Guide to Yeast Genetics and Molecular Biology
Edited by CHRISTINE GUTHRIE AND GERALD R. FINK

VOLUME 195. Adenylyl Cyclase, G Proteins, and Guanylyl Cyclase
Edited by ROGER A. JOHNSON AND JACKIE D. CORBIN

VOLUME 196. Molecular Motors and the Cytoskeleton
Edited by RICHARD B. VALLEE

VOLUME 197. Phospholipases
Edited by EDWARD A. DENNIS

VOLUME 198. Peptide Growth Factors (Part C)
Edited by DAVID BARNES, J. P. MATHER, AND GORDON H. SATO

VOLUME 199. Cumulative Subject Index Volumes 168–174, 176–194

VOLUME 200. Protein Phosphorylation (Part A: Protein Kinases: Assays,
Purification, Antibodies, Functional Analysis, Cloning, and Expression)
Edited by TONY HUNTER AND BARTHOLOMEW M. SEFTON

VOLUME 201. Protein Phosphorylation (Part B: Analysis of Protein
Phosphorylation, Protein Kinase Inhibitors, and Protein Phosphatases)
Edited by TONY HUNTER AND BARTHOLOMEW M. SEFTON

VOLUME 202. Molecular Design and Modeling: Concepts and Applications
(Part A: Proteins, Peptides, and Enzymes)
Edited by JOHN J. LANGONE

VOLUME 203. Molecular Design and Modeling: Concepts and Applications
(Part B: Antibodies and Antigens, Nucleic Acids, Polysaccharides, and Drugs)
Edited by JOHN J. LANGONE

VOLUME 204. Bacterial Genetic Systems
Edited by JEFFREY H. MILLER

VOLUME 205. Metallobiochemistry (Part B: Metallothionein and
Related Molecules)
Edited by JAMES F. RIORDAN AND BERT L. VALLEE

VOLUME 206. Cytochrome P450
Edited by MICHAEL R. WATERMAN AND ERIC F. JOHNSON

VOLUME 207. Ion Channels
Edited by BERNARDO RUDY AND LINDA E. IVERSON

VOLUME 208. Protein–DNA Interactions
Edited by ROBERT T. SAUER

VOLUME 209. Phospholipid Biosynthesis
Edited by EDWARD A. DENNIS AND DENNIS E. VANCE

VOLUME 210. Numerical Computer Methods
Edited by LUDWIG BRAND AND MICHAEL L. JOHNSON

VOLUME 211. DNA Structures (Part A: Synthesis and Physical Analysis of DNA)
Edited by DAVID M. J. LILLEY AND JAMES E. DAHLBERG

VOLUME 212. DNA Structures (Part B: Chemical and Electrophoretic Analysis of DNA)
Edited by DAVID M. J. LILLEY AND JAMES E. DAHLBERG

VOLUME 213. Carotenoids (Part A: Chemistry, Separation, Quantitation, and Antioxidation)
Edited by LESTER PACKER

VOLUME 214. Carotenoids (Part B: Metabolism, Genetics, and Biosynthesis)
Edited by LESTER PACKER

VOLUME 215. Platelets: Receptors, Adhesion, Secretion (Part B)
Edited by JACEK J. HAWIGER

VOLUME 216. Recombinant DNA (Part G)
Edited by RAY WU

VOLUME 217. Recombinant DNA (Part H)
Edited by RAY WU

VOLUME 218. Recombinant DNA (Part I)
Edited by RAY WU

VOLUME 219. Reconstitution of Intracellular Transport
Edited by JAMES E. ROTHMAN

VOLUME 220. Membrane Fusion Techniques (Part A)
Edited by NEJAT DÜZGÜNEŞ

VOLUME 221. Membrane Fusion Techniques (Part B)
Edited by NEJAT DÜZGÜNEŞ

VOLUME 222. Proteolytic Enzymes in Coagulation, Fibrinolysis, and Complement Activation (Part A: Mammalian Blood Coagulation Factors and Inhibitors)
Edited by LASZLO LORAND AND KENNETH G. MANN

VOLUME 223. Proteolytic Enzymes in Coagulation, Fibrinolysis, and Complement Activation (Part B: Complement Activation, Fibrinolysis, and Nonmammalian Blood Coagulation Factors)
Edited by LASZLO LORAND AND KENNETH G. MANN

VOLUME 224. Molecular Evolution: Producing the Biochemical Data
Edited by ELIZABETH ANNE ZIMMER, THOMAS J. WHITE, REBECCA L. CANN, AND ALLAN C. WILSON

VOLUME 225. Guide to Techniques in Mouse Development
Edited by PAUL M. WASSARMAN AND MELVIN L. DEPAMPHILIS

VOLUME 226. Metallobiochemistry (Part C: Spectroscopic and Physical Methods for Probing Metal Ion Environments in Metalloenzymes and Metalloproteins)
Edited by JAMES F. RIORDAN AND BERT L. VALLEE

VOLUME 227. Metallobiochemistry (Part D: Physical and Spectroscopic Methods for Probing Metal Ion Environments in Metalloproteins)
Edited by JAMES F. RIORDAN AND BERT L. VALLEE

VOLUME 228. Aqueous Two-Phase Systems
Edited by HARRY WALTER AND GÖTE JOHANSSON

VOLUME 229. Cumulative Subject Index Volumes 195–198, 200–227

VOLUME 230. Guide to Techniques in Glycobiology
Edited by WILLIAM J. LENNARZ AND GERALD W. HART

VOLUME 231. Hemoglobins (Part B: Biochemical and Analytical Methods)
Edited by JOHANNES EVERSE, KIM D. VANDEGRIFF, AND ROBERT M. WINSLOW

VOLUME 232. Hemoglobins (Part C: Biophysical Methods)
Edited by JOHANNES EVERSE, KIM D. VANDEGRIFF, AND ROBERT M. WINSLOW

VOLUME 233. Oxygen Radicals in Biological Systems (Part C)
Edited by LESTER PACKER

VOLUME 234. Oxygen Radicals in Biological Systems (Part D)
Edited by LESTER PACKER

VOLUME 235. Bacterial Pathogenesis (Part A: Identification and Regulation of Virulence Factors)
Edited by VIRGINIA L. CLARK AND PATRIK M. BAVOIL

VOLUME 236. Bacterial Pathogenesis (Part B: Integration of Pathogenic Bacteria with Host Cells)
Edited by VIRGINIA L. CLARK AND PATRIK M. BAVOIL

VOLUME 237. Heterotrimeric G Proteins
Edited by RAVI IYENGAR

VOLUME 238. Heterotrimeric G-Protein Effectors
Edited by RAVI IYENGAR

VOLUME 239. Nuclear Magnetic Resonance (Part C)
Edited by THOMAS L. JAMES AND NORMAN J. OPPENHEIMER

VOLUME 240. Numerical Computer Methods (Part B)
Edited by MICHAEL L. JOHNSON AND LUDWIG BRAND

VOLUME 241. Retroviral Proteases
Edited by LAWRENCE C. KUO AND JULES A. SHAFER

VOLUME 242. Neoglycoconjugates (Part A)
Edited by Y. C. LEE AND REIKO T. LEE

VOLUME 243. Inorganic Microbial Sulfur Metabolism
Edited by HARRY D. PECK, JR., AND JEAN LEGALL

VOLUME 244. Proteolytic Enzymes: Serine and Cysteine Peptidases
Edited by ALAN J. BARRETT

VOLUME 245. Extracellular Matrix Components
Edited by E. RUOSLAHTI AND E. ENGVALL

VOLUME 246. Biochemical Spectroscopy
Edited by KENNETH SAUER

VOLUME 247. Neoglycoconjugates (Part B: Biomedical Applications)
Edited by Y. C. LEE AND REIKO T. LEE

VOLUME 248. Proteolytic Enzymes: Aspartic and Metallo Peptidases
Edited by ALAN J. BARRETT

VOLUME 249. Enzyme Kinetics and Mechanism (Part D: Developments in Enzyme Dynamics)
Edited by DANIEL L. PURICH

VOLUME 250. Lipid Modifications of Proteins
Edited by PATRICK J. CASEY AND JANICE E. BUSS

VOLUME 251. Biothiols (Part A: Monothiols and Dithiols, Protein Thiols, and Thiyl Radicals)
Edited by LESTER PACKER

VOLUME 252. Biothiols (Part B: Glutathione and Thioredoxin; Thiols in Signal Transduction and Gene Regulation)
Edited by LESTER PACKER

VOLUME 253. Adhesion of Microbial Pathogens
Edited by RON J. DOYLE AND ITZHAK OFEK

VOLUME 254. Oncogene Techniques
Edited by PETER K. VOGT AND INDER M. VERMA

VOLUME 255. Small GTPases and Their Regulators (Part A: Ras Family)
Edited by W. E. BALCH, CHANNING J. DER, AND ALAN HALL

VOLUME 256. Small GTPases and Their Regulators (Part B: Rho Family)
Edited by W. E. BALCH, CHANNING J. DER, AND ALAN HALL

VOLUME 257. Small GTPases and Their Regulators (Part C: Proteins Involved in Transport)
Edited by W. E. BALCH, CHANNING J. DER, AND ALAN HALL

VOLUME 258. Redox-Active Amino Acids in Biology
Edited by JUDITH P. KLINMAN

VOLUME 259. Energetics of Biological Macromolecules
Edited by MICHAEL L. JOHNSON AND GARY K. ACKERS

VOLUME 260. Mitochondrial Biogenesis and Genetics (Part A)
Edited by GIUSEPPE M. ATTARDI AND ANNE CHOMYN

VOLUME 261. Nuclear Magnetic Resonance and Nucleic Acids
Edited by THOMAS L. JAMES

VOLUME 262. DNA Replication
Edited by JUDITH L. CAMPBELL

VOLUME 263. Plasma Lipoproteins (Part C: Quantitation)
Edited by WILLIAM A. BRADLEY, SANDRA H. GIANTURCO, AND JERE P. SEGREST

VOLUME 264. Mitochondrial Biogenesis and Genetics (Part B)
Edited by GIUSEPPE M. ATTARDI AND ANNE CHOMYN

VOLUME 265. Cumulative Subject Index Volumes 228, 230–262

VOLUME 266. Computer Methods for Macromolecular Sequence Analysis
Edited by RUSSELL F. DOOLITTLE

VOLUME 267. Combinatorial Chemistry
Edited by JOHN N. ABELSON

VOLUME 268. Nitric Oxide (Part A: Sources and Detection of NO; NO Synthase)
Edited by LESTER PACKER

VOLUME 269. Nitric Oxide (Part B: Physiological and Pathological Processes)
Edited by LESTER PACKER

VOLUME 270. High Resolution Separation and Analysis of Biological Macromolecules (Part A: Fundamentals)
Edited by BARRY L. KARGER AND WILLIAM S. HANCOCK

VOLUME 271. High Resolution Separation and Analysis of Biological Macromolecules (Part B: Applications)
Edited by BARRY L. KARGER AND WILLIAM S. HANCOCK

VOLUME 272. Cytochrome P450 (Part B)
Edited by ERIC F. JOHNSON AND MICHAEL R. WATERMAN

VOLUME 273. RNA Polymerase and Associated Factors (Part A)
Edited by SANKAR ADHYA

VOLUME 274. RNA Polymerase and Associated Factors (Part B)
Edited by SANKAR ADHYA

VOLUME 275. Viral Polymerases and Related Proteins
Edited by LAWRENCE C. KUO, DAVID B. OLSEN, AND STEVEN S. CARROLL

VOLUME 276. Macromolecular Crystallography (Part A)
Edited by CHARLES W. CARTER, JR., AND ROBERT M. SWEET

VOLUME 277. Macromolecular Crystallography (Part B)
Edited by CHARLES W. CARTER, JR., AND ROBERT M. SWEET

VOLUME 278. Fluorescence Spectroscopy
Edited by LUDWIG BRAND AND MICHAEL L. JOHNSON

VOLUME 279. Vitamins and Coenzymes (Part I)
Edited by DONALD B. MCCORMICK, JOHN W. SUTTIE, AND CONRAD WAGNER

VOLUME 280. Vitamins and Coenzymes (Part J)
Edited by DONALD B. MCCORMICK, JOHN W. SUTTIE, AND CONRAD WAGNER

VOLUME 281. Vitamins and Coenzymes (Part K)
Edited by DONALD B. MCCORMICK, JOHN W. SUTTIE, AND CONRAD WAGNER

VOLUME 282. Vitamins and Coenzymes (Part L)
Edited by DONALD B. MCCORMICK, JOHN W. SUTTIE, AND CONRAD WAGNER

VOLUME 283. Cell Cycle Control
Edited by WILLIAM G. DUNPHY

VOLUME 284. Lipases (Part A: Biotechnology)
Edited by BYRON RUBIN AND EDWARD A. DENNIS

VOLUME 285. Cumulative Subject Index Volumes 263, 264, 266–284, 286–289

VOLUME 286. Lipases (Part B: Enzyme Characterization and Utilization)
Edited by BYRON RUBIN AND EDWARD A. DENNIS

VOLUME 287. Chemokines
Edited by RICHARD HORUK

VOLUME 288. Chemokine Receptors
Edited by RICHARD HORUK

VOLUME 289. Solid Phase Peptide Synthesis
Edited by GREGG B. FIELDS

VOLUME 290. Molecular Chaperones
Edited by GEORGE H. LORIMER AND THOMAS BALDWIN

VOLUME 291. Caged Compounds
Edited by GERARD MARRIOTT

VOLUME 292. ABC Transporters: Biochemical, Cellular, and Molecular Aspects
Edited by SURESH V. AMBUDKAR AND MICHAEL M. GOTTESMAN

VOLUME 293. Ion Channels (Part B)
Edited by P. MICHAEL CONN

VOLUME 294. Ion Channels (Part C)
Edited by P. MICHAEL CONN

VOLUME 295. Energetics of Biological Macromolecules (Part B)
Edited by GARY K. ACKERS AND MICHAEL L. JOHNSON

VOLUME 296. Neurotransmitter Transporters
Edited by SUSAN G. AMARA

VOLUME 297. Photosynthesis: Molecular Biology of Energy Capture
Edited by LEE MCINTOSH

VOLUME 298. Molecular Motors and the Cytoskeleton (Part B)
Edited by RICHARD B. VALLEE

VOLUME 299. Oxidants and Antioxidants (Part A)
Edited by LESTER PACKER

VOLUME 300. Oxidants and Antioxidants (Part B)
Edited by LESTER PACKER

VOLUME 301. Nitric Oxide: Biological and Antioxidant Activities (Part C)
Edited by LESTER PACKER

VOLUME 302. Green Fluorescent Protein
Edited by P. MICHAEL CONN

VOLUME 303. cDNA Preparation and Display
Edited by SHERMAN M. WEISSMAN

VOLUME 304. Chromatin
Edited by PAUL M. WASSARMAN AND ALAN P. WOLFFE

VOLUME 305. Bioluminescence and Chemiluminescence (Part C)
Edited by THOMAS O. BALDWIN AND MIRIAM M. ZIEGLER

VOLUME 306. Expression of Recombinant Genes in Eukaryotic Systems
Edited by JOSEPH C. GLORIOSO AND MARTIN C. SCHMIDT

VOLUME 307. Confocal Microscopy
Edited by P. MICHAEL CONN

VOLUME 308. Enzyme Kinetics and Mechanism (Part E: Energetics of Enzyme Catalysis)
Edited by DANIEL L. PURICH AND VERN L. SCHRAMM

VOLUME 309. Amyloid, Prions, and Other Protein Aggregates
Edited by RONALD WETZEL

VOLUME 310. Biofilms
Edited by RON J. DOYLE

VOLUME 311. Sphingolipid Metabolism and Cell Signaling (Part A)
Edited by ALFRED H. MERRILL, JR., AND YUSUF A. HANNUN

VOLUME 312. Sphingolipid Metabolism and Cell Signaling (Part B)
Edited by ALFRED H. MERRILL, JR., AND YUSUF A. HANNUN

VOLUME 313. Antisense Technology (Part A: General Methods, Methods of Delivery, and RNA Studies)
Edited by M. IAN PHILLIPS

VOLUME 314. Antisense Technology (Part B: Applications)
Edited by M. IAN PHILLIPS

VOLUME 315. Vertebrate Phototransduction and the Visual Cycle (Part A)
Edited by KRZYSZTOF PALCZEWSKI

VOLUME 316. Vertebrate Phototransduction and the Visual Cycle (Part B)
Edited by KRZYSZTOF PALCZEWSKI

VOLUME 317. RNA–Ligand Interactions (Part A: Structural Biology Methods)
Edited by DANIEL W. CELANDER AND JOHN N. ABELSON

VOLUME 318. RNA–Ligand Interactions (Part B: Molecular Biology Methods)
Edited by DANIEL W. CELANDER AND JOHN N. ABELSON

VOLUME 319. Singlet Oxygen, UV-A, and Ozone
Edited by LESTER PACKER AND HELMUT SIES

VOLUME 320. Cumulative Subject Index Volumes 290–319

VOLUME 321. Numerical Computer Methods (Part C)
Edited by MICHAEL L. JOHNSON AND LUDWIG BRAND

VOLUME 322. Apoptosis
Edited by JOHN C. REED

VOLUME 323. Energetics of Biological Macromolecules (Part C)
Edited by MICHAEL L. JOHNSON AND GARY K. ACKERS

VOLUME 324. Branched-Chain Amino Acids (Part B)
Edited by ROBERT A. HARRIS AND JOHN R. SOKATCH

VOLUME 325. Regulators and Effectors of Small GTPases (Part D: Rho Family)
Edited by W. E. BALCH, CHANNING J. DER, AND ALAN HALL

VOLUME 326. Applications of Chimeric Genes and Hybrid Proteins (Part A: Gene Expression and Protein Purification)
Edited by JEREMY THORNER, SCOTT D. EMR, AND JOHN N. ABELSON

VOLUME 327. Applications of Chimeric Genes and Hybrid Proteins (Part B: Cell Biology and Physiology)
Edited by JEREMY THORNER, SCOTT D. EMR, AND JOHN N. ABELSON

VOLUME 328. Applications of Chimeric Genes and Hybrid Proteins (Part C: Protein–Protein Interactions and Genomics)
Edited by JEREMY THORNER, SCOTT D. EMR, AND JOHN N. ABELSON

VOLUME 329. Regulators and Effectors of Small GTPases (Part E: GTPases Involved in Vesicular Traffic)
Edited by W. E. BALCH, CHANNING J. DER, AND ALAN HALL

VOLUME 330. Hyperthermophilic Enzymes (Part A)
Edited by MICHAEL W. W. ADAMS AND ROBERT M. KELLY

VOLUME 331. Hyperthermophilic Enzymes (Part B)
Edited by MICHAEL W. W. ADAMS AND ROBERT M. KELLY

VOLUME 332. Regulators and Effectors of Small GTPases (Part F: Ras Family I)
Edited by W. E. BALCH, CHANNING J. DER, AND ALAN HALL

VOLUME 333. Regulators and Effectors of Small GTPases (Part G: Ras Family II)
Edited by W. E. BALCH, CHANNING J. DER, AND ALAN HALL

VOLUME 334. Hyperthermophilic Enzymes (Part C)
Edited by MICHAEL W. W. ADAMS AND ROBERT M. KELLY

VOLUME 335. Flavonoids and Other Polyphenols
Edited by LESTER PACKER

VOLUME 336. Microbial Growth in Biofilms (Part A: Developmental and Molecular Biological Aspects)
Edited by RON J. DOYLE

VOLUME 337. Microbial Growth in Biofilms (Part B: Special Environments and Physicochemical Aspects)
Edited by RON J. DOYLE

VOLUME 338. Nuclear Magnetic Resonance of Biological Macromolecules (Part A)
Edited by THOMAS L. JAMES, VOLKER DÖTSCH, AND ULI SCHMITZ

VOLUME 339. Nuclear Magnetic Resonance of Biological Macromolecules (Part B)
Edited by THOMAS L. JAMES, VOLKER DÖTSCH, AND ULI SCHMITZ

VOLUME 340. Drug–Nucleic Acid Interactions
Edited by JONATHAN B. CHAIRES AND MICHAEL J. WARING

VOLUME 341. Ribonucleases (Part A)
Edited by ALLEN W. NICHOLSON

VOLUME 342. Ribonucleases (Part B)
Edited by ALLEN W. NICHOLSON

VOLUME 343. G Protein Pathways (Part A: Receptors)
Edited by RAVI IYENGAR AND JOHN D. HILDEBRANDT

VOLUME 344. G Protein Pathways (Part B: G Proteins and Their Regulators)
Edited by RAVI IYENGAR AND JOHN D. HILDEBRANDT

VOLUME 345. G Protein Pathways (Part C: Effector Mechanisms)
Edited by RAVI IYENGAR AND JOHN D. HILDEBRANDT

VOLUME 346. Gene Therapy Methods
Edited by M. IAN PHILLIPS

VOLUME 347. Protein Sensors and Reactive Oxygen Species (Part A: Selenoproteins and Thioredoxin)
Edited by HELMUT SIES AND LESTER PACKER

VOLUME 348. Protein Sensors and Reactive Oxygen Species (Part B: Thiol Enzymes and Proteins)
Edited by HELMUT SIES AND LESTER PACKER

VOLUME 349. Superoxide Dismutase
Edited by LESTER PACKER

VOLUME 350. Guide to Yeast Genetics and Molecular and Cell Biology (Part B)
Edited by CHRISTINE GUTHRIE AND GERALD R. FINK

VOLUME 351. Guide to Yeast Genetics and Molecular and Cell Biology (Part C)
Edited by CHRISTINE GUTHRIE AND GERALD R. FINK

VOLUME 352. Redox Cell Biology and Genetics (Part A)
Edited by CHANDAN K. SEN AND LESTER PACKER

VOLUME 353. Redox Cell Biology and Genetics (Part B)
Edited by CHANDAN K. SEN AND LESTER PACKER

VOLUME 354. Enzyme Kinetics and Mechanisms (Part F: Detection and Characterization of Enzyme Reaction Intermediates)
Edited by DANIEL L. PURICH

VOLUME 355. Cumulative Subject Index Volumes 321–354

VOLUME 356. Laser Capture Microscopy and Microdissection
Edited by P. MICHAEL CONN

VOLUME 357. Cytochrome P450, Part C
Edited by ERIC F. JOHNSON AND MICHAEL R. WATERMAN

VOLUME 358. Bacterial Pathogenesis (Part C: Identification, Regulation, and Function of Virulence Factors)
Edited by VIRGINIA L. CLARK AND PATRIK M. BAVOIL

VOLUME 359. Nitric Oxide (Part D)
Edited by ENRIQUE CADENAS AND LESTER PACKER

VOLUME 360. Biophotonics (Part A)
Edited by GERARD MARRIOTT AND IAN PARKER

VOLUME 361. Biophotonics (Part B)
Edited by GERARD MARRIOTT AND IAN PARKER

VOLUME 362. Recognition of Carbohydrates in Biological Systems (Part A)
Edited by YUAN C. LEE AND REIKO T. LEE

VOLUME 363. Recognition of Carbohydrates in Biological Systems (Part B)
Edited by YUAN C. LEE AND REIKO T. LEE

VOLUME 364. Nuclear Receptors
Edited by DAVID W. RUSSELL AND DAVID J. MANGELSDORF

VOLUME 365. Differentiation of Embryonic Stem Cells
Edited by PAUL M. WASSAUMAN AND GORDON M. KELLER

VOLUME 366. Protein Phosphatases
Edited by SUSANNE KLUMPP AND JOSEF KRIEGLSTEIN

VOLUME 367. Liposomes (Part A)
Edited by NEJAT DÜZGÜNEŞ

VOLUME 368. Macromolecular Crystallography (Part C)
Edited by CHARLES W. CARTER, JR., AND ROBERT M. SWEET

VOLUME 369. Combinational Chemistry (Part B)
Edited by GUILLERMO A. MORALES AND BARRY A. BUNIN

VOLUME 370. RNA Polymerases and Associated Factors (Part C)
Edited by SANKAR L. ADHYA AND SUSAN GARGES

VOLUME 371. RNA Polymerases and Associated Factors (Part D)
Edited by SANKAR L. ADHYA AND SUSAN GARGES

VOLUME 372. Liposomes (Part B)
Edited by NEJAT DÜZGÜNEŞ

VOLUME 373. Liposomes (Part C)
Edited by NEJAT DÜZGÜNEŞ

VOLUME 374. Macromolecular Crystallography (Part D)
Edited by CHARLES W. CARTER, JR., AND ROBERT W. SWEET

VOLUME 375. Chromatin and Chromatin Remodeling Enzymes (Part A)
Edited by C. DAVID ALLIS AND CARL WU

VOLUME 376. Chromatin and Chromatin Remodeling Enzymes (Part B)
Edited by C. DAVID ALLIS AND CARL WU

VOLUME 377. Chromatin and Chromatin Remodeling Enzymes (Part C)
Edited by C. DAVID ALLIS AND CARL WU

VOLUME 378. Quinones and Quinone Enzymes (Part A)
Edited by HELMUT SIES AND LESTER PACKER

VOLUME 379. Energetics of Biological Macromolecules (Part D)
Edited by JO M. HOLT, MICHAEL L. JOHNSON, AND GARY K. ACKERS

VOLUME 380. Energetics of Biological Macromolecules (Part E)
Edited by JO M. HOLT, MICHAEL L. JOHNSON, AND GARY K. ACKERS

VOLUME 381. Oxygen Sensing
Edited by CHANDAN K. SEN AND GREGG L. SEMENZA

VOLUME 382. Quinones and Quinone Enzymes (Part B)
Edited by HELMUT SIES AND LESTER PACKER

VOLUME 383. Numerical Computer Methods (Part D)
Edited by LUDWIG BRAND AND MICHAEL L. JOHNSON

VOLUME 384. Numerical Computer Methods (Part E)
Edited by LUDWIG BRAND AND MICHAEL L. JOHNSON

VOLUME 385. Imaging in Biological Research (Part A)
Edited by P. MICHAEL CONN

VOLUME 386. Imaging in Biological Research (Part B)
Edited by P. MICHAEL CONN

VOLUME 387. Liposomes (Part D)
Edited by NEJAT DÜZGÜNEŞ

VOLUME 388. Protein Engineering
Edited by DAN E. ROBERTSON AND JOSEPH P. NOEL

VOLUME 389. Regulators of G-Protein Signaling (Part A)
Edited by DAVID P. SIDEROVSKI

VOLUME 390. Regulators of G-Protein Signaling (Part B)
Edited by DAVID P. SIDEROVSKI

VOLUME 391. Liposomes (Part E)
Edited by NEJAT DÜZGÜNEŞ

VOLUME 392. RNA Interference
Edited by ENGELKE ROSSI

VOLUME 393. Circadian Rhythms
Edited by MICHAEL W. YOUNG

VOLUME 394. Nuclear Magnetic Resonance of Biological Macromolecules (Part C)
Edited by THOMAS L. JAMES

VOLUME 395. Producing the Biochemical Data (Part B)
Edited by ELIZABETH A. ZIMMER AND ERIC H. ROALSON

VOLUME 396. Nitric Oxide (Part E)
Edited by LESTER PACKER AND ENRIQUE CADENAS

VOLUME 397. Environmental Microbiology
Edited by JARED R. LEADBETTER

VOLUME 398. Ubiquitin and Protein Degradation (Part A)
Edited by RAYMOND J. DESHAIES

VOLUME 399. Ubiquitin and Protein Degradation (Part B)
Edited by RAYMOND J. DESHAIES

VOLUME 400. Phase II Conjugation Enzymes and Transport Systems
Edited by HELMUT SIES AND LESTER PACKER

VOLUME 401. Glutathione Transferases and Gamma Glutamyl Transpeptidases
Edited by HELMUT SIES AND LESTER PACKER

VOLUME 402. Biological Mass Spectrometry
Edited by A. L. BURLINGAME

VOLUME 403. GTPases Regulating Membrane Targeting and Fusion
Edited by WILLIAM E. BALCH, CHANNING J. DER, AND ALAN HALL

VOLUME 404. GTPases Regulating Membrane Dynamics
Edited by WILLIAM E. BALCH, CHANNING J. DER, AND ALAN HALL

VOLUME 405. Mass Spectrometry: Modified Proteins and Glycoconjugates
Edited by A. L. BURLINGAME

VOLUME 406. Regulators and Effectors of Small GTPases: Rho Family
Edited by WILLIAM E. BALCH, CHANNING J. DER, AND ALAN HALL

VOLUME 407. Regulators and Effectors of Small GTPases: Ras Family
Edited by WILLIAM E. BALCH, CHANNING J. DER, AND ALAN HALL

VOLUME 408. DNA Repair (Part A)
Edited by JUDITH L. CAMPBELL AND PAUL MODRICH

VOLUME 409. DNA Repair (Part B)
Edited by JUDITH L. CAMPBELL AND PAUL MODRICH

VOLUME 410. DNA Microarrays (Part A: Array Platforms and Web-Bench Protocols)
Edited by ALAN KIMMEL AND BRIAN OLIVER

VOLUME 411. DNA Microarrays (Part B: Databases and Statistics)
Edited by ALAN KIMMEL AND BRIAN OLIVER

VOLUME 412. Amyloid, Prions, and Other Protein Aggregates (Part B)
Edited by INDU KHETERPAL AND RONALD WETZEL

VOLUME 413. Amyloid, Prions, and Other Protein Aggregates (Part C)
Edited by INDU KHETERPAL AND RONALD WETZEL

VOLUME 414. Measuring Biological Responses with Automated Microscopy
Edited by JAMES INGLESE

VOLUME 415. Glycobiology
Edited by MINORU FUKUDA

VOLUME 416. Glycomics
Edited by MINORU FUKUDA

VOLUME 417. Functional Glycomics
Edited by MINORU FUKUDA

VOLUME 418. Embryonic Stem Cells
Edited by IRINA KLIMANSKAYA AND ROBERT LANZA

VOLUME 419. Adult Stem Cells
Edited by IRINA KLIMANSKAYA AND ROBERT LANZA

VOLUME 420. Stem Cell Tools and Other Experimental Protocols
Edited by IRINA KLIMANSKAYA AND ROBERT LANZA

VOLUME 421. Advanced Bacterial Genetics: Use of Transposons and Phage for Genomic Engineering
Edited by KELLY T. HUGHES

VOLUME 422. Two-Component Signaling Systems, Part A
Edited by MELVIN I. SIMON, BRIAN R. CRANE, AND ALEXANDRINE CRANE

VOLUME 423. Two-Component Signaling Systems, Part B
Edited by MELVIN I. SIMON, BRIAN R. CRANE, AND ALEXANDRINE CRANE

VOLUME 424. RNA Editing
Edited by JONATHA M. GOTT

VOLUME 425. RNA Modification
Edited by JONATHA M. GOTT

VOLUME 426. Integrins
Edited by DAVID CHERESH

VOLUME 427. MicroRNA Methods
Edited by JOHN J. ROSSI

VOLUME 428. Osmosensing and Osmosignaling
Edited by HELMUT SIES AND DIETER HAUSSINGER

VOLUME 429. Translation Initiation: Extract Systems and Molecular Genetics
Edited by JON LORSCH

VOLUME 430. Translation Initiation: Reconstituted Systems and Biophysical Methods
Edited by JON LORSCH

VOLUME 431. Translation Initiation: Cell Biology, High-Throughput and Chemical-Based Approaches
Edited by JON LORSCH

VOLUME 432. Lipidomics and Bioactive Lipids: Mass-Spectrometry–Based Lipid Analysis
Edited by H. ALEX BROWN

VOLUME 433. Lipidomics and Bioactive Lipids: Specialized Analytical Methods and Lipids in Disease
Edited by H. ALEX BROWN

VOLUME 434. Lipidomics and Bioactive Lipids: Lipids and Cell Signaling
Edited by H. ALEX BROWN

VOLUME 435. Oxygen Biology and Hypoxia
Edited by HELMUT SIES AND BERNHARD BRÜNE

VOLUME 436. Globins and Other Nitric Oxide-Reactive Protiens (Part A)
Edited by ROBERT K. POOLE

VOLUME 437. Globins and Other Nitric Oxide-Reactive Protiens (Part B)
Edited by ROBERT K. POOLE

VOLUME 438. Small GTPases in Disease (Part A)
Edited by WILLIAM E. BALCH, CHANNING J. DER, AND ALAN HALL

VOLUME 439. Small GTPases in Disease (Part B)
Edited by WILLIAM E. BALCH, CHANNING J. DER, AND ALAN HALL

VOLUME 440. Nitric Oxide, Part F Oxidative and Nitrosative Stress in Redox
Regulation of Cell Signaling
Edited by ENRIQUE CADENAS AND LESTER PACKER

VOLUME 441. Nitric Oxide, Part G Oxidative and Nitrosative Stress in Redox
Regulation of Cell Signaling
Edited by ENRIQUE CADENAS AND LESTER PACKER

VOLUME 442. Programmed Cell Death, General Principles for Studying Cell
Death (Part A)
Edited by ROYA KHOSRAVI-FAR, ZAHRA ZAKERI, RICHARD A. LOCKSHIN,
AND MAURO PIACENTINI

VOLUME 443. Angiogenesis: *In Vitro* Systems
Edited by DAVID A. CHERESH

VOLUME 444. Angiogenesis: *In Vivo* Systems (Part A)
Edited by DAVID A. CHERESH

VOLUME 445. Angiogenesis: *In Vivo* Systems (Part B)
Edited by DAVID A. CHERESH

VOLUME 446. Programmed Cell Death, The Biology and Therapeutic
Implications of Cell Death (Part B)
Edited by ROYA KHOSRAVI-FAR, ZAHRA ZAKERI, RICHARD A. LOCKSHIN,
AND MAURO PIACENTINI

VOLUME 447. RNA Turnover in Bacteria, Archaea and Organelles
Edited by LYNNE E. MAQUAT AND CECILIA M. ARRAIANO

VOLUME 448. RNA Turnover in Eukaryotes: Nucleases, Pathways
and Anaylsis of mRNA Decay
Edited by LYNNE E. MAQUAT AND MEGERDITCH KILEDJIAN

VOLUME 449. RNA Turnover in Eukaryotes: Analysis of Specialized and Quality
Control RNA Decay Pathways
Edited by LYNNE E. MAQUAT AND MEGERDITCH KILEDJIAN

VOLUME 450. Fluorescence Spectroscopy
Edited by LUDWIG BRAND AND MICHAEL L. JOHNSON

VOLUME 451. Autophagy: Lower Eukaryotes and Non-Mammalian Systems (Part A)
Edited by DANIEL J. KLIONSKY

VOLUME 452. Autophagy in Mammalian Systems (Part B)
Edited by DANIEL J. KLIONSKY

VOLUME 453. Autophagy in Disease and Clinical Applications (Part C)
Edited by DANIEL J. KLIONSKY

VOLUME 454. Computer Methods (Part A)
Edited by MICHAEL L. JOHNSON AND LUDWIG BRAND

VOLUME 455. Biothermodynamics (Part A)
Edited by MICHAEL L. JOHNSON, JO M. HOLT, AND GARY K. ACKERS (RETIRED)

VOLUME 456. Mitochondrial Function, Part A: Mitochondrial Electron Transport Complexes and Reactive Oxygen Species
Edited by WILLIAM S. ALLISON AND IMMO E. SCHEFFLER

VOLUME 457. Mitochondrial Function, Part B: Mitochondrial Protein Kinases, Protein Phosphatases and Mitochondrial Diseases
Edited by WILLIAM S. ALLISON AND ANNE N. MURPHY

DEFINING THE MITOCHONDRIAL PROTEOME, INCLUDING PHOSPHORYLATED, ACETYLATED, AND PALMITOYLATED PROTEINS

CHAPTER ONE

THE MITOCHONDRIAL PROTEOME DATABASE: MITOP2

M. Elstner,*,‡ C. Andreoli,† T. Klopstock,‡ T. Meitinger,*,† and H. Prokisch*,†

Contents

1. Introduction	4
2. Yeast, Human, and Mouse Proteome Information	6
3. Mitochondrial Reference Set, Candidates, and SVM Prediction Score	8
4. Integrated Genome-Wide Approaches	9
4.1. Proteome, transcriptome, and sub-localization experiments in yeast	9
4.2. Proteome, transcriptome, and sub-localization experiments in human and mouse	10
4.3. In silico predictions	12
5. Other Search Options: Restrictions to Functional Categories, Gene Locus and Disease-Causing Proteins	12
6. Output Lists and Practical Examples	14
References	17

Abstract

Defining the mitochondrial proteome is a prerequisite for fully understanding the organelles function as well as mechanisms underlying mitochondrial pathology. The core functions of mitochondria include oxidative phosphorylation, amino acid metabolism, fatty acid oxidation, and ion homeostasis. In addition to these well-known functions, many crucial properties in cell signaling, cell differentiation and cell death are only now being elucidated, and with them the proteins involved. With the wealth of information arriving from single protein studies and sophisticated genome-wide approaches, MitoP2 was designed and is maintained to consolidate knowledge on mitochon-

* Institute of Human Genetics, Helmholtz Zentrum Munich—German Research Center for Environmental Health, Neuherberg, Germany
† Institute of Human Genetics, Technical University Munich, Munich, Germany
‡ Department of Neurology with Friedrich-Baur-Institute, Ludwig-Maximilians University, Munich, Germany

Methods in Enzymology, Volume 457
ISSN 0076-6879, DOI: 10.1016/S0076-6879(09)05001-0

3

drial proteins in one comprehensive database, thus making all pertinent data readily accessible (http://www.mitop2.de). Although the identification of the human mitochondrial proteome is ultimately the prime objective, integration of other species includes *Saccharomyces cerevisiae*, mouse, *Arabidopsis thaliana*, and *Neurospora crassa* so orthology between these species can be interrogated. Data from genome-wide studies can be individually retrieved and are also processed by a support vector machine (SVM) to generate a score that indicates the likelihood of a candidate protein having a mitochondrial location. Manually validated proteins constitute the reference set of the database that contains over 590 yeast, 920 human, and 1020 mouse entries, and that is used for benchmarking the SVM score. Multiple search options allow for the interrogation of the reference set, candidates, disease related proteins, chromosome locations as well as availability of mouse models. Taken together, MitoP2 is a valuable tool for basic scientists, geneticists, and clinicians who are investigating mitochondrial physiology and dysfunction.

1. INTRODUCTION

Systematic genome–wide studies have greatly facilitated progress toward defining the mitochondrial proteome (Meisinger *et al.*, 2008). Successful approaches include computational predictions of signaling sequences or sequence composition, sequence similarity and ancestry analysis, proteome mapping, mutant screening, expression profiling, protein–protein interaction, and cellular sub-localization studies (Gabaldón, 2006; Vo and Palsson, 2007). Since expression of certain genes may be restricted to developmental or pathological processes (i.e., apoptosis), a great many novel proteins can be expected from future studies looking at nonphysiological conditions.

The key advantage of these studies is an unbiased and comprehensive approach to the identification of mitochondrial candidates. Nevertheless they have the limitation of generating false negative and false positive results, that is, for biological and technical reasons, no such study will detect all mitochondrial proteins (i.e., reach 100% sensitivity) nor exclusively mitochondrial proteins either (i.e., reach 100% specificity). As an example, about 590 of the predicted 800 yeast proteins are certainly known. Conversely, high–throughput experiments have delivered more than 6000 candidates (Prokisch *et al.*, 2004). Likewise, with convincing evidence for only about 900 of an estimated 1500 proteins, the identification of all molecular components of human mitochondria is far from complete. In future, this could possibly be achieved by merging information coming from diverse approaches and species (Gabaldón, 2006). This is the fundamental rationale of creating and maintaining MitoP2 (Andreoli *et al.*, 2004; Prokisch *et al.*, 2006).

The study and integration of other species in the search for human mitochondrial proteins has pragmatic reasons. Due to the high conservation of core mitochondrial functions during evolution, detection of orthologous genes in model organisms is plausible. For example, yeast (*Saccharomyces cerevisiae*) is the most intensively studied eukaryotic model organism and the availability of full genome information based on open reading frames (ORFs), as well as convenient and inexpensive experimental conditions greatly simplifies its examination (Perocchi *et al.*, 2006). In elegant studies, yeast cultures have been mutated and manipulated to generate conditions that do justice to the intricate process of mitochondrial morphogenesis and function. In mouse, the homology to the human mitochondrial proteome is likely to be almost complete and the availability of mutant phenotypes available is growing steadily. Thereby, studies in different organisms mutually enhance knowledge about organelle structure and function. For this purpose, search functions in MitoP2 allow identification of the orthology between species including yeast, mouse, human, *Arabidopsis thaliana*, and *Neurospora crassa*.

Data coming from genome-wide studies on mitochondrial proteins are integrated into MitoP2 serving two main purposes. First, this information can be interrogated individually or in combination with other search parameters, depending on requirements. Secondly, by utilizing a support vector machine (SVM), these data serve to compile a combined score for each candidate, which provides a best estimate of mitochondrial localization. This score, together with information on the specific studies providing this evidence is given in the search output for any given protein. As a cautionary note it should be mentioned that in some cases integration of data into MitoP2 is not fully complete, depending on the data source and identifiers used.

Nevertheless, proteins that are being identified through genome-wide studies can only be considered candidates as their status needs to be tested in more detailed single protein studies. In MitoP2, proteins where thorough studies show convincing evidence of a mitochondrial location are provided as the reference set. This set results from a manual literature search of candidates and does not contain information from any of the integrated high-throughput approaches, thus allowing benchmarking of these against the reference set. It contains 597 yeast entries, as well as 924 human and 1022 mouse proteins and is easily retrievable through the input mask of MitoP2.

In combining the above aspects, any given yeast, human and mouse protein interrogated in MitoP2 will be distinguishable as a mitochondrial protein, a candidate or as a protein without any evidence for mitochondrial localization. Further database features include search functions according to chromosome location or biological function as well as numerous links to other databases. This chapter introduces the concept, structure and function of MitoP2.

2. YEAST, HUMAN, AND MOUSE PROTEOME INFORMATION

In its basic function MitoP2 is a protein database containing all entries available for human, mouse, and yeast, based on the Saccharomyces Genome Database (SGD) (Cherry *et al.*, 1997, 1998) and the "non-redundant" Swiss-Prot data sets (last release no. 56/2008, http://www.expasy.org). All consecutively described search functions are based upon this collection, and proteins can be extracted or excluded from the search by choosing the appropriate item according to the desired characteristics (Fig. 1.1). The yeast dataset contains all 6516 ORFs of the yeast genome and includes the subcellular localization and the functional annotation from SGD. It is complemented by the functional catalog entries from the Munich Information Centre for Protein Sequences (MIPS), which also provides the physical features, including the deoxyribonuleic acid (DNA) and protein sequence (Mewes *et al.*, 2008; Ruepp *et al.*, 2004). The human and mouse sections of the MitoP2 data sets comprise 38,453 and 32,833 protein entries, respectively. Human–mouse orthologs were determined by a bidirectional best BLAST hit or best BLAST hit with a score $<10^{-10}$. The orthologs are used equally for both the annotation of the human and mouse mitochondrial reference sets and the integration of high-throughput data sets. For each protein the description, the chromosomal localization, the subcellular localization, the cross-references and literature were extracted from the Swiss-Prot database. Functional annotations were compiled from Gene

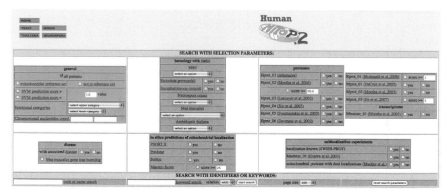

Figure 1.1   Search mask for the human section of the MitoP2 database (http://www.mitop2.de/). The mask is structured into several search boxes. Explanatory notes and the original literature can be accessed via the links. By choosing "all proteins" an output will be generated containing all available proteins without restrictions. The underlined text provides a link to the original reference or a pup up window with more details about the integrated data set or method used.

Ontology. The mitochondrial reference set proteins for mouse were also annotated according to the MIPS functional catalog and supplemented by physical features, DNA and protein sequence. In these cases, an explanation or PubMed link for the mitochondrial annotation is given. Human proteins already associated with Mendelian disorders are extracted from Swiss-Prot, linked to Online Mendelian Inheritance in Man (OMIM) and listed with the OMIM title (Hamosh *et al.*, 2005). In the output list, they are marked in red if they are mitochondrial or in orange if not (Fig. 1.2). In addition, the mouse entries are listed with phenotypes if a mouse model exists in the Mouse Genome Database (MGD) (Bult *et al.*, 2008). So far, more than 134 mouse models have been investigated that have mutations in genes coding

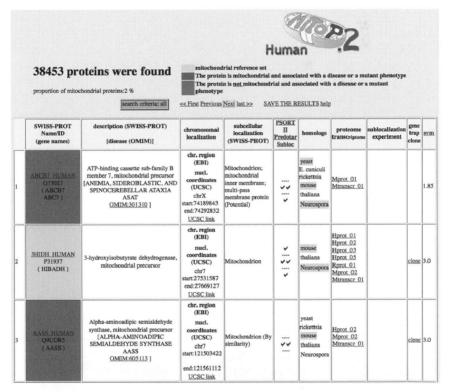

Figure 1.2 Part of the output list after query in the human section of the MitoP2 database. MitoP2 extracts 38,453 human proteins, 2% of which are known mitochondrial as related to the reference set. Proteins are color coded to indicate their mitochondrial and disease status. Also included are Swiss-Prot and OMIM descriptions, the chromosomal location, and its subcellular localization, as well as links to genome-wide studies providing evidence for its mitochondrial location. Links to the NCBI webpage provide quick access to information on gene trap clones where available. The SVM prediction score for mitochondrial localization provides an estimate.

mitochondrial proteins. For those users interested in creating mouse models, available mouse gene trap insertion cell lines from the International Gene Trap Consortium are listed as well (486 clones available in the human reference set, http://www.genetrap.org).

3. MITOCHONDRIAL REFERENCE SET, CANDIDATES, AND SVM PREDICTION SCORE

In the "general" box, the selection of searched proteins can be restricted to the mitochondrial reference set, and this selection can be further limited by adding other options. In yeast, the current reference set contains about 600 manually annotated proteins with experimental evidence for mitochondrial localization; in human the reference set contains more than 900 entries. The reference set is based on single gene studies only and does not contain information from any of the integrated high-throughput approaches. Thus, because no bias toward either of the high-throughput approaches was generated, it also allows for their benchmarking. In contrast, mitochondrial candidates can be evaluated by selecting all proteins that are not in the reference set (by clicking "not in reference set") and by requesting a computed evidence score. This score was implemented to improve predictions by utilizing a SVM (http://svmlight.joachims.org). The SVMs are learning machines based on a statistical learning theory used for solving classification tasks. They are trained by using a reference set of mitochondrial proteins and a set of proteins with a known localization to other cellular compartments. For each of the proteins, a multi-dimensional vector was identified using the datasets of available systematic studies. This results in a multi-dimensional input matrix which is used to train the SVM. After training, the SVM predicts mitochondrial proteins with a specificity and sensitivity considerably higher than the individual studies, specific for the cutoff and the organism chosen (Fig. 1.2). A threshold must be defined meeting individual demands, depending on the desired specificity and sensitivity of the output. The specificity to identify mitochondrial-localized proteins is estimated by the proportion of detected proteins (in percent) that are part of the mitochondrial reference set. The sensitivity of the output is calculated through the proportion of the reference set (in percent) that is covered. Thus, choosing a higher value for the SVM will result in a smaller set of proteins that are more likely to be mitochondrial. Conversely, choosing a lower value will result in a larger output with the risk of false positive results.

4. INTEGRATED GENOME-WIDE APPROACHES

4.1. Proteome, transcriptome, and sub-localization experiments in yeast

The right column of the entry mask contains options for restriction to, or exclusion of experimental data from proteome and transcriptome, as well as sub-localization experiments. In yeast, three systemic studies on subcellular localization are included under the alias Ysubloc_01–03 (Huh *et al.*, 2003; Kumar *et al.*, 2002; Matsuyama *et al.*, 2006). This approach used immunofluorescence imaging analysis for the detection of tagged proteins and directly tracked proteins in the specific organelles they are directed to. Methods based on transcriptome data take advantage of the ability of budding yeast to grow under anaerobic (fermentable) and aerobic (non-fermentable) conditions, by comparing gene expression in both states. The so-called "diauxic shift" occurs when multiple sets of genes are differently controlled to adjust to changing growth conditions, which can also be simulated by over-expressing the transcription factor Hap4. Three data sets contain transcriptome data of differentially expressed genes under fermentable and non-fermentable growth conditions that are regulated in response to the diauxic shift or by Hap4 (Ytransc_01–03) (DeRisi *et al.*, 1997; Lascaris *et al.*, 2003; Prokisch *et al.*, 2004). Proteome analysis of purified mitochondria includes seven studies (Yprot_01–06) (Ohlmeier *et al.*, 2004; Pflieger *et al.*, 2002; Prokisch *et al.*, 2004; Reinders *et al.*, 2006; Sickmann *et al.*, 2003; Zahedi *et al.*, 2006), one of which contains proteome data on *N. crassa* (Prokisch, unpublished). These studies are mostly based on highly purified mitochondrial fractions analyzed by SDS–PAGE based or tandem mass spectrometry techniques or a combination of both. Ghaemmaghami and colleagues introduced a novel method creating a *S. cerevisiae* fusion library where each ORF is tagged with a high-affinity epitope and expressed from its natural chromosomal location (Ghaemmaghami *et al.*, 2003). Through immunodetection of the common tag a census of proteins expressed during log-phase growth and measurements of their absolute levels were obtained. This information on protein abundance in copy numbers can be interrogated in MitoP2, as well as the confidence of interaction with mitochondrial proteins found in two other studies (Perocchi *et al.*, 2006; von Mering *et al.*, 2002). Here, high confidence is achieved when protein–protein interaction was shown with two independent methods, or with one method in different study datasets. In yeast, the "disease section" contains searchable null mutant phenotypes, which stem from two genome-wide investigations (Dimmer *et al.*, 2002; Steinmetz *et al.*, 2002). A null allele is a mutant copy of a gene that completely lacks that gene's normal function and at the phenotypic level is equivalent to a deletion of the entire locus. Yeast mutants of genes located in the nucleus and required for respiration competence (as defined by growing in fermentable and non-fermentable growth conditions) are commonly referred to as nuclear *petite* or *pet* mutants. Apart from these intrinsic features of mitochondrial oxidative metabolism, mutants with visible disruption of wild-type mitochondrial morphology can be further distinguished. Class I mutant genes encode proteins that are essential for establishing wild-type mitochondrial morphology. Class II mutants are respiratory competent and appear to exhibit mitochondrial morphology defects only under certain conditions. Class III mutants display a *pet* phenotype in addition to their mitochondrial morphology defect. From this information, conclusions can be drawn for protein function regarding their role in morphogenesis, membrane trafficking, fission, and fusion.

The agreement of the reference set with the datasets available for these studies can be expressed as specificity and sensitivity of the method in relation to the reference set. An overview of the yeast dataset is given in Fig. 1.3. Owing to the incompleteness of the reference set, the specificities and sensitivities given are conservative estimates.

4.2. Proteome, transcriptome, and sub-localization experiments in human and mouse

Studies in human tissue are more complicated due to restricted tissue availability but essentially the same techniques are being applied as in model organisms. In the effort to improve sensitivity and specificity, various techniques and combination of techniques are being employed to either improve separation of organelle fractions or protein detection. In human heart, proteome analysis was accomplished by a combination of gradient centrifugation and gel based purification techniques followed by mass spectrometry or SDS–PAGE detection (Hprot_01 + 06) (Devreese *et al.*, 2002; Gaucher *et al.*, 2004; Taylor *et al.*, 2003). Two-dimensional gel electrophoresis was used in studies on human placenta (Hprot_03) (Lescuyer *et al.*, 2003) and human lymphoblastoid cell lines (Hprot_04) (Xie *et al.*, 2005). These were augmented by matrix-assisted laser desorption/ionization-mass spectrometry in neuroblastoma cell lines (Hprot_05) (Fountoulakis and Schlaeger, 2003). Data coming from studies on rodent tissue are a valuable surrogate for studies in human tissue due to the above-mentioned orthology. In rat tissue, several techniques for organelle separation were applied to improve specificity, since impurity of fractions is the biggest interference factor (Rprot_01) (McDonald *et al.*, 2006). In mouse, proteins were extracted from highly purified liver mitochondrial inner membranes with organic acid and analyzed by two-dimensional liquid chromatography coupled to tandem mass spectrometry (Mprot_01) (Da Cruz *et al.*, 2003). Ozawa *et al.* presented an intriguing method that allows rapid identification of novel proteins compartmentalized in mitochondria by screening large-scale cDNA libraries combined with split-enhanced

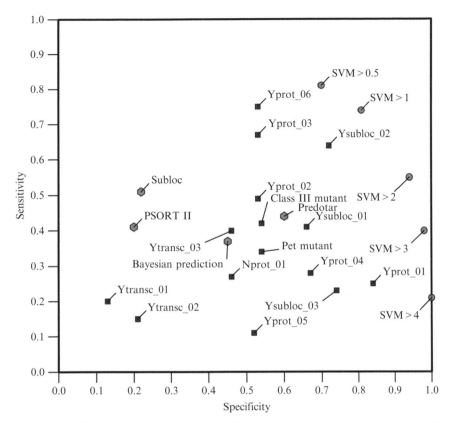

Figure 1.3 Plot shows sensitivity and specificity of all the approaches used to identify mitochondrial proteins. The yeast datasets were benchmarked against the mitochondrial reference set. Each point represents a dataset whose position is determined by benchmarking against the 590 reference proteins from MitoP2-yeast. The bioinformatics datasets (hexagon) are PSORT II (Nakai and Horton, 1999), Predotar (Small *et al.*, 2004), Subloc (Hua and Sun, 2001), and the Bayesian system in yeast (Drawid and Gerstein, 2000); the experimental datasets (square) are Yprot.01–06 (Ohlmeier *et al.*, 2004; Pflieger *et al.*, 2002; Prokisch *et al.*, 2004; Reinders *et al.*, 2006; Sickmann *et al.*, 2003; Zahedi *et al.*, 2006;), Ysubloc.01–03 (Huh *et al.*, 2003) (Kumar *et al.*, 2002; Matsuyama *et al.*, 2006), Ytransc.01–03 (DeRisi *et al.*, 1997; Lascaris *et al.*, 2003; Prokisch *et al.*, 2004), Nprot.01 and the yeast null mutant phenotypes (Dimmer *et al.*, 2002; Steinmetz *et al.*, 2002). The predictive score for a mitochondrial protein (SVM, circle) was based on the combination of the systematic datasets, calculated for different thresholds.

green fluorescent protein techniques (Msubloc_01) (Ozawa *et al.*, 2003). Jin *et al.* compared the mitochondrial protein profiles of MES cells (a mouse dopaminergic cell line) exposed to rotenone versus control, thus introducing an interesting functional component into the study (Mprot_03) (Jin *et al.*, 2007). They used a recently developed proteomic technology called stable isotope labeling by amino acids in cell culture (SILAC) combined with PAGE and liquid chromatography–tandem mass spectrometry. The studies presented by Moothas group consistently lead in sensitivity and specificity by using integrated approaches including targeted sequence prediction, protein domain enrichment, presence of *cis*-regulatory motifs, yeast homology, ancestry, tandem-mass spectrometry, co-expression and transcriptional induction during mitochondrial biogenesis (Hprot_02, Mprot_02) (Calvo *et al.*, 2006; Mootha *et al.*, 2003). The individual components of these studies display the same restrictions in sensitivity and specificity (Mtransc_01) (Mootha *et al.*, 2003) Taken together, as in yeast, studies differ in sensitivity and specificity, and combining the available information greatly increases the evidence (Fig. 1.4).

4.3. *In silico* predictions

A putative mitochondrial localization can be predicted by several *in silico* approaches and the most relevant computational strategies are integrated in MitoP2. One option is to search for homology to a known mitochondrial protein from another species (defined as bidirectional best BLAST hit of best BLAST hit with a score $<10^{-10}$) (Altschul *et al.*, 1997). For example, homologous proteins from *Rickettsia prowazekii*, a prokaryote believed to be closest to the evolutionary origin of mitochondria, can be included and those for *Encephallitozoon cuniculi*, a eukaryote lacking mitochondria, be excluded. Another option is the homology search to one or more of the remaining four organisms searchable in the database. A different *in silico* concept computes information on either the presence or absence of a mitochondrial targeting sequences or a predictive amino acid composition. In MitoP2 this is realized by the inclusion of various different computational approaches, namely PSORT (Nakai and Horton, 1999), Predotar (Small *et al.*, 2004), Subloc (Hua and Sun, 2001), and the Bayesian system in yeast (Drawid and Gerstein, 2000). Of particular interest is the option "Maestro score." A naïve Bayes classifier was implemented with a computer program called Maestro to integrate eight genome scale approaches. For a detailed explanation see Calvo *et al.* (2006). Working much like the SVM, raising the cut-off threshold will increase specificity at the expense of sensitivity (see Fig. 1.2). The most recent data integrated is a protein expression study of 14 mouse tissues (Pagliarini *et al.*, 2008).

5. Other Search Options: Restrictions to Functional Categories, Gene Locus and Disease-Causing Proteins

Categories can be chosen according to the major functional entities of mitochondria (e.g., respiratory chain complex, protein synthesis or apoptosis). In yeast, the search can be further restricted to the finer protein modules

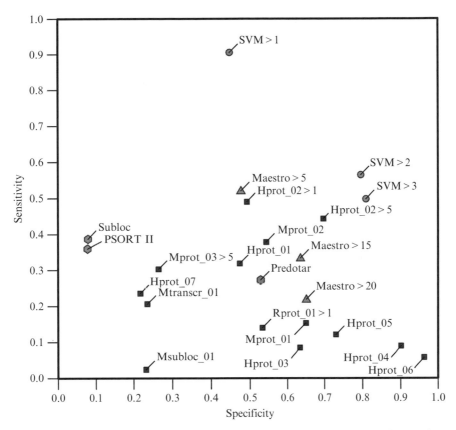

Figure 1.4 The plot shows sensitivity and specificity of all the approaches used to identify mitochondrial proteins. The human datasets were benchmarked against the mitochondrial reference set. Each point represents a dataset whose position is determined by benchmarking against the 922 reference proteins from MitoP2-human. The bioinformatics datasets (hexagon) are PSORT II (Nakai and Horton, 1999), Predotar (Small *et al.*, 2004), Subloc (Hua and Sun, 2001), and Maestro (Calvo *et al.*, 2006; Pagliarini *et al.*, 2008) (triangle, at different thresholds); the experimental datasets (square) are Hprot.01–07 (Hprot.02 at different thresholds) (Devreese *et al.*, 2002; Fountoulakis and Schlaeger, 2003; Gaucher *et al.*, 2004; Lescuyer *et al.*, 2003; Mootha *et al.*, 2003; Taylor *et al.*, 2003; Xie *et al.*, 2005), Mprot.01–03 (Da Cruz *et al.*, 2003; Jin *et al.*, 2007; Mootha *et al.*, 2003), Mtransc.01 (Mootha *et al.*, 2003), Msubloc.01 (Ozawa *et al.*, 2003), and Rprot.01 (McDonald *et al.*, 2006). The predictive score for a mitochondrial protein (SVM, circle) was based on the combination of the systematic datasets, calculated for different thresholds.

of similar function as described in Perocchi *et al.* (2006). In all organisms, the chromosomal region of coding genes can be restricted, which is particularly helpful when mitochondrial candidate genes in this region need to be identified. Another option is the limitation to proteins associated with human diseases. So far, more than 153 of these proteins have shown to be involved in human diseases and can be interrogated by selecting the reference set and disease association. Most of these proteins take part in the metabolism of amino acids, nucleic acid, lipids, heme, or coenzymes (MitoP2, OMIM). In addition, defects in the mitochondrial respiratory chain/oxidative phosphorylation system are responsible for a panoply of human disorders, ranging from sporadic myopathies to fatal encephalomyopathies. Recent epidemiological studies suggest that disorders of the mitochondrial respiratory chain, that is, the classical mitochondrial diseases, affect at least nine in 100,000 of the population, making these disorders a common genetically determined disease entity (Schaefer *et al.*, 2008). About half of the patients carry mutations in the mitochondrial DNA (mtDNA) and so far, more than 100 different pathogenic mtDNA point mutations and an even larger number of different mtDNA deletions have been identified. The causes of the disease in a large proportion of the remaining patients with nuclear mutations have yet to be determined. This points out the relevance of future mitochondrial proteome studies for diagnosis and treatment of human disease.

The last search option is for the availability of a mouse gene trap homolog (for human) and existing gene trap clones or mutant phenotypes (available from the German Gene Trap Consortium or listing in the MGD), or null mutant phenotypes in yeast.

6. Output Lists and Practical Examples

The result of a search is given in output lists, which may include annotated mitochondrial proteins from the reference set (labeled in green), as well as candidates. For example, in the yeast mask, when selecting mitochondrial subcellular localization according to Huh *et al.* (2003) (YSubloc_02) and an *in silico* prediction tool such as Predotar (Small *et al.*, 2004), MitoP2 extracts a list of 258 proteins (Fig. 1.5). This list contains 210 proteins from the mitochondrial reference set and 48 candidate proteins fulfilling the two selection criteria. The specificity in detecting mitochondrial proteins of this combined query is estimated from the percentage of the listed proteins ($n = 258$) that are present in the mitochondrial reference set ($n = 213$), in this case 81%. The sensitivity of the combined two approaches in detecting the known mitochondrial proteins (yeast reference set contains 546 entries) is 36% (213/596). Each individual protein list (given in a single line) contains the ORF description from SGD, the subcellular localization according to SGD or MitoP2 annotations, and the detailed information from high-throughput experiments and *in silico* calculations. In addition to the matrix provided by the columns with systematic experimental results and *in silico* data, the user can access detailed information for each single

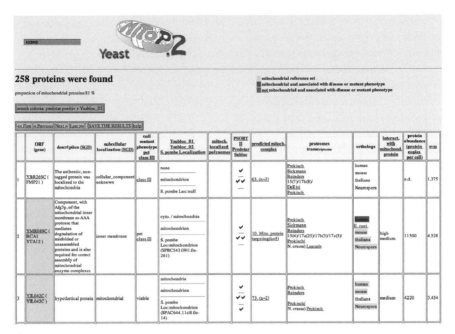

Figure 1.5 Part of the output list after query in the yeast section of the MitoP2 database. MitoP2 extracts 258 of 6516 yeast proteins predicted to be mitochondrial by the Predotar program (Small *et al.*, 2004) and by the high-throughput subcellular localization of Huh *et al.* (2003).

protein by selecting the ORF button. An extra page linked to each entry provides (1) a description of the individual protein compiled from SGD plus additional functional, genetic, and biophysical properties from MIPS (CYGD, Comprehensive Yeast Genome Database); (2) the corresponding entry from the Gene Ontology database (Ashburner *et al.*, 2000); (3) a list of homologs from other species generated by bidirectional best BLAST hits; and (4) a compilation of low-, medium-, or high-confidence protein–protein interactions weighted according to von Mering *et al.* (2002) in addition to protein–protein interactions from single experiments. The high-throughput interaction data sets are provided with links to the original literature.

A common question in human genetics that can be addressed by using MitoP2 is the identification of candidate genes. For example, Smeitink *et al.* (2006) investigated the genetic basis for a mitochondrial protein-synthesis defect associated with a combined oxidative phosphorylation enzyme deficiency in two patients, one of whom presented with encephalomyopathy and the other with hypertrophic cardiomyopathy. Single nucleotide polymorphism (SNP) analysis identified a long segment of consecutive homozygous SNPs on chromosome 2 spanning 22.1 Mb and another segment on chromosome 12 spanning 30.79 Mb. With use of MitoP2, two proteins with known function in mitochondrial translation were identified within these regions of several hundred genes: the mitochondrial ribosomal protein S35 and the mitochondrial translation elongation factor *TSFM* (Fig. 1.6). Sequence analysis of *TSFM* revealed the same homozygous C997T mutation in exon 7 in both patients. Figure 1.7 represents the entry for *TSFM* as an example for a single protein entry in the human section of MitoP2, which provides in addition to the known matrix lane (i) a list of homologies to sequences from other species, including the BLAST E value and the percentage of the aligned protein length, linked to the corresponding MitoP2 pages; (ii) if it exists, a Swiss-Prot description of an associated disease; (iii) the corresponding Gene Ontology annotations for molecular protein function, biological processes in which the protein is involved, and cellular components; (iv) any available literature about the protein and protein variants listed with authors and title; and (v) a table of cross references annotated in Swiss-Prot.

A last example of the application of MitoP2 has recently been reported by Dinur-Mills and colleagues (Dinur-Mills *et al.*, 2008). They divided the MitoP2 reference set of previously reported mitochondrial proteins into a group of proteins with dual distribution in more than one subcellular compartment and an exclusively mitochondrial group. They found that dually localized mitochondrial proteins comprise approximately one quarter of the predicted mitochondrial proteome, and have a weaker mitochondrial

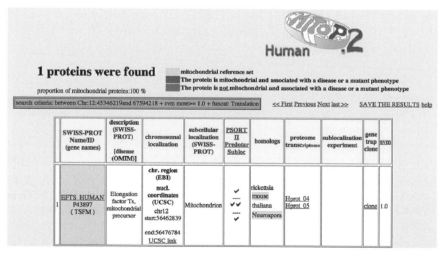

Figure 1.6   MitoP2 can be used to identify mitochondrial proteins and candidates in a specific chromosomal region. In this practical example, the search criteria were chosen as the chromosomal region (Chr:12:45346219–67594218), the functional category "protein synthesis/translation," and a SVM score >1.

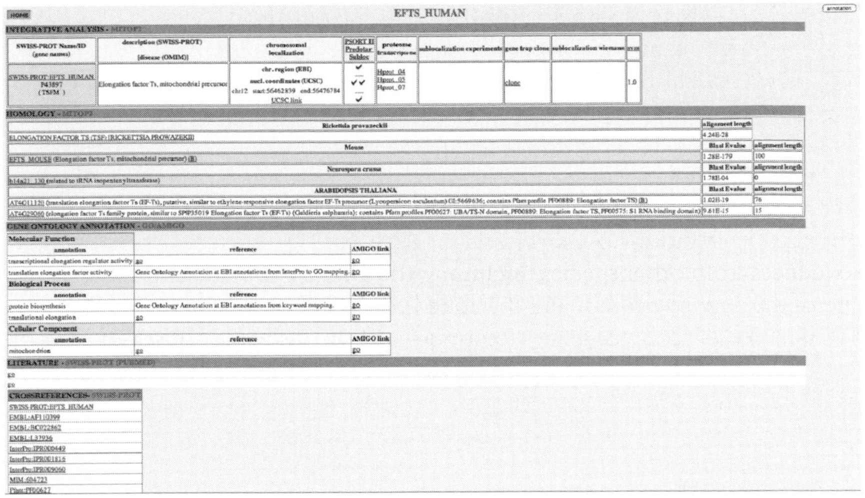

Figure 1.7 Single protein entry for the human translation elongation factor EFTS (TSFM) in the MitoP2 database. For each protein entry MitoP2 presents the Swiss-Prot name and description, the chromosomal location, the prediction results from mitochondrial prediction programs, homologous proteins in other organisms, gene ontology annotations and relevant literature.

targeting sequence score and a lower whole-protein net charge, when compared to exclusive mitochondrial proteins. These results establish dual targeting as a widely abundant phenomenon that should affect our concepts of gene expression and protein function.

REFERENCES

Altschul, S. F., Madden, T. L., Schäffer, A. A., Zhang, J., Zhang, Z., Miller, W., and Lipman, D. J. (1997). Gapped BLAST and PSI-BLAST: A new generation of protein database search programs. *Nucleic Acids Res.* **25,** 3389–3402.

Andreoli, C., Prokisch, H., Hörtnagel, K., Mueller, J. C., Münsterkötter, M., Scharfe, C., and Meitinger, T. (2004). MitoP2, an integrated database on mitochondrial proteins in yeast and man. *Nucleic Acids Res.* **32,** D459–D462.

Ashburner, M., Ball, C. A., Blake, J. A., Botstein, D., Butler, H., Cherry, J. M., Davis, A. P., Dolinski, K., Dwight, S. S., Eppig, J. T., Harris, M. A., Hill, D. P., *et al.* (2000). Gene ontology: Tool for the unification of biology. The gene ontology consortium. *Nat. Genet.* **25,** 25–29.

Bult, C. J., Eppig, J. T., Kadin, J. A., Richardson, J. E., and Blake, J. A.Mouse Genome Database Group (2008). The Mouse Genome Database (MGD): Mouse biology and model systems. *Nucleic Acids Res.* **36,** D724–D728.

Calvo, S., Jain, M., Xie, X., Sheth, S. A., Chang, B., Goldberger, O. A., Spinazzola, A., Zeviani, M., Carr, S. A., and Mootha, V. K. (2006). Systematic identification of human mitochondrial disease genes through integrative genomics. *Nat. Genet.* **38,** 576–582.

Cherry, J. M., Adler, C., Ball, C., Chervitz, S. A., Dwight, S. S., Hester, E. T., Jia, Y., Juvik, G., Roe, T., Schroeder, M., Weng, S., and Botstein, D. (1998). SGD: Saccharomyces Genome Database. *Nucleic Acids Res.* **26,** 73–79.

Cherry, J. M., Ball, C., Weng, S., Juvik, G., Schmidt, R., Adler, C., Dunn, B., Dwight, S., Riles, L., Mortimer, R. K., and Botstein, D. (1997). Genetic and physical maps of *Saccharomyces cerevisiae. Nature* **387,** 67–73.

Da Cruz, S., Xenarios, I., Langridge, J., Vilbois, F., Parone, P. A., and Martinou, J. C. (2003). Proteomic analysis of the mouse liver mitochondrial inner membrane. *J. Biol. Chem.* **278,** 41566–41571.

DeRisi, J. L., Iyer, V. R., and Brown, P. O. (1997). Exploring the metabolic and genetic control of gene expression on a genomic scale. *Science* **278,** 680–686.

Devreese, B., Vanrobaeys, F., Smet, J., Van Beeumen, J., and Van Coster, R. (2002). Mass spectrometric identification of mitochondrial oxidative phosphorylation subunits separated by two-dimensional blue-native polyacrylamide gel electrophoresis. *Electrophoresis* **23,** 2525–2533.

Dimmer, K. S., Fritz, S., Fuchs, F., Messerschmitt, M., Weinbach, N., Neupert, W., and Westermann, B. (2002). Genetic basis of mitochondrial function and morphology in *Saccharomyces cerevisiae. Mol. Biol. Cell.* **13,** 847–853.

Dinur-Mills, M., Tal, M., and Pines, O. (2008). Dual targeted mitochondrial proteins are characterized by lower MTS parameters and total net charge. *PLoS ONE* **3,** e2161.

Drawid, A., and Gerstein, M. (2000). A Bayesian system integrating expression data with sequence patterns for localizing proteins: Comprehensive application to the yeast genome. *J. Mol. Biol.* **301,** 1059–1075.

Fountoulakis, M., and Schlaeger, E. J. (2003). The mitochondrial proteins of the neuroblastoma cell line IMR-32. *Electrophoresis* **24,** 260–275.

Gabaldón, T. (2006). Computational approaches for the prediction of protein function in the mitochondrion. *Am. J. Physiol. Cell. Physiol.* **291,** C1121–C1128.

Gaucher, S. P., Taylor, S. W., Fahy, E., Zhang, B., Warnock, D. E., Ghosh, S. S., and Gibson, B. W. (2004). Expanded coverage of the human heart mitochondrial proteome using multidimensional liquid chromatography coupled with tandem mass spectrometry. *J. Proteome Res.* **3,** 495–505.

Ghaemmaghami, S., Huh, W. K., Bower, K., Howson, R. W., Belle, A., Dephoure, N., O'Shea, E. K., and Weissman, J. S. (2003). Global analysis of protein expression in yeast. *Nature* **425,** 737–741.

Hamosh, A., Scott, A. F., Amberger, J. S., Bocchini, C. A., and McKusick, V. A. (2005). Online Mendelian Inheritance in Man (OMIM), a knowledgebase of human genes and genetic disorders. *Nucleic Acids Res.* **33,** D514–D517.

Hua, S., and Sun, Z. (2001). Support vector machine approach for protein subcellular localization prediction. *Bioinformatics* **17,** 721–728.

Huh, W. K., Falvo, J. V., Gerke, L. C., Carroll, A. S., Howson, R. W., Weissman, J. S., and O'Shea, E. K. (2003). Global analysis of protein localization in budding yeast. *Nature* **425,** 686–691.

Jin, J., Davis, J., Zhu, D., Kashima, D. T., Leroueil, M., Pan, C., Montine, K. S., and Zhang, J. (2007). Identification of novel proteins affected by rotenone in mitochondria of dopaminergic cells. *BMC Neurosci.* **8,** 67.

Kumar, A., Cheung, K. H., Tosches, N., Masiar, P., Liu, Y., Miller, P., and Snyder, M. (2002). The TRIPLES database: A community resource for yeast molecular biology. *Nucleic Acids Res.* **30,** 73–75.

Lascaris, R., Bussemaker, H. J., Boorsma, A., Piper, M., van der Spek, H., Grivell, L., and Blom, J. (2003). Hap4p overexpression in glucose-grown *Saccharomyces cerevisiae* induces cells to enter a novel metabolic state. *Genome Biol.* **4,** R3.

Lescuyer, P., Strub, J. M., Luche, S., Diemer, H., Martinez, P., Van Dorsselaer, A., Lunardi, J., and Rabilloud, T. (2003). Progress in the definition of a reference human mitochondrial proteome. *Proteomics* **3,** 157–167.

Matsuyama, A., Arai, R., Yashiroda, Y., Shirai, A., Kamata, A., Sekido, S., Kobayashi, Y., Hashimoto, A., Hamamoto, M., Hiraoka, Y., Horinouchi, S., and Yoshida, M. (2006). ORFeome cloning and global analysis of protein localization in the fission yeast Schizosaccharomyces pombe. *Nat. Biotechnol.* **24,** 841–847.

McDonald, T., Sheng, S., Stanley, B., Chen, D., Ko, Y., Cole, R. N., Pedersen, P., and Van Eyk, J. E. (2006). Expanding the subproteome of the inner mitochondria using protein separation technologies: One- and two-dimensional liquid chromatography and two-dimensional gel electrophoresis. *Mol. Cell. Proteomics* **5,** 2392–2411.

Meisinger, C., Sickmann, A., and Pfanner, N. (2008). The mitochondrial proteome: From inventory to function. *Cell* **134,** 22–24.

Mewes, H. W., Dietmann, S., Frishman, D., Gregory, R., Mannhaupt, G., Mayer, K. F., Münsterkötter, M., Ruepp, A., Spannagl, M., Stümpflen, V., and Rattei, T. (2008). MIPS: Analysis and annotation of genome information in 2007. *Nucleic Acids Res.* **36,** D196–D201.

Mootha, V. K., Bunkenborg, J., Olsen, J. V., Hjerrild, M., Wisniewski, J. R., Stahl, E., Bolouri, M. S., Ray, H. N., Sihag, S., Kamal, M., Patterson, N., Lander, E. S., *et al.* (2003). Integrated analysis of protein composition, tissue diversity, and gene regulation in mouse mitochondria. *Cell* **115,** 629–640.

Nakai, K., and Horton, P. (1999). PSORT: A program for detecting sorting signals in proteins and predicting their subcellular localization. *Trends Biochem. Sci.* **24,** 34–36.

Ohlmeier, S., Kastaniotis, A. J., Hiltunen, J. K., and Bergmann, U. (2004). The yeast mitochondrial proteome, a study of fermentative and respiratory growth. *J. Biol. Chem.* **279,** 3956–3979.

Ozawa, T., Sako, Y., Sato, M., Kitamura, T., and Umezawa, Y. (2003). A genetic approach to identifying mitochondrial proteins. *Nat. Biotechnol.* **21,** 287–293.

Pagliarini, D. J., Calvo, S. E., Chang, B., Sheth, S. A., Vafai, S. B., Ong, S. E., Walford, G. A., Sugiana, C., Boneh, A., Chen, W. K., Hill, D. E., Vidal, M., *et al.* (2008). A mitochondrial protein compendium elucidates complex I disease biology. *Cell* **134,** 112–123.

Perocchi, F., Jensen, L. J., Gagneur, J., Ahting, U., von Mering, C., Bork, P., Prokisch, H., and Steinmetz, L. M. (2006). Assessing systems properties of yeast mitochondria through an interaction map of the organelle. *PLoS Genet.* **2,** e170.

Pflieger, D., Le Caer, J. P., Lemaire, C., Bernard, B. A., Dujardin, G., and Rossier, J. (2002). Systematic identification of mitochondrial proteins by LC-MS/MS. *Anal. Chem.* **74,** 2400–2406.

Prokisch, H., Andreoli, C., Ahting, U., Heiss, K., Ruepp, A., Scharfe, C., and Meitinger, T. (2006). MitoP2: The mitochondrial proteome database—Now including mouse data. *Nucleic Acids Res.* **34,** D705–D711.

Prokisch, H., Scharfe, C., Camp, D. G., Xiao, W., David, L., Andreoli, C., Monroe, M. E., Moore, R. J., Gritsenko, M. A., Kozany, C., Hixson, K. K., Mottaz, H. M., *et al.* (2004). Integrative analysis of the mitochondrial proteome in yeast. *PLoS Biol.* **2,** e160.

Reinders, J., Zahedi, R. P., Pfanner, N., Meisinger, C., and Sickmann, A. (2006). Toward the complete yeast mitochondrial proteome: Multidimensional separation techniques for mitochondrial proteomics. *J. Proteome Res.* **5,** 1543–1554.

Ruepp, A., Zollner, A., Maier, D., Albermann, K., Hani, J., Mokrejs, M., Tetko, I., Güldener, U., Mannhaupt, G., Münsterkötter, M., and Mewes, H. W. (2004). The FunCat, a functional annotation scheme for systematic classification of proteins from whole genomes. *Nucleic Acids Res.* **32,** 5539–5545.

Schaefer, A. M., McFarland, R., Blakely, E. L., He, L., Whittaker, R. G., Taylor, R. W., Chinnery, P. F., and Turnbull, D. M. (2008). Prevalence of mitochondrial DNA disease in adults. *Ann. Neurol.* **63,** 35–39.

Sickmann, A., Reinders, J., Wagner, Y., Joppich, C., Zahedi, R., Meyer, H. E., Schönfisch, B., Perschil, I., Chacinska, A., Guiard, B., Rehling, P., Pfanner, N., *et al.* (2003). The proteome of *Saccharomyces cerevisiae* mitochondria. *Proc. Natl. Acad. Sci. USA* **100,** 13207–13212.

Small, I., Peeters, N., Legeai, F., and Lurin, C. (2004). Predotar: A tool for rapidly screening proteomes for N-terminal targeting sequences. *Proteomics* **4,** 1581–1590.

Smeitink, J. A., Elpeleg, O., Antonicka, H., Diepstra, H., Saada, A., Smits, P., Sasarman, F., Vriend, G., Jacob-Hirsch, J., Shaag, A., Rechavi, G., Welling, B., *et al.* (2006). Distinct clinical phenotypes associated with a mutation in the mitochondrial translation elongation factor EFTs. *Am. J. Hum. Genet.* **79,** 869–877.

Steinmetz, L. M., Scharfe, C., Deutschbauer, A. M., Mokranjac, D., Herman, Z. S., Jones, T., Chu, A. M., Giaever, G., Prokisch, H., Oefner, P. J., and Davis, R. W. (2002). Systematic screen for human disease genes in yeast. *Nat. Genet.* **31,** 400–404.

Taylor, S. W., Fahy, E., Zhang, B., Glenn, G. M., Warnock, D. E., Wiley, S., Murphy, A. N., Gaucher, S. P., Capaldi, R. A., Gibson, B. W., and Ghosh, S. S. (2003). Characterization of the human heart mitochondrial proteome. *Nat. Biotechnol.* **21,** 281–286.

Vo, T. D., and Palsson, B. O. (2007). Building the power house: Recent advances in mitochondrial studies through proteomics and systems biology. *Am. J. Physiol. Cell. Physiol.* **292,** C164–C177.

von Mering, C., Krause, R., Snel, B., Cornell, M., Oliver, S. G., Fields, S., and Bork, P. (2002). Comparative assessment of large-scale data sets of protein–protein interactions. *Nature* **417,** 399–403.

Xie, J., Techritz, S., Haebel, S., Horn, A., Neitzel, H., Klose, J., and Schuelke, M. (2005). A two-dimensional electrophoretic map of human mitochondrial proteins from immortalized lymphoblastoid cell lines: A prerequisite to study mitochondrial disorders in patients. *Proteomics* **5,** 2981–2999.

Zahedi, R. P., Sickmann, A., Boehm, A. M., Winkler, C., Zufall, N., Schönfisch, B., Guiard, B., Pfanner, N., and Meisinger, C. (2006). Proteomic analysis of the yeast mitochondrial outer membrane reveals accumulation of a subclass of preproteins. *Mol. Biol. Cell.* **17,** 1436–1450.

CHAPTER TWO

PREDICTING PROTEOMES OF MITOCHONDRIA AND RELATED ORGANELLES FROM GENOMIC AND EXPRESSED SEQUENCE TAG DATA

Daniel Gaston, Anastasios D. Tsaousis, *and* Andrew J. Roger

Contents

1. Introduction	22
1.1. Evolution and diversity of mitochondria and mitochondrion-derived organelles	23
1.2. Hydrogenosomes	24
1.3. Mitosomes	24
1.4. Mitochondrial proteomic and genomic diversity	25
1.5. Organellar proteome characterization	25
1.6. Mitochondrial targeting and import	26
1.7. Computational methods for predicting mitochondrial localization	27
2. Data Preparation	28
3. Screening Large Datasets for Mitochondrial Proteins	29
3.1. CBOrg	29
3.2. Predotar	32
3.3. MITOPRED	32
4. Detailed Analysis	33
4.1. Programs for N-terminal targeting sequence prediction for single sequences	33
4.2. Sequence feature-based prediction for prediction of mitochondrial localization	35
4.3. Mixture approaches	37
4.4. Meta-analysis	39
4.5. Assignment of localization	41

Centre for Comparative Genomics and Evolutionary Bioinformatics, Department of Biochemistry and Molecular Biology, Dalhousie University, Halifax, Nova Scotia, Canada

Methods in Enzymology, Volume 457
ISSN 0076-6879, DOI: 10.1016/S0076-6879(09)05002-2

5. Comparison of Methods   41
    5.1. Conclusion   43
Acknowledgments   44
References   44

Abstract

In eukaryotes, determination of the subcellular location of a novel protein encoded in genomic or transcriptomic data provides useful clues as to its possible function. However, experimental localization studies are expensive and time-consuming. As a result, accurate *in silico* prediction of subcellular localization from sequence data alone is an extremely important field of study in bioinformatics. This is especially so as genomic studies expand beyond model system organisms to encompass the full diversity of eukaryotes. Here we review some of the more commonly used programs for prediction of proteins that function in mitochondria, or mitochondrion-related organelles in diverse eukaryotic lineages and provide recommendations on how to apply these methods. Furthermore, we compare the predictive performance of these programs on a mixed set of mitochondrial and non-mitochondrial proteins. Although N-terminal targeting peptide prediction programs tend to have the highest accuracy, they cannot be effectively used for partial coding sequences derived from high-throughput expressed sequence tag surveys where data for the N-terminus of the encoded protein is often missing. Therefore methods that do not rely on the presence of an N-terminal targeting sequence alone are extremely useful, especially for expressed sequence tag data. The best strategy for classification of unknown proteins is to use multiple programs, incorporating a variety of prediction strategies, and closely examine the predictions with an understanding of how each of those programs will likely handle the data.

1. INTRODUCTION

Eukaryotic cells have subcellular compartments such as the nucleus, an endomembrane system, mitochondria, peroxisomes, and chloroplasts as well as a host of other cellular and molecular features. Each of these subcellular compartments houses characteristic biochemical processes carried out by specific sets of proteins. In endosymbiont–derived organelles with autonomous genomes, such as mitochondria and chloroplasts, some of these proteins are encoded by the organellar genome. However, the vast majority of proteins in these organelles are encoded by the host nuclear genome, translated into polypeptides in the cytosol and are subsequently targeted to the organelle and imported via specialized protein import apparatuses (Dolezal *et al.*, 2006; Pfanner *et al.*, 2004). Determining the subcellular location of a protein can provide useful information about its metabolic or biochemical role in a given subcellular compartment;

however, experimental localization of a protein using immunolocalization, subcellular fractionation, proteomics or other techniques can be expensive in terms of both time and resources. Indeed experimental localization studies cannot keep pace with the ever more rapid accumulation of high-throughput genomic data. For this reason, computational prediction methods have become increasingly important for annotating newly sequenced genes and for providing testable hypotheses regarding localization and function that may be followed up by experimental studies. In this chapter, we shall focus on current computational methods that are designed to predict whether proteins are localized in mitochondria or mitochondrion-related organelles. Before examining prediction methods in detail, we will review what is currently known regarding the evolutionary history of mitochondria, and current knowledge of their diversity in protein content and function across the eukaryotic realm. It is this diversity that will ultimately determine how well mitochondrial proteome prediction methods will perform as the field progresses from characterizing mitochondria of model systems to investigating these organelles in a much broader array of eukaryotic systems.

1.1. Evolution and diversity of mitochondria and mitochondrion-derived organelles

Mitochondria are the result of the endosymbiosis and subsequent enslavement of an α-proteobacterium in a common ancestor of all known extant eukaryotes (Gray et al., 1999). After the initial phase of endosymbiosis, the mitochondrial endosymbiont eventually ceased to exist as an independent organism and instead evolved into the organelle that we observe today. "Classical" aerobically functioning mitochondria, such as those present in mammals, plants, and fungi, are the typical text-book examples of mitochondria that generate ATP by oxidative phosphorylation and carry out other biochemical functions including replication of mitochondrial DNA, transcription and translation, iron–sulfur (Fe/S) cluster biogenesis, metabolism of lipids, interconversion of amino acids, and so on. However, these classical mitochondria comprise only a portion of the diversity found in organelles derived from this endosymbiotic event. For instance, there are also anaerobic mitochondria that generate ATP by respiration, but utilize terminal electron acceptors other than oxygen (e.g., fumarate, nitrate, etc.) in the electron transport chain. This type of anaerobic ATP production has been observed in a variety of eukaryotic lineages but is arguably best understood in parasitic animals (e.g., *Ascaris*) (Finlay et al., 1983; Howe, 2008; Kobayashi et al., 1996; Takaya et al., 1999; Tielens and van Hellemond, 1998; Tielens et al., 2002; van Hellemond et al., 1998). Among the unicellular eukaryotes (protists) living in low oxygen conditions, even stranger mitochondrion-related organelles have been discovered (Embley 2006), including hydrogenosomes (Cerkasovová *et al.*, 1973; Lindmark and Müller, 1973) and mitosomes (Tovar *et al.*, 1999) that are discussed briefly below.

1.2. Hydrogenosomes

Hydrogenosomes are genome-lacking double-membrane bounded orga-nelles characterized by the production of molecular hydrogen as a by-product of anaerobic ATP generation (Müller, 1993). They catabolize pyruvate and malate and, via substrate-level phosphorylation, produce ATP anaerobically using a series of enzymes, many of which are not generally found in the mitochondria of aerobic eukaryotes (e.g., pyruvate: ferredoxin oxidoreductase (PFO), [FeFe]-hydrogenase and acetate:succi-nate CoA transferase) (Embley 2006). These organelles were first described in parasitic parabasalid protists (e.g., *Tritrichomonas* and *Trichomonas*) and it was initially unclear whether they represented new, non-endosymbiont-derived organelles, an endosymbiotic organelle of unique origin, or were related to mitochondria. The matter was resolved in the 1990s when nuclear-encoded mitochondrial marker proteins, such as mitochondrial-type chaperonin 60 (CPN60) and Hsp70 (mtHSP70) were discovered within the hydrogenosomal proteome and phylogenetic analysis indicated that they were most closely related to homologs existing in other mito-chondria (Bui *et al.*, 1996; Germot *et al.*, 1996; Horner *et al.*, 1996; Roger *et al.*, 1996; Palmer, 1997). More recently, the hydrogenosomal localization of a number of systems including: a mitochondrial carrier family (MCF) transporter, mitochondrial iron–sulfur (Fe/S) proteins, and the presence of two subunits of complex I confirm that hydrogenosomes are in fact highly modified mitochondria (Dolezal *et al.*, 2005; Hrdy *et al.*, 2004; Sutak *et al.*, 2004; Tachezy *et al.*, 2001; van der Giezen *et al.*, 2002). Hydrogenosomes have evolved from mitochondria multiple times in a number of distantly related eukaryotic lineages including parabasalids, ciliates and anaerobic fungi (Barberà *et al.*, 2007; Embley 2006). Although their energy generating pathways may represent a convergent adaptation to anaerobiosis, it is likely that hydrogenosomes in these different lineages (like mitochondria of different eukaryotic lineages) have distinct properties.

1.3. Mitosomes

Mitosomes are a poorly characterized and heterogeneous category of mitochondrion-like organelles (MLOs) that possess two membranes, lack cristae and a genome, and, as far as is currently known, play no role in energy generation. They were first discovered (and named) in the amoeboid human parasite *Entamoeba histolytica* (Mai *et al.*, 1999; Tovar *et al.*, 1999) and were shown to contain CPN60 homologs clearly related to mitochondrial homologs in their proteomes (Dolezal *et al.*, 2005; Mai *et al.*, 1999) suggesting a shared origin with mitochondria. More recently, mitosomes have been shown to exist in Microsporidia such as *Trachipleistophora hominis* (Williams *et al.*, 2002) and the diplomonad *Giardia intestinalis* (Tovar *et al.*, 2003). Like hydrogenosomes, mitosomes have distinct origins in these diverse eukaryotic protists resulting from the parallel loss of many typical aerobic mitochondrial pathways in these different lineages (Embley 2006). Although the exact functional roles of mitosomes are still unclear, they are thought to consume, but not produce, ATP (Tsaousis *et al.*, 2008) and, as a shared common function, carry out iron–sulfur cluster assembly (Goldberg *et al.*, 2008; Roger and Silberman, 2002).

1.4. Mitochondrial proteomic and genomic diversity

As a hallmark of their endosymbiotic origin, "classical" mitochondria contain genomes of their own, that are most closely related to those of α-proteobacteria. Although the structure and content of mitochondrial genomes varies widely between the organelles of different lineages (Gray *et al.*, 1999, 2004; Marande and Burger, 2007; Smith *et al.*, 2007). Over evolutionary time, different genes have been transferred from the mitochondrial genome to the host nucleus, and their products were then re-targeted back to the organelle in a process known as endosymbiotic gene transfer (EGT) (Timmis *et al.*, 2004). In the case of hydrogenosomes and mitosomes where organellar genomes have likely been lost entirely, all essential organellar genes must be located in the host genome. However, the MLOs of *Nyctotherus* (Boxma *et al.*, 2005) and *Blastocystis* (Stechmann *et al.*, 2008) fall somewhere between classical mitochondria and hydrogenosomes. In both of these organisms, their MLOs retain genomes, and are predicted to house part of the tricarboxylic acid (TCA) cycle, parts of the electron transport chain as well as a variety of other mitochondrial pathways. Curiously, both MLOs also possess a [FeFe] hydrogenase enzyme and other proteins typical of hydrogenosomes. These data, and new information regarding organelles in a variety of previously unstudied anaerobic protists, are beginning to reveal that MLOs fall along a continuum of biochemical function (van der Giezen and Tovar, 2005) ranging from classical aerobically-respiring mitochondria to the mitosome. MLOs, such as those found in *Blastocystis* sp. (Stechmann *et al.*, 2008), *Mastigamoeba balamuthi* (Gill *et al.*, 2007), and *Trimastix pyriformis* (Hampl *et al.*, 2008) blur the distinctions between mitochondria, hydrogenosomes and mitosomes and illustrate this continuum of diversity.

1.5. Organellar proteome characterization

Until recently, the most in depth characterizations of mitochondrial proteomes have been limited to those of plants (Heazlewood *et al.*, 2004), animals (Pagliarini *et al.*, 2008), and fungi (e.g., yeast) (Prokisch *et al.*, 2004;

Reinders *et al.*, 2006; Sickmann *et al.*, 2003) while those of protists have not been as well studied. However, these studies have shown that mitochondria are evolutionary mosaics containing proteins derived both from the endosymbiont, the eukaryotic host cell, bacterial—but not alpha-proteobacterial—origin, and lineage-specific proteins of unknown/uncertain origin (Gray *et al.*, 1998; Smith *et al.*, 2007). The recent mass-spectrometry-based characterization of the mitochondrial proteome of the ciliate *Tetrahymena thermophila* exemplifies the tremendous diversity within even classical aerobic mitochondria (Smith *et al.*, 2007). A large number of putative mitochondrial proteins in *Tetrahymena* have no known homologs, or have homologs only in *Paramecium*, and are therefore possibly ciliate specific. It remains to be determined whether these are truly novel protein sequences or they are so highly divergent as to be unrecognizably related to orthologs in other eukaryotes. In any case, the apparent fluidity of the organellar proteome across eukaryotic diversity suggests that predictive methods for protein localization within mitochondria should be "trained" on a broad range of eukaryotic groups if they are to be used to make accurate predictions in non-model organisms.

1.6. Mitochondrial targeting and import

As many mitochondrial protein prediction tools rely on detecting mitochondrial targeting sequences on proteins translated in the cytosol, it is worthwhile to review briefly what is known regarding targeting and protein import into these organelles. Mitochondrial proteins may be targeted to the mitochondrial matrix, inner membrane, outer membrane, or the intermembrane space (Dolezal *et al.*, 2006). There are two main import pathways, one mediate by N-terminal targeting sequences (presequences) and the other, known as the carrier protein import pathway, mediated by internal targeting sequences. Import into the organelle is mediated by the translocase of the outer membrane (TOM) complex, molecular chaperone proteins, and targeting sequences. The translocase of the inner membrane (TIM) complexes mediate transit from the intermembrane space into the mitochondrial matrix as well as embedding of proteins into the inner membrane. The carrier translocase (TIM22) embeds carrier proteins and the TIM23 complex embeds N-terminal targeted inner membrane proteins (Bolender *et al.*, 2008). Proteins destined for import into the organelle via the presequence import pathway typically carry an N-terminal targeting sequence capable of forming an amphipathic helix. The amphipathic helix structure is recognized by the import machinery and is ultimately cleaved off after the protein enters the mitochondrial matrix by a mitochondrial processing peptidase. For more details on protein import into mitochondria several comprehensive review papers (e.g., Bolender *et al.*, 2008; Dolezal *et al.*, 2006; Gabriel and Pfanner, 2007; Pfanner *et al.*, 2004) should be consulted.

Recent studies have shown that a significant percentage of mitochondrial proteins do not carry an N-terminal targeting sequence and carry only an internal or C-terminal sequence if they have a detectable localization sequence at all (Bolender et al., 2008; Marcotte et al., 2000; Sickmann et al., 2003). Unfortunately, from the point of view of prediction, mitochondrial targeting sequences also tend to be poorly conserved across proteins, with the exact features being only loosely understood. In general, mitochondrial targeting sequences are enriched in positive, hydrophobic, and hydroxylated amino acids with acidic residues avoided. Import requirements have only rigorously been examined in a small subset of eukaryotes such as yeast and *Neurospora crassa* (Gabriel and Pfanner, 2007). While some proteins may lack a targeting sequence, others are "dual targeted" (Peeters and Small, 2001); that is, they are targeted to the mitochondrion and to one or more additional subcellular compartments. It has been argued that up to 25% of yeast mitochondrial proteins may be dual-targeted in this fashion (Dinur-Mills et al., 2008). These factors can make characterization of the requirements for mitochondrial import difficult and bioinformatics-based prediction strategies even more so.

Hydrogenosomes and mitosomes, like classical mitochondria, also require host nucleus-encoded proteins to be targeted back to the organelle and often possess N-terminal targeting sequences on these proteins. Furthermore, parts of their import/refolding machinery have been shown to be homologous to those of classical mitochondria (Embley and Martin, 2006).

1.7. Computational methods for predicting mitochondrial localization

Computational methods designed for predicting mitochondrial localization of proteins fall into several broad categories: (1) methods that rely on comparisons with well-characterized mitochondrial proteomes to detect homologs of mitochondrial proteins in these systems, (2) methods that predict N-terminal signal/targeting peptides, (3) full protein sequence feature-based methods, and (4) approaches that utilize a mixture of the previous discussed strategies. N-terminal signal/targeting peptide classification is the most widely used type of subcellular prediction strategy, with the majority of available programs relying on the presence of such a signal exclusively, or using the presence/absence of a signal/targeting peptide as one of several kinds of evidence to base their predictions upon.

Methods not only differ in which mitochondrial features they detect, but also in the computational classification methods they are based upon. Machine learning techniques such as Artificial Neural Networks (ANNs), Hidden Markov Models (HMMs), and Support Vector Machines (SVM) have all been used for the task of prediction, along with so-called "expert systems." The details of machine learning algorithms are outside of the scope of this review. Briefly, all of these approaches are "trained" on data where target-ing has been confirmed experimentally (or is very certain) and an attempt is made to optimize the discriminatory power of classification based on that training data. In many cases, training data is split into training and testing data sets, so that performance of the trained method can be evaluated using separate test data sets. In all of these approaches, features of the input data are extracted and then clustered in some way into one or more groups, such as subcellular location, based on specified criteria. ANNs are mathematical models that simulate biological neural networks while SVMs attempt to find a hyperplane that maximally separates input data into two groups. In some cases more than one of these machine-learning classifiers are used in multi-ple layers. Individual classifiers work on one component of a larger problem and a final "master" classifier combines the outputs to make a global classi-fication. This master classifier may itself be implemented as any sort of useful classifier including expert voting systems or clustering algorithms. For more detailed reviews of these algorithms see Cristianini and Shawe-Taylor (2000), Gurney (1997), Kecman (2001), and Rabiner (1989).

Most of the programs discussed below have been tested on protein data sets compiled from Swiss-Prot. The experimental localization information found in that database is extracted and used to train the programs. However, as discussed above, the organellar proteomes of diverse eukaryotes including protists are quite variable. Since complete, experimentally determined proteomes remain few and those proteomes that have been so determined are biased toward classical mitochondria, it is unclear how well the programs tested by using annotation information in databases will perform on these organisms. Furthermore, no comprehensive proteomic studies of anaerobic mitochondria, hydrogenosomes, mitosomes, and other MLOs have yet been published. Given that many of the proteins found in hydrogenosomes and mitosomes are relatively unique and not found in classical mitochondria, these proteins may be especially difficult to detect by programs designed to predict the subcellular location of, for example, mammalian proteins. In the following sections we present a selection of subcellular location prediction programs useful for more detailed analysis and characterization of unknown protein sequences.

2. DATA PREPARATION

Most mitochondrial prediction programs take, as input, protein sequences in FASTA format which features a definition line beginning with the ">" symbol. This definition line typically contains a label for the sequence but may contain other information. The instructions regarding the precise format of the definition line of a FASTA file should be consulted for each prediction program, as the format may vary slightly. The sequence itself begins on the next line. A properly formatted FASTA sequence file may contain any number of sequences as in the following example:

>Sequence 1
MKLESVV...
>Sequence 2
MKLESVI...

Most, but not all, of the programs described below allow for the input of multiple sequences in this FASTA format except where noted below.

 ## 3. SCREENING LARGE DATASETS FOR MITOCHONDRIAL PROTEINS

While most mitochondrial prediction programs will operate on multiple protein sequences, few are optimized for large genomic or EST datasets which are becoming increasingly common as genome sequencing becomes easier, quicker, and cheaper. In many cases, the researcher with open reading frames predicted from genomic or EST sequences will want to screen these large datasets for putative mitochondrial proteins to focus further analysis in an efficient manner and to minimize false positives. Several programs are described below that have been designed as screening tools or are optimized for genome-scale analyses. These tools should be used to construct a list of putative mitochondrial proteins that can then be examined in more detail with a wider suite of mitochondrial prediction tools.

3.1. CBOrg

Comparative BLAST for Organelles (CBOrg) is a homology-based method for the prediction of subcellular localization. Specifically designed for screening of large EST and/or genomic datasets, it has been used by our group for the analysis of the MLOs of *Blastocystis* sp. (Stechmann *et al.*, 2008) and *M. balamuthi* (Gill *et al.*, 2007). It does not rely on N-terminal targeting peptide predictions and is therefore especially useful for analysis of EST data where the sequences of the 5′-ends of genes can be incomplete. CBOrg utilizes the experimentally determined or predicted mitochondrial (or hydrogenosomal) proteome databases of humans, yeast, *Trichomonas vaginalis*, and *T. thermophila* and can easily accommodate databases from any number of organisms. For each organism, CBOrg has pre-built data-bases of their organellar proteome(s) (mitochondria or hydrogenosome) and a "subtractive" cellular proteome. The latter is derived from the complete protein predictions from the nuclear genome, with the proteins found in the organellar proteome removed. Input sequences are analyzed using

BLAST against each organelle and subtractive cellular proteome database for each organism. Then, for each organism, and each database, a score for the "best hit" in the database sequence is recorded; the compartment that has the best hit is putatively assigned as the best guess as to the subcellular location of the protein for that organism. Figure. 2.1 is a schematic representation of the CBOrg method.

CBOrg is usable on Linux or Mac OSX systems and can be run either on the command line as a series of analysis scripts or as a fully graphical environment. CBOrg requires and makes use of features found in Perl version 5.10 as well as the Gtk2-Perl libraries for the graphical user interface. At its core CBOrg utilizes the power of BLAST to identify homologs to the query sequences. Input must be formatted, as above, in proper FASTA format and any number of sequences can be accommodated. Because CBOrg uses BLAST for homology searching it requires a separate, stand-alone installation of NCBI BLAST be present. This allows CBOrg to accommodate both nucleotide and protein sequences by using either the blastx or blastp algorithms, respectively.

Sequences can then be sorted based on the number of organisms "agreeing," and the degree of agreement, on the subcellular location. Because CBOrg is intended for use as a filtering program for large datasets and not as a definitive classifier, we suggest it be used in a way that minimizes false negatives at the expense of an excess of false positives which can later be eliminated by careful analysis. By default, proteins with at least one hit to an MLO database are sorted into a file of putative mitochondrial proteins. Because CBOrg relies on homology, mitochondrial proteins unique to an experimenter's organism (or proteins found outside of the organelle in other organisms) will not be detected by this method and so at least one other non-homology-based screening method should be used on large data sets. Prediction of Organelle Targeting Sequences (Predotar) (Small *et al.*, 2004) and MITOPRED (Guda *et al.*, 2004a,b) are subcellular localization tools that are both optimized for genome-scale datasets. Predotar does N-terminal targeting sequence detection while MITOPRED bases its predictions on a variety of other factors. Both of these programs, described in more detail below, are able to assign proteins to subcellular locations other than mitochondria. If follow-up studies are to be carried out after the initial screening procedure, proteins predicted to contain a targeting sequence for the chloroplast or other subcellular compartments should be included as putative mitochondrial proteins because targeting sequences for different compartments such as chloroplasts and mitochondria often have some similarities. As well as using the results of programs such as Predotar and MITOPRED, proteins found in the "no hit" sorted file of CBOrg should also be more closely examined. This file contains all query proteins with no significant BLAST hits in any of

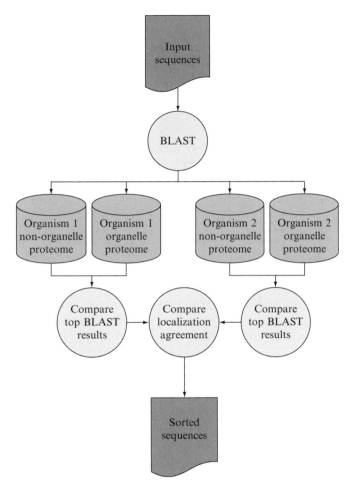

Figure 2.1 Schematic representation of the Comparative BLAST for Organelles (CBOrg) method. Non-organelle proteome refers to the set of nucleus encoded proteins with the organellar (mitochondrial) sequenced subtracted. Top scoring results of BLAST are compared between databases for localization within that organism. Localization data are then compared between organisms to sort input sequences. Where localization is not clear (non-agreement between organisms) input sequences are sorted based on level of agreement for mitochondrial localization. A standard rule that maximizes true positive mitochondrial predictions (at the expense of a high false positive rate) is to flag a query as potentially mitochondrial if at least one organism has a best hit to its mitochondrial proteome.

the used databases and is likely to contain unique proteins, some of which may be mitochondrial. CBOrg is available for download and local installation from the Roger Lab Web site (http://rogerlab.biochem.dal.ca/ Software/Software.htm).

3.2. Predotar

Predotar utilizes a neural network to detect and classify N-terminal targeting sequences. Predotar, unlike most N-terminal sequence prediction programs, was designed and optimized for analysis of large numbers of input sequences and whole genome/proteome data, making it suitable as a tool for initial screening. The neural network operates on three basic parameters: (1) charge of the amino acid side chain for each residue in the putative N-terminal signal/targeting peptide region, (2) the hydrophobicity score for each amino acid in this same region, and (3) the amino acid composition in two sections of the sequence. The final output of Predotar, like in other N-terminal prediction programs, is a final probability that the input sequence contains a targeting sequence. The authors of Predotar claim this program is an improvement over other methods for distinguishing between those proteins targeted to the chloroplast or to the mitochondria (Small *et al.*, 2004). The training set, like many others, was taken from annotated Swiss-Prot data. Proteins encoded on organellar genomes, targeted to multiple locations, and/or lacking N-terminal methionines were removed from the final test dataset. Predotar is usable both through a Web interface (http://urgi.versailles.inra.fr/predotar/predotar.html) and as a local installation (see http://urgi.versailles.inra.fr/predotar/) by request from the authors.

Any additional query sequences with probable N-terminal targeting sequences detected by Predotar should be added to the initial list of putative mitochondrial proteins (as well as any "no hit" proteins) compiled by CBOrg and subjected to more detailed analysis using some, or all, of the programs discussed in Section 4.

3.3. MITOPRED

MITOPRED is designed for high-throughput analysis and is not based solely on the prediction of N-terminal targeting sequences, making it an excellent program for use in initial screening of large datasets. It is a mixed classifier (for more information see Section 4.3) based on (1) the presence (and occurrence pattern) of "subcellular location-specific" Pfam domains in the protein, (2) the amino acid composition, and (3) the isoelectric point (pI) of the protein (Guda *et al.*, 2004a,b). As with many other programs, Swiss-Prot data were separated into subcellular localization categories based on annotation information and used for training. Average amino acid composition, pI, and Pfam domain occurrences could then be determined for each location category. Query sequences can then be compared to each subcellular location and scores are calculated individually for the amino acid composition/pI and the Pfam domain occurrence. MITOPRED is available as a Web service (http://bioapps.rit.albany.edu/MITOPRED/)

and source code is available from the authors upon request. MITOPRED accepts multiple input sequences in FASTA format as well as Swiss-Prot/ TrEMBL accession numbers. Probable mitochondrial proteins detected by MITOPRED should also be added to the list of likely mitochondrial proteins from CBOrg and Predotar. Proteins predicted as mitochondrial by all three of these programs are almost certainly correctly classified as these methods rely on different kinds of evidence to make their predictions.

4. Detailed Analysis

Ideally, initial screening with the programs described above generates a list of putative mitochondrial proteins. While these programs alone provide a good indication of subcellular localization, more detailed analysis using a variety of subcellular prediction programs is recommended to refine the predictions and remove false positives.

4.1. Programs for N-terminal targeting sequence prediction for single sequences

Signal/targeting peptide prediction has proved to be quite powerful and highly accurate. The presence of a known N-terminal signal/targeting sequence is, generally, a straightforward and unambiguous means of deter- mining subcellular location with the caveat that some organellar proteins have cryptic targeting sequences, internal targeting sequences, or lack targeting sequences altogether. Although the sequence of N-terminal targeting peptides can be quite variable across eukaryotic diversity, they do tend to require certain key features such as the amphipathic helix discussed previously (Smith *et al.*, 2007). Because of these shared character- istics, N-terminal targeting peptide prediction has proved to be of broad use across all eukaryotes. Although both Predotar and MITOPRED make predictions fully or partially based on N-terminal targeting peptide predic- tions, once a set of putative organellar proteins have been obtained, it is useful to follow these up with other prediction programs many of which use different classification algorithms and/or have been trained on different data sets. Discussed below are several useful programs for the detection/predic- tion of N-terminal targeting sequences.

4.1.1. MitoProt

MitoProt is one of the earliest programs designed to predict the presence of a mitochondrial targeting sequence (Claros and Vincens, 1996). MitoProt is freely available from a Web interface (http://ihg2.helmholtz-muenchen.de/ ihg/mitoprot.html), and the source code is freely available (ftp://ftp.

biologie.ens.fr/pub/molbio/) for local installation. To detect and calculate the statistical support of a mitochondrial targeting sequence, MitoProt considers the following parameters (1) length of the putative N-terminal extension region, (2) number of acidic residues (D or E) in the putative N-terminal extension, (3) number of positively charged residues (K or R) in the putative N-terminal extension, (4) amino acid composition within the first 20 residues of the N-terminal region, (5) amino acid composition within the whole N-terminal region, (6) presence of a putative cleavage site, (7) the net charge over the entire sequence, and (8) several factors that evaluate the impact of hydrophobic regions within the protein. The hydrophobic factors are broken down into several different categories such as the maximum hydrophobic moment and the maximum hydrophobicity. To avoid bias in these calculations, several different hydrophobicity scales are used for each of these categories. A total of 47 parameters are used to calculate the final score of any given sequence. Linear discriminant analysis was used in the development of MitoProt to obtain a discriminant function for differentiating between mitochondrial and non-mitochondrial proteins. Due to the similarity of chloroplast and mitochondrial proteins, especially in the properties of their targeting peptides, chloroplast proteins may be misclassified as mitochondrial, which is not unusual in many programs designed solely to classify sequences as either mitochondrial or non-mitochondrial. MitoProt primarily classifies proteins based on the presence or absence, according to its 47 parameters, of an N-terminal mitochondrial targeting sequence. From the Web interface only one sequence can be input and analyzed at a time.

4.1.2. TargetP

Like MitoProt, TargetP is based on identifying N-terminal targeting sequences. However, TargetP is not exclusively designed to predict mitochondrial targeting sequences but also chloroplast and secretory system targeting peptides (Emanuelsson *et al.*, 2000; Nielsen *et al.*, 1997). Here we will restrict attention to the mitochondrial targeting sequence prediction although misclassification, in this case, of putative mitochondrial proteins as being targeted to a chloroplast or for secretion can also be informative. TargetP was originally based on a method that used two different neural networks, one for predicting cleavage sites and the other for predicting signal peptides (Nielsen *et al.*, 1997). The most recent version of TargetP (Emanuelsson *et al.*, 2000) still utilizes a neural network but now classifies proteins into subcellular localization categories: mitochondrial, chloroplastic, secretory, or "other" and the classification is still based on the presence of an N-terminal targeting sequence and a cleavage site. The current version of TargetP features a dual-layer neural network, with the first layer holding dedicated networks for predictions for each subcellular localization while a second layer integrates the outputs of the first layer networks to make a final prediction. The neural networks were trained on data from Swiss-Prot and the experimental localization annotation found in the database entries. TargetP is usable both as a Web interface (http://www.cbs.dtu.dk/services/TargetP/) and as a stand-alone local installation (http://www.cbs.dtu.dk/cgi-bin/nph-sw_request?targetp).

4.1.3. PProwler

PProwler (Bodén and Hawkins, 2005; Hawkins and Bodén, 2006) is a prediction program based on TargetP but, unlike TargetP, prediction of an N-terminal targeting sequence is not restricted to a simple linear analysis of the N-terminal region of the input protein (the first 100 amino acids in the case of TargetP). PProwler attempts to examine the N-terminal region in a nonlinear manner known as a recurrent neural network. This recurrent network is structured such that the classification of any given residue as belonging to a signal peptide depends on the state of the amino acid residues immediately up- and downstream. These residues in turn are dependent on the state of the residues up- and downstream of them, and so on in a recursive pattern. This recursion is designed to overcome limitations of looking at the linear sequence alone, providing information on longer range interactions important in the three-dimensional structure, such as the amphipathic helix of the signal/targeting peptide itself. A network is developed for each subcellular compartment or destination (mitochondria, chloroplast, secretory system, and other) and prediction is performed on each residue in the N-terminal 100 residues as a sliding window under each prediction network. The output of each detection network must then be "sorted" and a final classification performed. In the current version of PProwler (version 1.2), a SVM is used for this final classification. PProwler is available for analysis on the Web (http://pprowler.itee.uq.edu.au/pprowler_webapp_1-2/).

4.2. Sequence feature-based prediction for prediction of mitochondrial localization

As discussed previously, not all mitochondrial proteins contain a canonical N-terminal targeting sequence and may carry only an internal targeting sequence or no such sequence at all; in these cases targeting peptide-based predictions will undoubtedly fail. However, there are other features that can be used to identify mitochondrial proteins. Feature-based prediction programs employ a broad range of strategies that may be implemented either singly or collectively. They may utilize information that can be obtained directly from the protein sequence entry in a database, such as

Gene Ontology (GO) terms, sequence annotation, and/or structural data. However, most of these classifiers make use of data which can be extracted from the sequence alone including amino acid composition, di- or tri-peptide compositions, presence/absence of particular domains, secondary structure, hydrophobicity, isoelectric point (pI), and so on. As well as direct sequence features, this class of predictors also encompasses sequence comparison methods that use homology or phylogenetic data to make classifications. Homology-based prediction programs are limited by presence of sequence similar homologs in the searched databases. Although the existence of known homologs certainly improves the prediction accuracy, examples from the full mass-spectrometry-based characterization of the *T. thermophila* mitochondrion (Smith *et al.*, 2007), and from our experience on the characterization of mitochondria and mitochondrion-derived organelles (Gill *et al.*, 2007; Stechmann *et al.*, 2008), have demonstrated that many hypothetical mitochondrial proteins of unknown function could be organism specific or be phylogenetically restricted to a particular group of eukaryotes. Broader taxon sampling of organellar diversity may reduce the number of lineage-specific proteins with unknown homologs, expanding the repertoire of mitochondrial proteins and increasing our understanding of the properties of those proteins. Composition-based classifiers also come in a wide variety of flavors, from simple ones that consider only the amino acid composition over the entire sequence, to classifiers that break the sequence into sub-sequences or that also evaluate parameters such as the composition and frequency of di-, tri-, tetra-peptides, and so on.

4.2.1. CELLO

SubCELlular LOcalization (CELLO) is a prediction program originally designed for analysis of Gram-negative bacterial proteins but has since been extended to Gram-positive bacteria and eukaryotes (Yu *et al.*, 2006). This program uses a SVM as its machine learning technique and is a protein sequence feature-based classifier that analyzes amino acid composition vectors of varying lengths (n-peptide composition vectors). When $n = 1$ the vector reflects the amino acid composition of the entire sequence, when $n = 2$ it reflects the composition of di-peptides in the sequence, and so on. Like some other prediction programs CELLO utilizes two SVM layers. The first is composed of classifiers that operate on individual sequence composition vectors while the second layer acts as a master classifier. CELLO is accessible as a Web-based tool (http://cello.life.nctu.edu.tw/).

There are other programs discussed below which also rely heavily on these types of full sequence features; however, many also use N-terminal targeting sequence detection or other data, such as literature references, to classify protein localizations. For this reason these programs are discussed in the context of mixture-based classification.

4.3. Mixture approaches

"Mixed" approaches typically use several different types of protein sequence feature-based classifiers and are frequently coupled with N-terminal signal peptide and cleavage site prediction strategies. The results are combined, usually by use of a voting system or master classifier, into a final predicted subcellular location. These mixed approaches are useful for overcoming the shortcomings of individual localization programs, significantly improving performance as shown recently with yimLOC (Shen and Burger, 2007). Integrating the results from separate programs can be a challenge, especially if missing data from some programs are to be allowed. Flexibility is often preferred to minimize the impact of the limitations of individual classifiers. If, for instance, the input data are a large number of EST sequences, then missing N-terminal sequences that could contain targeting peptides should not remove sequences from consideration by classifiers that are not dependent on this information.

4.3.1. PSORT

PSORT was first developed in 1990 and has since been updated to yield several different program variants such as WoLF PSORT (Horton *et al.*, 2006, 2007), PSORT II (Nakai and Horton, 1999), and iPSORT (Bannai *et al.*, 2002). PSORT II, the newest version, utilizes a comprehensive algorithm that analyzes several features of the protein sequence: (1) the presence of an N-terminal signal peptide for the endomembrane system, (2) presence of a mitochondrial targeting signal, (3) nuclear localization signal, (4) ER-lumen-retention signal, (5) ER-membrane-retention signal, (6) Peroxisomal-retention signal, (7) Vacuolar targeting signal, (8) Golgi-transport signal, tyrosine-containing motif (number of tyrosines in cytoplasmic tail), (9) di-leucine motif (in cytoplasmic tail), (10) presence of transmembrane segments, (11) RNA-binding motif, (12) actin-binding motifs, (13) isoprenyl motif, (14) GPI-anchor, (15) N-myristoylation motif, (16) DNA-binding motifs, (17) ribosomal-protein motifs, (18) prokaryotic DNA-binding motifs, (19) amino acid composition, (20) number of residues in predicted coiled-coil secondary structure, and (21) the overall length of the sequence. Most of these features are compiled from structural/sequence motifs that provide good indications of protein function, such as RNA/DNA binding, actin binding, and so on. Some of these features are more direct, such as the presence of targeting signals (mitochondrial), while others are indirect, such as amino acid composition. A k-nearest neighbors machine learning classifier is used to determine the probability of localization to several different subcellular compartments (except in iPSORT which uses a decision list). All of the PSORT programs are available online at the PSORT WWW-Server (http://psort.ims.u-tokyo.ac.jp). PSORT II was initially trained on yeast data but can be trained on any user-supplied data set, which increases its flexibility. It is available both as a Web service and as a stand-alone program for local installation.

4.3.2. PA-SUB

PA-SUB (Lu *et al.*, 2004) is a collection of five different classifiers for subcellular localization. Each classifier is trained and designed to classify proteins from a particular group of organisms: animals, plants, fungi, Gram-negative bacteria, and Gram-positive bacteria. The authors of this program report a prediction accuracy of 81% for fungi and 92–94% for the other categories. PA-SUB was designed by increasing the taxonomic diversity of training sequences (especially by including bacteria) and optimizing predictive accuracy. PA-SUB is integrated into a larger suite (Proteome Analyst) that also predicts the general function of an input protein and allows training of custom predictive classifiers. In brief PA-SUB performs classification based on presence/absence of annotated features in the Swiss-Prot entries of the top-scoring BLAST hits to the query sequence. PA-SUB is unable to classify sequences if no homologous sequences can be found in Swiss-Prot or if no suitable feature annotations can be found in the Swiss-Prot entries of homologs. PA-SUB is available as a Web-based prediction program (http://pasub.cs.ualberta.ca:8080/pa/Subcellular). The user must select which classifier to use on the query sequences (Animal, Plant, Fungi, Gram-negative Bacteria, or Gram-positive Bacteria).

4.3.3. SherLoc

SherLoc (Höglund *et al.*, 2006a; Shatkay *et al.*, 2007) is a Web-based program (http://www-bs.informatik.uni-tuebingen.de/Services/SherLoc/) designed to be comprehensive for a broad range of both organisms and subcellular compartments. SherLoc implements a novel mixture-based approach that combines both sequence features, such as amino acid composition, and text features from the literature to predict the subcellular location; classification is done using an SVM. Instead of using location references directly from the database entry, the authors of SherLoc designed it to use textual references extracted from titles and abstracts associated with the database entry that they suggest are correlated with subcellular location such as textual references regarding functional roles, interacting proteins, and so on. SherLoc is an example of a multi-tiered classifier with four sequence-based classifiers and one text-based classifier; the predictions of these individual classifiers can then be integrated into a final result. The sequence-based classifiers are based on those in MultiLoc (Höglund *et al.*, 2006b) and are each based on a different sequence feature: amino acid composition, N-terminal targeting sequence, internal anchor sequences, and sorting sequence motifs. For signal/targeting peptide detection, SherLoc predicts chloroplast (plant only), mitochondrial, secretory pathway, or "other" subcellular locations. Presence of a signal anchor is a presence/absence property important only for some proteins destined for the secretory system. The text-based classifier operates on data extracted from the titles and abstracts, available on Pubmed, referenced by the Swiss-Prot entry for the query sequence. For query sequences without a Swiss-Prot entry, an attempt is made to find close homologs in the database and use the textual information associated with those entries. If no Pubmed entries can be found, the text classifier is not used for that query, with the prediction then based solely upon the four sequence feature classifiers.

4.4. Meta-analysis

Meta-analysis is an extension of the mixture-based approach which, in the context of computational prediction and classification, is often termed a "mixture-of-experts" strategy. In this strategy, the results of several different types of classifiers are combined and analyzed by one global "expert" classifier to come up with a global prediction. In the case of meta-analysis the outputs of separate programs are integrated and a master classifier is used to make a final prediction. These meta-analyses are useful for objectively deciding for/against mitochondrial localization based on detailed analyses with a wide variety of programs.

4.4.1. YimLOC

YimLOC is a meta-analysis tool (Shen and Burger, 2007) that is available through the Web-server AnaBench (http://anabench.bcm.umontreal.ca/anabench/index.jsp). The source code is also freely available in the supplementary material of the manuscript. This program combines the results of several different independent programs and their predictions into one integrated platform as an attempt to avoid the pitfalls/weaknesses of any given program. All subcellular localization programs use different machine learning algorithms trained on different datasets and are optimized to meet the needs of their individual developers. Many (but not all) of the individual programs whose output is used for yimLOC have been discussed already in this review and all of the programs used by yimLOC are available for use as Web services. Those programs which have already been discussed have their major features summarized in Table 2.1. After integrating this data, the program will classify input sequences into their probable subcellular location. yimLOC was tested by Shen and Burger (2007) using proteins of known subcellular localization from yeast, human, and *Arabidopsis*. Both majority-win voting and decision trees were assessed as possible integration methods, with the decision tree showing the best overall performance. If the program is installed locally, then the decision tree can be trained on the users data and a selection of primary classifiers (LOC-tools) can be implemented. In the standard implementation, the programs that detect N-terminal mitochondrial targeting sequences were combined into

Table 2.1   Major features of reviewed subcellular localization software

Program	Predicts N-terminal targeting sequence	Classifier	Trained on	Available as Web service
TargetP	Exclusively	Neural network	Swiss-Prot	Yes
SherLoc	Yes, mixed	SVM	Swiss-Prot	Yes
Predotar	Exclusively	Neural network	Swiss-Prot	Yes
MitoProt	Exclusively	Discriminant function	Swiss-Prot	Yes
PA-SUB	No	Naïve Bayes	Swiss-Prot/ Gram-negative bacteria/ Custom	Yes
PProwler	Exclusively	Recurrent ANN/ SVM	Same as TargetP	Yes
MITOPRED	Yes, mixed	Matthews correlation coefficient	Swiss-Prot	Yes
CELLO	No	SVM	Swiss-Prot	Yes
WoLF PSORT	Yes, mixed	K-nearest neighbor	UniProt/ Swiss-Prot	Yes
PSORT II	Yes, mixed	K-nearest neighbor	Yeast/ Custom	Yes
iPSORT	Yes, mixed	Decision list	Same as TargetP	Yes
yimLOC	No	Decision tree	Yeast/ Human/ *Arabidopsis*	Yes
CBOrg	No	None	–	No

one decision tree, with the output of that tree combined with the outputs of the other programs (based on sequence features) in what is known as a stacked decision tree. To improve prediction of mitochondrial membrane proteins, which are typically more poorly predicted than matrix proteins, several tools for the prediction of transmembrane domains were integrated into the decision trees (e.g., TMHMM). Unlike most programs discussed here, the input to yimLOC is not the raw sequence data. Instead it takes a specially formatted summary of the predictions made by the individual prediction programs.

4.5. Assignment of localization

After completion of screening and detailed analysis, as well as meta-analysis (if desired), the experimenter is left with a table of predictions similar to that shown in Table 2.2. This series of predictions provides in impression of the degree of evidence for/against a mitochondrial location of a protein. In most cases a "majority rules" type evaluation will be a sufficient (if conservative) way of deciding on the location. If individual programs disagree a great extent then predictions from yimLOC will be especially helpful.

5. COMPARISON OF METHODS

Below, we have compared the performance of several mitochondrial prediction methods on sequences obtained during the EST survey and organelle characterization of *Blastocystis* sp. (Stechmann *et al.*, 2008) and the results are shown in Table 2.2. Where possible, intact N-terminal targeting sequences are present and the data are relatively evenly split with eight mitochondrial and seven non-mitochondrial sequences. While this is not an exhaustive analysis, we have deliberately included proteins such as mitochondrial-type heat shock protein 70 (mtHsp70) which has a sequence-similar cytosolic homolog (Hsp70) to make prediction difficult. For comparative purposes, all programs were operated with their default/ standard options to emulate a naive user.

All of these prediction programs have their own individual strengths and weaknesses. Programs such as yimLOC are designed to overcome the individual weaknesses of various types of classifiers by uniting their predictions into one global classification using a rules-based decision tree system. The results of yimLOC's prediction (Table 2.2) were consistent with those programs based on signal peptide detection and classification while overcoming misidentifications provided by programs such as PProwler and PA-SUB. This flexibility produces a highly powerful prediction program.

Those programs which incorporate N-terminal targeting sequence prediction, either directly or indirectly, such as MitoProt (prediction accuracy: 0.867), yimLOC (0.933), and SherLoc (0.800), performed the best overall in terms of predictive accuracy defined as #true predictions/#total predictions. Since, most of these organellar sequences did contain targeting sequences with "classical" targeting peptide properties, this result was expected. While the overall predictive accuracy of CBOrg (0.667) is not as good as programs such as MitoProt, yimLOC, or Predotar, it was designed primarily as a screening tool and not as a definitive classifier.

Table 2.2 Results of subcellular localization prediction on selected *Blastocystis* sequences

Sequence	TargetP	SherLoc	Predotar	MitoProt	CELLO	PProwler	PA-SUB	WoLF PSORT	PSORT II	iPSORT	Mitopred	yimLoc	CBOrg
[FeFe]-hydrogenase	M	M	M	M	C	M	M	M	M	M	M	M	C[a]
PFO	M	M	M	M	C	O	ER/M?	M	M	N	M	M	M[a]
ND4	S	S	ER	M	PM	Signal	M	M	PM	M	N	M	M
Serine dehydrogenase	O	C	O	M	C/M	O	C	M	M	M	M	N/M?	C
mIF2Like[b]	O	M	–	–	C	O	C	C	C	N	N	N	M
[2Fe-2S] Ferredoxin	M	M	M	M	S	M	M	M	M	M	N	M	M
mtHsp70	M	M	M	M	M	M	C	M	M	M	M	M	M
MCP	O	M	O	N	PM	O	M	M	C	N	N	M	M
RNAse-L-inhibitor	O	P	O	N	C	O	M	C	C	M	N	N	M
6PGD	O	C	O	N	C	O	N	C	C	N	N	N	C
LCFA-CoA ligase	O	P	O	N	M	O	N	C	C	N	N	N	M
Acyl-CoA binding protein	O	P	O	N	C	O	N	M	C	N	N	N	M
FGAR synthase	O	Ncl.	O	N	Ncl.	O	C	C	Ncl.	N	N	N	C
Hydroxypyruvate reductase	O	C	O	N	C	O	N	C	C	N	N	N	C
Sodium/chloride transporter	O	M	O	N	PM	O	N	M	ER	N	N	N	C

[a] Hydrogenosomal protein, close BLAST matches in both Organellar and Non-organellar databases.
[b] Missing N-terminal methionine residue, some predictions not available.
M, Mitochondrial; N, Non-Mitochondrial; C, Cytoplasm; O, Other; S, Secretory; PM, Plasma Membrane; ER, Endoplasmic Reticulum; Ncl, Nuclear; ?, Ambiguous localization; –, No prediction.

CBOrg is most useful as a method when the gene sequences to be screened are partial and may therefore be missing N-terminal targeting regions. Furthermore, using it in the way we suggest above maximizes the number of true mitochondrial predictions at the expense of generating false positive predictions. Indeed, all but two organellar sequences were flagged as putatively organellar for at least one of the organisms queried and is flagged for further investigation by CBOrg. One of those sequences, [FeFe]-Hydrogenase, is found in hydrogenosomes but not "classical" mitochondria and had an extremely high-scoring hit to *Trichomonas* sequences in both the organellar and non-organellar databases and so would normally be investigated further. Restricting attention to mitochondrial predictions alone, CBOrg's accuracy improves to 0.75.

5.1. Conclusion

A wide variety of subcellular localization programs have proved to be useful in protein and proteome analysis and some, such as MitoProt and TargetP are both widely known and used by molecular biologists. Prediction of an N-terminal targeting sequence has proved to be quite effective and can be used to classify subcellular localization with a high degree of accuracy in methods that rely only on these kinds of data and is incorporated directly (PSORT) or indirectly (yimLOC) in most "mixture-based" classification schemes. High-throughput analysis of sequence data is a pressing concern in bioinformatics and some of the most popular programs, such as MitoProt in its online Web server, can only accommodate a single query sequence at a time, requiring local installation for larger scale analysis. A new generation of tools including Predotar and PA-SUB, have been specifically designed for high-throughput studies and can handle large volumes of input sequences. These high throughput tools will become more prevalent as genomic data continues to accumulate at a rapid pace. For current studies the use of a mixture of the tools described above will provide the most thorough predictions possible. For genomic and transcriptomic data, we recommend the use of screening tools such as CBOrg, Predotar, and MITOPRED as a first pass to generate a list of candidate proteins for follow-up investigation using the other programs that we have discussed (e.g., yimLOC, MitoProt, WoLF PSORT, SherLoc, and PA-SUB). Predictions made by programs that primarily rely on N-terminal targeting sequence information can be combined with predictions by methods that use other features, such as homology and amino acid composition, to generate localization hypotheses that can be followed up by targeted experimental approaches.

ACKNOWLEDGMENTS

The work is supported by a grant MOP-62809 from the Canadian Institutes of Health Research awarded to AJR and a Centre for Comparative Genomics and Evolutionary Bioinformatics postdoctoral fellowship from the Tula Foundation held by ADT.

REFERENCES

Bannai, H., Tamada, Y., Maruyama, O., Nakai, K., and Miyano, S. (2002). Extensive feature detection of N-terminal protein sorting signals. *Bioinformatics* **18,** 298–305.

Barberà, M. J., Ruiz-Trillio, I., Leigh, J., Hug, L. A., and Roger, A. J. (2007). The diversity of mitochondrion-related organelles amongst eukaryotic microbes. *In* "Origin of Mitochondria and Hydrogenosomes" (W. F. Martin and M. Müller, eds), pp. 239–275. Springer, Berlin.

Bodén, M., and Hawkins, J. (2005). Prediction of subcellular localisation using sequence-biased recurrent networks. *Bioinformatics* **21,** 2279–2286.

Bolender, N., Sickmann, A., Wagner, R., Meisinger, C., and Pfanner, N. (2008). Multiple pathways for sorting mitochondrial precursor proteins. *EMBO Rep.* **9,** 42–49.

Boxma, B., de Graaf, R. M., van der Staay, G. W., van Alen, T. A., Ricard, G., Gabaldón, T., van Hoek, A. H., Moon-van der Staay, S. Y., Koopman, W. J., van Hellemond, J. J., Tielens, A. G., Friedrich, T., *et al.* (2005). An anaerobic mitochondrion that produces hydrogen. *Nature* **434,** 74–79.

Bui, E. T., Bradley, P. J., and Johnson, P. J. (1996). A common evolutionary origin for mitochondria and hydrogenosomes. *Proc. Natl. Acad. Sci. USA* **93,** 9651–9656.

Cerkasovová, A., Lukasová, G., Cerkasòv, J., and Kulda, J. (1973). Biochemical characterization of large granule fraction of *Tritrichomonas foetus* (strain KV1). *J. Protozool.* **20,** 525.

Claros, M. G., and Vincens, P. (1996). Computational method to predict mitochondrially imported proteins and their targeting sequences. *Eur. J. Biochem.* **241,** 779–786.

Cristianini, N., and Shawe-Taylor, J. (2000). "An Introduction to Support Vector Machines and Other Kernel-Based Learning Methods" Cambridge University Press, Cambridge.

Dinur-Mills, M., Tal, M., and Pines, O. (2008). Dual targeted mitochondrial proteins are characterized by lower MTS parameters and total Net charge. *PLoS One* **3,** e2161.

Dolezal, P., Likic, V., Tachezy, J., and Lithgow, T. (2006). Evolution of the molecular machines for protein import into mitochondria. *Science* **313,** 314–318.

Dolezal, P., Smid, O., Rada, P., Zubacova, Z., Bursac, D., Sutak, R., Nebesarova, J., Lithgow, T., and Tachezy, J. (2005). *Giardia* mitosomes and trichomonad hydrogenosomes share a common mode of protein targeting. *Proc. Natl. Acad. Sci. USA* **102,** 10924–10929.

Emanuelsson, O., Nielsen, H., Brunak, S., and von Heijne, G. (2000). Predicting subcellular localization of proteins based on their N-terminal amino acid sequence. *J. Mol. Biol.* **300,** 1005–1016.

Embley, T. M. (2006). Multiple secondary origins of the anaerobic lifestyle in eukaryotes. *Philos. Trans. R. Soc. Lond. B Biol. Sci.* **361,** 1055–1067.

Embley, T. M., and Martin, W. (2006). Eukaryotic evolution, changes and challenges. *Nature* **440,** 623–630.

Finlay, B. J., Span, A. S. W., and Harman, J. M. P. (1983). Nitrate respiration in primitive eukaryotes. *Nature* **303,** 333–336.

Gabriel, K., and Pfanner, N. (2007). The mitochondrial machinery for import of precursor proteins. *Methods Mol. Biol.* **390,** 99–117.

Germot, A., Phillipe, H., and Le Guyader, H. (1996). Presence of a mitochondrial-type 70-kDa heat shock protein in *Trichomonas vaginalis* suggests a very early mitochondrial endosymbiosis in eukaryotes. *Proc. Natl. Acad. Sci. USA* **93**, 14614–14617.

Gill, E. E., Diaz-Triviño, S., Barbera, M. J., Silberman, J. D., Stechmann, A., Gaston, D., Tamas, I., and Roger, A. J. (2007). Novel mitochondrion-related organelles in the anaerobic amoeba *Mastigamoeba balamuthi*. *Mol. Microbiol.* **66**, 1306–1320.

Goldberg, A. V., Molik, S., Tsaousis, A. D., Neumann, K., Kuhnke, G., Delbac, F., Vivares, C. P., Hirt, R. P., Lill, R., and Embley, T. M. (2008). Localization and functionality of microsporidian iron–sulphur cluster assembly proteins. *Nature* **452**, 624–628.

Gray, M. W., Burger, G., and Lang, B. F. (1999). Mitochondrial evolution. *Science* **283**, 1476–1481.

Gray, M. W., Lang, B. F., and Burger, G. (2004). Mitochondria of protists. *Annu. Rev. Genet.* **38**, 477–524.

Gray, M. W., Lang, B. F., Cedergren, R., Golding, G. B., Lemieux, C., Sankoff, D., Turmel, M., Brossard, N., Delage, E., Littlejohn, T. G., Plante, I., Rioux, P., *et al.* (1998). Genome structure and gene content in protist mitochondrial DNAs. *Nucl. Acids Res.* **26**, 865–878.

Guda, C., Fahy, E., and Subramaniam, S. (2004a). MITOPRED: A genome-scale method for prediction of nuclear-encoded mitochondrial proteins. *Bioinformatics* **20**, 1785–1794.

Guda, C., Guda, P., Fahy, E., and Subramaniam, S. (2004b). MITOPRED: A web server for the prediction of mitochondrial proteins. *Nucleic Acids Res.* **32**, W372–W374.

Gurney, K. (1997). "An Introduction to Neural Networks" Taylor & Francis, Bristol, PA.

Hampl, V., Silbermann, J. D., Stechmann, A., Diaz-Triviño, S., Johnson, P. J., and Roger, A. J. (2008). Genetic evidence for a mitochondriate ancestry in the 'amito-chondriate' flagellate *Trimastix pyriformis*. *PLoS ONE* **3**, e1383.

Hawkins, J., and Bodén, M. (2006). Detecting and sorting targeting peptides with recurrent networks and support vector machines. *J. Bioinform. Comput. Biol.* **4**, 1–18.

Heazlewood, J. L., Tonti-Fillippini, J. S., Gout, A. M., Day, D. A., Whelan, J., and Millar, A. H. (2004). Experimental analysis of the *Arabidopsis* mitochondrial proteome highlights signaling and regulatory components, provides assessment of targeting prediction programs, and indicates plant-specific mitochondrial proteins. *Plant Cell* **16**, 241–256.

Höglund, A., Blum, T., Brady, S., Dönnes, P., Miguel, J. S., Rocheford, M., Kohlbacher, O., and Shatkay, H. (2006a). Significantly improved prediction of subcellular localization by integrating text and protein sequence data. *Proceedings of the Pacific Symposium on Biocomputing (PSB 2006)*, pp. 16–27.

Höglund, A., Dönnes, P., Blum, T., Adolph, H., and Kohlbacher, O. (2006b). MultiLoc: Prediction of protein subcellular localization using N-terminal targeting sequences, sequence motifs, and amino acid composition. *Bioinformatics* **22**, 1158–1165.

Horner, D. S., Hirt, R. P., Kilvington, S., Lloyd, D., and Embley, T. M. (1996). Molecular data suggest an early acquisition of the mitochondrion endosymbiont. *Proc. R. Soc. B Biol. Sci.* **263**, 1053–1059.

Horton, P., Park, K. J., Obayashi, T., Fujita, N., Harada, H., Adams-Collier, C. J., and Nakai, K. (2007). WoLF PSORT: Protein localization predictor. *Nucl. Acids Res.* **35**, w585–w587.

Horton, P., Park, K. J., Obayashi, T., and Nakai, K. (2006). Protein subcellular localization prediction with WoLF PSORT. *Proceedings of the 4th Annual Asia Pacific Bioinformatics Conference APBC06*, Taipei, Taiwan pp. 39–48.

Howe, C. J. (2008). Cellular evolution: What's in a mitochondrion? *Curr. Biol.* **18**, R429–R431.

Hrdy, I., Hirt, R. P., Dolezal, P., Bardonova, L., Foster, P. G., Tachezy, J., and Embley, T. M. (2004). *Trichomonas* hydrogenosomes contain the NADH dehydrogenase module of mitochondrial complex I. *Nature* **432**, 618–622.

46 Daniel Gaston *et al.*

Kecman, V. (2001). "Learning and Soft Computing—Support Vector Machines, Neural Networks, Fuzzy Logic Systems" The MIT Press, Cambridge, MA.
Kobayashi, M., Matsuo, Y., Takimoto, A., Suzuki, S., Maruo, F., and Shoun, H. (1996). Denitrification, a novel type of respiratory metabolism in fungal mitochondrion. *J. Biol. Chem.* **271**, 16263–16267.
Lindmark, D. G., and Müller, M. (1973). Hydrogenosome, a cytoplasmic organelle of the anaerobic flagellate *Tritrichomonas foetus*, and its role in pyruvate metabolism. *J. Biol. Chem.* **248**, 7724–7728.
Lu, Z., Szafron, D., Greiner, R., Lu, P., Wishart, D. S., Poulin, B., Anvik, J., Macdonell, C., and Eisner, R. (2004). Predicting protein subcellular localization of proteins using machine-learned classifiers. *Bioinformatics* **20**, 547–556.
Mai, Z., Ghosh, S., Frisardi, M., Rosenthal, B., Rogers, R., and Samuelson, J. (1999). Hsp60 is targeted to a cryptic mitochondrion-derived organelle ("crypton") in the microaerophilic protozoan parasite *Entamoeba histolytica*. *Mol. Cell Biol.* **19**, 2198–2205.
Marande, W., and Burger, G. (2007). Mitochondrial DNA as a genomic jigsaw puzzle. *Science* **318**, 415.
Marcotte, E. M., Xenarios, I., van Der Bliek, A. M., and Eisenberg, D. (2000). Localizing proteins in the cell from their phylogenetic profiles. *Proc. Natl. Acad. Sci. USA* **97**, 12115–12120.
Müller, M. (1993). The hydrogenosome. *J. Gen. Microbiol.* **139**, 2879–2889.
Nakai, K., and Horton, P. (1999). PSORT: A program for detecting the sorting signals of proteins and predicting their subcellular localization. *Trends Biochem. Sci.* **24**, 34–35.
Nielsen, H., Engelbrecht, J., Brunak, S., and von Heijne, G. (1997). Identification of prokaryotic and eukaryotic signal peptides and prediction of their cleavage sites. *Protein Eng.* **10**, 1–6.
Pagliarini, D. J., Calvo, S. E., Chang, B., Sheth, S. A., Vafai, S. B., Ong, S. E., Walford, G. A., Sugiana, C., Boneh, A., Chen, W. K., Hill, D. E., Vidal, M., *et al.* (2008). A mitochondrial protein compendium elucidates complex I disease biology. *Cell* **134**, 112–123.
Palmer, J. D. (1997). Organelle genomes: Going, going, gone! *Science* **275**, 790–791.
Peeters, N., and Small, I. (2001). Dual targeting to mitochondria and chloroplasts. *Biochem. Biophys. Acta* **1541**, 54–63.
Pfanner, N., Wiedemann, N., Meisinger, C., and Lithgow, T. (2004). Assembling the mitochondrial outer membrane. *Nat. Struct. Mol. Biol.* **11**, 1044–1048.
Prokisch, H., Scharfe, C., Camp, D. G. 2nd, Xiao, W., David, L., Andreoli, C., Monroe, M. E., Moore, R. J., Gritsenko, M. A., Kozany, C., Hixson, K. K., Mottaz, H. M., *et al.* (2004). Integrative analysis of the mitochondrial proteome in yeast. *PLoS Biol.* **2**, e160.
Rabiner, L. R. (1989). A tutorial on hidden Markov models and selected applications in speech recognition. *Proceedings of the IEEE.* **77**, 257–286.
Reinders, J., Zahedi, R. P., Pfanner, N., Meisinger, C., and Sickmann, A. (2006). Toward the complete yeast mitochondrial proteome: Multidimensional separation techniques for mitochondrial proteomics. *J. Proteome Res.* **5**, 1543–1554.
Roger, A. J., Clark, C. G., and Doolittle, W. F. (1996). A possible mitochondrial gene in the early-branching amitochondriate protist *Trichomonas vaginalis*. *Proc. Natl. Acad. Sci. USA* **93**, 14618–14622.
Roger, A. J., and Silberman, J. D. (2002). Cell evolution: Mitochondria in hiding. *Nature* **418**, 827–829.
Shatkay, H., Höglund, A., Brady, S., Blum, T., Dönnes, P., and Kohlbacher, O. (2007). SherLoc: High-accuracy prediction of protein subcellular localization by integrating text and protein sequence data. *Bioinformatics* **23**, 1410–1417.
Shen, Y. Q., and Burger, G. (2007). 'Unite and conquer': Enhanced prediction of protein subcellular localization by integrating multiple specialized tools. *BMC Bioinformatics* **8**, 420–430.

Sickmann, A., Reinders, J., Wagner, Y., Joppich, C., Zahedi, R., Meyer, H. E., Schönfisch, B., Perschil, I., Chacinska, A., Guiard, B., Rehling, P., Pfanner, N., *et al.* (2003). The proteome of *Saccharomyces cerevisiae* mitochondria. *Proc. Natl. Acad. Sci. USA* **100,** 13207–13212.

Small, I., Peeters, N., Legeai, F., and Lurin, C. (2004). Predotar: A tool for rapidly screening proteomes for N-terminal targeting sequences. *Proteomics* **4,** 1581–1590.

Smith, D. G., Gawryluk, R. M., Spencer, D. E., Pearlman, R. E., Siu, K. W., and Gray, M. W. (2007). Exploring the mitochondrial proteome of the ciliate protozoon *Tetrahymena thermophila*: Direct analysis by tandem mass spectrometry. *J. Mol. Biol.* **374,** 837–863.

Stechmann, A., Hamblin, K., Perez-Brocal, V., Gaston, D., Richmond, G. S., van der Giezen, M., Clark, C. G., and Roger, A. J. (2008). Organelles in *Blastocystis* that blur the distinction between mitochondria and hydrogenosomes. *Curr. Biol.* **18,** 580–585.

Sutak, R., Dolezal, P., Fiumera, H. L., Hrdy, I., Dancis, A., Delgadillo-Correa, M., Johnson, P. J., Müller, M., and Tachezy, J. (2004). Mitochondrial-type assembly of FeS centers in the hydrogenosomes of the amitochondriate eukaryote *Trichomonas vaginalis*. *Proc. Natl. Acad. Sci. USA* **101,** 10368–10373.

Takaya, N., Suzuki, S., Kuwazaki, S., Shoun, H., Maruo, F., Yamaguchi, M., and Takeo, K. (1999). Cytochrome P450nor, a novel class of mitochondrial cytochrome P450 involved in nitrate respiration in the fungus *Fusarium oxysporum*. *Arch. Biochem. Biophys.* **372,** 340–346.

Tachezy, J., Sanchez, L. B., and Müller, M. (2001). Mitochondrial type iron–sulfur cluster assembly in the amitochondriate eukaryotes *Trichomonas vaginalis* and *Giardia intestinalis*, as indicated by the phylogeny of IscS. *Mol. Biol. Evol.* **18,** 1919–1928.

Tielens, A. G., Rotte, C., Van Hellemond, J. J., and Martin, W. (2002). Mitochondria as we don't know them. *Trends Biochem. Sci.* **27,** 564–572.

Tielens, A. G., and Van Hellemond, J. J. (1998). The electron transport chain in anaerobically functioning eukaryotes. *Biochim. Biophys. Acta* **1365,** 71–78.

Timmis, J. N., Ayliffe, M. A., Huang, C. Y., and Martin, W. (2004). Endosymbiotic gene transfer: Organelle genomes forge eukaryotic chromosomes. *Nat. Rev. Genet.* **5,** 123–135.

Tovar, J., Fischer, A., and Clark, C. G. (1999). The mitosome, a novel organelle related to mitochondria in the amitochondrial parasite *Entamoeba histolytica*. *Mol. Microbiol.* **32,** 1013–1021.

Tovar, J., Leon-Avila, G., Sanchez, L. B., Sutak, R., Tachezy, J., van der Giezen, M., Hernandez, M., Müller, M., and Lucocq, J. M. (2003). Mitochondrial remnant organelles of *Giardia* function in iron–sulphur protein maturation. *Nature* **426,** 172–176.

Tsaousis, A. D., Kunji, E. R., Goldberg, A. V., Lucocq, J. M., Hirt, R. P., and Embley, T. M. (2008). A novel route for ATP acquisition by the remnant mitochondria of *Encephalitozoon cuniculi*. *Nature* **453,** 553–556.

van der Giezen, M., Slotboom, D. J., Horner, D. S., Dyal, P. L., Harding, M., Xue, G. P., Embley, T. M., and Kunji, E. R. (2002). Conserved properties of hydrogenosomal and mitochondrial ADP/ATP carriers: A common origin for both organelles. *EMBO J.* **21,** 572–579.

van der Giezen, M., and Tovar, J. (2005). Degenerate mitochondria. *EMBO Rep.* **6,** 525–530.

Van Hellemond, J. J., Opperdoes, F. R., and Tielens, A. G. (1998). Trypanosomatidae produce acetate via a mitochondrial acetate:succinate CoA transferase. *Proc. Natl. Acad. Sci. USA* **95,** 3036–3041.

Williams, B. A., Hirt, R. P., Lucocq, J. M., and Embley, T. M. (2002). A mitochondrial remnant in the microsporidian *Trachipleistophora hominis*. *Nature* **418,** 865–869.

Yu, C. S., Chen, Y. C., Lu, C. H., and Hwang, J. K. (2006). Prediction of protein subcellular localization. *Proteins*. **64,** 643–651.

CHAPTER THREE

PROTEOME CHARACTERIZATION OF MOUSE BRAIN MITOCHONDRIA USING ELECTROSPRAY IONIZATION TANDEM MASS SPECTROMETRY

X. Fang *and* Cheng S. Lee

Contents

1. Introduction	50
2. Analysis of Mitochondrial Proteome by Two-Dimensional PAGE and LC–MS Techniques	51
3. Selective Proteome Enrichment by Capillary Isotachophoresis (CITP)-Based Separations	51
4. Analysis of Mouse Brain Mitochondria Using the CITP-Based Proteome Platform	52
4.1. Isolation of synaptic mouse brain mitochondria	52
4.2. Mitochondria proteome sample preparation	53
4.3. Coupling transient CITP/CZE-based multidimensional separations with electrospray ionization (ESI)–MS	53
4.4. Proteome data analysis	54
5. Proteome Characterization of Mouse Brain Mitochondria	55
References	60

Abstract

Mitochondria are responsible for several essential functions, including apoptosis, iron–sulfur cluster assembly, and several important catabolic pathways in mammalian cells. Characterization of the mitochondrial proteome may therefore be extremely useful for identification and validation of targets for therapeutic treatment of human diseases, including aging and age-related disorders, diabetes, and obesity.

While capillary isotachophoresis (CITP) has been introduced for on-capillary analyte preconcentration prior to electrophoretic separation, the application of transient CITP/capillary zone electrophoresis (CZE) to selectively enrich and resolve trace amounts of proteins/peptide is employed in an effort to expand

Department of Chemistry and Biochemistry, University of Maryland, College Park, Maryland, USA

Methods in Enzymology, Volume 457
ISSN 0076-6879, DOI: 10.1016/S0076-6879(09)05003-4

mitochondrial proteome coverage in mammals. In this study, a total of 2095 distinct mouse SwissProt protein entries are identified from synaptic mitochondria isolated from mouse brain, corresponding to 50% and 76% coverage of the Maestro and MitoP2 mitochondrial reference sets, respectively. The increased coverage of mouse mitochondrial proteome further reveals the top 17 biosynthetic and metabolic pathways in synaptic mitochondria using the Ingenuity Pathway Analysis™.

1. INTRODUCTION

Mitochondria are a major source and target of free radicals (Hirano *et al.*, 2001; Leonard and Schapira, 2000; Lopez and Melov, 2002; Melov, 2000), and the collapse of the mitochondrial transmembrane potential can initiate the signaling cascades involved in programmed cell death or apoptosis (Kuwana *et al.*, 1998; Nicholls and Budd, 2000; Susin *et al.*, 1999). Mitochondria play a crucial role in the regeneration of antioxidants through the production of reducing equivalents (Kagan and Tyurina, 1998), and are responsible for the vast majority of ATP production within most cells and higher organisms. Mitochondria also infer a major role in cell signaling, as well as biosynthesis (e.g., heme) and degradation (e.g., urea cycle) (Gunter *et al.*, 1998). Mitochondrial mutations and dysfunction have been implicated in numerous diseases including aging, cancer, heart disease, neurological diseases such as Parkinson's, Alzheimer's, and Huntington's, and various neuromuscular syndromes (Beckman and Ames, 1998; Hirano *et al.*, 2001; Kovacic and Jacintho, 2001; Verma *et al.*, 2003). The mitochondrial connection to these major degenerative diseases has consequently driven significant research efforts to define the mitochondrial proteome and to discover new molecular targets for drug development and therapeutic intervention.

In addition to their biological and medical significance, mitochondria are ideal targets for global proteome analysis because they have a manageable level of complexity as a consequence of their apparent prokaryotic ancestry. Their endosymbiotic origins have been preserved in their double membrane structure, and they possess their own circular genome with mitochondria-specific transcription, translation, and protein assembly systems. Mammalian mitochondrial DNA encodes for 13 essential protein components of the oxidative phosphorylation system, small and large rRNAs, and 22 tRNAs. The remaining mitochondrial proteins of the oxidative phosphorylation system, metabolic enzymes, membrane channel proteins, and any other proteins regulating mitochondrial function are derived from the nuclear genome. Import of these proteins into the mitochondria is highly regulated through the function of protein chaperones and the inner and outer transmembrane peptide import complexes (Koehler, 2000; Lithgow, 2000).

2. ANALYSIS OF MITOCHONDRIAL PROTEOME BY TWO-DIMENSIONAL PAGE AND LC–MS TECHNIQUES

While two-dimensional (2-D) polyacrylamide gel electrophoresis (2-D PAGE) has been employed to define the components of the mitochondrial proteome (Chang et al., 2003; Hanson et al., 2001; Lopez and Melov, 2002; Lopez et al., 2000; McDonald et al., 2006; Reinders et al., 2006; Sickmann et al., 2003), mitochondria contain many membrane-bound and very basic proteins that are quite difficult to be analyzed by 2-D PAGE. Thus, considerable efforts have been devoted to the application of various liquid chromatography (LC) techniques in single or multidimensional separation format prior to mass spectrometry (MS) detection for the analysis of mito-chondrial proteins (Forner et al., 2006; Johnson et al., 2007; Kislinger et al., 2006; McDonald et al., 2006; Mootha et al., 2003; Prokisch et al., 2004; Reinders et al., 2006; Rezaul et al., 2005; Sickmann et al., 2003; Taylor et al., 2002, 2003b). In particular, the peptide-based shotgun proteomic studies fully exploit the resolution and sensitivity achievable with multidimensional LC–MS or sodium dodecyl sulfate (SDS)-PAGE/LC–MS approaches, allowing many additional mitochondrial proteins to be identified.

Still, proteins cataloged from reported proteomic studies comprise only small part of the estimated 1500 (DiMauro and Schon, 1998; Taylor et al., 2003a) and 697–4532 (Calvo et al., 2006) total human mitochondrial pro-teins. The actual number of mitochondrial proteins is difficult to determine since these estimations can have a false discovery rate (FDR) of up to 68% (Calvo et al., 2006). Even with reduced protein complexity in the subcellular proteome, the goal of achieving comprehensive proteome analysis is still challenged by the large variation of protein relative abundances.

3. SELECTIVE PROTEOME ENRICHMENT BY CAPILLARY ISOTACHOPHORESIS (CITP)-BASED SEPARATIONS

Besides protein complexity, the greatest bioanalytical challenge facing comprehensive proteome analysis, particularly in the discovery of low abundance proteins, is related to the large variation of protein relative abundances. Thus, an important aspect of any multidimensional separation platform is its ability to improve the MS detection of analytes present in low quantities during the analyses of complex protein/peptide mixtures. As opposed to universally enriching all proteins by a similar degree in most sample concentration techniques, the CITP stacking process specifi-cally targets trace amounts of proteins (Foret et al., 1993; Gebauer and Bocek, 2000; Stegehuis et al., 1991) and thus drastically reduces the range of relative protein abundances for providing unparallel advantages towards the identification of low abundance proteins.

In order to obtain excellent separation resolution in CITP, the operational electrolyte system should be optimized to give appropriate differences among the effective mobilities of analytes. As the pH and ionic strength of the stacked zones in CITP are different from each other, optimization of the separation in CITP is not straightforward in comparison with capillary zone electrophoresis (CZE). Thus, on-column transition of CITP to CZE has been performed for the optimization of selective enrichment of trace peptides in CITP and the subsequent separation resolution achieved using CZE (An et al., 2006).

By comparing with displacement chromatography, the only reported approach in the literature for the selective enrichment of low abundance peptides in the absence of any bioaffinity interactions (Wilkins et al., 2002; Xiang et al., 2003), CITP offers the benefits of speed, easy manipulation/ switching between the stacking and separation modes in transient CITP/ CZE, and no need for column regeneration, including the removal of bound displacers. Transient CITP/CZE further provides seamless combination with nano-reversed phase LC (nano-RPLC) as two highly resolving and completely orthogonal separation techniques critically needed for analyzing complex proteomes (Fang et al., 2007, 2008).

4. Analysis of Mouse Brain Mitochondria Using the CITP-Based Proteome Platform

4.1. Isolation of synaptic mouse brain mitochondria

Mice (Male wild-type C57BL/6) are decapitated, and brains are removed and homogenized in ice-cold isolation medium containing 225 mM mannitol, 75 mM sucrose, 5 mM N-2-hydroxyethylpiperazine-N'-2-ethanesulfonic acid, and 1 mM ethylene glycol tetraacetic acid, pH 7.4, at 4 °C. As described previously (Chandrasekaran et al., 2006), a slightly modified procedure by Dunkley et al. (1988) is employed to separate synaptosomes and nonsynaptic mitochondria. Briefly, after low-speed spin (1300g for 3 min) of the brain homogenate, the supernatant is centrifuged at 21,000g for 10 min. The pellet is suspended in 3% Percoll and layered on top of a preformed gradient of 24% (3.5 ml), 15% (2 ml), 10% (1.5 ml) of Percoll and centrifuged at 32,000g for 8 min. Nonsynaptic mitochondria sediment at the bottom of the tube and the synaptosomes accumulate at the interface of 24% and 15% Percoll and also at the interface of 15% and 10% Percoll.

The synaptosomes are collected and the mitochondria are released from synaptosomes using a nitrogen cavitation technique (Brown et al., 2004; Kristian et al., 2006). The suspension is layered onto a preformed gradient of

40% (1.5 ml) and 24% (3.5 ml) Percoll. The mitochondria accumulate at the interface of 40% and 24% Percoll gradient. After the mitochondria fraction is collected and diluted with isolation medium, it is centrifuged to pellet purified synaptic mitochondria. Synaptic mitochondria are suspended in isolation medium containing 1 mg/ml bovine serum albumin to remove free fatty acid from mitochondrial membranes. The purity of synaptosome and mitochondria isolated from the Percoll gradient is examined with electron microscopy (Chandrasekaran et al., 2006; Kristian et al., 2006).

4.2. Mitochondria proteome sample preparation

After centrifugation at 7500g, the mitochondrial pellet is suspended and treated with a solution containing 1% SDS, 50 mM Tris(hydroxymethyl) aminomethane (Tris) at pH 8.0, and 1 mM dithiothreitol (DTT). After centrifugation at 20,000g for 30 min, the supernatant is collected as the SDS-solubilized mitochondrial proteins. The SDS-solubilized mitochondrial proteins are placed in a dialysis cup and dialyzed overnight at 4 °C against 100 mM Tris at pH 8.0. The dialyzed proteins are denatured, reduced, and alkylated by sequentially adding urea, DTT, and iodoacetamide with final concentrations of 8 M, 10 mg/ml, and 20 mg/ml, respectively. The solution is incubated at 37 °C for 1 h in the dark and then diluted eightfold with 100 mM ammonium acetate at pH 8.0. Trypsin is added at a 1:40 (w/w) enzyme to substrate ratio and the solution is incubated at 37 °C overnight. Tryptic digests are desalted using a Peptide MacroTrap column, lyophilized to dryness using a SpeedVac, and then stored at −80 °C. Approximately, 100–300 μg of mitochondrial proteins are obtained from each of the whole mouse forebrain (brain without the cerebellum).

4.3. Coupling transient CITP/CZE-based multidimensional separations with electrospray ionization (ESI)–MS

The CITP apparatus is constructed in-house using a CZE 1000R high-voltage power supply. An 80-cm long CITP capillary (100 μm i.d./365 μm o.d.) coated with hydroxypropyl cellulose (Chen et al., 2003; Shen et al., 2000) is initially filled a background electrophoresis buffer of 0.1 M acetic acid at pH 2.8. The sample containing mitochondrial protein digests is prepared in a 2% pharmalyte 3–10 solution and is hydrodynamically injected into the capillary (Fang et al., 2007; Mohan and Lee, 2002; Mohan et al., 2003). A positive electric voltage of 24 kV is then applied to the inlet reservoir, which is filled with a 0.1 M acetic acid solution.

UV absorbance detection at 214 nm is placed at 14 cm from the cathodic end of the capillary. The cathodic end of the capillary is housed inside a stainless steel needle using a coaxial liquid sheath flow configuration

(Yang *et al.*, 1998). A sheath liquid composed of 0.1 *M* acetic acid is delivered at a flow rate of 1 μl/min using a syringe pump. The stacked and resolved peptides in the CITP/CZE capillary are sequentially fractionated and loaded into individual wells on a moving microtiter plate.

To couple transient CITP/CZE with nano-RPLC, peptides collected in individual wells are sequentially injected into individual trap columns (3 cm $\times$ 200 μm i.d. $\times$ 365 μm o.d.) packed with 5 μm porous C_{18} particles. Each peptide fraction is subsequently analyzed by nano-RPLC equipped with a dual-quaternary pump and a dual nano-flow splitter connected to two pulled-tip fused-silica capillaries (50 μm i.d. $\times$ 365 μm o.d.). These two 15-cm long capillaries are packed with 3-μm C_{18} particles.

Nano-RPLC separations are performed in parallel in which a dual-quaternary pump delivered two identical 2-h organic solvent gradients with an offset of 1 h (Balgley *et al.*, 2008; Wang *et al.*, 2007). Peptides are eluted at a flow rate of 200 nl/min using a 5–45% linear acetonitrile gradient over 100 min with the remaining 20 min for column regeneration and equilibration. The peptide eluants are monitored using a linear ion-trap mass spectrometer operated in a data dependent mode. Full scans are collected from 400–1400 *m/z* and 5 data dependent MS/MS scans are collected with dynamic exclusion set to 30 s. A moving stage housing two nano-RPLC columns is employed to provide electrical contacts for applying ESI voltages, and most importantly to position the columns in-line with the ESI inlet at each chromatography separation and data acquisition cycle.

4.4. Proteome data analysis

Raw mass spectra data are converted to peak list files. OMSSA developed at the National Center for Biotechnology Information (Geer *et al.*, 2004) is used to search the peak list files against a decoyed *Mus musculus* subset of the UniProt sequence library (April 20, 2006). This database is constructed by reversing all 41,178 sequences and appending them to the end of the sequence library. Searches are performed using the following parameters: fully tryptic, 1.5 Da precursor ion mass tolerance, 0.4 Da fragment ion mass tolerance, 1 missed cleavage, alkylated Cys as a fixed modification, and variable modification of Met oxidation. Searches are run in parallel on a 12-node, 24-CPU cluster.

FDRs are determined using a target-decoy search strategy introduced by Elias and co-workers (Elias *et al.*, 2005). Briefly, FDRs are calculated by multiplying the number of false positive identifications (hits to the reversed sequences scoring below a given threshold) by 2 and dividing by the number of total identifications. An *E*-value threshold corresponding to a 1% FDR in total peptide identifications is used as a cutoff as this correlates with the maximum sensitivity versus specificity (Balgley *et al.*, 2007). No other cutoffs, such as requiring a minimum number of distinct peptides per protein identification, are utilized. After generation of search data, the OMSSA XML result files are parsed using a Java parser and loaded into an Oracle 10g database.

5. PROTEOME CHARACTERIZATION OF MOUSE BRAIN MITOCHONDRIA

By using a target–decoy search strategy (Balgley et al., 2007; Elias et al., 2005), an E-value threshold of 0.0013, corresponding to 1% FDR of total peptide identifications, is chosen as the cutoff and leads to the identification of 12,110 fully tryptic peptides covering 1795 distinct mouse SwissProt protein entries. OMSSA E-value is the expectation value that a tandem MS event represents the predicted peptide given a number of factors, including the quality of the match between the experimental and theoretical fragment ion spectra, the search space as determined by the sequence library, the precursor and fragment ion mass accuracy settings, enzyme specificity, and missed cleavages and modifications (Geer et al., 2004). The use of the target–decoy determined FDR and E-value threshold not only accurately reduces false negative identifications, but also controls the degree of false positives, while improving the predictive power of a typical OMSSA search. Due to the differences in the complexity of various proteome samples, the sample preparation procedures, the proteome measurements, and the search parameters, a decoyed database search is always performed at our laboratory to determine specific score thresholds employed for peptide and protein identifications.

The percentage of overlapping proteins obtained from repeated CITP/ CZE-nano-RPLC runs is around 86%. A total of 2095 distinct mouse SwissProt protein entries are identified from two CITP/CZE-nano-RPLC runs of a single mouse brain mitochondria sample. Combined proteomic results are further compared to the mitochondrial protein reference set predicted by Maestro (Calvo et al., 2006). By using a threshold score of 4.69 (corresponding to a 10% FDR), the Maestro reference set contains a list of 1207 mouse mitochondrial proteins. Combined CITP/ CZE analyses cover 598 and 413 proteins of the Maestro mitochondrial reference set on the basis of proteins identified by a single and at least two distinct peptides, respectively. In contrast, only 178 out of 1075 mouse brain mitochondrial proteins identified by Kislinger et al. (2006) overlap with the Maestro mitochondrial reference set.

The combined CITP/CZE dataset is further compared to the mitochondrial protein reference set of the MitoP2-database (Prokisch et al., 2006) containing a list of 731 mouse mitochondrial proteins. Out of 2095 distinct proteins, 553 overlap with the MitoP2-database corresponding to

76% mouse mitochondrial proteome coverage. The proteomic results of mouse brain mitochondria achieved by Kislinger *et al.* (2006) are also compared to the mitochondrial protein reference set of the MitoP2-database. Only 146 out of 1075 cataloged proteins overlap with the MitoP2-database, corresponding to less than 20% coverage. It is interesting to note that the entirety of these 146 proteins and 90% of proteins collected from the work of Kislinger *et al.* (2006) are included in the combined CITP/CZE dataset.

By employing the mitochondrial proteome dataset obtained from combined CITP/CZE analyses, the top 17 biosynthetic and metabolic pathways are mapped out by the Ingenuity Pathway Analysis^{TM} and are shown in Fig. 3.1. The ratio is calculated as the number of proteins in a given pathway which meet cutoff criteria divided by total number of proteins which make up that pathway. The significance value associated with the functional analysis for a dataset is a measure for how likely the proteins from the dataset under investigation participate in that function. The significance is expressed as a p-value, which is calculated using the right-tailed Fisher's Exact Test. The p-value is calculated by comparing the number of user-specified proteins of interest that participate in a given function or pathway, relative to the total number of occurrences of these proteins in all functional/pathway annotations stored in the Ingenuity Pathway Knowledge Base.

The most significant metabolic pathway revealed by our mitochondrial dataset is clearly the oxidative phosphorylation pathway. As energy producers

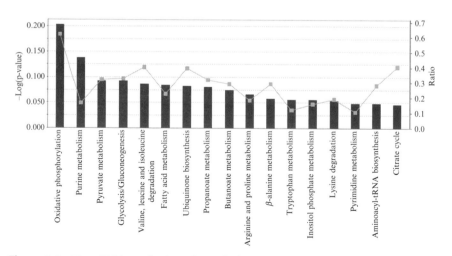

Figure 3.1 Top 17 biosynthetic and metabolic pathways constructed from combined mitochondrial proteome dataset using the Ingenuity Pathway Analysis^{TM}. The significance and the ratio of the proteome dataset associated with individual pathways are represented by the bar and the line, respectively (Fang *et al.*, 2008).

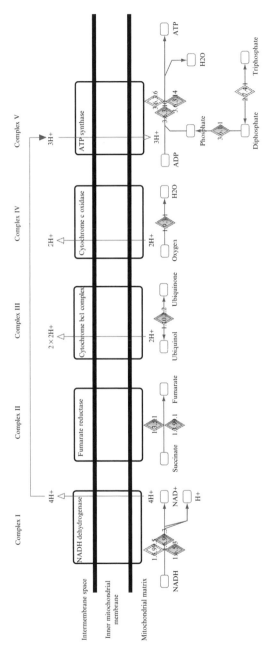

Figure 3.2 Simplified illustration of the oxidative phosphorylation pathway containing five complexes (Fang *et al.*, 2008).

to sustain the living cell activities, mitochondria constantly convert the glycolysis products, pyruvate and NADH, into energy in the form of ATP through the cellular respiration process. In order to generate ATP in the efficient way, mitochondria require numerous transport proteins and enzymes to perform the oxidation reactions of the respiratory chain, ATP synthesis, and regulation of metabolite and substance transport.

A total of 100 proteins engaged in the oxidative phosphorylation pathway, corresponding to ~62% pathway coverage, are identified. Five complexes involved in the oxidative phosphorylation pathway are shown in Fig. 3.2, including NADH dehydrogenase, fumarate reductase, cytochrome $bc1$ complex, cytochrome c oxidase, and ATP synthase families. The coverage of different complexes are analyzed as 87% (40/46) for complex I, 80% (4/5) for complex II, 73% (8/11) for complex III, 85% (11/13) for complex IV, and 94% (15/16) for complex V. The overall complex coverage is around 86% (78/91) and higher than 62% pathway coverage. As shown in Fig. 3.3, tandem MS spectra of single unique peptide hits leading to the identifications of NDUA1, NU3M, and NU4LM as representative subunits of complex I exhibit excellent E-values of 5.8×10^{-10}, 6.6×10^{-12}, and 8.3×10^{-16}, respectively, from a target-decoy OMSSA search.

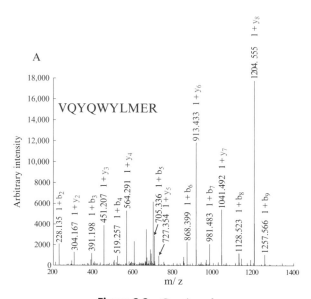

Figure 3.3   Continued

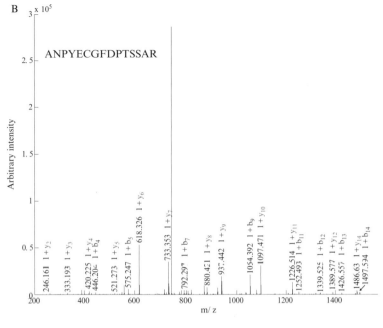

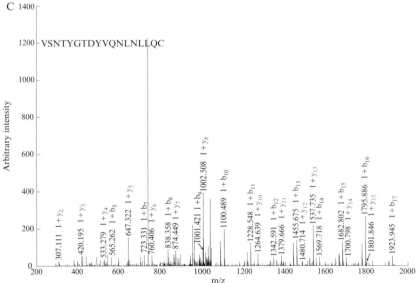

Figure 3.3 Tandem MS spectra of single unique peptide hits leading to the identifications of (A) NDUA1, (B) NU3M, and (C) NU4LM (Fang *et al.*, 2008).

REFERENCES

An, Y., Cooper, J. W., Balgley, B. M., and Lee, C. S. (2006). Selective enrichment and ultrasensitive identification of trace peptides in proteome analysis using transient capillary isotachophoresis/zone electrophoresis coupled with nano-ESI-MS. *Electrophoresis* **27,** 3599–3608.

Balgley, B. M., Laudeman, T., Yang, L., Song, T., and Lee, C. S. (2007). Comparative evaluation of tandem MS search algorithms using a target-decoy search strategy. *Mol. Cell. Proteomics* **6,** 1599–1608.

Balgley, B. M., Wang, W., Song, T., Fang, X., Yang, L., and Lee, C. S. (2008). Evaluation of confidence and reproducibility in quantitative proteomics performed by a capillary isoelectric focusing-based proteomic platform coupled with a spectral counting approach. *Electrophoresis* **29,** 3047–3054.

Beckman, K. B., and Ames, B. N. (1998). The free radical theory of aging matures. *Physiol. Rev.* **78,** 547–581.

Brown, M. R., Sullivan, P. G., Dorenbos, K. A., Modafferi, E. A., Geddes, J. W., and Steward, O. (2004). Nitrogen disruption of synaptoneurosomes: An alternative method to isolate brain mitochondria. *J. Neurosci. Methods* **137,** 299–303.

Calvo, S., Jain, M., Xie, X., Sheth, S. A., Chang, B., Goldberger, O. A., Spinazzola, A., Zeviani, M., Carr, S. A., and Mootha, V. K. (2006). Systematic identification of human mitochondrial disease genes through integrative genomics. *Nat. Genet.* **38,** 576–582.

Chandrasekaran, K., Hazelton, J. L., Wang, Y., Fiskum, G., and Kristian, T. (2006). Neuron-specific conditional expression of a mitochondrially targeted fluorescent protein in mice. *J. Neurosci.* **26,** 13123–13127.

Chang, J., Van Remmen, H., Cornell, J., Richardson, A., and Ward, W. F. (2003). Comparative proteomics: Characterization of a two-dimensional gel electrophoresis system to study the effect of aging on mitochondrial proteins. *Mech. Ageing Dev.* **124,** 33–41.

Chen, J., Balgley, B. M., DeVoe, D. L., and Lee, C. S. (2003). Capillary isoelectric focusing-based multidimensional concentration/separation platform for proteome analysis. *Anal. Chem.* **75,** 3145–3152.

DiMauro, S., and Schon, E. A. (1998). Nuclear power and mitochondrial disease. *Nat. Genet.* **19,** 214–215.

Dunkley, P. R., Heath, J. W., Harrison, S. M., Jarvie, P. E., Glenfield, P. J., and Rostas, J. A. (1988). A rapid Percoll gradient procedure for isolation of synaptosomes directly from an S1 fraction: Homogeneity and morphology of subcellular fractions. *Brain Res.* **441,** 59–71.

Elias, J. E., Haas, W., Faherty, B. K., and Gygi, S. P. (2005). Comparative evaluation of mass spectrometry platforms used in large-scale proteomics investigations. *Nat. Methods* **2,** 667–675.

Fang, X., Wang, W., Yang, L., Chandrasekaran, K., Kristian, T., Balgley, B. M., and Lee, C. S. (2008). Application of capillary isotachophoresis-based multidimensional separations coupled with electrospray ionization-tandem mass spectrometry for characterization of mouse brain mitochondrial proteome. *Electrophoresis* **29,** 2215–2223.

Fang, X., Yang, L., Wang, W., Song, T., Lee, C., Devoe, D., and Balgley, B. (2007). Comparison of electrokinetics-based multidimensional separations coupled with electrospray ionization-tandem mass spectrometry for characterization of human salivary proteins. *Anal. Chem.* **79,** 5785–5792.

Foret, F., Szoko, E., and Karger, B. L. (1993). Trace analysis of proteins by capillary zone electrophoresis with on-column transient isotachophoretic preconcentration. *Electrophoresis* **14,** 417–428.

Forner, F., Foster, L. J., Campanaro, S., Valle, G., and Mann, M. (2006). Quantitative proteomic comparison of rat mitochondria from muscle, heart, and liver. *Mol. Cell. Proteomics* **5,** 608–619.

Gebauer, P., and Bocek, P. (2000). Recent progress in capillary isotachophoresis. *Electrophoresis* **21,** 3898–3904.

Geer, L. Y., Markey, S. P., Kowalak, J. A., Wagner, L., Xu, M., Maynard, D. M., Yang, X., Shi, W., and Bryant, S. H. (2004). Open mass spectrometry search algorithm. *J. Proteome Res.* **3,** 958–964.

Gunter, T. E., Buntinas, L., Sparagna, G. C., and Gunter, K. K. (1998). The Ca^{2+} transport mechanisms of mitochondria and Ca^{2+} uptake from physiological-type Ca^{2+} transients. *Biochim. Biophys. Acta* **1366,** 5–15.

Hanson, B. J., Schulenberg, B., Patton, W. F., and Capaldi, R. A. (2001). A novel subfractionation approach for mitochondrial proteins: A three-dimensional mitochondrial proteome map. *Electrophoresis* **22,** 950–959.

Hirano, M., Davidson, M., and DiMauro, S. (2001). Mitochondria and the heart. *Curr. Opin. Cardiol.* **16,** 201–210.

Johnson, D. T., Harris, R. A., French, S., Blair, P. V., You, J., Bemis, K. G., Wang, M., and Balaban, R. S. (2007). Tissue heterogeneity of the mammalian mitochondrial proteome. *Am. J. Physiol. Cell Physiol.* **292,** 689–697.

Kagan, V. E., and Tyurina, Y. Y. (1998). Recycling and redox cycling of phenolic antioxidants. *Ann. N. Y. Acad. Sci.* **854,** 425–434.

Kislinger, T., Cox, B., Kannan, A., Chung, C., Hu, P., Ignatchenko, A., Scott, M. S., Gramolini, A. O., Morris, Q., Hallett, M. T., Rossant, J., Hughes, T. R., et al. (2006). Global survey of organ and organelle protein expression in mouse: Combined proteomic and transcriptomic profiling. *Cell* **125,** 173–186.

Koehler, C. M. (2000). Protein translocation pathways of the mitochondrion. *FEBS Lett.* **476,** 27–31.

Kovacic, P., and Jacintho, J. D. (2001). Reproductive toxins: Pervasive theme of oxidative stress and electron transfer. *Curr. Med. Chem.* **8,** 863–892.

Kristian, T., Hopkins, I. B., McKenna, M. C., and Fiskum, G. (2006). Isolation of mitochondria with high respiratory control from primary cultures of neurons and astrocytes using nitrogen cavitation. *J. Neurosci. Methods* **152,** 136–143.

Kuwana, T., Smith, J. J., Muzio, M., Dixit, V., Newmeyer, D. D., and Kornbluth, S. (1998). Apoptosis induction by caspase-8 is amplified through the mitochondrial release of cytochrome c. *J. Biol. Chem.* **273,** 16589–16594.

Leonard, J. V., and Schapira, A. H. (2000). Mitochondrial respiratory chain disorders I: Mitochondrial DNA defects. *Lancet* **355,** 299–304.

Lithgow, T. (2000). Targeting of proteins to mitochondria. *FEBS Lett.* **476,** 22–26.

Lopez, M. F., Kristal, B. S., Chernokalskaya, E., Lazarev, A., Shestopalov, A. I., Bogdanova, A., and Robinson, M. (2000). High-throughput profiling of the mitochondrial proteome using affinity fractionation and automation. *Electrophoresis* **21,** 3427–3440.

Lopez, M. F., and Melov, S. (2002). Applied proteomics: Mitochondrial proteins and effect on function. *Circ. Res.* **90,** 380–389.

McDonald, T., Sheng, S., Stanley, B., Chen, D., Ko, Y., Cole, R. N., Pedersen, P., and Van Eyk, J. E. (2006). Expanding the subproteome of the inner mitochondria using protein separation technologies: One- and two-dimensional liquid chromatography and two-dimensional gel electrophoresis. *Mol. Cell. Proteomics* **5,** 2392–2411.

Melov, S. (2000). Mitochondrial oxidative stress. Physiologic consequences and potential for a role in aging. *Ann. N. Y. Acad. Sci.* **908,** 219–225.

Mohan, D., and Lee, C. S. (2002). On-line coupling of capillary isoelectric focusing with transient isotachophoresis-zone electrophoresis: A two-dimensional separation system for proteomics. *Electrophoresis* **23,** 3160–3167.

Mohan, D., Pasa-Tolic, L., Masselon, C. D., Tolic, N., Bogdanov, B., Hixson, K. K., Smith, R. D., and Lee, C. S. (2003). Integration of electrokinetic-based multidimensional separation/concentration platform with electrospray ionization-Fourier transform ion cyclotron resonance-mass spectrometry for proteome analysis of *Shewanella oneidensis*. *Anal. Chem.* **75,** 4432–4440.

Mootha, V. K., Bunkenborg, J., Olsen, J. V., Hjerrild, M., Wisniewski, J. R., Stahl, E., Bolouri, M. S., Ray, H. N., Sihag, S., Kamal, M., Patterson, N., Lander, E. S., et al. (2003). Integrated analysis of protein composition, tissue diversity, and gene regulation in mouse mitochondria. *Cell* **115,** 629–640.

Nicholls, D. G., and Budd, S. L. (2000). Mitochondria and neuronal survival. *Physiol. Rev.* **80,** 315–360.

Prokisch, H., Andreoli, C., Ahting, U., Heiss, K., Ruepp, A., Scharfe, C., and Meitinger, T. (2006). MitoP2: The mitochondrial proteome database—Now including mouse data. *Nucleic Acids Res.* **34,** 705–711.

Prokisch, H., Scharfe, C., Camp, D. G., Xiao, W., David, L., Andreoli, C., Monroe, M. E., Moore, R. J., Gritsenko, M. A., Kozany, C., Hixson, K. K., Mottaz, H. M., et al. (2004). Integrative analysis of the mitochondrial proteome in yeast. *PLoS. Biol.* **2,** e160.

Reinders, J., Zahedi, R. P., Pfanner, N., Meisinger, C., and Sickmann, A. (2006). Toward the complete yeast mitochondrial proteome: Multidimensional separation techniques for mitochondrial proteomics. *J. Proteome Res.* **5,** 1543–1554.

Rezaul, K., Wu, L., Mayya, V., Hwang, S.-I., and Han, D. (2005). A systematic characterization of mitochondrial proteome from human T leukemia cells. *Mol. Cell. Proteomics* **4,** 169–181.

Shen, Y., Berger, S. J., Anderson, G. A., and Smith, R. D. (2000). High-efficiency capillary isoelectric focusing of peptides. *Anal. Chem.* **72,** 2154–2159.

Sickmann, A., Reinders, J., Wagner, Y., Joppich, C., Zahedi, R., Meyer, H. E., Schonfisch, B., Perschil, I., Chacinska, A., Guiard, B., Rehling, P., Pfanner, N., et al. (2003). The proteome of *Saccharomyces cerevisiae* mitochondria. *Proc. Natl. Acad. Sci. USA* **100,** 13207–13212.

Stegehuis, D. S., Irth, H., Tjaden, U. R., and Van der Greef, J. (1991). Isotachophoresis as an on-line concentration pretreatment technique in capillary electrophoresis. *J. Chromatogr.* **538,** 393–402.

Susin, S. A., Lorenzo, H. K., Zamzami, N., Marzo, I., Snow, B. E., Brothers, G. M., Mangion, J., Jacotot, E., Costantini, P., Loeffler, M., Larochette, N., Goodlett, D. R., et al. (1999). Molecular characterization of mitochondrial apoptosis-inducing factor. *Nature* **397,** 441–446.

Taylor, S. W., Fahy, E., and Ghosh, S. S. (2003a). Global organellar proteomics. *Trends Biotechnol.* **21,** 82–88.

Taylor, S. W., Fahy, E., Zhang, B., Glenn, G. M., Warnock, D. E., Wiley, S., Murphy, A. N., Gaucher, S. P., Capaldi, R. A., Gibson, B. W., and Ghosh, S. S. (2003b). Characterization of the human heart mitochondrial proteome. *Nat. Biotechnol.* **21,** 281–286.

Taylor, S. W., Warnock, D. E., Glenn, G. M., Zhang, B., Fahy, E., Gaucher, S. P., Capaldi, R. A., Gibson, B. W., and Ghosh, S. S. (2002). An alternative strategy to determine the mitochondrial proteome using sucrose gradient fractionation and 1D PAGE on highly purified human heart mitochondria. *J. Proteome Res.* **1,** 451–458.

Verma, M., Kagan, J., Sidransky, D., and Srivastava, S. (2003). Proteomic analysis of cancer-cell mitochondria. *Nat. Rev. Cancer* **3,** 789–795.

Wang, W., Guo, T., Rudnick, P. A., Song, T., Li, J., Zhuang, Z., Zheng, W., Devoe, D. L., Lee, C. S., and Balgley, B. M. (2007). Membrane proteome analysis of microdissected ovarian tumor tissues using capillary isoelectric focusing/reversed-phase liquid chromatography-tandem MS. *Anal. Chem.* **79,** 1002–1009.

Wilkins, J. A., Xiang, R., and Horvath, C. (2002). Selective enrichment of low-abundance peptides in complex mixtures by elution-modified displacement chromatography and their identification by electrospray ionization mass spectrometry. *Anal. Chem.* **74,** 3933–3941.

Xiang, R., Horvath, C., and Wilkins, J. A. (2003). Elution-modified displacement chromatography coupled with electrospray ionization-MS: On-line detection of trace peptides at low-femtomole level in peptide digests. *Anal. Chem.* **75,** 1819–1827.

Yang, L., Lee, C. S., Hofstadler, S. A., Pasa-Tolic, L., and Smith, R. D. (1998). Capillary isoelectric focusing-electrospray ionization Fourier transform ion cyclotron resonance mass spectrometry for protein characterization. *Anal. Chem.* **70,** 3235–3241.

CHAPTER FOUR

32P LABELING OF PROTEIN PHOSPHORYLATION AND METABOLITE ASSOCIATION IN THE MITOCHONDRIA MATRIX

Angel M. Aponte,* Darci Phillips,[†] Robert A. Harris,[‡] Ksenia Blinova,[†] Stephanie French,* D. Thor Johnson,*,[‡] and Robert S. Balaban[†]

Contents

1. Introduction	64
2. Methods	65
2.1. Mitochondria isolation and phosphate loading	66
2.2. 32P labeling of intact mitochondria	68
2.3. Sample preparation for two-dimensional gel electrophoresis	68
2.4. Image acquisition and analysis	71
2.5. Mass spectrometry	71
3. Results and Discussion	72
3.1. 32P dose and time course	72
3.2. 32P assignments	73
3.3. Chase experiments with cold phosphate	74
3.4. Suppression of PDH	75
3.5. Purified 32P labeled proteins and complexes	76
3.6. Blue native gel electrophoresis	77
4. Summary	78
References	79

Abstract

Protein phosphorylations, as well as phosphate metabolite binding, are well characterized post-translational mechanisms that regulate enzyme activity in the cytosol, but remain poorly defined in mitochondria. Recently extensive

* Proteomics Core Facility, National Heart Lung and Blood Institute, DHHS, Bethesda, Maryland, USA
[†] Laboratory of Cardiac Energetics, National Heart Lung and Blood Institute, DHHS, Bethesda, Maryland, USA
[‡] Department of Biochemistry and Molecular Biology, Indiana University School of Medicine, Indianapolis, Indiana, USA

Methods in Enzymology, Volume 457
ISSN 0076-6879, DOI: 10.1016/S0076-6879(09)05004-6

matrix protein phosphorylation sites have been discovered but their functional significance is unclear. Herein we describe methods of using ^{32}P labeling of intact mitochondria to determine the dynamic pools of protein phosphorylation as well as phosphate metabolite association. This screening approach may be useful in not only characterizing the dynamics of these pools, but also provide insight into which phosphorylation sites have a functional significance. Using the mitochondrial ATP synthetic capacity under appropriate conditions, inorganic ^{32}P was added to energized mitochondria to generate high specific activity γ-P[32]-ATP in the matrix. In general, SDS denaturing and gel electrophoresis was used to primarily follow protein phosphorylation, whereas native gel techniques were used to observe weaker metabolite associations since the structure of mitochondrial complexes was minimally affected. The protein phosphorylation and metabolite association within the matrix was found to be extensive using these approaches. [32]P labeling in 2D gels was detected in over 40 proteins, including most of the complexes of the cytochrome chain and proteins associated with intermediary metabolism, biosynthetic pathways, membrane transport, and reactive oxygen species metabolism. [32]P pulse-chase experiments further revealed the overall dynamics of these processes that included phosphorylation site turnover as well as phosphate–protein pool size alterations. The high sensitivity of ^{32}P resulted in many proteins being intensely labeled, but not identified due to the sensitivity limitations of mass spectrometry. These low concentration proteins may represent signaling proteins within the matrix. These results demonstrate that the mitochondrial matrix phosphoproteome is both extensive and dynamic. The use of this, *in situ*, labeling approach is extremely valuable in confirming protein phosphorylation sites as well as examining the dynamics of these processes under near physiological conditions.

1. INTRODUCTION

Protein phosphorylation and phosphate containing metabolite association plays a critical role in the regulation of mitochondrial function (Bender and Kadenbach, 2000; Denton *et al.*, 1975; Gabriel *et al.*, 1986; Lee *et al.*, 2005; Ludwig *et al.*, 2001; Randle, 1981; Steenaart and Shore, 1997; Walker, 1994). However, the actual enzymatic and non-enzymatic mechanisms and kinetics for most of these interactions are poorly understood. For example, mass spectrometry and dye screening methods have revealed numerous of protein phosphorylation sites (Ballif *et al.*, 2004; Hopper *et al.*, 2006; Schulenberg *et al.*, 2003, 2004; Struglics *et al.*, 2000; Villen *et al.*, 2007) (http://www.phosphosite.org) in the matrix, while the kinase/phosphatase system within the matrix is poorly defined. It is also unclear how many of these phosphorylation sites are functionally active in the protein and how many are simply decorations resulting from 'non-specific' kinase interactions or events that occurred early in protein folding processes to statically control structure and are no longer dynamically controlled. With regard to non-specific phosphorylation by kinases, the kinase selectivity is based on both the phosphorylation site sequence (Yaffe et al., 2001) as well as docking region for the kinase on the target protein (Biondi and Nebreda, 2003), but does not result in complete exclusion. Thus, it is likely that some of the mass spectrometry or dye detected protein phosphorylations, especially determined after enhancement of phosphopeptides (Ballif et al., 2004; Villen and Gygi, 2008; Villen et al., 2007), may have no functional significance and may also only represent a small fraction of the total protein pool, resulting in little functional impact.

Phosphate metabolite associations may come in two general classes, covalent interactions, that include ADP ribosylation (Scovassi, 2004), and numerous allosteric and substrate binding sites that are present for GTP and ATP as well as the di- and monophosphate metabolites of these molecules along with Pi, NADH/NAD (Berger et al., 2005; Grunicke et al., 1975), NADPH/NADP and FAD/FADH.

One of the gold standards in determining protein phosphorylation is the direct labeling of proteins with ^{32}P from γ-^{32}P-ATP in vitro (Hopper et al., 2006) or intact systems (Edes and Kranias, 1990; Kiss et al., 1997). The advantages of ^{32}P labeling include its high sensitivity and specificity as well as its required association site turnover and kinetics of binding. We have developed a ^{32}P labeling strategy for monitoring protein phosphorylation in intact isolated mitochondria. The purpose of this approach was to screen for protein phosphorylation as well as phosphate metabolite associations in matrix proteins.

2. METHODS

Mitochondria are essentially impermeable to extramitochondrial ATP with exception of the modest net ATP import via the Mg-ATP transporter (Aprille, 1993). This is due to the fact that the membrane potential drives ATP out of the matrix while ADP is driven into the matrix through the electrogenic adenylate translocase (LaNoue et al., 1978). Thus, γ-^{32}P-ATP cannot be added externally, but must be generated in the matrix to monitor protein phosphorylation. Our strategy to generate matrix γ-^{32}P-ATP is outlined in Fig. 4.1, where mitochondria are directly incubated in ^{32}P inorganic phosphate (Pi). In energized mitochondria the labeled Pi is converted to γ-^{32}P-ATP, via oxidative phosphorylation as well as succinyl-CoA-synthetase (SCS). Exchanging of the ^{32}P label could occur primarily via adenylate kinase (AK) between β and γ ATP phosphates. Though adenosine kinase (ADK) has not been directly associated with the mitochondria matrix, this reaction could also result in the conversion of label at γ to α. Previous studies have demonstrated that ^{32}P incorporation occurs most rapidly in γ-ATP followed by β-ADP via AK (Tokumitsu and Ui, 1973).

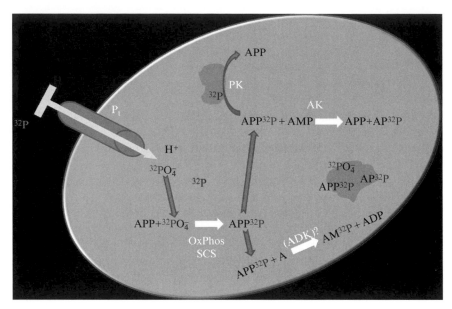

Figure 4.1 Schematic diagram of mitochondrial matrix generating γ-^{32}P-ATP from extramitochondrial ^{32}P as well as other phosphate labeled metabolites. Pt: phosphate transport protein. ADK: adenosine kinase, it is not clear that this enzyme is present and active in the matrix. AK: adenylate kinase. PK: protein kinase. OxPhos: Oxidative phosphorylation. SCS: Succinate CoA Synthetase. Grey masses represent proteins that have metabolites bound or are phosphorylated.

2.1. Mitochondria isolation and phosphate loading

We conducted these ^{32}P labeling experiments on intact porcine heart mitochondria. Due to unique aspects of preventing Pi depletion in the isolation process we will briefly review our mitochondria isolation methodology. Pig heart mitochondria were isolated from tissue that was cold-perfused *in situ* to remove blood and extracellular Ca^{2+} as well as prevent any warm ischemia. Pigs were anesthetized and euthanized using potassium chloride. The heart was removed immediately. The coronaries were flushed retrograde through the aorta with 0.5 L cold buffer A (0.28 M sucrose, 10 mM HEPES, 1 mM EDTA, 1 mM EGTA pH 7.1). The atria, right ventricle, fat and connective tissue were meticulously removed. The left ventricle was weighed, chopped, and added to 500 mL cold buffer A. After mincing in a food processor (Oskar, Sunbeam), attached to a rheostat set to 30%, for 10 min the mixture was centrifuged at 600g for 10 min at 4 °C. The pellet was resuspended in buffer A with 0.5 mg trypsin per gram of tissue and incubated for 15 min at 4 °C. Trypsin inhibitor (2.6 mg/g of tissue) and BSA (1 mg/mL buffer A) were added and the suspension was homogenized two times with a loose fitting tissue grinder and five times with a tight fitting tissue grinder. The puree was centrifuged at 600g for 10 min at 4 °C. The supernatant was immediately centrifuged at 8000g for 10 min at 4 °C to yield a pellet of mitochondria. One notable modification was that 1 mM K_2PO_4 was added to buffer A for the first re-suspension of the mitochondrial pellets to assure that matrix phosphate depletion was not occurring in the isolation process. Mitochondria were then washed two times with buffer A alone, and finally in buffer B (137 mM KCl, 10 mM HEPES, 2.5 mM $MgCl_2$, 0. 5 mM K_2EDTA) for storage on ice. All experiments were conducted on the same day as isolation.

We had some original concerns that the trypsin treatment used to remove the mitochondria from the heart muscle fibers may be contributing to proteolysis within the sample. However, preparing heart mitochondria in the absence of trypsin did not reveal any change in lower molecular weight proteolytic products, though some differences in proteins were detected likely due to different populations of mitochondria enhanced in the non-trypsin sample (Palmer et al., 1977).

Mitochondria preparations were tested for viability by measuring the respiratory control ratio (RCR) by taking the ratio of the rate of oxygen consumption in the presence and absence of ADP (1 mM) at 37 °C. To accept a mitochondrial preparation, the RCR had to exceed 8 at 37 °C to assure a well coupled system for generating ATP.

In addition, we also confirmed the generation of matrix ATP with re-energization of the mitochondria using standard ATP assay procedures. ATP assays were conducted after different incubation conditions by taking 1 mL of the suspension and placing it into 2 mL of 6% perchloric acid (PCA) on ice. After 10 min on ice, 3 M potassium carbonate (~300 μL) was used to titrate to neutral pH. An alkaline overshoot during titration was avoided by adding a universal pH indicator (20 μL) directly into the extract and monitoring the pH during the neutralization process. After neutralization the sample was centrifuged and the supernatant collected for ATP analysis. The supernatant was stored at −80 °C and analyzed within 48 h. Total ATP content was determined using a luminescence assay with luciferin–luciferase (Invitrogen, Carlsbad, CA), against a standard curve of ATP.

Note that we pre-incubated the mitochondria in phosphate media during the ice-cold isolation procedure. We have found, as have others (Siess et al., 1984), that standard isolation procedures lead to the depletion of matrix Pi. This hampered the ability to energize the mitochondria (Bose et al., 2003) as well as the overall generation of ATP when picomolar amounts of Pi were added to the suspension, simulating a ³²P labeling experiment. We have found that this pre-incubation increased matrix Pi of the mitochondria and stabilized the ATP matrix content as well as the ³²P labeling patterns that were highly variable in the absence of the Pi pre-incubation. We believe that low matrix Pi hampered ATP synthesis as well as other Pi dependent metabolic reactions (Bose et al., 2003), which resulted in variable ^{32}P labeling. Indeed, others have made similar observations on the maximum velocity of oxidative phosphorylation and other processes with Pi pre-incubations (Siess *et al.*, 1984). We highly recommend that this pre-incubation with cold phosphate be used in all mitochondria preparations designed to monitor ^{32}P associations as well as other metabolic studies.

2.2. ^{32}P labeling of intact mitochondria

The standard condition for ^{32}P incubation (outlined in Table 4.1) was to incubate 5 nmol of cytochrome oxidase in 5 mL of oxygenated buffer C (125 mM KCl, 15 mM NaCl, 20 mM HEPES, 1 mM EGTA, 1 mM EDTA, 5 mM MgCl$_2$, 5 mM potassium-glutamate, 5 mM potassium-malate at pH 7.1) at 37 °C. Studies were performed in a 50 mL chamber with 100% O$_2$ passed over the top while rocking in a water bath at 37 °C. Glutamate and malate were added to energize the mitochondria to generate a membrane potential and support oxidative phosphorylation. We have also been successful using other substrates, such as pyruvate, α-ketoglutarate or fatty acids. Free calcium was assumed to be nominally zero under these conditions; calcium was added in some experimental conditions by calculating the free calcium using the known buffering properties of the ions in this media. ^{32}P labeled free inorganic phosphate was used as the label. In general, ^{32}P labeled phosphate was the only exogenous phosphate added to the preparation to optimize the specific activity of ^{32}P in the matrix. This was added directly to the incubation medium under different experimental conditions or incubation times. Samples could be extracted at any time to follow the kinetics of ^{32}P incorporation.

2.3. Sample preparation for two-dimensional gel electrophoresis

In preparation for 2D gel electrophoresis, mitochondrial incubations were stopped by adding an equal sample volume of 10% trichloroacetic acid (TCA) solution and placing the samples on ice overnight at 4 °C to achieve maximum protein precipitation. The precipitated proteins were centrifuged at 10,000*g* for 30 min at 4 °C, and the supernatant was aspirated. The pellet was washed twice with cold (−20 °C) 100% acetone, broken apart using a pipette tip and vortexing, and then centrifuged for 15 min at 10,000*g* 4 °C in-between wash steps. After the second wash step with acetone the sample was centrifuged for 30 min at 4 °C to keep the pellet intact. The supernatant was removed with a vacuum tipped probe and the pellet was air dried for 5 min. If the pellet is dried for longer periods of time it will be very difficult to get it back into suspension, a cup–horn sonicator will be needed. The pellet was re-suspended in 200 μL of lysis buffer containing 7 M urea, 2 M thiourea, and 4% CHAPS (w/v). Protein concentration was determined by

Table 4.1 Standard incubation procedure for ³²P labeled intact mitochondria

Reagents	Stock	Storage
Buffer C	125 mM KCl, 15 mM NaCl, 20 mM HEPES, 1 mM EGTA, 1 mM EDTA, 5mM MgCl$_2$, pH 7.1 at 37 °C	4 °C
Glutamate/ Malate	0.5 M potassium-glutamate/0.5 M potassium-malate, pH 7	−20 °C
32-inorganic Phosphorus	Perkin Elmer (NEX053H005MC) 5mCi vial	−20 °C
TCA	10% TCA (w/v) in deionized water	4 °C
Acetone	100% Acetone	−20 °C
Lysis Buffer	7 M Urea, 2 M Thiourea, 4% CHAPS	−20 °C
Bradford Dye	USB #(30098)	Room Temp.

Procedure for 5 mg Mito Prep

1. Freshly isolated mitochondria should be left on ice in isolation buffer until ready to be used.
2. Place 20 mL of Buffer C in a warm water bath 37 °C and oxygenate with 100% O$_2$.
3. Thaw previously frozen stock Glutamate/Malate solution.
4. Thaw ³²P 5 min. before incubation procedure starts.
5. In a 50 mL tube add 50 μL of Glutamate/Malate solution, 250 μCi ³²P per nmol cyto a, adjust final volume to 5 mL with Buffer C.
6. Mix gently before adding 5 mg of Mitochondria directly into the 50 mL tube.
7. The 50 mL tube should be in a water bath rocker set to 37 °C, rocking for 20 min.
8. Stop the reaction by adding 5 mL of 10% TCA, place tube on ice for overnight precipitation in 4 °C.
9. Next day, centrifuge 50 mL tube at 10,000×g for 30 min, 4 °C.
10. Decant TCA solution be careful not lose pellet.
11. Add 5 mL of ice cold acetone to the pellet and vortex to disrupt pellet.
12. Centrifuge 50 mL tube at 10,000×g for 15 min, 4 °C.
13. Decant acetone solution.
14. Repeat steps 11–13 but extend the centrifugation for 30 min to get a tighter pellet.
15. Air dry pellet for 5 min.
16. Resuspend pellet with 200 μL Lysis buffer and transfer to 1.5 mL vial.
17. The pellet needs to be fully resuspended in lysis buffer, and sit on ice for 5 min.
18. Clarify the sample by centrifuging 10,000×g for 5 min at 4 °C.
19. Transfer supernatent to a new tube.
20. Perform Bradford protein assay, follow manufacturer's protocol, USB.

the Bradford protein assay method (USB Quant kit, USB Corp., Cleveland, OH). We have tested the use of phosphatase inhibitors in this protocol and found no increase in detected ^{32}P labeling under these extraction conditions. Most likely the high dilution of the matrix volume (over 1000-fold) and the TCA treatment of our extraction procedure reduced the activity of intrinsic phosphatases without requiring inhibitors.

Three hundred micrograms of cleaned ^{32}P labeled mitochondrial lysate (10 μg/μL concentration) were then mixed with rehydration solution (7 M urea, 2 M thiourea, 4% CHAPS (w/v), 13 mM DTT, 1% (pH 3–10 NL) Pharmalyte (v/v), trace bromophenol blue), to a final volume of 443 μL. The mixture was vortexed, centrifuged at 10,000g, 4 °C for 1 min and placed on ice for 5 min. Seven microlitters of Destreak reagent (GE Healthcare) were added to the sample rehydration buffer, vortexed, centrifuged at 10,000g, 4 °C for 1 min and placed on ice before loading onto a 24 cm Immobiline DryStrip gels (pH 3–10 NL) (GE Healthcare). Isoelectric focusing was achieved by active rehydration for 12 h at 30 V followed by stepwise application of 250 V (1 h), 500 V (1 h), 1000 V (1 h), gradient to 8000 V (1 h) and final step at 8000 V (8 h) for a total of ~72,000 V h (Ettan IPGPhor2, GE Healthcare).

Immobiline DryStrip gels were equilibrated in 10 mL of SDS equilibration solution (50 mM Tris–HCl (pH 8.8), 6 M urea, 30% glycerol, and 2% SDS) for 10 min, first containing 100 mg of DTT. The strip was then transferred to another tube containing 10 mL of SDS equilibration solution with 250 mg of iodoacetemide for 10 min. Gel strips were applied to 10–15% Tris–HCl pH 8.8, gradient gels (Nextgen Sciences, Ann Arbor, MI) and sealed with 0.5% agarose containing bromophenol blue. Electrophoresis was performed in an Ettan DALT-12 tank (GE Healthcare) in 20 °C electrophoresis buffer consisting of 25 mM Tris (pH 8.3), 192 mM glycine, and 0.2% SDS until the dye front advanced completely (~2000 V h).

In preparation for drying radio-labeled gels, each gel was placed in 400 mL of saturated Coomassie blue stain which consisted of 50% methanol, 3% phosphoric acid, and 0.5% colloidal Coomassie G250 (Bio-Rad Laboratories, Hercules, CA), agitating for 2 h. The gels were visually inspected to verify total protein staining. To reduce the background staining they were first destained with 400 mL of 50% methanol and 3% phosphoric acid, agitating for 15 min, and then transferred to another destain solution containing 30% methanol and 3% phosphoric acid, agitating for 2 h. The gels were rehydrated for 2 min with deionized water (longer periods of rehydration will cause the gel to swell abnormally) and placed on filter paper backing (cut a little larger then the size of the gel) (#165–0962, Bio-Rad) that had been sprayed with deionized water (spray enough water to give a generous coating of water but not saturated). The gels were then sprayed with deionized water and covered with plastic wrap (Glad^{TM} wrap) before placing in a large format dryer (model 583, Bio-Rad) for 2 h set to gradient made to slowly reach 70 °C under a vacuum. We do not advise using glycerol in the gel rehydration solution due to the sticky nature of the gel after drying. Dried gels were taped in a screen development cassette (GE Healthcare) to prevent the gel from curling or moving in the cassette. A marker solution was made by mixing 1 μL of radioisotope to 49 μL of Coomassie blue stain (50% methanol and 0.1% (w/v) colloidal Coomassie blue G250). One microlitter of the marker solution was spotted on each corner of the dried Coomassie blue stained gel and let dry at room temperature for 10 min or until spot is completely dried. The fiducial markers were used to serve as points of reference for Coomassie blue and ³²P image overlays. A phosphor screen (GE Healthcare) was placed on top of the dried gels for an exposure time of 72 h.

2.4. Image acquisition and analysis

The phosphor screens were scanned on a Typhoon 9410 variable mode imager (GE Healthcare) at a resolution of 100 μm. The scanner was set in phosphor mode with the selection of maximum sensitivity. The dried Coomassie blue 2D gel was placed in the AlphaImager HP imaging station (Alpha Innotech, San Leandro, CA) set to visible white light mode and the image was captured. To achieve proper alignment the image dimensions of both Coomassie blue and ³²P images need to be close in size. The phosphor screens allowed us to achieve a dynamic range of up to 5 orders of magnitude compared to film. If using film, take an image of the film with the same system used for taking the image of the Coomassie blue stained gel, to achieve proper alignment.

The 2D images from Coomassie blue stain and ³²P autoradiograms were aligned using Progenesis SameSpot or PG240 software (Fig. 4.4) (Nonlinear Dynamics, Newcastle upon Tyne, UK). All four fiducial points seen on the Coomassie blue image were aligned to the fiducial points on the phosphor-screen image only. This is important to prevent any bias aligning to the images. Alternatively, image alignment can be performed using Adobe Photoshop software, but it is not as accurate or flexible as Progenesis SameSpot or PG240 software.

2.5. Mass spectrometry

Protein identification was performed on a non–radiolabeled preparative pick gel. Using Progenesis SameSpot software we were first able to align the ³²P labeled image to total protein Coomassie blue image according to the fiducial markers. In the same Progenesis image analysis experiment the preparative pick gel (stained with Coomassie blue) was added to the analysis. The preparative pick gel was then aligned to the Coomassie blue total protein image. The group of spots that was selected on the ³²P alignment was also selected on the preparative pick gel. The selected spots were picked using an automated spot picking system that also digested each protein with trypsin and spotted onto a MALDI mass spectrometry plate for further analysis.

3. RESULTS AND DISCUSSION

3.1. ^{32}P dose and time course

Dose and time course studies for ^{32}P labeling were conducted on porcine heart mitochondria. The ^{32}P dosing studies at 1, 25, 50, and 250 μCi ^{32}P per nmol cytochrome oxidase protein revealed that the 250 μCi ^{32}P dose provided the best overall labeling pattern for a majority of proteins (Fig. 4.2). Based on these studies we used the 250 μCi ^{32}P dose for the

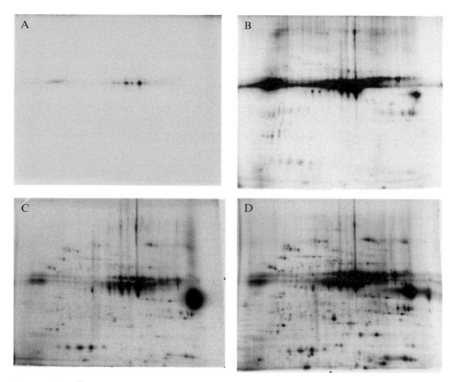

Figure 4.2 ^{32}P Dose–response in heart mitochondria. Porcine heart mitochondria were incubated in standard conditions for 20 min where (A) 1 μCi (B) 25 μCi, (C) 50 μCi, and (D) 250 μCi of ^{32}P was added. Proteins were separated in the horizontal direction by isoelectric focusing point (pI), from pH $\sim$4 to 10, and vertically by molecular weight, from $\sim$150 to 10 kDa. The relative amplitude for each image was arbitrarily set.

majority of experiments. We did not use higher levels of ³²P since we seemed to be approaching the limit of the dynamic range and resolution of the 2D gel autoradiogram system at that concentration.

In the time course studies, samples were collected at different times from the same agitating sample chamber, after initiating the ³²P labeling. These studies revealed that most associations reached a maximum at 20 min at 37 °C. Based on this data we used 20 min as our standard incubation time.

3.2. ³²P assignments

Figure 4.3 presents a partially identified ³²P labeling pattern in a 2D gel. Assignments are in the figure legend. Assignments were made using the procedures outlined in Section 2.5. Note that many labeled proteins are not assigned; these are primarily the result of low protein content or poor dispersion of the protein preventing a unique identification.

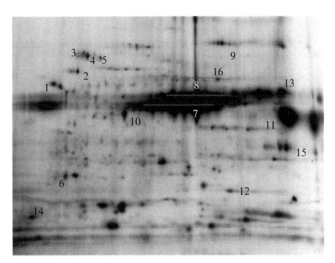

Figure 4.3   ³²P labeling assignments in isolated heart mitochondria. Mitochondria were incubated for 20 min with 250 μCi of ³²P. Chromatography is the same at in Fig. 4.2. Protein assignments: (1) ATP synthase, mitochondrial F1 complex, β subunit, (2) 60 kDa heat shock protein, mitochondrial precursor, (3) NADH dehydrogenase (ubiquinone) Fe–S protein 1 (75 kDa), (4) 70 kDa heat shock protein, mitochondrial precursor, (5) Pyruvate dehydrogenase complex, E2 subunit, (6) ATP synthase d-chain, mitochondrial precursor, (7) Pyruvate dehydrogenase complex, E1 subunit, (8) Elongation factor Tu, mitochondrial precursor, (9) Aconitase hydratase, mitochondrial precursor, (10) Isocitrate dehydrogenase (NAD) subunit alpha, mitochondrial precursor, (11) Succinyl-CoA ligase (GDP-forming) alpha–chain, mitochondrial precursor, (12) Mn superoxide dismutase, (13) Creatine kinase, sarcomeric mitochondrial precursor, (14) Cytochrome c oxidase polypeptide Va, mitochondrial precursor, (15) Voltage-dependent anion channel 1, and (16) ATP synthase, mitochondrial F1 complex, α subunit.

3.3. Chase experiments with cold phosphate

The extent of ^{32}P incorporation is a function of both the turnover of existing phosphoproteins and the generation of new phosphorylated proteins in the presence of ^{32}P. The initial pool of phosphorylated protein (PE_i), of the total protein E, will increase ^{32}P labeling by enzymatic turnover that could be modeled $PE_t = PE_i(1-e^{-kt})$ where t is time while k is the exchange rate for turnover. This assumes that all of the free matrix phosphate is ^{32}P labeled. While the generation of new pools of phosphorylated protein with ^{32}P can be modeled in the extreme case where all protein will be phosphorylated as $PE_n = (E-PE_i)(1-e^{-gt})$ where g is the rate constant of generating, or regenerating, the phospho-protein pool. Using these models the total ^{32}P protein signal will equal $PE_i(1-e^{kt})$ (i.e., the turnover component) $+ (E-PE_i)(1-e^{-gt})$ (i.e., the newly synthesized pool). Depending on the pool sizes and the relative rate constants, either of these processes could dominate the ^{32}P labeling process in the matrix.

As discussed above, ^{32}P labeling alone cannot distinguish turnover of sites from the creation of newly labeled protein. For example, it has been previously established that pyruvate dehydrogenase (PDH) dephosphorylates under standard isolation conditions and rapidly rephosphorylates upon warming and re-energization (Kerbey *et al.*, 1976). To estimate the relative role of phosphoprotein pool building versus pure exchange in our incubation conditions, we performed a series of pulse-chase experiments with cold phosphate. An example of these studies is presented in Fig. 4.4. It is clear that the majority of ^{32}P is not competed off of its sites if cold phosphate is added after ^{32}P, while the initial addition of cold phosphate eliminates most of the ^{32}P labeling. This was clearly the case for PDH labeling, as predicted by the previous work. These results imply that many of the proteins were phospho-depleted during the de-energization associated with mitochondria isolation, likely reflecting a decrease in matrix ATP, followed by an increase in phosphorylation associated with the warming and re-energization of the mitochondria. This may be fortuitous since the building of the phosphor-protein pools upon re-enegerzation will enhance the overall ^{32}P incorporation into the proteins. However, the rates of incorporation cannot be interpreted as turnover, due to the non-steady state condition of the phosphor-protein levels and phospho-protein pool building seems to dominate the labeling patterns detected in our standard protocol. To avoid these non-steady state complications, a 20 min incubation with low cold Pi to achieve steady state before adding ^{32}P may enhance the turnover contribution to the ^{32}P labeling. We have not explored this possibility extensively, since we have been focusing on phosphorylation site identification relying on both terms to generate the largest ^{32}P labeled pool.

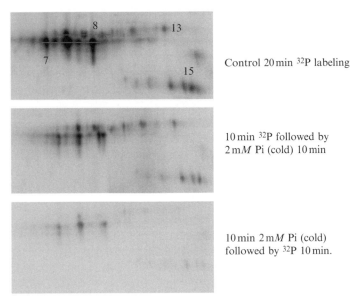

Control 20 min ³²P labeling

10 min ³²P followed by
2 mM Pi (cold) 10 min

10 min 2 mM Pi (cold)
followed by ³²P 10 min.

Figure 4.4 ³²P Pulse-chase experiments. Control, ³²P was added to porcine heart mitochondria and incubated for 20 min. ³²P was added to heart mitochondria and incubated for 10 min then cold inorganic phosphate was added to the incubation media for an additional 10 min. Heart mitochondria incubated with cold inorganic phosphate for 10 min then ³²P was added to the incubation media for an additional 10 min. Proteins were separated in the horizontal direction by isoelectric focusing point (pI), from pH ~4 to 10, and vertically by molecular weight, from ~150 to 10 kDa. All exposures were identical along with window level values in the display.

3.4. Suppression of PDH

There are a few mitochondrial proteins that have strong ³²P incorporation such as PDH, which limits the dynamic range of the autoradiogram. We believe that this large incorporation is more due to the re-generation of PDH during the re-energization process rather than simply a high turnover, as demonstrated by the pulse-chase experiments. In mammalian tissue, PDH exists in two forms: active (dephosphorylated) or inactive (phosphorylated) and is regulated by its own kinase/phosphatase system. An approach to selectively suppress PDH rephosphorylation would be to inhibit the kinase by using the halogenated organic acid dichloroacetic acid (DCA) (Whitehouse et al., 1974) and/or pyruvate (Denton et al., 1975). To test this concept, heart mitochondria were incubated for 1 h on ice with 0.1 mM DCA and 5 mM pyruvate before the standard 20 min ³²P incubation at 37 °C in the presence of glutamate and malate, substrates not dependent on PDH for oxidation. These inhibitors decreased the labeling of PDH as shown in Fig. 4.5, as predicted from the inhibition of PDHK. Similar results were obtained with DCA or pyruvate alone. However, we

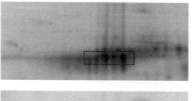
Control

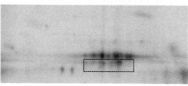
DCA/Pyruvate: 1 h pre-incubation
with 0.1 mM DCA 5 mM pyruvate
then ^{32}P 20 min

Figure 4.5 The effects of dichloroacetate (DCA) and Pyruvate on ^{32}P labeled heart mitochondria. Control: Heart mitochondria incubated in standard conditions for 20 min with ^{32}P. DCA/Pyruvate: Heart mitochondria treated with 0.1 mM DCA and 5 mM pyruvate for 1 h on ice then ^{32}P was added during the standard 20 min incubation at 37°. Box indicates region of the phosphorylated pyruvate dehydrogenase E1 protein.

noted that ^{32}P labeling was surprising suppressed in many other proteins. We speculated that this global effect of DCA and pyruvate may be due to a suppression of ATP production even though these mitochondria were provided glutamate and malate as alternative substrates. We assayed mitochondria matrix ATP content in the presence of DCA and pyruvate. A dose-dependent decrease in matrix ATP content with both DCA and pyruvate was observed. The mechanism for this global inhibition of ATP generation by DCA and pyruvate is unknown, but this inhibition could explain the global decrease in ^{32}P labeling with these agents.

3.5. Purified ^{32}P labeled proteins and complexes

To further validate the assignment of ^{32}P labeling sites, we have found that purifying proteins after *in situ* ^{32}P labeling is a useful approach. After incubating intact mitochondria, as described in the ^{32}P labeling section, Complex V was isolated with an immunocapture kit (Mitosciences, Eugene, OR) according to the manufacturer's instructions. Briefly, after incubation, mitochondria were pelleted and solubilized to a final concentration of 5 mg/mL in 1× PBS with 1% lauryl maltoside detergent. Mitochondria were then mixed with a pipette and placed on ice for 30 min, before centrifuging at 10,000 rpm for 30 min at 4 °C. One milliliter aliquots of solubilized mitochondrial supernatant were transferred to fresh 1.5 mL vials containing the following mixture: 5 mM potassium-fluoride, 10 μL of protease inhibitor cocktail (Sigma–Aldrich, St Louis, MO), and 50 μL of antibody loaded G-agarose beads. This mixture was allowed to mix overnight at 4 °C using a vial rotator. Beads were then washed three times in 1× PBS with 0.05% lauryl maltoside, and Complex V was eluted by resuspending in two volumes of 4 M urea, pH 7.5.

Isolating Complex V provided confirmation for the ³²P labeled subunits seen in the total mitochondria labeled gels as well as revealed sites that were masked by other phosphorylated proteins. Complex V isolation demonstrated ³²P labeling of the β, α, γ, and d-chain subunits, with some residual contamination from the intensely labeled PDH (Fig. 4.6). It should be noted that the location of the β subunit labeling was slightly above the Commassie stained gel. This implied that other posttranslational modifications may be associated with this labeling.

We have also been successful in isolated individual complexes using Blue native approaches (Schagger and von, 1991) (see below). Blue native revealed better maintained subunit stoichiometry than immuncapture methods, likely due to partial dissociation of the complexes prior to or during the immune-capture procedure, while only complete complexes properly migrate in the Blue Native gels.

3.6. Blue native gel electrophoresis

Native electrophoresis was used to maintain mitochondrial protein complexes in their intact form (Schagger and von, 1991) and permit the detection of weak ³²P metabolite association with mitochondrial proteins as well as covalent protein phosphorylation events within complete protein complexes. Blue native gel electrophoresis (BN-PAGE) was performed according to an Invitrogen protocol for NativePAGE Novex Bis-Tris Gel System, except for the destaining step where cold water was used to destain and detect metabolite association. Four–16% 1 mm bis-tris gels were used

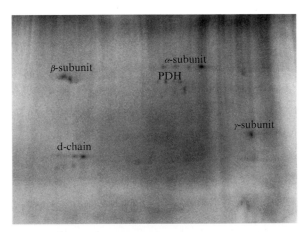

Figure 4.6 Purified labeled complex V from porcine heart mitochondria. Subunit identifications were obtained by mass spectrometry. Purified proteins were separated by two-dimensional gel electrophoresis, first in the horizontal direction by isoelectric focusing point (pI), from pH ~4 to 10, and then vertically by molecular weight, from ~150 to 10 kDa.

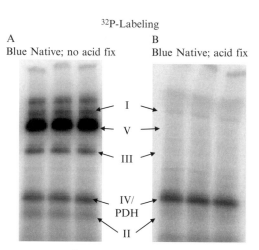

^{32}P-Labeling

A   B
Blue Native; no acid fix   Blue Native; acid fix

I
V
III
IV/
PDH
II

Figure 4.7 Native PAGE of ^{32}P labeled porcine heart mitochondrial complexes. (A) cold water destain, no acid for 2 min (B) BN-PAGE gel was fixed in an hot fixative followed by hot acid destain for additional 15 min.

for BN-PAGE and electrophoresis was performed at 4 °C for 1 h at 150 V then for 1.2 h at 250 V. The BN-PAGE gels were dried the same way as the 2D gels except the total drying time was 45 min on a gradient cycle.

Using only cold water to destain the BN-PAGE gels, a large number of ^{32}P associations were detected that did not quantitatively correlate with the 2D gels. These weak metabolite associations were removed when more standard organic solvents and acids were used to fix/destain the BN-PAGE gels. When fixed for 15 min in hot fixative solution (40% methanol, 10% acetic acid) and destained in hot 8% acetic acid solution many of the non-covalent metabolite associations to proteins were lost, especially in complexes I and V (Fig. 4.7). These results demonstrate a new property of the Blue Native approach in monitoring the association of metabolites to individual enzyme complexes as well as protein phosphorylation. This approach, in combination with in-gel assay approaches, should provide a useful tool in analyzing the effect of metabolite association and protein phosphorylation on enzyme complex activity. The specific metabolites weakly associated with the proteins in the BN-PAGE gels have not been fully characterized. However, preliminary studies suggest that ATP is the dominate metabolite in these weak interactions.

4. SUMMARY

The methodology for performing *in situ* ^{32}P labeling of matrix proteins in mitochondria was described. This approach reveals a large number of dynamic protein phosphorylations and metabolite associations that are only just beginning to be understood and characterized. The approach was validated by confirming many of the classical observations on PDH, while providing new insights on numerous other protein systems. This approach may provide an excellent screening tool to select the numerous protein phosphorylation sites detected by mass spectroscopy that maybe linked to dynamic modulation of enzyme or protein function in the matrix. Many of the phosphorylations detected were expressed at levels below mass spectrometry identification limits, suggesting many more new discoveries of protein phosphorylation sites are still to be made on proteins that may play a significant role in matrix signaling networks.

REFERENCES

Aprille, J. R. (1993). Mechanism and regulation of the mitochondrial ATP-Mg/P(i) carrier. *J. Bioenerg. Biomembr.* **25,** 473–481.

Ballif, B. A., Villen, J., Beausoleil, S. A., Schwartz, D., and Gygi, S. P. (2004). Phospho-proteomic analysis of the developing mouse brain. *Mol. Cell Proteomics* **3,** 1093–1101.

Bender, E., and Kadenbach, B. (2000). The allosteric ATP-inhibition of cytochrome c oxidase activity is reversibly switched on by cAMP-dependent phosphorylation. *FEBS Lett.* **466,** 130–134.

Berger, F., Lau, C., Dahlmann, M., and Ziegler, M. (2005). Subcellular compartmentation and differential catalytic properties of the three human nicotinamide mononucleotide adenylyltransferase isoforms. *J. Biol. Chem.* **280,** 36334–36341.

Biondi, R. M., and Nebreda, A. R. (2003). Signalling specificity of Ser/Thr protein kinases through docking-site-mediated interactions. *Biochem. J.* **372,** 1–13.

Bose, S., French, S., Evans, F. J., Joubert, F., and Balaban, R. S. (2003). Metabolic network control of oxidative phosphorylation: Multiple roles of inorganic phosphate. *J. Biol. Chem.* **278,** 39155–39165.

Denton, R. M., Randle, P. J., Bridges, B. J., Cooper, R. H., Kerbey, A. L., Pask, H. T., Severson, D. L., Stansbie, D., and Whitehouse, S. (1975). Regulation of mammalian pyruvate dehydrogenase. *Mol. Cell Biochem.* **9,** 27–53.

Edes, I., and Kranias, E. G. (1990). Phospholamban and troponin I are substrates for protein kinase C *in vitro* but not in intact beating guinea pig hearts. *Circ. Res.* **67,** 394–400.

Gabriel, J. L., Zervos, P. R., and Plaut, G. W. (1986). Activity of purified NAD-specific isocitrate dehydrogenase at modulator and substrate concentrations approximating con-ditions in mitochondria. *Metabolism* **35,** 661–667.

Grunicke, H., Keller, H. J., Puschendorf, B., and Benaguid, A. (1975). Biosynthesis of nicotinamide adenine dinucleotide in mitochondria. *Eur. J. Biochem.* **53,** 41–45.

Hopper, R. K., Carroll, S., Aponte, A. M., Johnson, D. T., French, S., Shen, R. F., Witzmann, F. A., Harris, R. A., and Balaban, R. S. (2006). Mitochondrial matrix phosphoproteome: Effect of extra mitochondrial calcium. *Biochemistry* **45,** 2524–2536.

Kerbey, A. L., Randle, P. J., Cooper, R. H., Whitehouse, S., Pask, H. T., and Denton, R. M. (1976). Regulation of pyruvate dehydrogenase in rat heart. Mechanism of regulation of proportions of dephosphorylated and phosphorylated enzyme by oxida-tion of fatty acids and ketone bodies and of effects of diabetes: Role of coenzyme A, acetyl-coenzyme A and reduced and oxidized nicotinamide-adenine dinucleotide. *Biochem. J.* **154,** 327–348.

Kiss, E., Edes, I., Sato, Y., Luo, W., Liggett, S. B., and Kranias, E. G. (1997). Beta-Adrenergic regulation of cAMP and protein phosphorylation in phospholamban-knockout mouse hearts. *Am. J. Physiol.* **272,** H785–H790.

LaNoue, K. F., Mizani, S. M., and Klingenberg, M. (1978). Electrical unbalance of adenosine nucelotide transport across the mitochondrial membrane. *J. Biol. Chem.* **253,** 191–198.

Lee, I., Salomon, A. R., Ficarro, S., Mathes, I., Lottspeich, F., Grossman, L. I., and Huttemann, M. (2005). cAMP-dependent tyrosine phosphorylation of subunit I inhibits cytochrome c oxidase activity. *J. Biol. Chem.* **280,** 6094–6100.

Ludwig, B., Bender, E., Arnold, S., Huttemann, M., Lee, I., and Kadenbach, B. (2001). Cytochrome C oxidase and the regulation of oxidative phosphorylation. *Chembiochem.* **2,** 392–403.

Palmer, J. W., Tandler, B., and Hoppel, C. L. (1977). Biochemical properties of subsarco-lemmal and interfibrillar mitochondria isolated from rat cardiac muscle. *J. Biol. Chem.* **252,** 8731–8739.

Randle, P. J. (1981). Phosphorylation–dephosphorylation cycles and the regulation of fuel selection in mammals. *Curr. Top. Cell Regul.* **18,** 107–129.

Schagger, H., and von, J. G. (1991). Blue native electrophoresis for isolation of membrane protein complexes in enzymatically active form. *Anal. Biochem.* **199,** 223–231.

Schulenberg, B., Aggeler, R., Beechem, J. M., Capaldi, R. A., and Patton, W. F. (2003). Analysis of steady-state protein phosphorylation in mitochondria using a novel fluores-cent phosphosensor dye. *J. Biol. Chem.* **278,** 27251–27255.

Schulenberg, B., Goodman, T. N., Aggeler, R., Capaldi, R. A., and Patton, W. F. (2004). Characterization of dynamic and steady-state protein phosphorylation using a fluorescent phosphoprotein gel stain and mass spectrometry. *Electrophoresis* **25,** 2526–2532.

Scovassi, A. I. (2004). Mitochondrial poly(ADP-ribosylation): From old data to new per-spectives. *FASEB J.* **18,** 1487–1488.

Siess, E. A., Kientsch-Engel, R. I., Fahimi, F. M., and Wieland, O. H. (1984). Possible role of Pi supply in mitochondrial actions of glucagon. *Eur. J. Biochem.* **141,** 543–548.

Steenaart, N. A., and Shore, G. C. (1997). Mitochondrial cytochrome c oxidase subunit IV is phosphorylated by an endogenous kinase. *FEBS Lett.* **415,** 294–298.

Struglics, A., Fredlund, K. M., Konstantinov, Y. M., Allen, J. F., and Moller, I. M. (2000). Protein phosphorylation/dephosphorylation in the inner membrane of potato tuber mitochondria. *FEBS Lett.* **475,** 213–217.

Tokumitsu, Y., and Ui, M. (1973). The incorporation of 32 P i into intramitochondrial ADP fraction dependent on the substrate-level phosphorylation. *Biochim. Biophys. Acta* **292,** 310–324.

Villen, J., Beausoleil, S. A., Gerber, S. A., and Gygi, S. P. (2007). Large-scale phosphoryla-tion analysis of mouse liver. *Proc. Natl. Acad. Sci. USA* **104,** 1488–1493.

Villen, J., and Gygi, S. P. (2008). The SCX/IMAC enrichment approach for global phosphorylation analysis by mass spectrometry. *Nat. Protoc.* **3,** 1630–1638.

Walker, J. E. (1994). The regulation of catalysis in ATP synthase. *Curr. Opin. Struct. Biol.* **4,** 912–918.

Whitehouse, S., Cooper, R. H., and Randle, P. J. (1974). Mechanism of activation of pyruvate dehydrogenase by dichloroacetate and other halogenated carboxylic acids. *Biochem. J.* **141,** 761–774.

Yaffe, M. B., Leparc, G. G., Lai, J., Obata, T., Volinia, S., and Cantley, L. C. (2001). A motif-based profile scanning approach for genome-wide prediction of signaling path-ways. *Nat. Biotechnol.* **19,** 348–353.

CHAPTER FIVE

SELECTIVE ENRICHMENT IN PHOSPHOPEPTIDES FOR THE IDENTIFICATION OF PHOSPHORYLATED MITOCHONDRIAL PROTEINS

Gabriella Pocsfalvi

Contents

1. Introduction   82
2. Off-Line Phosphopeptide Enrichment Methods   84
   2.1. Immobilized metal affinity chromatography (IMAC)   84
   2.2. Single-tube and microtip methods using metal oxides   85
   2.3. Chemical reactions prior enrichment   88
   2.4. Immunoaffinity purification   89
3. On-Line 2D-LC Phosphopeptide Enrichment Methods   90
   3.1. 2D-TiO$_2$-RP-LC-MS using biphasic column   91
   3.2. 2D-SCX-TiO$_2$-RP-LC-MS   91
4. Mass Spectrometry-Based Methods   92
   4.1. Maldi-MS and MS/MS analysis   92
   4.2. ESI-MS and data dependent MS/MS strategies   93
5. Quantitative Phosphoproteomics   94
References   94

Abstract

In vivo protein phosphorylation is a reversible and dynamic process controlled by protein kinases and phosphatases for the addition and removal of the phosphate group to serine, threonine, and tyrosine residues. In addition to other regulation events, increasing evidence indicates that reversible protein phosphorylation plays an important role in regulating mitochondrial function. Detecting changes in the state of protein phosphorylation is a difficult task since phosphorylation on a specific protein is typically transient and usually presents in substoichiometric concentration. Moreover, what makes mass spectrometry (MS)-based phosphopeptide analysis difficult is that ionization efficiency of phosphopeptides is lower than their nonphosphorylated analogues.

Istituto di Biochimica delle Proteine–Consiglio Nazionale delle Ricerche, Naples, Italy

Methods in Enzymology, Volume 457
ISSN 0076-6879, DOI: 10.1016/S0076-6879(09)05005-8

Different strategies have been proposed for the selective enrichment of low abundance phosphoproteins and phosphopeptides present in biological samples. As a result of recent advances, phosphoproteomics has become one of the most rapidly developing areas of proteomics. In the last few years, a number of studies have focused on the analysis of phosphoproteome purified from yeast, liver and heart mitochondria, as well as on the identification of endogenously phosphorylated subunits of mitochondrial oxidative phosphorylation system complexes. This chapter describes methods for the selective enrichment of phosphoserine, phosphothreonine, and phosphotyrosine containing peptides and proteins and phosphate-specific MS strategy generally applicable to phosphoprotein analysis but focusing specifically on mitochondrial samples.

1. INTRODUCTION

Mitochondria are associated with various vital and complex cellular functions. Besides its central role in energy metabolisms, mitochondria have also been implicated in a wide range of normal and disease processes, including metabolic disorders, neurodegenerative disease, cancer and aging. Proteins have fundamental roles in the functioning of mitochondria. Most mitochondrial proteins are nuclear encoded, synthesized on cytosolic ribosomes and imported posttranslationally into the organelles, whereas a small fraction of proteins are encoded by the mitochondrial genome. Over the past years, mitochondrion targeted proteomics has provided a significant amount of experimental data for deciphering the mitochondrial proteome. Experimental datasets obtained by extensive large-scale proteomics on isolated mitochondrial samples have aided the validation of computationally predicted mitochondrial proteins collected in several public databases, including MiGenes (http://www.pharm.stonybrook.edu/migenes) (Basu *et al.*, 2006), AMPDB (http://www.ampdb.bcs.uwa.edu.au/) (Heazlewood and Millar, 2005), MitoP2 (http://www.mitop2.de) (Andreoli *et al.*, 2004; Elstner *et al.*, 2008), MitoProteome (http://www.mitoproteome.org) (Cotter *et al.*, 2004), and Human Mitochondrial Protein Database (http://bioinfo.nist.gov:8080/examples/servlets/index.html). Until recently, only half of the ~1500 proteins predicted to function in mitochondria could have been identified due to sample complexity, the dynamic nature of the mitochondrial proteome, the presence of posttranslational protein modifications, and the difficulties in identification of low-copy number-proteins. Creating an inventory was complicated even more by the significant numbers of co-purifying "false positives." More recently, by applying a combined approach based on subtractive proteomics, GFP microscopy and machine learning, Pagliriani *et al.* have overcome some of these difficulties and gathered an inventory of 1098 proteins with a strong support of mitochondrial localization and their expressions across different mouse tissues (Pagliarini *et al.*, 2008).

Apart from the creation of detailed catalogues of mitochondrial protein components, it is expected that a broader understanding of mitochondrial function in normal and diseased conditions can be obtained from proteome analyses (Meisinger *et al.*, 2008). Comparative and quantitative proteomics has already provided valuable insights into the role of some mitochondrial proteins in several disease processes including cardiovascular (Melanie, 2008) and neurodegenerative diseases (Frank, 2006), cancer (Verma *et al.*, 2003), liver dysfunction (Liu *et al.*, 2008), diabetes type II and obesity (Højlund *et al.*, 2008), and aging (Chakravarti *et al.*, 2008). Further challenges in mitochondrial proteomics include the identification of reversible posttranslational modifications (PTMs) (Distler *et al.*, 2007) and the determination of their role in the regulation of protein activity (Distler *et al.*, 2007). Mitochondrial proteins may undergo numerous reversible PTMs like lysine acetylation (Kim *et al.*, 2006; Schwer *et al.*, 2006), tyrosine nitration (Souza *et al.*, 2008), *S*-glutathionylation (Hurd *et al.*, 2005), *N*-myristoylation (Toshihiko *et al.*, 2003), ubiquitination (Sutovsky *et al.*, 2000), and phosphorylation (Reinders *et al.*, 2007).

Reversible protein phosphorylation mediated by protein kinases and phosphatases is the most studied PTM in general, and in mitochondria as well (Pagliarini and Dixon, 2006). The most common amino acid residues modified by phosphorylation are serine (90%), threonine (10%), and tyrosine (0.05%). In the last few years the steadily increasing number of reported mitochondrial phosphoproteins, kinases and phosphatases suggests that reversible protein phosphorylation is likely to be an important regulatory event in mitochondria. Traditional *in vitro* phosphorylation studies use ^{32}P radiolabeling added ATP and protein kinases to visualize phosphorylated proteins. Analysis is combined with Edman degradation for the identification of phosphoproteins. Although *in vitro* phosphorylation studies provide valuable information on phosphorylable proteins, these phosphorylation sites are not necessarily phosphorylated in living systems. Therefore, direct assessment of the endogenous phosphorylation state of mitochondrial proteins under a variety of physiologic conditions is particularly important. ^{32}P radiolabeling is being gradually replaced by the recently developed gel-based and in-solution proteomics strategies. Gel-based phosphoprotein analysis separates the proteins first by polyacrylamide gel electrophoresis, visualizes phosphoproteins by the use of specific dyes in gels, and then identifies them by mass spectrometry–based methods. High sensitivity phosphoprotein specific fluorescence-dye can be used with standard SDS–PAGE and 2D gels and is compatible with downstream mass spectrometry analysis. Though this method does enable comparative profiling, it does not always allow identification of the phosphoproteins due to the generally low amount of phosphoprotein per band or per spot. To successfully detect low abundance endogenously phosphorylated proteins from complex mixture and to identify the protein phosphorylation site, MS-based in-solution proteomics strategy is the method of choice. Complex protein mixtures such as cell or mitochondrial lysates are proteolytically digested in the first step resulting in an even more complex mixture containing peptides. In this strategy it is fundamental to remove abundant nonphosphorylated peptides, and thus to enrich phosphopeptides prior MS analysis. This chapter presents the various ways to enrich phosphoproteins and phosphopeptides followed by a description of different phosphor-specific MS acquisition methods.

2. OFF-LINE PHOSPHOPEPTIDE ENRICHMENT METHODS

Isolation and purification of intact mitochondria from various sources such as cell lines, liver, brain, heart, and muscle tissues can be performed using established protocols compatible with downstream MS analysis. For successful phosphoprotein analysis it is fundamental that isolation and purification is carried out in the presence of protease and phosphatase inhibitors.

2.1. Immobilized metal affinity chromatography (IMAC)

2.1.1. Preparation of IMAC media

IMAC using ferric ions immobilized on iminodiacetate-agarose gel was first introduced by Andersson and Porath as a generic phosphoprotein purification protocol (Andersson and Porath, 1986). IMAC is the most used method to isolate phosphopeptides. The enrichment process is based on the interaction between the immobilized metal ion and the target phosphoproteins or phosphopeptides. The classical IMAC system employs a carrier matrix complexed with iminodiacetic (IDA) or nitrilotriacetic acid (NTA) chelating group and loaded with metal ions such as Fe^{3+}, Cu^{2+}, Ni^{2+}, and Ga^{3+}. IMAC media can be prepared using 50% slurry of metal-chelating resins like POROS 20MC (Applied Biosystems), IPAC (Eprogen) or Propac-IMAC (Dionex) prepared in water. Resin is washed twice in water, stripped by 50 mM EDTA, washed in water, and then washed in 1% acetic acid. Metal loading is performed using 100 mM metal salt solution (like iron chloride ($FeCl_3$), copper sulfate ($CuSO_4$), nickel chloride ($NiCl_2$), gallium nitrate [Ga$(NO_3)_3$]) in 1% acetic acid for 15 min with agitation. To remove unbound metal ions, resin is rinsed in 30% ACN, 1% acetic acid and then again in 1% acetic acid. Activated IMAC resin then can be used directly in batchwise phosphopeptide enrichment protocols (see Section 2.2) with the convenience that one can easily change the amount of the resin as a function of sample quantity (scaling-up or scaling-down). Moreover, different ready to use metal ion loaded resins, which are usually commercialized in gravity or spin column formats, are also available. In downstream mass spectrometry, applications spin columns are usually preferred over gravity columns due to the lower sample volumes and significantly higher sample throughput. Another IMAC solution broadly used in phosphoproteomic applications is ZipTip$_{MC}$ pipette tips (MilliPore) containing IDA resin which can be charged with an appropriate metal ion prior to use.

2.1.2. Application of spin column IMAC for phosphopeptide enrichment

The mitochondrial protein sample (100 μg) is reduced, alkylated, enzymatically, or chemically digested, desalinated and lyophilized according to standard protocols used in proteomics. In a typical spin column arrangement, lyophilized peptide sample is reconstituted in 50 μl of 0.1% formic acid, 30% ACN and combined with an equal volume of IMAC Binding Buffer (5% acetic acid). IMAC spin column, containing typically 25–50 μl of metal–chelated resin is preconditioned by washing two times with IMAC Binding Buffer followed by spinning to remove the liquid. The peptide solution is then loaded onto the column by centrifugation at 700g for 1 min, followed by incubation for 30 min at room temperature. The resin is then washed two times by 50 μl of Wash Buffer 1 (0.1% acetic acid), two times by 50 μl of Wash Buffer 2 (0.1% acetic acid and 30% ACN), and once by 50 μl of water followed by centrifugation. Flow-through washing solutions can be combined and analyzed by MS if one is interested in nonphosphorylated peptides. The IMAC-retained peptides then subjected to three replicate 50 μl elution steps using 100 mM NH$_4$HCO$_3$ in 5% acetonitrile at pH 9. Elutes are combined into the same tube and finally 50 μl 10% formic acid is used to neutralize pH.

Drawbacks of the IMAC procedure are (i) the relatively low specificity due to nonspecific binding of nonphosphorylated peptides, especially acidic peptides, and (ii) the metal ion leaching during fractionation and storage. To improve separation and phosphopeptide recovery different binding, washing and eluting solutions and conditions have been proposed by different groups. Among these is a batchwise IMAC procedure which uses Poros 20 MC beads was recently optimized to study the mitochondrial phosphoproteome. This procedure led to the identification of 84 phosphorylation sites in 62 mouse liver mitochondrial proteins (Lee *et al.*, 2007). More recently, Ga^{3+}-IMAC was used to probe the mammalian mitochondrial ribosomal Death-associated protein 3 (DAP3) which was shown to be endogenously phosphorylated. Six residues were identified by mass spectrometry (Miller *et al.*, 2008).

2.2. Single-tube and microtip methods using metal oxides

Affinity of some metal oxides such as titanium dioxide (TiO$_2$) (Pinkse *et al.*, 2008; Pocsfalvi *et al.*, 2007), zirconium dioxide (ZrO$_2$) (Cuccurullo *et al.*, 2007; Kweon and Hakansson, 2006), tin dioxide (SnO$_2$) (Sturm *et al.*, 2008), and niobium pentoxide (Nb$_2$O$_5$) (Ficarro *et al.*, 2008) for phosphate groups have also been exploited for the enrichment of phosphopeptides from complex mixtures. Titania and zirconia are the most widely employed in phosphopeptide enrichment experiments due to their excellent chemical, mechanical, and thermal stability. Since this method does not require chelation of metal ion, it overcomes the metal ion leaching problem one can encounter in the IMAC method. Titania is an amphoteric ion-exchanger and can react either as a Lewis acid or base depending on the pH of the solution: having anion exchange properties at acidic pH and cation exchange properties at alkaline pH. Analyte absorption and desorption on the metal oxide surface is influenced by a number of factors, like the type of the Lewis base buffer anion, pH and ionic strength of the eluent, bulkiness of counter-ion and the presence of organic modifier (Hu *et al.*, 2002). Metal oxide-based affinity separation can be performed by employing the bare micro- or nano-sized particles in single-tube experiments (Cuccurullo *et al.*, 2007; Pan *et al.*, 2008; Pocsfalvi *et al.*, 2007) or alternatively they can be packed in fused silica capillaries (Pinkse *et al.*, 2008) or into microtips. Metal oxides can also be bound to polymer (Feng *et al.*, 2007), magnetic beads (Li *et al.*, 2008) and chip surface.

2.2.1. Single-tube experiment applying bare metal oxide particles

The single-tube, batch-wise phosphopeptide enrichment method consists of four principal steps (Fig. 5.1): (1) wetting and conditioning of particles, (2) sample loading, (3) washing to remove unbound peptides, and (4) eluting of bound peptides (Cuccurullo *et al.*, 2007). The quantities of metal oxide powder and protein sample used are highly depend on the crystal properties of the metal oxide, its particle size, the complexity of the sample and the amount of phosphorylated peptides to be separated. Advantages of the single-tube method include the simplicity, scalability and the feasibility to control experimental conditions. In a typical setup metal oxide particles (3 mg) are placed into an eppendorf tube and washed three times with 100 μl 80% acetonitrile in 0.1% TFA. The slurry is centrifuged at 3000g and the solvent is removed. Particles are conditioned by 100 μl 0.1% TFA. Enzymatic mitochondrial protein digest (20 μg/mg ZrO$_2$ at 1 mg/ml concentration in 0.1% TFA) is added, mixed, and incubated under gentle shaking for 30 min at room temperature. Sample is centrifuged, and the

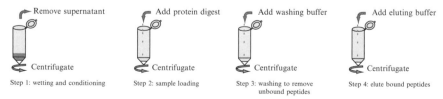

Remove supernatant   Add protein digest   Add washing buffer   Add eluting buffer

Centrifugate   Centrifugate   Centrifugate   Centrifugate

Step 1: wetting and conditioning   Step 2: sample loading   Step 3: washing to remove   Step 4: elute bound peptides
                                                              unbound peptides

Figure 5.1 Schematic representation of phosphopeptide enrichment method using metal oxide particles in a single-tube experimental setup.

supernatant is removed. Non-bound peptides are eluted by 10 μl 80% acetonitrile in 0.1% TFA. The particles are washed five times by 200 μl 80% acetonitrile in 0.1% TFA. Bound phosphopeptides are eluted by adding 20 μl of diammonium phosphate, 100 mM at pH 9 for 5 min. In the elution step phosphate acts a strong Lewis-base anion and ammonium as buffer counterion. After centrifugation, the supernatant containing the bound phosphopeptides is collected and acidified by adding 2 μl TFA 10%.

The single-tube TiO$_2$-based method in combination with shotgun proteomics was shown to be useful for the study of steady-state phosphorylation state of mitochondrial oxidative phosphorylation complexes. Using this method phosphorylation of B14.5a and 42 kDa subunits of complex I from bovine heart have been identified and the relating phosphorylation sites have been determined (Pocsfalvi et al., 2007).

2.2.2. Single-tube experiment applying magnetic particles-bound metal oxides

A novel, elegant single-tube method which employs Fe$_3$O$_4$-@TiO$_2$ magnetic microspheres for the purification has recently been reported (Li et al., 2008). Conventional single-tube enrichment uses centrifugation force in contrast with the magnetic particles-bound metal oxides-based experiment which uses magnetic force to mechanically separate the particles from the liquid phase. Magnetic handling provides easy washing and separation without need of centrifugation. Magnetic bead-based phosphopeptide enrichment technique is therefore a rapid and efficient procedure. The basic enrichment steps are the same as used in the conventional single-tube method. After the removal of non-bound peptides, a small volume of slurry containing the magnetic beads with the bound phosphopeptides can directly be loaded onto MALDI target plate for on-bead analysis. Alternatively, bound peptides are eluted with 10 μl of 12.5% aqueous NH$_3$ solution for 5 min, sample is neutralized with formic acid, and analyzed by ESI-MS.

Magnetic particle-bound TiO$_2$ was successfully applied in the characterization of rat liver phosphoproteome identifying 56 phosphopeptides with 65 phosphorylation sites in a single nano-LC-MS/MS analysis (Li et al., 2008).

2.2.3. Single-tube experiment applying Zr^{4+}- and Ti^{4+}-IMAC beads

Two novel polymer beads charged with Zr^{4+} (Feng et al., 2007) and Ti^{4+} (Zhou et al., 2008) cations through chelating interactions have been recently synthesized and used in single-tube phosphopeptide enrichment experiments. The mechanism of phosphopeptide isolation by Zr^{4+}-IMAC and Ti^{4+}-IMAC is similar to that of zirconia and titania. Therefore conditions and enrichment protocol are basically the same as the one described for the metal oxides. Ti^{4+}-IMAC was shown to have superior sensitivity, specificity, and efficiency compared to other enrichment methods in use including Fe^{3+}-IMAC, Zr^{4+}-IMAC, TiO$_2$, and ZrO$_2$. This was demonstrated by phosphoproteome analyses of mouse liver lysates (100 μg) (Zhou *et al.*, 2008). Ti^{4+}-IMAC resulted in the identification of 156 phosphopeptides and 251 phosphorylation sites. Applications of Ti^{4+}-IMAC or Zr^{4+}-IMAC in mitochondrial phosphoprotein characterization are promising, though still pending.

2.2.4. Enrichment using microtip-loaded metal oxides

TiO$_2$-based microtip-loaded metal oxides for the purification of phospho-peptides was introduced and optimized by MALDI-MS application by Larsen *et al.* (2005). Today there are a number of titania and zirconia pipette tips commercially available in standard 10–200 μl tip sizes with the advantage of automatic processing. The purification process, which is similar to the single-tube method, includes four main steps: (1) tip wetting and conditioning, (2) sample loading, (3) washing steps, and (4) sample elution. The main advantage of this technique is that it is easily and quickly performed. It is especially suited to MALDI-TOF-MS application. For MALDI-TOF-MS sample preparation in the phosphopeptide elution step 2 μl of 2, 5–dihydroxybenzoic acid (DHB) matrix solution (25 mg/ml in 70% ACN) containing 1% H$_3$PO$_4$ (v/v) is added and 0.5 μl of resulting solution is deposited on the MALDI target. Besides DHB, other additives like 1-octanesulfonic acid, lactic acid, formic acid at relatively high concentration (2–5%) have also been shown to decrease nonspecific binding during sample loading.

2.3. Chemical reactions prior enrichment

2.3.1. Methyl esterification to increase selectivity

Both IMAC and metal oxide-based phosphopeptide enrichment methods bind phosphopeptides under acidic conditions and elute under basic conditions. Some nonphosphorylated peptides, particularly acidic peptides have similar behavior under these conditions. Eluting nonphosphorylated peptides increases sample complexity and thus compromises successful phosphopeptide analysis. Improvements in the selectivity of phosphopeptide enrichment protocols can be achieved by the conversion of peptide carboxylates to their corresponding methyl esters. Methyl esterification of a peptide converts carboxylic acids present on the side chains of aspartic (D) and glutamic acid (E) as well as the free carboxyl group at C-terminus to their corresponding methyl esters. Methanolic HCl solution is prepared *in situ* by the dropwise addition of 32 μl acetyl chloride to 200 μl of dry methanol generating HCl and methyl acetate. Lyophilized sample containing the proteolytic digest (up to 250 pmol) is redissolved in 250 μl of methanolic HCl solution using a polypropylene micro-centrifuge tube. Methyl esterification is allowed to proceed for 2–3 h at room temperature. Solvent is removed by lyophilization. For simple peptide mixtures, the esterification reaction was reported to increase selectivity of both the

IMAC and the metal oxide-based phosphopeptide enrichments by at least two orders of magnitude. However, the reaction is not free of side reactions such as deamination of asparagine and glutamine which make this method troublesome for the analysis of complex protein mixtures.

2.3.2. Phosphopeptide derivatization by β-elimination and Michael addition to increase sensitivity

Alkaline-induced beta-elimination of phosphate groups from phosphoserine (pS) and phosphothreonine (pT) residues followed by addition of an affinity tag has been introduced as a strategy for phosphopeptide enrichment from complex mixtures. β-Elimination of the phosphate moiety on pS and pT residues (note pY cannot be derivatized using this method) is performed under strongly alkaline conditions leading to the formation of dehydroalanine or dehydroamino-2-butyric acid residues, respectively. Several bases including barium hydroxide, sodium hydroxide, and ammonium bicarbonate can be employ in the β-elimination reaction, saturated barium hydroxide in organic solvents such as ethanol or acetonitrile is the most widely used. The unsaturated residue used as Michael acceptor can react with different nucleophiles (like ethanedithiol, DTT, mercaptoethylpyridine). Substitution of the negatively charged phosphate group by a positively charged group by β-elimination and Michael addition can greatly improve the ionization of the derivatized peptides. The main drawbacks of derivatization procedures are the yields of derivatization and the occurrence of side reactions during β-elimination step.

An elegant way to increase the sensitivity of phosphopeptide enrichment was recently introduced by Guo et al. (2007). The new chemical derivatization strategy uses a soluble polyamide dendrimer support to derivatize and immobilize phosphopeptides in a single step. Methylated phosphopeptides are captured directly on the dendrimer via a carbodiimide-activated reaction between amine and phosphate groups. Phosphopeptides bound to dendrimer then can be easily separated from the nonphosphorylated ones using size-exclusion chromatography or simple molecular filtration using a membrane with 5 kDa cutoff. Modified phosphopeptides are released from the dendrimer via acid hydrolysis followed by molecular filtration (5 kDa cutoff). The dendrimer-based enrichment can be combined with stabile isotope labeling during the methyl esterification reaction which grants quantitative characteristic to the approach.

2.4. Immunoaffinity purification

2.4.1. Immunoaffinity-based enrichment by the use of phosphotyrosine antibody

Specific antibodies directed against pS, pT, and pY can be used in immunoprecipitation experiments to capture proteins or peptides containing these residues. Available pS and pT antibodies generally have lower specificity than pY antibodies. Therefore pY antibodies are the most frequently used in immunoaffinity purification (IAP). A large-scale procedure employing pY antibody to enrich phosphotyrosine containing peptides from proteolytically digested cellular extract combined with LC–MS/MS has recently been reported for the identification of phosphotyrosine-containing proteins (Rush *et al.*, 2005). Generally, affinity purified mouse monoclonal antibodies are immobilized to protein G agarose beads through non–covalent binding according to standard protocols. Immobilized pY antibody can be directly used in single-tube IAP experiments or can be loaded into columns. Mixture of immobilized antibody and sample containing protease digest is incubated in IAP buffer containing 20 mM Tris/HCl pH 7.2, 10 mM sodium phosphate, 50 mM NaCl overnight at 4 °C with gentle shaking. The beads are washed three times with IAP buffer and twice with water. Bound peptides are eluted from beads by 0.1% TFA for 10 min. This elution step also releases the antibody from protein G which should be removed before downstream analysis.

Presence of protein tyrosine kinases (members of the Src kinase family), and tyrosine phosphatases (like Shp2 and PTPM-1) in mitochondria has been demonstrated. Tyrosine phosphorylation, in spite of its relatively rare occurrence, is particularly important because of its regulatory role in cell signal transduction. Tyrosine phosphorylation is generally switched off under normal conditions due to the high activity of tyrosine phosphatases and to the general inactivation state of the kinases in the absence of an appropriate signal. Among identified tyrosine phosphorylated mitochondrial proteins are the 39-kDa subunit of complex I, F_oF_1-ATP synthase α chain, subunit Va of cytochrome c oxidase, cytochrome c, and adenine nucleotide translocator-1. Moreover, a large-scale analysis using immunoaffinity enrichment of phosphotyrosine containing peptides from rat brain mitochondria extract led to the identifications of novel tyrosine phosphorylated mitochondrial proteins (Urs *et al.*, 2008). In this work, the use of peroxovanadate was shown to be key importance in the inhibition of tyrosine phosphatases prior analysis.

3. ON-LINE 2D-LC PHOSPHOPEPTIDE ENRICHMENT METHODS

In-solution enzymatic digestion of a complex cell lysate or a mitochondrial protein extract leads to an extremely complex peptide mixture. The two-dimensional (2D) chromatographic approach for the selectively enrichment and separation of phosphorylated peptides from such complex digests is a key element for successful phosphoprotein characterization. Both the IMAC and the metal oxide-based phosphopeptide enrichment methods can be carried out "online" with MS using appropriately packed LC columns and LC systems. The best performing nanoflow 2D-LC-systems today employ a combination of TiO_2-based selective phosphopeptide enrichment, reversed phase (RP) C18 and strong cation exchange (SCX) chromatography.

3.1. 2D-TiO$_2$-RP-LC-MS using biphasic column

One of the most suited techniques for large-scale phosphoproteome analysis used today is the on-line two-dimensional HPLC system using titania in the first dimension for phosphopeptide enrichment followed by monolithic or reversed phase C18 separation in the second dimension. Self-pulled or a commercially available column needle can be packed with two different chromatographic resins to construct a biphasic analytical column which functions as a sample introduction needle in an on-line ESI-MS setup. A TiO_2-C18 biphasic column containing C18 beads (10 cm length) at the needle end followed by titania beads (5 cm length) can be directly used in 2D-nano-LC setup after loading approximately 25 μg of proteolitic digest dissolved in 0.5% acetic acid. Nonphosphorylated peptides are eluted in 0.5% acetic acid and 80% acetonitrile, whereas phosphopeptides are bound under these conditions. Elution of phosphopeptides is performed by injecting 5 μl of 0.5% ammonium hydroxide into the column. Separation and analysis of phosphopeptides are followed by RP nano-HPLC-ESI-MS/MS using established conditions. 671 phosphorylation sites, derived from 512 phosphopeptides of digested HeLa cells, were successfully identified using a calcined rutile-form titania/C18 biphasic column in a 2D-nano-LC-MS/MS experiment (Imami *et al.*, 2008).

3.2. 2D-SCX-TiO$_2$-RP-LC-MS

A highly efficient automated 2D online phosphopeptide enrichment/separation was recently introduced by Pinkse *et al.* using a triple-stage "sandwich" C18-TiO_2-C18 precolumn (Pinkse *et al.*, 2008). Separation in the first dimension is achieved by off-line SCX chromatography which separates peptides in a number of fractions depending on sample complexity. In the second dimension, peptides in the single SCX fractions are first trapped onto the first stage of the C18 column. Nonphosphorylated peptides are then eluted and separated with RP chromatography using gradient elution at a nano-flow rate (100 nl/min) achieved by flow splitting. Under this condition, phosphorylated peptides are efficiently trapped on the TiO_2 precolumn. Elution of phosphorylated peptides is performed by the injection of 30 μl of 250 mM ammonium hydrogen bicarbonate, pH 9.0 (adjusted with ammonia) containing 10 mM sodium phosphate, 5 mM sodium orthovanadate, and 1 mM potassium fluoride, followed by an injection of 20 μl of 5% formic acid. Finally, phosphorylated peptides are separated by a second run of conventional RP chromatography coupled on-line to the mass spectrometer.

4. MASS SPECTROMETRY-BASED METHODS

Various mass spectrometry methods have been exploited for the identification of phosphoproteins and the determination of their phosphorylation sites. Nevertheless, the task remains technically challenging due to the intrinsic physicochemical characteristics of phosphopeptides. In the positive ion mode, phosphopeptides typically have one charge less than their corresponding nonphosphorylated analogues due the presence of negatively charged phosphate groups and, therefore, are characterized by limited ionization efficiency. This suggests that the negative ion mode should be the ionization mode of choice for phosphopeptide analysis. In fact, MALDI- and ESI-MS in the negative ion mode result in relatively the abundant phosphopeptide molecular ion, $[M-H]^-$. Single-stage MALDI- and ESI-MS however are limited to molecular mass information and to indicate a possible phosphorylation event. To determine the site of phosphorylation, tandem mass spectrometry (MS/MS) has to be performed. In the negative ion mode, difficulties in spectral interpretation make phosphopeptide sequencing quite challenging. Due to these difficulties, positive ion mode is most commonly used today in high-throughput mass spectrometry-based phosphorylation studies.

4.1. Maldi-MS and MS/MS analysis

MALDI-MS is well suited to phosphorylation studies which target a specific protein or moderately complex mixtures. Comparing MALDI-TOF spectra of mixtures containing both phosphorylated and nonphosphorylated peptides acquired in positive and negative modes is useful for identifying the phosphopeptides based on the relatively higher ion intensities generated by the phosphorylated peptides in negative ion mode. In addition, singly charged serine and threonine phosphorylated peptides can be identified in the positive ion mode based on the presence of characteristic $[MH-H_3PO_4]^+$ and $[MH-HPO_3]^+$ fragment ions that appear at MH-98 and MH-80 daltons, respectively, in MALDI reflectron spectra. Phosphotyrosine containing peptides commonly loses $[MH-HPO_3]^+$. However, depending on sequence, phosphoric acid loss ($[MH-H_3PO_4]^+$) has also been experienced. The most commonly used matrix for MALDI-MS detection of phosphopeptides is 2,5-DHB with 0.1% phosphoric acid (DHB-PA). Ionization efficiency can be enhanced in the positive ion mode by chemical reaction (methyl esterification and β-elimination/ Michael addition) or the supplement of additives (phosphoric acid, ammonium phosphate (mono- and dibasic), diammonium citrate, lactic acid, and so on). In the negative ion mode, tyrosine, serine, and threonine phosphorylated peptides all gave essentially identical reflectron spectra, exhibiting an ion at $[M-H-97]^-$.

To produce sequence specific fragment ions and to assign the site of phosphorylation MALDI, tandem mass spectrometry using TOF/TOF or quadrupole time of flight (QTOF) analyzers can be performed on the isolated single charged phosphopeptide molecular ion.

4.2. ESI-MS and data dependent MS/MS strategies

Electrospray ionization mass spectrometry (ESI-MS) is the most widely used technique in protein phosphorylation studies. ESI-MS coupled to liquid chromatography (LC-ESI-MS) is well suited to facilitate automation of robust large-scale phosphoproteomics. To obtain sequence information, collision induced fragmentation is required using collision induced dissociation (CID) and tandem mass spectrometry (MS/MS). There are a number of possible instrument configurations and MS/MS scanning techniques exploited for phosphopeptide characterization. Besides the most commonly used product ion scanning, precursor ion scanning in the negative ion mode for the phosphate-derived anion m/z 79 on a triple-quadrupole instrument has also been shown to be useful. During product ion scans, phosposerine and phosphothreonine containing peptides undergo β-elimination and lose neutral phosphoric acid yielding highly abundant MH-98 Da fragment ions. The neutral loss of phosphoric acid can be monitored in the positive ion mode for pS and pT containing peptides using in-source or collision cell fragmentation. Whereas for pY containing peptides, the formation of characteristic immonium ion at m/z 216 and the neutral loss of HPO_3 (-80 Da) are the fragmentation processes to be monitored.

To generate additional sequence specific fragment ions created by peptide backbone cleavage, neutral loss triggered MS3 can be performed. In this procedure, the ion that has lost phosphoric acid is identified and subsequently fragmented. The utility of this method was demonstrated in analysis the mouse liver phosphoproteome using a hybrid triple-quadrupole linear ion trap mass spectrometer operated in a data-dependent acquisition mode (Lee et al., 2007). In this type of experiment, the single-stage MS scan (MS1) is followed by four pairs MS2/MS3 scans. The second stage MS2 scan is a CID product ion scan (MS/MS) performed on the eight most intense ions observed in the MS1 scan. In the case of phosphopeptides, MS2 generates the characteristic $[M + H-98]^+$, $[M + 2H-98]^{2+}$, and $[M + 3H-98]^{3+}$ ions at $-98\ m/z$, $49\ m/z$, and $32.7\ m/z$, respectively, due to phosphoric acid loss from the single, double, and triple charged molecular ions. Third stage MS3

(MS/MS/MS) scan is automatically triggered when one of the peaks of phosphoric acid loss is detected among the top eight most intense peaks in the MS2 spectrum. Using this strategy in combination with IMAC enrichment and methyl esterification, Lee *et al.* has reported one of the most comprehensive mitochondrial phosphoprotemic work today which identified 84 phosphorylation sites from 62 phosphoproteins in mouse liver (Lee *et al.*, 2007).

5. QUANTITATIVE PHOSPHOPROTEOMICS

Reversible protein phosphorylation in a living cell is a dynamic process. Kinetics and stochiometry of phosphorylation is regulated by protein kinases and phosphatases, whereby the phosphorylation status of a protein may vary with time and with stimuli. The aim of quantitative phosphoproteomics is to examine relative changes occurring in phosphorylation levels. The two most frequently used quantitative proteomics methods are (i) stable-isotope labeling by amino acids in cell culture (SILAC) and (ii) isobaric tagging for relative and absolute quantitation (iTRAQ). While SILAC have been successfully applied in studies with cultured human cells and in yeast, iTRAQ is based on the labeling of the N termini and lysine residues of extracted (phospho-) peptides. Quantitative analysis is becoming even more important because it permits the study of *in vivo* phosphorylation events, as recently highlighted by the comparison of control and phosphatase inhibited mouse liver cell line using SILAC analysis (Pan *et al.*, 2008). We still lack comprehensive quantitative analysis for the mitochondrial phosphoproteomes.

REFERENCES

Andersson, L., and Porath, J. (1986). Isolation of phosphoproteins by immobilized metal (Fe^{3+}) affinity chromatography. *Anal. Biochem.* **154,** 250–254.
Andreoli, C., Prokisch, H., Hortnagel, K., Mueller, J. C., Munsterkotter, M., Scharfe, C., and Meitinger, T. (2004). MitoP2, an integrated database on mitochondrial proteins in yeast and man. *Nucleic Acids Res.* **32,** D459–D462.
Basu, S., Bremer, E., Zhou, C., and Bogenhagen, D. F. (2006). MiGenes: A searchable interspecies database of mitochondrial proteins curated using gene ontology annotation. *Bioinformatics* **22,** 485–492.
Chakravarti, B., Oseguera, M., Dalal, N., Fathy, P., Mallik, B., Raval, A., and Chakravarti, D. N. (2008). Proteomic profiling of aging in the mouse heart: Altered expression of mitochondrial proteins. *Arch. Biochem. Biophys.* **474,** 22–31.
Cotter, D., Guda, P., Fahy, E., and Subramaniam, S. (2004). MitoProteome: Mitochondrial protein sequence database and annotation system. *Nucleic Acids Res.* **32,** D463–D467.

Cuccurullo, M., Schlosser, G., Cacace, G., Malorni, L., and Pocsfalvi, G. (2007). Identification of phosphoproteins and determination of phosphorylation sites by zirconium dioxide enrichment and SELDI-MS/MS. *J. Mass Spectrom.* **42,** 1069–1078.

Distler, A. M., Kerner, J., and Hoppel, C. L. (2007). Post-translational modifications of rat liver mitochondrial outer membrane proteins identified by mass spectrometry. *Biochim. Biophys. Acta (BBA) Proteins Proteomics* **1774,** 628–636.

Elstner, M., Andreoli, C., Ahting, U., Tetko, I., Klopstock, T., Meitinger, T., and Prokisch, H. (2008). MitoP2: An integrative tool for the analysis of the mitochondrial proteome. *Molec. Biotechnol.* **40,** 306–315.

Feng, S., Ye, M., Zhou, H., Jiang, X., Jiang, X., Zou, H., and Gong, B. (2007). Immobilized zirconium ion affinity chromatography for specific enrichment of phosphopeptides in phosphoproteome analysis. *Mol. Cell Proteomics* **6,** 1656–1665.

Ficarro, S. B., Parikh, J. R., Blank, N. C., and Marto, J. A. (2008). Niobium(V) Oxide (Nb2O5): Application to phosphoproteomics. *Anal. Chem.* **80,** 4606–4613.

Frank, G. (2006). Differential mitochondrial protein expression profiling in neurodegenerative diseases. *Electrophoresis* **27,** 2814–2818.

Guo, M., Galan, J., and Tao, W. A. (2007). Soluble nanopolymer-based phosphoproteomics for studying protein phosphatase. *Methods* **42,** 289–297.

Heazlewood, J. L., and Millar, A. H. (2005). AMPDB: The arabidopsis mitochondrial protein database. *Nucleic Acids Res.* **33,** D605–D610.

Højlund, K., Mogensen, M., Sahlin, K., and Beck-Nielsen, H. (2008). Mitochondrial dysfunction in type 2 diabetes and obesity. *Endocrinol. Metab. Clin. North Am.* **37,** 713–731.

Hu, Y., A Yang, X., and A Carr, P. W. (2002). Mixed-mode reversed-phase and ion-exchange separations of cationic analytes on polybutadiene-coated zirconia. *J. Chromatogr. A* **968,** 17–29.

Hurd, T. R., Costa, N. J., Dahm, C. C., Beer, S. M., Brown, S. E., Filipovska, A., and Murphy, M. P. (2005). Glutathionylation of mitochondrial proteins. *Antioxid. Redox Signal.* **7,** 999–1010.

Imami, K., Sugiyama, N., Kyono, Y., Tomita, M., and Ishihama, Y. (2008). Automated phosphoproteome analysis for cultured cancer cells by two-dimensional NanoLC-MS using a calcined titania/C18 biphasic column. *Anal. Sci.* **24,** 161–166.

Kim, S. C., Sprung, R., Chen, Y., Xu, Y., Ball, H., Pei, J., Cheng, T., Kho, Y., Xiao, H., Xiao, L., Grishin, N. V., White, M., et al. (2006). Substrate and functional diversity of lysine acetylation revealed by a proteomics survey. *Mol. Cell* **23,** 607–618.

Kweon, H. K., and Hakansson, K. (2006). Selective zirconium dioxide-based enrichment of phosphorylated peptides for mass spectrometric analysis. *Anal. Chem.* **78,** 1743–1749.

Larsen, M. R., Thingholm, T. E., Jensen, O. N., Roepstorff, P., and Jorgensen, T. J. D. (2005). Highly selective enrichment of phosphorylated peptide mixture using titanium dioxide microcolumns. *Mol. Cell. Proteomics* **4,** 873–886.

Lee, J., Xu, Y., Chen, Y., Sprung, R., Kim, S. C., Xie, S., and Zhao, Y. (2007). Mitochondrial phosphoproteome revealed by an improved IMAC method and MS/MS/MS. *Mol. Cell. Proteomics* **6,** 669–676.

Li, Y., Xu, X., Qi, D., Deng, C., Yang, P., and Zhang, X. (2008). Novel Fe3O4@TiO2 core-shell microspheres for selective enrichment of phosphopeptides in phosphoproteome analysis. *J. Proteome Res.* **7,** 2526–2538.

Liu, Y., He, J., Ji, S., Wang, Q., Pu, H., Jiang, T., Meng, L., Yang, X., and Ji, J. (2008). Comparative studies of early liver dysfunction in senescence-accelerated mouse using mitochondrial proteomics approaches. *Mol. Cell. Proteomics* **7,** 1737–1747.

Meisinger, C., Sickmann, A., and Pfanner, N. (2008). The mitochondrial proteome: From inventory to function. *Cell* **134,** 22–24.

Melanie, Y. W. (2008). Mitochondria: A mirror into cellular dysfunction in heart disease. *Proteomics: Clin. Appl.* **2,** 845–861.

Miller, J. L., Koc, H., and Koc, E. C. (2008). Identification of phosphorylation sites in mammalian mitochondrial ribosomal protein DAP3. *Protein Sci.* **17,** 251–260.

Pagliarini, D. J., Calvo, S. E., Chang, B., Sheth, S. A., Vafai, S. B., Ong, S.-E., Walford, G. A., Sugiana, C., Boneh, A., Chen, W. K., Hill, D. E., Vidal, M., *et al.* (2008). A mitochondrial protein compendium elucidates complex I disease biology. *Cell* **134,** 112–123.

Pagliarini, D. J., and Dixon, J. E. (2006). Mitochondrial modulation: Reversible phosphorylation takes center stage? *Trends Biochem. Sci.* **31,** 26–34.

Pan, C., Gnad, F., Olsen, J. V., and Mann, M. (2008). Quantitative phosphoproteome analysis of a mouse liver cell line reveals specificity of phosphatase inhibitors. *Proteomics* **8,** 4534–4546.

Pinkse, M. W. H., Mohammed, S., Gouw, J. W., van Breukelen, B., Vos, H. R., and Heck, A. J. R. (2008). Highly robust, automated, and sensitive online TiO2-based phosphoproteomics applied to study endogenous phosphorylation in *Drosophila melanogaster. J. Proteome Res.* **7,** 687–697.

Pocsfalvi, G., Cuccurullo, M., Schlosser, G., Scacco, S., Papa, S., and Malorni, A. (2007). Phosphorylation of B14.5a subunit from bovine heart complex I identified by titanium dioxide selective enrichment and shotgun proteomics. *Mol. Cell. Proteomics* **6,** 231–238.

Reinders, J., Wagner, K., Zahedi, R. P., Stojanovski, D., Eyrich, B., van der Laan, M., Rehling, P., Sickmann, A., Pfanner, N., and Meisinger, C. (2007). Profiling phosphoproteins of yeast mitochondria reveals a role of phosphorylation in assembly of the ATP synthase. *Mol. Cell. Proteomics* **6,** 1896–1906.

Rush, J., Moritz, A., Lee, K. A., Guo, A., Goss, V. L., Spek, E. J., Zhang, H., Zha, X. -M., Polakiewicz, R. D., and Comb, M. J. (2005). Immunoaffinity profiling of tyrosine phosphorylation in cancer cells. *Nat. Biotech.* **23,** 94–101.

Schwer, B., Bunkenborg, J., Verdin, R. O., Andersen, J. S., and Verdin, E. (2006). Reversible lysine acetylation controls the activity of the mitochondrial enzyme acetyl-CoA synthetase 2. *Proc. Nat. Acad. Sci. USA* **103,** 10224–10229.

Souza, J. M., Peluffo, G., and Radi, R. (2008). Protein tyrosine nitration—Functional alteration or just a biomarker? *Free Rad. Biol. Med.* **45,** 357–366.

Sturm, M., Leitner, A., Jan-Henrik, S., Mika, S., and Lindner, L. W. (2008). Tin dioxide microspheres as a promising material for phosphopeptide enrichment prior to liquid chromatography-(tandem) mass spectrometry analysis. *Adv. Funct. Mater.* **18,** 2381–2389.

Sutovsky, P., Moreno, R. D., Ramalho-Santos, J., Dominko, T., Simerly, C., and Schatten, G. (2000). Ubiquitinated sperm mitochondria, selective proteolysis, and the regulation of mitochondrial inheritance in mammalian embryos. *Biol. Reprod.* **63,** 582–590.

Toshihiko, U., Nagisa, S., Kengo, N., and Rumi, I. (2003). C-terminal 15 kDa fragment of cytoskeletal actin is posttranslationally *N*-myristoylated upon caspase-mediated cleavage and targeted to mitochondria. *FEBS Lett.* **539,** 37–44.

Urs, L., Albert, S., Luca, C., Anna Maria, B., Antonio, T., and Mauro, S. (2008). Identification of new tyrosine phosphorylated proteins in rat brain mitochondria. *FEBS Lett.* **582,** 1104–1110.

Verma, M., Kagan, J., Sidransky, D., and Srivastava, S. (2003). Proteomic analysis of cancer-cell mitochondria. *Nat. Rev. Cancer* **3,** 789–795.

Zhou, H., Ye, M., Dong, J., Han, G., Jiang, X., Wu, R., and Zou, H. (2008). Specific phosphopeptide enrichment with immobilized titanium ion affinity chromatography adsorbent for phosphoproteome analysis. *J. Proteome Res.* **7,** 3957–3967.

CHAPTER SIX

POST-TRANSLATIONAL MODIFICATIONS OF MITOCHONDRIAL OUTER MEMBRANE PROTEINS

Anne M. Distler,*,‡ Janos Kerner,*,‡ Kwangwon Lee,*,‡ and Charles L. Hoppel*,†,‡

Contents

1. Introduction	98
2. Isolation of Mitochondria and Mitochondrial Membranes	99
2.1. Preparation of percoll purified mitochondria	100
2.2. Preparation of mitochondrial outer membranes	100
3. Detection of Post-translational Modifications Using One-dimensional Gel Electrophoresis	101
3.1. One-dimensional gel electrophoresis and electroelution	102
3.2. Reduction, alkylation, and enzymatic digestion of the protein	103
3.3. Detection of post-translational modifications using mass spectrometry	104
4. Focused Proteomics Using Immunological Approaches	105
4.1. Antibody preparation and immobilization	105
4.2. Isolation of CPT-I	105
4.3. Mass spectrometric analysis of the isolated CPT-I	107
5. Shotgun Methods for Identification of Post-translational Modifications	109
5.1. Digestion of proteins	110
5.2. Mass spectrometric analysis of digestion	110
6. Conclusion	111
Acknowledgments	111
References	112

* Department of Pharmacology, Case Western Reserve University, Cleveland, Ohio, USA
† Department of Medicine, Case Western Reserve University, Cleveland, Ohio, USA
‡ Center for Mitochondrial Disease, Case Western Reserve University, Cleveland, Ohio, USA

Methods in Enzymology, Volume 457
ISSN 0076-6879, DOI: 10.1016/S0076-6879(09)05006-X

Abstract

In recent years, a wide variety of proteomic approaches using gel electrophoresis and mass spectrometry has been developed to detect post-translational modifications. Mitochondria are often a focus of these studies due to their important role in cellular function. Many of their crucial transport and oxidative-phosphorylation functions are performed by proteins residing in the inner and outer membranes of the mitochondria. Although proteomic technologies have greatly enhanced our understanding of regulation in cellular processes, analysis of membrane proteins has lagged behind that of soluble proteins. Herein, we present techniques to facilitate the detection of post-translational modifications of mitochondrial membrane proteins including the isolation of resident membranes as well as electrophoretic and immunological-based methods for identification of post-translational modifications.

1. INTRODUCTION

The field of proteomics plays a cardinal role in advancing the understanding of physiological and cellular processes. Using a variety of techniques, cellular systems can be studied in either a global or more targeted approach. Recently, many proteomic studies have focused on sub-cellular systems such as lysosomes (Bagshaw *et al.*, 2005), Golgi complex (Bell *et al.*, 2001; Taylor *et al.*, 2000; Wu *et al.*, 2004), endoplasmic reticulum (Knoblach *et al.*, 2003), peroxisomes (Marelli *et al.*, 2004), and mitochondria (Crouser *et al.*, 2006; Distler *et al.*, 2008; Mootha *et al.*, 2003; Prokisch *et al.*, 2002; Sickmann *et al.*, 2003; Taylor *et al.*, 2003). Mitochondria are of particular interest due to their involvement in energy production as well as their proposed role in apoptosis (Green and Kroemer, 2004; Green and Reed, 1998; Spierings *et al.*, 2005).

The majority of mitochondrial proteomic studies identify, or focus, on soluble proteins from the matrix of the mitochondria and only a few studies detail integral membrane proteins of the inner membrane or outer membrane (Bernardi, 1999; Da Cruz and Martinou, 2008; Da Cruz *et al.*, 2003; Schmitt *et al.*, 2006; Zahedi *et al.*, 2006). Although there are relatively few studies focused on the mitochondrial outer membrane, the proteins in this membrane play crucial roles in mitochondria. For example, long-chain acyl–CoA synthetase (ACS), voltage dependent anion channel (VDAC), and carnitine palmitoyltransferase-I (CPT-I) play an obligatory role in the mitochondrial uptake of long-chain fatty acids, the major substrate for energy production in many tissues (Turkaly *et al.*, 1999). Additionally, mitochondrial outer membrane proteins are involved in energy exchange and metabolite trafficking (VDAC) (Benz, 1994), apoptosis (VDAC, CPT-I)

(Cheng *et al.*, 2003; Paumen *et al.*, 1997; Shimizu *et al.*, 2001), and serve as a docking site for cytosolic proteins (VDAC) (Linden and Karlsson, 1996; Vyssokikh and Brdiczka, 2004).

Numerous proteins are post-translationally modified. While many of the identified modifications are functionally significant, the physiological relevance of many protein modifications has not been elucidated. The post-translational modification of proteins in mitochondrial membranes is a field that still needs to be further explored. There is a variety of covalent modifications that play an important role in cellular function. For example, protein phosphorylation is considered as the primary means of altering the activity of a protein rapidly and, for this reason, is seen as a key event in many signal transduction pathways. While less common than phosphorylation, tyrosine nitration also is an important modification in cellular function. Nitration of tyrosine residues in a protein can induce diverse physiological and pathological responses. Protein acetylation is another frequent protein modification and can occur co-translationally at the N-terminus and post-translationally at lysine residues. Although N-terminal acetylation is widespread in eukaryotes the biological relevance of this modification is known for only a few proteins (Caesar *et al.*, 2006; Perrier *et al.*, 2005; Polevoda *et al.*, 1999). In contrast, lysine acetylation is reversible with a wide range of functional consequences (Bruserud *et al.*, 2006; Yang, 2004).

While important to cellular function, post-translational modifications are difficult to detect using analytical techniques such as gel electrophoresis and mass spectrometry. These difficulties arise for a number of reasons. Protein modifications often are unstable at conditions used during sample preparation. Also, these modifications are usually present sub-stoichiometrically making detection difficult. For these reasons, high sequence coverage and a sensitive mass spectrometric technique are required for the detection and localization of post-translational modifications.

Here we present several methods developed to examine the post-translational modifications of mitochondrial outer membrane proteins. While these studies are focused on the CPT-I protein, this protein serves as model for all outer membrane proteins, and the approaches described herein are easily adapted to other membrane proteins in mitochondria (Distler *et al.*, 2006, 2007).

2. Isolation of Mitochondria and Mitochondrial Membranes

Mitochondria associate *in vivo* with other cytoplasmic components with physiological relevance (Bionda *et al.*, 2004; Linden and Karlsson, 1996; Pizzo and Pozzan, 2007). Depending on the tissue, mitochondria isolated by conventional differential centrifugation often are contaminated with non-mitochondrial organelles such as peroxisomes, endoplasmic reticulum or microsomes, Golgi complexes, myofibrillar proteins, cytoskeletal proteins, and plasma membranes. To eliminate or significantly reduce these contaminants, mitochondria isolated by differential centrifugation can be further purified using density gradient centrifugation with Percoll, Nycodenz, or Iodixanol and their purity assessed by marker enzyme activity, electron microscopy, and immunoblot analysis of the organelles using markers for mitochondria as well as for contaminating organelles (Hoppel *et al.*, 1998, 2001, 2002).

2.1. Preparation of percoll purified mitochondria

The livers from male Sprague–Dawley rats are minced and rinsed with MSM isolation buffer (220 mM mannitol, 70 mM sucrose, 5 mM 3-(N-morpholino)propanesulfonic acid, pH 7.4). The isolation buffer can be supplemented with commercially available protein phosphatase inhibitor cocktails (Roche, Sigma) or a mixture of 10 mM sodium fluoride, 1 mM sodium orthovanadate, 1 mM sodium pyrophosphate, and 4 mM β-glycerophosphate in the isolation buffer (Vavvas *et al.*, 1997). A 10% homogenate is prepared in MSM with 2 mM EDTA using two strokes of a Potter–Elvehjem loose-fitting Teflon pestle. The homogenate is centrifuged (10 min, 300g) and the supernatant is saved. The mitochondria are isolated from the supernatant by differential centrifugation (10 min, 7000g), washed twice in MSM isolation buffer, and resuspended in 0.2 mL of isolation buffer per gram of liver (80–100 mg mitochondria/mL) (Hoppel *et al.*, 1979). This crude preparation of mitochondria still contains other organelles such as peroxisomes, lysosomes, and mitochondria associated membranes (MAMs).

Further purification of mitochondria is achieved using self-forming Percoll gradient centrifugation. One milliliter aliquots of mitochondrial suspensions are layered on top of 20 mL of 30% Percoll in MSM and centrifuged (111,406g) for 30 min (Beckman 50.2 Ti, 35,000 rpm). The mitochondria are separated into two distinct layers separated by a clear zone. The upper layer has been shown to be peroxisomes and lysosomes by electron microscopy, marker enzymes, and immunoblot analysis. The peroxisomes and the clear zone above the mitochondria are removed, and the lower brownish layer consisting of mitochondria is harvested. Following a twofold dilution with isolation buffer, the mitochondria are collected by centrifugation (10 min, 7000g) and washed once with isolation buffer.

2.2. Preparation of mitochondrial outer membranes

The washed Percoll purified mitochondria are slowly resuspended to 2 mg/mL using 20 mM potassium phosphate/0.2% defatted BSA (pH 7.2) and incubated on ice for 20 min with gentle stirring to induce swelling of

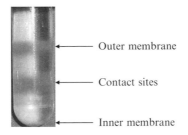

Outer membrane

Contact sites

Inner membrane

Figure 6.1 Photograph of the outer membrane, contact site, and inner membrane layers formed in the discontinuous sucrose gradient after centrifugation.

the mitoplast resulting in rupture of the mitochondrial outer membrane. To shrink the mitochondria, ATP and $MgCl_2$ are then added to a final concentration of 1.0 mM each and the suspension is incubated on ice with stirring for an additional 5 min. The swollen/shrunken mitochondria are centrifuged for 20 min at 4 °C at 22,550g_{av} (Sorvall RC5C, SS-34 rotor) and the pellet is slowly resuspended to the original volume (around 2 mg/mL) in 20 mM potassium phosphate/0.2% defatted BSA, pH 7.2. The mitochondrial suspension is treated with two strokes of a tight-fitting pestle in a glass homogenizer (Potter–Elvjehem, setting 3) and centrifuged for 15 min at 4 °C with 1,930g (Sorvall RC5C, SS-34 rotor). The supernatant is carefully removed and centrifuged (34,000g) for 20 min (Sorvall RC5C, SS-34 rotor) to obtain the crude outer membrane pellet. Using a hand-driven Potter homogenizer, the pellet is resuspended in 2 mL of 20 mM potassium phosphate, pH 7.2 (13–17 mg protein/mL), and separated into outer membranes, contact site fraction, and inner membranes by discontinuous sucrose gradient centrifugation (see Fig. 6.1), consisting of 1.2 mL of each 51.3, 37.7, and 25.2% sucrose in 20 mM potassium phosphate, pH 7.2 (Hoppel *et al.*, 2001; Parsons *et al.*, 1966).

3. DETECTION OF POST-TRANSLATIONAL MODIFICATIONS USING ONE-DIMENSIONAL GEL ELECTROPHORESIS

To study complex protein mixtures from biological systems, two-dimensional polyacrylamide gel electrophoresis (2D-PAGE) is used in conjunction with mass spectrometric analysis (Barnouin, 2004; Issaq and Veenstra, 2008; Rabilloud, 2002). Despite recent advances (reviewed in Braun *et al.*, 2007), 2D-PAGE separation methods often result in underrepresentation of membrane proteins (Santoni *et al.*, 2000) due to the low solubility of hydrophobic membrane proteins, low abundances of membrane proteins compared to soluble proteins, and difficulty of focusing large membrane proteins in a 2D-gel. When analyzing hydrophobic membrane proteins, one-dimensional gel electrophoresis is a more viable option. Although in-gel digestions are successful for analyzing some membrane proteins, only partial cleavage of the protein occurs, especially for those proteins with many membrane-spanning regions (Buhler *et al.*, 1998; Hellmann *et al.*, 1995). In-gel digestion often results in low sequence coverage for the proteins making identification of post-translational modifications difficult.

To increase sequence coverage and identify post-translational modifications, we have described a method for the targeted analysis of membrane proteins (Distler *et al.*, 2006). The method is based on the isolation of the proteins by semi-preparative SDS–PAGE, electroelution of the proteins from the gel, and digestion of the proteins in-suspension. While the protein complexity in each gel band is greater in 1D as compared to 2D electrophoresis, enrichment of the proteins by isolation of the mitochondrial outer membranes (as described above) simplifies the mixture of proteins. The use of liquid chromatography (LC) coupled to mass spectrometry further reduces the complexity of the digestion products. This focused technique allows for high coverage of the proteins of interest without the use of detergents. Use of this method allows protein coverage from 82–99% for the hydrophobic integral mitochondrial outer membrane proteins, as well as the identification of post-translational modifications including phosphorylation, acetylation, and nitration (Distler *et al.*, 2006). Furthermore, the electroeluted proteins can be subjected to analytical SDS–PAGE and immunoblotting for post-translational modifications (Fig. 6.2).

3.1. One-dimensional gel electrophoresis and electroelution

Proteins are isolated from Percoll-purified rat liver mitochondrial outer membranes by semi-preparative SDS–PAGE as described earlier (Hoppel *et al.*, 2002; Kerner *et al.*, 2004). One milligram of mitochondrial outer membranes is subjected to semi-preparative (100 mm × 72 mm × 1.5 mm) SDS–PAGE using a 10% separating gel and a 4% concentrating gel. The separated proteins are visualized by a brief staining with Coomassie Brilliant Blue R250. The bands of interest are then excised from the gel and the proteins in the excised gel piece are electroeluted for further analysis. The electroeluted proteins are precipitated with 80% ethanol and the precipitated proteins washed twice with 80% ethanol to remove excess protein stain and SDS. After each wash, the solution is centrifuged (13,200*g*) and the supernatant discarded. Once the membrane proteins are precipitated, resolubilization is difficult and requires the use of detergents that will interfere in the mass spectrometric analysis. As a result, the remainder of the protocol is carried out on the precipitated protein pellet.

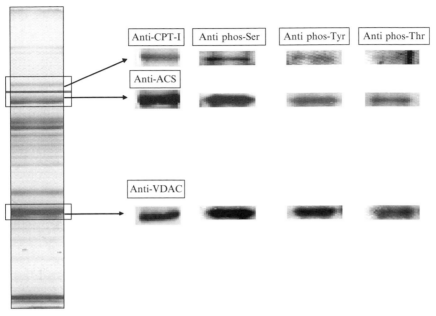

Figure 6.2 Analysis of CPT-I, ACS, and VDAC phosphorylation by Western blotting. Rat liver mitochondrial outer membranes were separated by semi-preparative SDS–PAGE (Coomassie Brilliant Blue staining—left) and CPT-I, ACS1, and VDAC isolated by electro-elution. Analytical SDS–PAGE/Western blot analysis of electroeluted proteins using CPT-I, ACS, and VDAC1 antibodies and anti-phospho-serine, anti-phospho-threonine, and anti-phospho-tyrosine antibodies (right). Antibodies against ACS (Hoppel *et al.*, 2001) and VDAC1 (Distler *et al.*, 2006) were prepared as referenced.

3.2. Reduction, alkylation, and enzymatic digestion of the protein

Since the electroeluted and denatured hydrophobic membrane proteins could not be resolubilized, the reduction, alkylation, and digestion of the proteins is performed on the precipitated proteins with constant gentle stirring using a magnetic stir plate to keep the proteins in-suspension. The precipitated, washed proteins are resuspended in 100 μL of 50 mM ammonium bicarbonate. To the suspension, 10 μL of 20 mM dithiothreitol in 50 mM ammonium bicarbonate is added and incubated for 1 h at 55 °C. The suspension is centrifuged for 10 min (13,200g) and the supernatant removed. Again, the pellet is resuspended in 100 μL of 50 mM ammonium bicarbonate and 20 μL of 100 mM iodoacetamide in 50 mM ammonium bicarbonate is added. This suspension is incubated in the dark for 30 min at room temperature. The suspension is centrifuged and the supernatant is removed. The pellet is washed with 80% ethanol to remove any excess alkylating agent.

The in-suspension tryptic digestion of the isolated proteins is carried out at 37 °C in 50 mM ammonium bicarbonate/1.0 mM CaCl$_2$ in a final volume of 100 μL with 1 μg trypsin added. After 3 h of digestion, another 1 μg of trypsin was added and the digestion continued overnight. After overnight digestion, the reaction mixture is centrifuged (13,200g) for 10 min and the supernatant is removed. Frequently in proteomic analyses, trypsin is used to digest proteins before mass spectrometric analysis. Trypsin cleaves peptide chains predominantly at the carboxyl side of the amino acids lysine and arginine. While appropriate for many proteomic applications, a less specific protease enhances the protein digestion especially for hydrophobic regions that contain few basic residues. To increase chain coverage, after the supernatant containing the tryptic peptides is removed for mass spectrometric analysis, the undigested protein pellet is further digested with the less specific proteinase K enzyme. The latter is a nonspecific enzyme useful for analyzing membrane proteins and detecting post-translational modifications (MacCoss *et al.*, 2002; Wu *et al.*, 2003). The remaining, partially digested pelleted protein is further treated with proteinase K. Fifty microliters of water are added to the protein pellet and a suspension made using constant stirring. Five micrograms of proteinase K are added and the solution incubated for 4 h at room temperature. Again, the digestion of the protein pellet is facilitated by constant, gentle stirring. After 4 h, the solution is centrifuged (13,200g) and the supernantant removed for mass spectrometric analysis.

3.3. Detection of post-translational modifications using mass spectrometry

Electrospray spectra are acquired on a ThermoElectron (San Jose, CA) LTQ linear quadrupole ion trap mass spectrometer coupled to a ThermoElectron Surveyor liquid chromatograph equipped with a PepMap C18 column from LC Packings (Sunnyvale, CA). The flow rate is approximately 200 nl/min. The samples are diluted to concentrations of 100–500 fmol/μL and 2 μL of each sample are injected. The mobile phases are 0.1% formic acid in HPLC grade water and 0.1% formic acid in HPLC grade acetonitrile. Sample analyses are performed using data-dependent/dynamic exclusion with a repeat count of one in order to maximize coverage. The LTQ method employs eight full MS/MS scans following one full scan mass spectrum (400–2,000 Da). All data files are searched against the NR database using Bioworks (ThermoElectron) as described previously (Distler *et al.*, 2006, 2007; Kerner *et al.*, 2004). Phosphorylation (serine, threonine, and tyrosine), nitration (tyrosine), oxidation (methionine), and acetylation (N-terminus of protein, lysine) are selected as possible differential modifications.

4. Focused Proteomics Using Immunological Approaches

If membrane proteins could be isolated without using gel electrophoresis, the detection of post-translational modifications might be increased. To isolate a mitochondrial protein, an immunological approach can be developed. While this method can be adapted for any protein, we will here demonstrate the technique using antibodies raised against the liver isoform of the CPT-I protein.

4.1. Antibody preparation and immobilization

To efficiently isolate CPT-I from the limited amount of mitochondria available, the following approach is used. Peptides corresponding to the amino acid regions 230–245 (NYVSDWWEEYIYLRGR) and 620–631 (MDPKSTAEQRLK) of the liver CPT-I protein are custom synthesized with a carboxy-terminal cysteine added to provide a thiol linkage to keyhole limpet hemocyanin for immunization and coupling to SulfoLink resin for affinity purification of specific antibodies. For immunization, the peptides (5 mg) are coupled to maleimide-activated keyhole limpet hemocyanin (5 mg) as recommended by the manufacturer (Pierce). The antigens are emulsified with Freund's complete adjuvant and injected into rabbits (two/ antigen). Booster injections in Freund's incomplete adjuvant are given 5 weeks following immunization. To affinity purify the polyclonal antibodies, the peptides are covalently linked to Sulfolink (Pierce) via a stable thioether linkage. The CPT-I-peptide specific antibodies are then isolated from the antiserum by means of immunoaffinity chromatography, with 100 mM glycine/HCl, pH 2.7 as the eluent. The collected fractions are immediately neutralized with a predetermined volume of 1 M Tris (pH 8.0).

4.2. Isolation of CPT-I

To isolate the CPT-I protein, both immunocapture and immunoprecipitation are used. For these immunological approaches to protein purification, the same starting material is used. Rat liver mitochondria are isolated by Percoll-purification and the outer membranes isolated in high purity by the swell/shrink technique and discontinuous sucrose gradient centrifugation as described above. The outer membrane (1 mg) is lysed with 200 μL of lysis buffer containing 2% Lubrol-PX, 25 mM sodium phosphate, 500 mM NaCl, 1 tablet each of Roche protease inhibitor cocktail and phosphatase inhibitor cocktail in 7 mL (pH 7.1) for 30 min on ice in phosphate buffer. After 30 min, the outer membranes are centrifuged in an airfuge (Beckman) with 30 psi of pressure (160,000g) for 10 min. The pellet is discarded and the supernatant used for immunoprecipitation and immunocapture.

Immunoprecipitation. For immunoprecipitation, 20 μL GammaBind Plus slurry (GE Healthcare) is washed two times with a total of 800 μL binding buffer (10 mM sodium phosphate, 150 mM sodium chloride, and 10 mM EDTA). After each wash, the slurry is centrifuged for 1 min (100g) and the supernatant is discarded. The slurry is incubated with the affinity-purified CPT-I antibodies using end-over-end rotation for 2 h in a spin column (Pierce). The beads are washed with a total of 900 μL of binding buffer three times and then incubated with the extracted mitochondrial outer membrane proteins for 2 h at room temperature. The beads are washed three times with a total of 900 μL binding buffer containing 0.1% Lubrol-PX and an additional three times with a total of 900 μL binding buffer without detergent. The immunoprecipitated proteins are incubated with 100 μL of 0.1 M glycine (pH 2.8) at room temperature for 10 min while shaking. To elute the immunoprecipitated proteins, the Pierce spin column is placed in a collection tube containing 10 μL 1 M Tris (pH 8.0) to neutralize the pH of the eluted fraction. The spin column is centrifuged for 1 min (100g) and the eluted fractions are kept on ice.

Immunocapture. For immunocapture, CPT-I antibodies are treated with mercaptoethylamine to selectively reduce the disulfide bridges in the hinge region to yield half IgGs ("rIgG"). To form a stable thioether linkage, the reduced antibodies are then immobilized onto UltraLink Iodoacetyl gel via the free sulfhydryls of IgG proteins as recommended by the manufacturer (Pierce). The mitochondrial outer membrane proteins are incubated with gentle end-over-end mixing on a rotator for 2 h to capture CPT-I and associated proteins by the immobilized antibodies. The gel containing the captured proteins is washed three times with a total of 1 mL of 0.5 M NaCl containing 0.05% Lubrol-PX to remove nonspecifically bound proteins. The gel is additionally washed three times with a total of 900 μL of PBS (pH 7). The captured proteins are eluted with 100 μL of 0.1 M glycine (pH 2.8) as described above for immunoprecipitation. Again, 1 M Tris (pH 8.0) is used to neutralize the glycine solution and the eluted fractions are kept on ice.

Gel electrophoresis and Western-blotting. To determine the efficiency and purity of the immunocaptured and immunoprecipitated proteins, gel electrophoresis and Western-blotting are performed. The eluents are mixed with 4× SDS–PAGE sample buffer and heated for 5 min at 85 °C before loading onto a 10% acrylamide gel. The gel electrophoresis is carried out at 180 V for 50 min. The separated proteins are then silver stained or transferred to a polyvinylidene difluoride membrane (BioRad) for Western blot analysis (Fig. 6.3). The membrane is blocked with 5% non-fat milk in Tris buffered saline (25 mM Tris, 500 mM sodium chloride, pH 7.5) containing 0.2% Tween-20 and probed with the affinity-purified CPT-I antibody

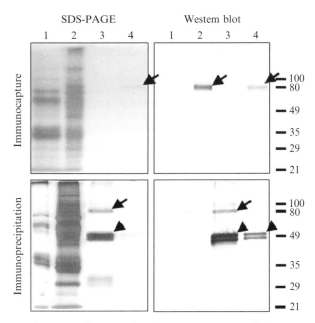

Figure 6.3 Purification of CPT-I by immunocapture and immunoprecipitation. The images shown here are the SDS–PAGE gels with silver staining on the left and the Western blot on the right. (1) Mitochondrial outer membrane pellet after Lurol-PX extraction. (2) Lubrol-PX extract. (3) First elution of immunoaffinity beads and immunoprecipitation. (4) Second elution of immunoaffinity beads and immunoprecipitation. The arrowheads and arrows represent CPT-I antibody and CPT-I, respectively. Molecular weight markers are shown at the right.

(1:2000) raised against the amino acid sequence from 230 to 245. The membrane is then incubated with alkaline phosphatase-conjugated secondary goat anti-rabbit antibody (1:4000) and developed with 5-bromo-4-chloro-3-indolyl phosphate and nitroblue tetrazolium solution (BioRad) for visualization (Fig. 6.3).

4.3. Mass spectrometric analysis of the isolated CPT-I

For analysis of the CPT-I protein by mass spectrometry, the protein is subjected to enzymatic digestion with trypsin. The digestions are carried out as described above in Section 3.2. Tryptic digestion is carried out at 37 °C in 50 mM ammonium bicarbonate overnight with additional trypsin added after 4 h for a total of 1 μg. The digested sample is evaporated to dryness in a vacuum evaporator. The tryptic peptides are reconstituted in 0.1% trifluoroacetic acid and purified with a C-18 ZipTip (Millipore) according to the manufacturer's instructions. Prior to LC-MS/MS analysis, the samples were analyzed using MALDI-TOF MS. For the

```
  1 MAEAHQAVAF QFTVTPDGID LRLSHEALKQ ICLSGLHSWK KKFIRFKNGI
 51 ITGVFPANPS SWLIVVVGVI SSMHAKVDPS LGMIAKISRT LDTTGRMSSQ
101 TKNIVSGVLF GTGLWVAVIM TMRYSLKVLL SYHGWMFAEH GKMSRSTKIW
151 MAMVKVLSGR KPMLYSFQTS LPRLPVPAVK DTVSRYLESV RPLMKEEDFQ
201 RMTALAQDFA VNLGPKLQWY LKLKSWWATN YVSDWWEEYI YLRGRGPLMV
251 NSNYYAMEML YITPTHIQAA RAGNTIHAIL LYRRTLDREE LKPIRLLGST
301 IPLCSAQWER LFNTSRIPGE ETDTIQHIKD SRHIVVYHRG RYFKVWLYHD
351 GRLLRPRELE QQMQQILDDP SEPQPGEAKL AALTAADRVP WAKCRQTYFA
401 RGKNKQSLDA VEKAAFFVTL DESEQGYREE DPEASIDSYA KSLLHGRCFD
451 RWFDKSITFV VFKNSKIGIN AEHSWADAPV VGHLWEYVMA TDVFQLGYSE
501 DGHCKGDTNP NIPKPTRLQW DIPGECQEVI DASLSSASLL ANDVDLHSFP
551 FDSFGKGLIK KCRTSPDAFI QLALQLAHYK DMGKFCLTYE ASMTRLFREG
601 RTETVRSCTM ESCNFVQAMM DPKSTAEQRL KLFKIACEKH QHLYRLAMTG
651 AGIDRHLFCL YVVSKYLAVD SPFLKEVLSE PWRLSTSQTP QQQVELFDFE
701 KNPDYVSCGG GFGPVADDGY GVSYIIVGEN FIHFHISSKF SSPETDSHRF
751 GKHLRQAMMD IITLFGLTIN SKK
```

Figure 6.4 Coverage of the immunopurified CPT-I protein by LC-MS/MS and MALDI-TOF MS.

MALDI experiments, a 10 mg/mL solution of α-cyano-4-hydroxycinnamic acid (CHCA) was made using 50% acetonitrile and 0.1% TFA. The solution containing the purified tryptic peptides (0.5 µL) was combined with the matrix solution (0.5 µL) and the mixture was spotted on the MALDI sample plate. The samples are then analyzed by MALDI-TOF MS using a prOTOF 2000 (Perkin Elmer) MALDI mass spectrometer. After analysis by MALDI, the solution containing the tryptic peptides was analyzed by LC-MS/MS using a LTQ-OrbiTrap (ThermoFisher) coupled to an Ultimate 3000 (Dionex) using the gradient described in Section 3.3. When the tryptic peptides were analyzed by both MALDI-TOF MS and ESI-LC-MS/MS, 38.6% of the protein is covered by the two techniques, Fig. 6.4.

Isolation of phosphopeptides. Many post-translational modifications are present in sub-stoichiometric quantities. To enhance the detection of phosphopeptides, they must be selectively purified from the more abundant, unmodified peptides. With this approach the negatively charged phosphorylated peptides are purified by their affinity to metal ions like Fe^{3+} or Ga^{3+}. However, non-phosphorylated peptides, including those containing multiple acidic residues, frequently are also enriched by this method (Ficarro *et al.*, 2002). Recently, titanium dioxide (Klemm *et al.*, 2006) and zirconium dioxide columns are used to bind phosphorylated peptides. Detection of mono-phosphorylated peptides is improved with the use of ZrO_2 and multiply phosphorylated peptides have improved detection with the use of TiO_2 tips (Kweon and Håkansson, 2006).

The porous titanium and zirconium oxide microtips are purchased from Glygen (Columbia, MD). The samples are prepared in 1% formic acid. The sample volume is then split into two parts for purification by either the TiO_2 or ZrO_2 tips. The microtip is conditioned by aspirating and expelling 50 µL of 1% formic acid. This is repeated four times. The sample (in 1% formic acid)

Table 6.1 Potential phosphopeptides identified by MALDI-TOF MS using immunopurified rat liver mitochondrial CPT-I

Experimental [M + H]$^+$	Theoretical [M + H]+	Missed cleavage	Position	Sequence
1110.5243	1110.5265	0	77–86	(K)VDPSLGMIAK(I)
1870.8198	1870.8346	0	128–142	(K)VLLSYHGWMF AEHGK(M)
1992.0016	1991.9582	0	756–772	(R)QAMMDIITLF GLTINSK(K)
2014.8364	2014.8718	1	640–655	(K)HQHLYRLAMT GAGIDR(H)
2347.0524	2347.0107	1	23–40	(R)LSHEALKQICLS GLHSWK(K)
2462.1385	2462.1737	1	753–772	(K)HLRQAMMDII TLFGLTINSK(K)
2504.1191	2504.1284	0	103–123	(K)NIVSGVLFGTG LWVAVIMTMR(Y)

is aspirated and expelled 20 times in order to adsorb the phosphopeptides to the metal oxide surface. To wash the sample, 50 μL of 1% formic acid is aspirated and expelled to waste. This washing is repeated 10 times to eliminate salts, lipids, detergents, and non–phosphorylated peptides. To elute the phosphorylated peptides from the metal oxide surface, 10 μL of 1% ammonium hydroxide is aspirated and expelled at least 10 times into a clean tube. The solution is immediately neutralized with formic acid to prevent the hydrolysis of the phosphate group from the peptide. When the immunopurified CPT-I was purified using a zirconium oxide column, several potentially phosphorylated peptides were identified as shown in Table 6.1.

5. SHOTGUN METHODS FOR IDENTIFICATION OF POST-TRANSLATIONAL MODIFICATIONS

To decrease the amount of sample handling, mass spectrometry is used to directly analyze membrane proteins. Shotgun methods involving direct digestion and analysis of a protein mixture (without any prior separation) are a viable option for membrane protein analysis (Washburn et al., 2001; Wu et al., 2003). Shotgun methods commonly use LC to separate peptides before MS analysis circumventing the need for gel electrophoresis and keeping the proteins/peptides in solution. Shotgun methods can easily be adapted for the detection of post-translational modifications.

Purification methods designed to selectively purify modified peptides can be employed. For example, in a recent study, the acetylation of proteins was examined by using an affinity-purified anti-acetyllysine antibody to purify those peptides containing an acetylated lysine (Kim *et al.*, 2006). Lysine acetylation was found to be abundant in the mitochondria of mouse liver with 277 unique acetylation sites in 133 proteins in mitochondria of fed and fasted mice.

5.1. Digestion of proteins

For these experiments, the Percoll-purified mitochondria are isolated and reconstituted in the presence of protein phosphatase inhibitors as described above. The mitochondria (1 mg) are treated with 10 μg of trypsin with constant gentle stirring on a magnetic stir plate. After 4 h, another 10 μg of trypsin is added to the mitochondria and incubated overnight at 37 °C. The mitochondria are centrifuged (10 min at 7,000g) and the supernatant is removed and frozen.

5.2. Mass spectrometric analysis of digestion

The phosphorylated peptides from the mitochondrial digestion are purified using titanium and zirconium columns as described above. The supernatant from Section 5.1 is divided into two parts with half of the sample being purified using a titanium column and the other half using a zirconium column. The peptides retained and subsequently eluted off the titanium and zirconium columns were analyzed by LC-MS/MS using the mass spectrometry method described above. While there were several peptides identified without sites of phosphorylation, there were peptides identified as phosphorylated from several known mitochondrial proteins including the five phosphorylated peptides from CPT-I, listed in Table 6.2. An example of an MS/MS spectrum from one of the peptides is shown in Fig. 6.5.

Table 6.2 Phosphorylated rat liver CPT-I peptides detected using a global proteomics approach

Sequence	Purification method
[295]LLGS(phos)TIPLCSAQWER[309]	Titanium
[379]LAALT(phos)AADR[309]	Titanium
[455]SIT(phos)FVVFK[309]	Titanium
[505]GDT(phos)NPNIPKPTR[516]	Titanium
[606]SCT(phos)MESCNFVQAMMDPK[622]	Zirconium

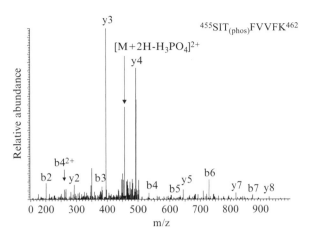

Figure 6.5 MS/MS spectrum of the peptide $^{455}SIT_{(phos)}FVVFK^{462}$ purified using a titanium oxide microcolumn as shown in Table 6.1.

6. CONCLUSION

Despite the state-of-the-art technologies currently in use, there is no single technique that can be used for all studies of mitochondrial membrane proteins. Here we report several methods to enhance the analysis of mitochondrial outer membrane proteins. Detection of hydrophobic mitochondrial membrane proteins is facilitated by the removal of the soluble matrix and intermembrane space proteins by isolation of the resident membranes. To determine sites of post-translational modifications on the membrane proteins, electrophoretic, immunological, and global proteomic approaches are presented here. To improve the coverage of mitochondrial membrane proteins, further advances are needed that use multiple proteases, fractionation of the hydrophobic proteins, and affinity purification methods. While post-translational modifications of mitochondrial membrane proteins can be identified using the approaches presented here, the functional impact these modifications have on the biological processes in the mitochondria is of great relevance. The future offers great promise in deepening our understanding of the effects of post-translational modifications on mitochondrial membrane proteins.

ACKNOWLEDGMENTS

We thank William Parland, Ph.D. for providing the mitochondrial fractions and Bernard Tandler, Ph.D. for his editorial assistance. This work was partially supported by NIH grants DK-066107 and PO1 AG-15885 as well as the Medical Research Service of the Department of Veterans Affairs.

REFERENCES

Bagshaw, R. D., Mahuran, D. J., and Callahan, J. W. (2005). A proteomics analysis of lysosomal integral-membrane proteins reveals the diverse composition of the organelle. *Mol. Cell Proteomics* **4,** 133–143.

Barnouin, K. (2004). Two-dimensional gel electrophoresis for analysis of protein complexes. *Methods Mol. Biol.* **261,** 479–498.

Bell, A. W., Ward, M. A., Blackstock, W. P., Freeman, H. N., Choudhary, J. S., Lewis, A. P., Chotai, D., Fazel, A., Gashue, J. N., Paiement, J., Palcy, S., Chevet, E., *et al.* (2001). Proteomics characterization of abundant Golgi membrane proteins. *J. Biol. Chem.* **276,** 5152–5165.

Benz, R. (1994). Permeation of hydrophilic solutes through mitochondrial outer membranes: Review on mitochondrial porins. *Biochim. Biophys. Acta* **1197,** 167–196.

Bernardi, P. (1999). Mitochondrial transport of cations: Channels, exchangers, and permeability transition. *Physiol. Rev.* **79,** 1127–1155.

Bionda, C., Portoukalian, J., Schmitt, D., Rodriguez-Lafrasse, C., and Ardail, D. (2004). Subcellular compartmentalization of ceramide metabolism: MAM (mitochondria-associated membrane) and/or mitochondria? *Biochem. J.* **382,** 527–533.

Braun, R. J., Kinkl, N., Beer, M., and Ueffing, M. (2007). Two-dimensional electrophoresis of membrane proteins. *Anal. Bioanal. Chem.* **389,** 1033–1045.

Bruserud, O., Stapnes, C., Tronstad, K. J., Ryningen, A., Anensen, N., and Gjertsen, B. T. (2006). Protein lysine acetylation in normal and leukaemic haematopoiesis: HDACs as possible therapeutic targets in adult AML. *Expert Opin. Ther. Targets* **10,** 51–68.

Buhler, S., Michels, J., Wendt, S., Rück, A., Brdiczka, D., Welte, W., and Przybylski, M. (1998). Mass spectrometric mapping of ion channel proteins (porins) and identification of their supramolecular membrane assembly. *Proteins* **30**(Suppl 2), 63–73.

Caesar, R., Warringer, J., and Blomberg, A. (2006). Physiological importance and identification of novel targets for the N-terminal acetyltransferase NatB. *Eukaryot. Cell* **5,** 368–378.

Cheng, E. H.-Y., Sheiko, T., Fisher, J. K., Craigen, W. J., and Korsmeyer, S. J. (2003). VDAC2 inhibits BAK activation and mitochondrial apoptosis. *Science* **301,** 513–517.

Crouser, E. D., Julian, M. W., Huff, J. E., Mandich, D. V., and Green-Church, K. B. (2006). A proteomic analysis of liver mitochondria during acute endotoxemia. *Intensive Care Med.* **32,** 1252–1262.

Da Cruz, S., and Martinou, J. C. (2008). Purification and proteomic analysis of the mouse liver mitochondrial inner membrane. *Methods Mol. Biol.* **432,** 101–116.

Da Cruz, S., Xenarios, I., Langridge, J., Vilbois, F., Parone, P. A., and Martinou, J. C. (2003). Proteomic analysis of the mouse liver mitochondrial inner membrane. *J. Biol. Chem.* **278,** 41566–41571.

Distler, A. M., Kerner, J., and Hoppel, C. L. (2007). Post-translational modifications of rat liver mitochondrial outer membrane proteins identified by mass spectrometry. *Biochim. Biophys. Acta* **1774,** 628–636.

Distler, A. M., Kerner, J., and Hoppel, C. L. (2008). Proteomics of mitochondrial inner and outer membranes. *Proteomics* **8,** 4066–4082.

Distler, A. M., Kerner, J., Peterman, S. M., and Hoppel, C. L. (2006). A targeted proteomic approach for the analysis of rat liver mitochondrial outer membrane proteins with extensive sequence coverage. *Anal. Biochem.* **356,** 18–29.

Ficarro, S. B., McCleland, M. L., Stukenberg, P. T., Burke, D. J., Ross, M. M., Shabanowitz, J., Hunt, D. F., and White, F. M. (2002). Phosphoproteome analysis by mass spectrometry and its application to *Saccharomyces cerevisiae*. *Nat. Biotech.* **20,** 301–305.

Green, D. R., and Kroemer, G. (2004). The pathophysiology of mitochondrial cell death. *Science* **305,** 626–629.

Green, D. R., and Reed, J. C. (1998). Mitochondria and apoptosis. *Science* **281**, 1309–1312.

Hellman, U., Wernstedt, C., Gonez, J., and Heldin, C. H. (1995). Improvement of an "In-Gel" digestion procedure for the micropreparation of internal protein fragments for amino acids sequencing. *Anal. Biochem.* **224**, 451–455.

Hoppel, C. L., DiMarco, J. P., and Tandler, B. (1979). Riboflavin and rat hepatic cell structure and function: Mitochondrial oxidative metabolism in deficiency states. *J. Biol. Chem.* **254**, 4164–4170.

Hoppel, C., Kerner, J., Turkaly, P., Minkler, P., and Tandler, B. (2002). Isolation of hepatic mitochondrial contact sites: Previously unrecognized inner membrane components. *Anal. Biochem.* **302**, 60–69.

Hoppel, C., Kerner, J., Turkaly, P., and Tandler, B. (2001). Rat liver mitochondrial contact sites and carnitine palmitoyltransferase-I. *Arch. Biochem. Biophys.* **392**, 321–325.

Hoppel, C. L., Kerner, J., Turkaly, P., Turkaly, J., and Tandler, B. (1998). The malonyl-CoA-sensitive form of carnitine palmitoyltransferase is not localized exclusively in the outer membrane of rat liver mitochondria. *J. Biol. Chem.* **273**, 23495–23503.

Issaq, H., and Veenstra, T. (2008). Two-dimensional polyacrylamide gel electrophoresis (2D-PAGE): Advances and perspectives. *Biotechniques* **44**, 697–700.

Kerner, J., Distler, A. M., Minkler, P., Parland, W., Peterman, S. M., and Hoppel, C. L. (2004). Phosphorylation of rat liver mitochondrial carnitinepalmitoyltransferase-I: Effect on the kinetic properties of the enzyme. *J. Biol. Chem.* **279**, 41104–41113.

Kim, S. C., Sprung, R., Chen, Y., Xu, Y., Ball, H., Pei, J., Cheng, T., Kho, Y., Xiao, H., Xiao, L., Grishin, N. V., White, M., *et al.* (2006). Substrate and functional diversity of lysine acetylation revealed by a proteomics survey. *Mol. Cell* **23**, 607–618.

Klemm, C., Otto, S., Wolf, C., Haseloff, R. F., Beyermann, M., and Krause, E. (2006). Evaluation of the titanium dioxide approach for MS analysis of phosphopeptides. *J. Mass Spectrom.* **41**, 1623–1632.

Knoblach, B., Keller, B. O., Gronendyk, J., Aldred, S., Zheng, J., Lemire, B. D., Li, L., and Michalak, M. (2003). Erp19 and Erp46, new membrers of the thioredoxin family of endoplasmic reticulum proteins. *Mol. Cell. Proteomics* **2**, 1104–1119.

Kweon, H. K., and Håkansson, K. (2006). Selective zirconium dioxide-based enrichment of phosphorylated peptides for mass spectrometric analysis. *Anal. Chem.* **78**, 1743–1749.

Linden, M., and Karlsson, G. (1996). Identification of porin as a binding site for MAP2. *Biochem. Biophys. Res. Commun.* **218**, 833–836.

MacCoss, M. J., McDonald, W. H., Saraf, A., Sadygov, R., Clark, J. M., Tasto, J. J., Gould, K. L., Wolters, D., Washburn, M., Weiss, A., Clark, J. I., and Yates, J. R. 3rd. (2002). Shotgun identification of protein modifications from protein complexes and lens tissue. *Proc. Natl. Acad. Sci. USA* **99**, 7900–7905.

Marelli, M., Smith, J. J., Jung, S., Yi, E., Nesvizhskii, A. I., Christmas, R. H., Saleem, R. A., Tam, Y. Y., Fagarasanu, A., Goodlett, D. R., Aebersold, R., Rachubinski, R. A., *et al.* (2004). Quantitative mass spectrometry reveals a role for the GTPase Rho1p in actin organization on the peroxisome membrane. *J. Cell. Biol.* **167**, 1099–1112.

Mootha, V. K., Bunkenborg, J., Olsen, J. V., Hjerrild, M., Wisniewski, J. R., Stahl, E., Bolouri, M. S., Ray, H. N., Sihag, S., Kamal, M., Patterson, N., Lander, E. S., *et al.* (2003). Integrated analysis of protein composition, tissue diversity, and gene regulation in mouse mitochondria. *Cell* **115**, 629–640.

Parsons, D. F., Williams, G. R., and Chance, B. (1966). Characteristics of isolated and purified preparations of the outer and inner membranes of mitochondria. *Ann. NY Acad. Sci.* **137**, 643–666.

Paumen, M. B., Ishida, Y., Han, H., Muramatsu, M., Eguchi, Y., Tsujimoto, Y., and Honjo, T. (1997). Direct interaction of the mitochondrial membrane protein carnitine palmitoyltransferase I with Bcl-2. *Biochem. Biophys. Res. Commun.* **231**, 523–525.

Perrier, J., Durand, A. G., Hiardina, T., and Puigserver, A. (2005). Catabolism of intracellular N-terminal acetylated proteins: Involvement of acylpeptide hydrolase and acylase. *Biochimie* **87**, 673–685.

Pizzo, P., and Pozzan, T. (2007). Mitochondria–endoplasmic reticulum choreography: Structure and signaling dynamics. *Trends Cell Biol.* **17**, 512–517.

Polevoda, B., Norbeck, J., Takakura, H., Blomberg, A., and Sherman, F. (1999). Identification and specificities of N-terminal acetyltransferases from *Saccharomyces cerevisiae*. *EMBO J.* **18**, 6155–6168.

Prokisch, H., Scharfe, C., Camp, D. G., Xiao, W., David, L., Andreoli, C., Monroe, M. E., Moore, R. J., Gritsenko, M. A., Kozany, C., Hixson, K. K., Mottaz, H. M., et al. (2004). Integrative analysis of the mitochondrial proteome in yeast. *PLoS Biol.* **2**, e160.

Rabilloud, T. (2002). Two-dimensional gel electrophoresis in proteomics: Old, old fashioned, but it still climbs up the mountains. *Proteomics* **2**, 3–10.

Santoni, V., Molloy, M., and Rabilloud, T. (2000). Membrane proteins and proteomics: *Un amour* impossible? *Electrophoresis* **21**, 1054–1070.

Schmitt, S., Prokisch, H., Schlunck, T., Camp, D. G., Ahting, U., Waizenegger, T., Scharfe, C., Meitinger, T., Imhof, A., Neupert, W., Oefner, P. J., and Rapaport, D. (2006). Proteome analysis of mitochondrial outer membrane from *Neurospora crassa*. *Proteomics* **6**, 72–80.

Shimizu, S., Matsuoka, Y., Shinohara, Y., Yoneda ., and Tsujimoto, Y. (2001). Essential role of voltage-dependent anion channel in various forms of apoptosis in mammalian cells. *J. Cell Biol.* **152**, 237–250.

Sickmann, A., Reinders, J., Wagner, Y., Joppich, C., Zahedi, R., Meyer, H. E., Schonfisch, B., Perschil, I., Charcinska, A., Guiard, B., Rehling, P., Pfanner, N., et al. (2003). The proteome of *Saccharomyces cereviase* mitochondria. *Proc. Natl. Acad. Sci. USA* **100**, 13207–13212.

Spierings, D., McStay, G., Saleh, M., Bender, C., Chipuk, J., Maurer, U., and Green, D. R. (2005). Connected to death: The (unexpurgated) mitochondrial pathway of apoptosis. *Science* **310**, 66–67.

Taylor, S. W., Fahy, E., Zhang, B., Glenn, G. M., Warnock, D. E., Wiley, S., Murphy, A. N., Gaucher, S. P., Capaldi, R. A., Gibson, B. W., and Ghosh, S. S. (2003). Characterization of the human heart mitochondrial proteome. *Nat. Biotechnol.* **21**, 281–286.

Taylor, R. S., Wu, C. C., Hays, L. G., Eng, J. K., Yates, J. R. 3rd, and Howell, K. E. (2000). Proteomics of rat liver Golgi complex: Minor proteins are identified through sequential fractionation. *Electrophoresis* **21**, 3441–3459.

Turkaly, P., Kerner, J., and Hoppel, C. L. (1999). A 22 kDa polyanion inhibits carnitine-dependent fatty acid oxidation in rat liver mitochondria. *FEBS Lett.* **460**, 241–245.

Vavvas, D., Apazidis, A., Saha, A. K., Gamble, J., Patel, A., Kemp, B. E., Witters, L. A., and Ruderman, N. B. (1997). Contraction-induced changes in acetyl-CoA carboxylase and 5′-AMP-activated kinase in skeletal muscle. *J. Biol. Chem.* **272**, 13255–13261.

Vyssokikh, M., and Brdiczka, D. (2004). VDAC and peripheral channelling complexes in health and disease. *Mol. Cell. Biochem.* **256–257**, 117–126.

Washburn, M. P., Wolters, D., and Yates, J. R. 3rd. (2001). Large-scale analysis of the yeast proteome by multidimensional protein identification technology. *Nat. Biotech.* **19**, 242–305.

Wu, C. C., MacCoss, M. J., Howell, K. E., and Yates, J. R. 3rd. (2003). A method for the comprehensive proteomic analysis of membrane proteins. *Nat. Biotech.* **21**, 532–538.

Wu, C. C., MacCoss, M. J., Mardones, G., Finnigan, C., Mogelsvang, S., Yates, J. R. 3rd, and Howell, K. E. (2004). Organellar proteomics reveals golgi arginine dimethylation. *Mol. Biol. Cell* **15**, 2907–2919.

Yang, X. J. (2004). Lysine acetylation and the bromodomain: A new partnership for signaling. *Bioessays* **26,** 1076–1087.

Zahedi, R. P., Sickmann, A., Boehm, A. M., Winkler, C., Zufall, N., Schönfisch, B., Guiard, B., Pfanner, N., and Meisinger, C. (2006). Proteomic analysis of the yeast mitochondrial outer membrane reveals accumulation of a subclass of preproteins. *Mol. Biol. Cell* **17,** 1436–1450.

CHAPTER SEVEN

Analysis of Tyrosine-Phosphorylated Proteins in Rat Brain Mitochondria

Urs Lewandrowski,* Elena Tibaldi,† Luca Cesaro,†
Anna M. Brunati,† Antonio Toninello,† Albert Sickmann,‡
and Mauro Salvi†

Contents

1. Introduction   118
2. Identification of Src Tyrosine Kinase Substrates in Rat
   Brain Mitochondria   119
   2.1. Purification of Src family tyrosine kinase Lyn and Fgr   119
   2.2. Isolation and purification of mitochondria from rat brain   122
   2.3. Tyrosine phosphorylation of mitochondrial proteins   123
3. Mass Spectrometric Analysis of Tyrosine-Phosphorylated Peptides
   in Mitochondria   125
   3.1. Enrichment of phosphopeptides   125
   3.2. Mass spectrometric detection   128
4. Bioinformatic Tools to Analyze the Potential Role of Tyrosine
   Phosphorylation   131
Acknowledgment   134
References   134

Abstract

Mitochondrial protein phosphorylation is emerging as a central event in mitochondrial signaling. In particular, tyrosine phosphorylation is proving to be an unappreciated mechanism involved in regulation of mitochondrial functions. Tyrosine kinases and phosphatases have been identified in mitochondrial compartments and there is a steadily increasing number of new identified tyrosine-phosphorylated proteins implicated in a wide spectrum of mitochondrial functions. The deciphering of the tyrosine phoshorylation signaling in mitochondria is strictly linked to the definition of the entire mitochondrial tyrosine phosphoproteome.

* ISAS–Institute for Analytical Sciences, Dortmund, Germany
† Department of Biological Chemistry, University of Padova, Padova, Italy
‡ ISAS–Institute for Analytical Sciences, Dortmund, Germany and Medizinisches Proteom-Center (MPC),
  Ruhr-Universitaet Bochum, Bochum, Germany

Methods in Enzymology, Volume 457
ISSN 0076-6879, DOI: 10.1016/S0076-6879(09)05007-1

This chapter describes methods to analyze tyrosine phosphorylation in brain mitochondria: identification of new substrates by biochemical and mass spectrometry approaches and bioinformatic tools to analyze the potential effect of tyrosine phosphorylation on the structure/activity of a protein.

1. INTRODUCTION

Overlooked until recently, mitochondrial protein phosphorylation is emerging as a central event in mitochondrial signaling and in particular Ser/Thr phosphorylation seems to be implicated in different mitochondrial processes (Pagliarini and Dixon, 2006). Tyrosine phosphorylation is also emerging as an important mechanism in regulating mitochondrial functions, and the involvement of tyrosine phosphorylation in mitochondrial signaling is discussed in depth in a previous review (Salvi *et al.*, 2005).

Thus far only tyrosine kinases belonging to the Src family have been described in mitochondrial compartments such as Lyn, c-Src, Fyn, and Fgr (Itoh *et al.*, 2005; Livigni *et al.*, 2006; Miyazaki *et al.*, 2003; Salvi *et al.*, 2002) and two different mitochondrial anchoring proteins for c-src, AKAP121 (Livigni *et al.*, 2006), and Dok-4 (Itoh *et al.*, 2005) have been described. Tyrosine phosphatases have also been identified: shp-2 has been described in mitochondrial compartments (Salvi *et al.*, 2004) and recently, the PTPM-1 (PTP localized to the Mitochondrion 1), has been discovered as the first tyrosine phosphatase with an almost exclusive mitochondrial localization (Pagliarini *et al.*, 2005).

Tyrosine phosphorylation has a primary role in regulating electron transport chain (ETC) activity. Several components of the ETC are subjected to tyrosine phosphorylation and the activity of ETC is modulated by the tyrosine phosphorylation level (Itoh *et al.*, 2005; Lee *et al.*, 2005; Livigni *et al.*, 2006; Miyazaki *et al.*, 2003; Pagliarini *et al.*, 2005).

We previously identified flavoprotein of succinate dehydrogenase and aconitase as specific substrates for Fgr tyrosine kinase by identifying the phosphorylation sites (Salvi *et al.*, 2007) and recently by a proteomic approach we identified new tyrosine-phosphorylated proteins strictly implicated in cell energy metabolism and in cell death thereby reporting a first draft of a mitochondrial tyrosine phosphoproteome (Lewandrowski *et al.*, 2008). This represents the tip of the iceberg and the definition of a mitochondrial tyrosine phosphoproteome is only at the very beginning. However, this inspection shows us that tyrosine-phosphorylated proteins are distributed to all mitochondrial compartments and may be involved in a wide range of mitochondrial functions from the ETC, Krebs cycle, fatty acids β-oxidation cycle, urea cycle, metabolite transport, and so on and could indeed represent a new frontier in mitochondrial signaling.

The deciphering of the tyrosine phoshorylation signaling in mitochondria is strictly linked to the definition of the entire mitochondrial tyrosine phosphoproteome.

This chapter describes methods for the identification of tyrosine-phosphorylated proteins, highlighting mass spectrometry requirements for analysis of tyrosine phosphorylated peptides and simple bioinformatic approaches useful to analyze the potential effect of tyrosine phosphorylation on the structure/activity of a protein.

2. IDENTIFICATION OF SRC TYROSINE KINASE SUBSTRATES IN RAT BRAIN MITOCHONDRIA

Several methods are available to identify phosphoproteins from large scale phosphoproteomics approaches to more specific ones.

First we report here an assay for the identification of tyrosine-phosphorylated mitochondrial proteins by just one specific tyrosine kinase. This method has been adapted from one developed by Knebel and Cohen to identify specific Ser/Thr kinase substrates in whole cell lysate (Cohen and Knebel, 2006; Knebel et al., 2001). This approach is based on the incubation of an exogenous kinase to a proteic lysate followed by the identification of the phoshorylated proteins. To reduce the complexity of the samples, the lysate is fractionated by classical chromatographic methods and each eluted fraction is assayed for the presence of specific substrates. The method requires $[\gamma\text{-}^{32}P]ATP$ of high specific radioactivity, high concentrations of an active kinase, and a short time of incubation (2–5 min) to reduce the phosphorylation background. Analyses are carried out using two closely related protein kinases with same activity to find specifically phosphorylated substrates by just one kinase (Cohen and Knebel, 2006; Knebel et al., 2001).

In the following method the highly active tyrosine kinases Fgr and Lyn, both belonging to Src kinases family, are added to fractions of mitochondrial lysate obtained by chromatographic procedures, and the tyrosine-phosphorylated proteins are analyzed. Bacterial-produced tyrosine kinases generally retain only modest activity. For this reason highly active tyrosine kinases are purified from rat spleen. We therefore describe here the method for purification of Fgr and Lyn tyrosine kinases, the isolation and purification of rat brain mitochondria and their fractionating, and finally the phosphorylation process and the analysis of the tyrosine-phosphorylated proteins.

2.1. Purification of Src family tyrosine kinase Lyn and Fgr

Twenty rat spleen are promptly removed and homogenized in the following buffer: 50 mM Tris/HCl (pH 7.5), 0.25 M sucrose, 5 mM EDTA, 1 mM EGTA, 0.1 mM vanadate, 0.1 mM phenylmethylsulfonyl fluoride (PMSF), using a Potter homogenizer with a Teflon pestle in an ice-water bath. Spleen homogenate is centrifuged at 10,000*g* for 10 min and the supernatant is newly centrifuged at 105,000*g* for 1 h to pellet the tyrosine kinase activity. The pellet is then extracted with 50 m*M* Tris/HCl (pH 7.5) including 10% glycerol, 1 m*M* EDTA, 10 m*M* β-mercaptoethanol, 3% Triton X-100 and 0.1 m*M* PMSF for 1 h. The lysate is then clarified to remove detergent-insoluble material by centrifuging at 105,000*g* for 1 h.

The supernatant is subjected to chromatography through a DEAE-Sepharose column (40 ml), equilibrated with buffer A: 25 m*M* Hepes (pH 7.0), 10% glycerol, 0.1% Triton X-100, 1 m*M* EDTA, 10 m*M* β-mercaptoethanol, and 0.1 m*M* PMSF. Proteins are eluted with a linear 0.0–0.5 *M* NaCl gradient in buffer A and 1-ml fractions are collected. Each fraction is assayed for tyrosine protein kinase (TPK) activity using the peptide cdc-2 as substrate (see below). Using this procedure, all tyrosine kinase activity is collected in three peaks (TPK-I, TPK-II, and TPK-III) eluting with 0.1, 0.18, and 0.26 *M* NaCl, respectively (Fig. 7.1).

2.1.1. Purification of Lyn tyrosine kinase

TPK-I and TPK-II fractions are combined, dialyzed against buffer A and subjected to affinity chromatography through a column of Heparin-Sepharose equilibrated with the same buffer.

The column (20 ml) is eluted with a linear 0–0.6 *M* NaCl gradient in buffer A; 1-ml fractions are collected and analyzed for TPK activity. Using this procedure, three peaks are obtained, the first one containing Lyn kinase, with a profile of elution of 0.3 *M* NaCl, is further purified on a polylysine-agarose column (10 ml), equilibrated in buffer A, and eluted with a linear 0–0.5 *M* NaCl gradient in buffer A; 0.4-ml fractions are collected. The peak

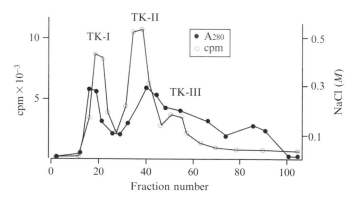

Figure 7.1 DEAE-Sepharose chromatography of rat spleen extracts. TK (tyrosine kinase activity).

eluting at about 0.18–0.2 M NaCl, is finally resolved on a ResourceQ/FPLC chromatography column.

The column, equilibrated with 20 mM Tris/HCl, (pH 7.5), 10% glycerol, 0.1 mM PMSF, 5 mM NaCl and 10 mM β-mercaptoethanol, is eluted at a flow rate of 0.5 ml/min with a linear NaCl gradient (0–0.5 M). A sharp peak of absorbance at 280 nm, overlapped by TPK activity, and corresponding to Lyn tyrosine kinase, is eluted at 0.15 M NaCl. Twenty microlitters of the eluted samples loaded onto SDS–PAGE gel and coomassie stained reveal a prevailing double band of about 53 and 56 kDa corresponding to Lyn tyrosine kinase.

2.1.2. Purification of Fgr tyrosine kinase

TPK-III, resolved by DEAE-Sepharose, is dialyzed against buffer A and subjected to affinity chromatography through a column of Heparin-Sepharose equilibrated with the same buffer.

The column is eluted with a linear 0–0.5 M NaCl gradient; 1 ml fractions are collected and analyzed for TPK activity. Two peaks are obtained, the second one, with a profile of elution of 0.35 M NaCl is further purified on a polylysine-agarose column (5 ml), equilibrated in buffer A, and eluted with a linear 0–0.5 M NaCl gradient; 0.2 ml fractions are collected. The peak eluting at 0.32–0.38 M NaCl is finally resolved on a Superdex75/FPLC chromatography column. The Superdex75 column is equilibrated with 20 mM Tris/HCI pH 7.5, 10% glycerol, 10 mM β-mercaptoethanol, 0.1 mM PMSF and 0.5 M NaCl, and eluted at a flow rate of 0.5 ml/min. Fgr elutes as a sharp peak of absorbance at 280 nm, overlapped by TPK activity. Twenty microlitters of the eluted samples loaded onto SDS–PAGE gel and coomassie stained reveals a major single band of about 55 kDa corresponding to Fgr.

From Western blot experiments, both purified enzymes only react with anti-Lyn or anti-Fgr antibodies, whereas no binding is observed with other tyrosine kinases such as Src, Fyn, Jak2, Jak3, and Syk. Both kinases retain their activity stored in aliquots at $-80\ °C$ for some years.

2.1.3. Assay of tyrosine kinase activity

To identify specific substrates for Fgr or Lyn kinases, we use the two enzymes at the same units measured with the synthetic peptide cdc-2 (KVE-KIGEGTYGVVYK). Different amounts of the enzymes are incubated in the radioactive mixture: 50 mM Tris–HCl, 10 mM MnCl$_2$, 100 μM EDTA, 0.01% β-mercaptoethanol, 0.1 mM [γ^{33}P-ATP] (500–1000 cpm/pmol) in presence or absence of 0.25 mM cdc-2 for 10 min at 30 °C. Twenty-five microliters are spotted onto P81 phosphocellulose paper. After washing three times with 0.75% phosphoric acid and once with acetone for 5 min each, the papers are dried and the radioactivity incorporated into the P81 paper is counted using a scintillation counter. One unit of activity, U, is that amount of protein kinase which catalyzed the phosphorylation of 1 nmol of the synthetic peptide substrate in 1 min.

2.2. Isolation and purification of mitochondria from rat brain

Mitochondria are routinely isolated from different rat organs such as liver, heart, and brain. The choice of the source of mitochondria should be linked to the type of analysis to be performed. Therefore, we previously checked the presence of tyrosine kinase activity in purified mitochondria obtained from liver, heart, and brain using the generic substrate Poly(Glu-Tyr)$_{4:1}$ which provided evidence that most of tyrosine kinase activity is present in brain mitochondria, less in heart mitochondria and virtually absent in liver mitochondria at least at our isolation conditions. For this reason we suggest the use of brain mitochondria as the best model to investigate tyrosine phosphorylation. Therefore the isolation and purification process of mitochondria from brain is here reported.

Rat brains (cerebral cortex) are promptly removed and immediately immersed in an ice-cold isolation medium containing 5 mM Hepes (pH 7.4), 320 mM sucrose, 0.5 mM EDTA. BSA (0.3% (w/v)) is added at this stage and during the first step of purification to protect mitochondrial membranes. Brains are minced, thoroughly rinsed three times with ice-cold medium and homogenized in the same buffer solution using a Potter homogenizer with a Teflon pestle in an ice-water bath. Mitochondria are then isolated by conventional differential centrifugation and purified by Ficoll discontinuous gradients. First, cell debris, nuclei, and other heavy components are removed by centrifugation at 900g for 5 min. The supernatant is centrifuged at 12,000g for 10 min to precipitate crude mitochondrial pellets. The pellets are resuspended in isolation medium plus 1 mM ATP and layered on top of a discontinuous gradient, composed of 2 ml of isolation medium containing 16% (w/v) Ficoll, 2 ml of isolation medium containing 14% (w/v) Ficoll, 3 ml of isolation medium containing 12% (w/v) Ficoll, and 3 ml of isolation medium containing 7% (w/v) Ficoll. The gradient is centrifuged for 30 min at 75,000g. Mitochondrial pellets are suspended in isolation medium and centrifuged for 10 min at 12,000g. Again the pellets are suspended in isolation medium without EDTA. Protein content is measured by the biuret method with bovine serum albumin as standard.

Mitochondria are isolated without using phosphatase inhibitors to favor the dephosphorylation of the proteins. Generally 20–25 mg of mitochondrial proteins is yielded from four rat brains.

2.2.1. Chromatography of mitochondrial extracts

Mitochondrial proteins (300 mg) are solubilized for 1 h at 4 °C in the following medium: 30 mM Tris/HCl (pH 7.5), 2 mM EGTA, 2 mM EDTA, 2 mM DTT, 3% Triton X-100, and protease inhibitor cocktail (Roche).

Solubilization is expedited by sonication: three times for 15 s in an ice bath with a Fisher Sonic Dismembrator 300, set at 60% maximal energy. Half of the extract is diluted 1:7 in extraction buffer where Triton is at 0.1%, filtered and chromatographed on 8-ml Source Q/FPLC, equilibrated in extraction buffer (0.1% Triton), using a linear salt gradient (0–1 M NaCl). Fractions, 4.5 ml each, are collected and analyzed as described below. Half of the flow through of the Source Q is put directly on a 1 ml Source S/FPLC column, equilibrated in extraction buffer (0.1% Triton), using a linear salt gradient (0–1 M NaCl). 0.8 ml fractions are collected and analyzed as described below.

2.3. Tyrosine phosphorylation of mitochondrial proteins

Aliquots of each fraction of chromatographed mitochondrial proteins are diluted with nine volumes of 30 mM Tris–HCl pH 7.5, 10 mM MnCl$_2$, 0.1 mM EGTA, 0.1% (v/v) β-mercaptoethanol. Aliquots, 25 μl each, are then incubated for 5 min at 30 °C with or without 10 mU of Fgr or Lyn and 2.5 $\times$ 10^6 c.p.m. of 20 nM [γ-^{32}P]ATP in a total volume of 30 μl. The reactions were stopped by the addition of 2X Laemmli buffer and heated for 2 min at 95 °C. Phosphorylated fractions were run on SDS-gels, subjected to Western blotting on a PVDF membrane. The high amount of kinase activity, the high radioactivity and the low time of incubation are all important factors to reduce the phosphorylation background and to ensure the resolution of specific radioactive spots.

To decrease endogenous Ser/Thr phosphorylation and thus reduce background phosphorylation, the PVDF membranes are subjected to alkali-treatment. Phosphotyrosine residues are stable at alkaline pH, whereas the O-linked phosphoamino acids phosphoserine and phosphothreonine are not. A nitrocellulose membrane cannot be used because it is decomposed by this treatment. PVDF blots are dipped into 2 M NaOH at 55 °C for 1 h followed by exposition to X-ray films.

All the fractions obtained from Source S and Source Q are screened looking for proteins that are phosphorylated by just one kinase. Figure 7.2 illustrates some of these fractions incubated without enzymes (background phosphorylation) and in presence of Lyn or Fgr. The incubation of Fgr and Lyn alone shows the autophosphorylative bands.

Specific radioactive spots should be further purified. Different methods could be employed to do this, such as standard chromatographic procedures or 2D bidimensional analysis.

Here we describe the purification of the two phosphorylated proteins found in the flow through fractions of Source Q and S. To purify these substrates, the flow through of Source S is put directly on a 5 ml HiTrap Heparin-column, equilibrated in extraction buffer (0.1% Triton), using a linear salt gradient (0–1 M NaCl). Fractions, 3 ml each, are collected and

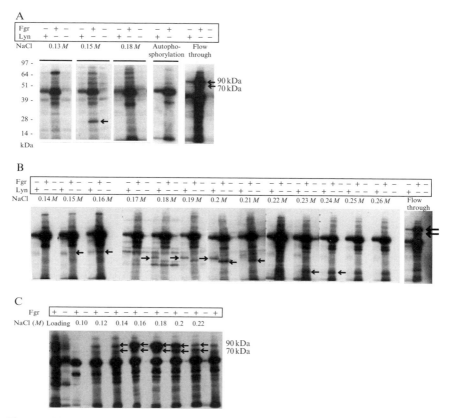

Figure 7.2 Detection of Fgr and Lyn substrates in mitochondrial extracts from rat brain. (A) Source Q fractions. (B) Source S fractions. (C) Heparin fractions. Reprinted from FEBS letters (Salvi *et al.*, 2007).

analyzed as described above. The two labeled proteins eluted together between 0.16 and 0.2 M NaCl (Fig. 7.2C). The two fractions are combined and concentrated by using Vivaspin 6 centrifugal concentrators. Proteins are then exhaustively phosphorylated by Fgr tyrosine kinase for 1 h at 30 °C in presence of 0.1 mM [γ^{33}P-ATP] (1000 cpm/pmol), and separated by SDS–PAGE gels and coomassie stained. The radioactive spots are cut, tryptically digested and identified by mass spectrometry as mitochondrial aconitase and flavoprotein of succinate dehydrogenase. The tyrosine phosphorylation sites are identified as Y71, Y544, Y665 for aconitase and *as* Y535 and Y596 for the flavoprotein.

Methods to enrich and analyze tyrosine-phosphorylated peptides for mass spectrometry are described below.

3. Mass Spectrometric Analysis of Tyrosine-Phosphorylated Peptides in Mitochondria

Although mass spectrometry has become a common tool for protein identification due to the constant development of proteomic workflows, the exact determination of posttranslational modifications still remains a major challenge. In case of phosphorylations, these challenges are due to low abundance of phosphorylated protein species, their susceptibility to phosphatase activity, as well as their instabilities under basic conditions. Furthermore, quenching effects can hamper detection in the presence of unmodified peptide species and phosphopeptides can spontaneously decompose under mass spectrometric conditions. Due to those obstacles, dedicated strategies for phosphopeptide analysis have been developed which integrate specific phosphopeptide enrichment procedures as well as special mass spectrometric scan events for detection and identification.

Over the years, several methods have been applied for generalized phosphopeptide enrichment, such as immobilized metal ion affinity chromatography (Ficarro et al., 2002), strong cation exchange chromatography (Beausoleil et al., 2004) or titanium dioxide affinity chromatography (Pinkse et al., 2004). Especially the latter technique has recently attracted a lot of attention thanks to the high quality of purification. However, as judged by results from several large scale datasets, phosphotyrosine phosphorylation (pY) is of low abundance even with respect to pS or pT. While Hunter (1998) reported a ratio of 1800:200:1 (pS:pT:pY) by amino acid analysis, large phosphoproteome studies by Olsen et al. (2006) (HeLa cells) and Zahedi et al. (2008) (resting platelets) resulted in ratios of 43:6:1 and 35:6:2, respectively. So even in phosphopeptide enriched fractions, pY–containing peptides might be underrepresented. This can be improved either by dedicated pY enrichment procedures based on specific antibodies, or by applying specialized mass spectrometric techniques, aiming specifically at identification of pY–peptides. Therefore, we use an integrated workflow for the analysis of mitochondrial pY–peptides as shown in Fig. 7.3A.

3.1. Enrichment of phosphopeptides

3.1.1. Specific immunoaffinity-based enrichment of pY-containing peptides

While antibodies directed against pS and pT rarely enable specific enrichment of those phosphorylated species unless combined with a special target sequence, antibodies for pY can very well recognize the larger epitope of a single phosphorylated tyrosine residue. Therefore, pY–antibodies can be

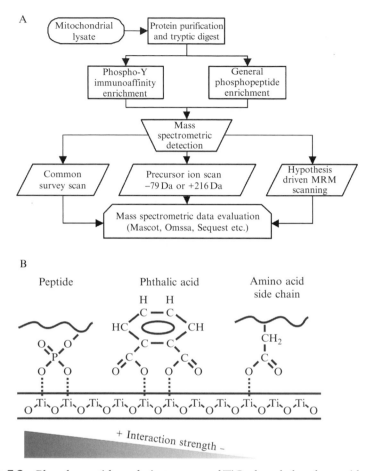

Figure 7.3   Phosphopeptide analysis strategy and TiO$_2$-based phosphopeptide enrichment. (A) After isolation and lysis of mitochondria as well as proteolytic digestion of proteins, phosphopeptides can be enriched specifically for pY-containing species by immunoprecipitation or generalized by TiO$_2$-based affinity chromatography. Enriched phosphopeptide fractions are analyzed by mass spectrometry using different survey scan types among which specialized techniques such as precursor ion scanning are prominent. For individual proteins of known sequence, hypothesis driven multiple reaction monitoring (MRM) is feasible as well. (B) General enrichment of phosphopeptides may be achieved by selective affinity of phosphate groups to the solid support surface of titanium dioxide. Weak unspecific binding may occur in case of acidic amino acid side chains. By *using* competitors like 2,5-dihydroxybenzoic acid or phthalic acid, unspecific binding of peptides can be reduced by competition for binding sides.

used for specific enrichment of pY-containing peptides out of crude peptide mixtures (Rush *et al.*, 2005).

   Initially, crude peptide mixtures are generated from the protein pool to be screened for pY-sites. After removal of salts, impurities as well as undigested proteins by solid phase extraction, tryptic peptides are lyophilized. Upon reconstitution in resuspension buffer (1 ml of 20 mM Tris/HCl pH 7.2, 10 mM sodium phosphate, 50 mM sodium chloride) insoluble matter is removed by centrifugation at 2,000g for 5 min. Although they may be used independently, the combined use of different pY-directed antibodies can potentially increase the coverage of pY-containing peptides. Therefore, the monoclonal phosphotyrosine antibodies 4G10 (Upstate Biotechnology) and PY20 (ICN) can be coupled non-covalently in a 5:1 ratio to protein G agarose (Roche) beads by incubation overnight at 4 °C with gentle shaking. Thereafter, the antibody resin is washed and equilibrated extensively with PBS and resuspension buffer. Peptides are incubated with the immobilized antibodies for 16 h at 4 °C under slight agitation. After washing with PBS, peptides are eluted from the beads by applying a pH shift using 0.1% trifluoroacetic acid at 25 °C for 10 min. Eluted peptides can be concentrated by solid phase extraction using C18-microcolumns (Millipore or equivalent) prior to mass spectrometric analysis.

3.1.2. General enrichment of phosphopeptides by TiO$_2$−affinity chromatography

The enrichment of phosphopeptides by titanium dioxide is based on the specific interaction of the phosphate group with the solid support surface as shown in Fig. 7.3B. Although being weak interactors in comparison to phosphates, modifiers such as 2,5-dihydroxybenzoic acid (Larsen *et al.*, 2005) or phthalic acid (Bodenmiller *et al.*, 2007) nevertheless suppress unspecific binding e.g., of acidic side chains of peptides like aspartic or glutamic acid. Elution is performed by short-term shifts to basic pH (see below). However, attention should be paid to contaminant removal prior enrichment. Phospholipids may bind to titanium dioxide as well and should be removed prior to phosphopeptide enrichment, e.g., by precipitation or solid phase extraction. Acidic oligosaccharides can bind to the stationary phase as well (Larsen *et al.*, 2007). However, these modifications are not to be expected during analysis of mitochondrial fractions.

For procedures outlined below, only highly pure chemicals were used, since contaminants can severely interfere with mass spectrometric analysis.

Initially, 4 mg TiO$_2$-material (Titansphere, 5 μm particle diameter, GL Sciences) per 1 mg of protein digest is equilibrated in 500 μl loading buffer (80% acetonitrile, 2.5% (v/v) trifluoroacetic acid, saturated with phthalic acid) for 5 min, and the supernatant is discarded after centrifugation. Desalted and lyophilized peptides are dissolved in 500 μl loading buffer (optionally dilute further) and incubated with the equilibrated beads for 15 min at RT under slight agitation. After brief centrifugation, the supernatant is discarded and the beads are washed twice with 500 μl loading buffer and twice with washing buffer (80% (v/v), 0.1% (v/v) trifluoroacetic acid) for 1 min each. Finally, the beads are washed twice with 500 μl

0.1% (v/v) trifluoroacetic acid. Elution is performed by short-term alkaline pH shifts. Since phosphate-diester bonds are labile under those conditions, the pH of the eluted fractions needs to be immediately acidified (~pH 3). For elution, the beads are incubated with 100 μl 30% (v/v) acetonitrile, 200 mM ammonium hydroxide for 30 s. After a quick spin-down the supernatant is transferred to a reaction tube with already deposited 2.5 μl formic acid. Subsequently, 5 μl formic acid is added to the same tube to prepare for a second elution from the beads by 100 μl 30% (v/v) acetonitrile, 300 mM ammonium hydroxide for 30 s. The two supernatants are combined and further 10 μl formic acid is added. Upon elution of remaining phosphopeptides by 100 μl 30% (v/v) acetonitrile, 400 mM ammonium hydroxide for 30 s all supernatants are combined. Check the pH to be below pH 3 and lyophilize the samples to remove the acetonitrile prior to mass spectrometric analysis.

3.2. Mass spectrometric detection

For analysis of phosphorylations and other posttranslational modifications, specialized mass spectrometric detection techniques have been developed. Most often, mass spectrometers in proteomic experiments employ data dependant fragmentation of precursor ions, which elute from the up-front liquid chromatographic separation. Usually, ions are picked and analyzed in decreasing order of their relative intensity. Since phosphopeptides often exhibit only moderate ion intensities present in mixtures, it is advisable to specifically identify precursor ions which correspond to potential phosphopeptides and hence focus the analysis time on these alone by reducing the sample complexity on the mass spectrometric level. This can be done either by screening for marker ions of phosphorylation or by a hypothetically driven workflow when dealing with a few, known proteins. Commonly, triple quadrupole mass spectrometers are used for this purpose. Mass spectrometric conditions as outlined below are usually preceded by a nano-scale liquid chromatography step for peptide separation as, e.g., shown by Schindler *et al.* (2008).

3.2.1. Common mass spectrometric identification

For a first survey of isolated phosphopeptide mixtures, a common survey scan may be used in combination with subsequent fragmentation of precursor ions (see Fig. 7.4). Derived tandem mass spectrometric data (MS/MS) can then be searched and identified by comparison with protein sequence databases using algorithms like Mascot (Perkins *et al.*, 1999), Sequest, or Omssa (Geer *et al.*, 2004). This analysis can offer conclusions on enrichment effectiveness as well as a first round of phosphopeptide identifications. Although triple quadrupole mass spectrometers allow for unique scan events, their MS/MS sensitivity is often inferior to those of ion trap instruments.

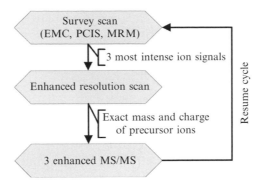

Figure 7.4 Scan events for mass spectrometric phosphopeptide analysis. Mass spectrometric detection of phosphopeptides and peptides in general is quite commonly based
on a three-step procedure using triple-quadrupole linear ion trap mass spectrometric
equipment. First, peptide ions are detected either unbiased by a common enhanced
multiple charge scan or selectively by scans for phosphopeptides. The latter comprise
precursor ion scanning (PCIS) or hypothesis driven MRM approaches. Following
selection of the three most abundant peptide ions in the first scan, an enhanced resolution scan is performed for exact mass and charge determination. Lastly, MS/MS
fragmentation spectra of the three peptide ions are acquired for peptide sequencing and
phosphorylation site identification.

In contrast, hybrid systems like triple quadrupole linear ion traps combine
scan events like true precursor ion scanning (see Section 3.2.2) with
improved MS/MS sensitivity.

   For the analysis of mitochondrial pY peptides by a Qtrap4000 triple
quadrupole linear ion trap (Applied Biosystems), we successfully employed
an enhanced multiple charge scan followed by an enhanced resolution scan
(4000 amu/s) of the three most intensive precursor ions (for exact m/z
determination, scan rate 250 amu/s). Scan range is set to m/z 380–1500;
within this range most tryptic peptides are detectable as doubly or triply
charged ions, which are most suitable for collision induced dissociation. Singly
charged ions are not selected for fragmentation during the subsequent three
MS/MS spectra (scan range m/z 115–1500 at 4000 amu/s scan rate). Hence
collision energy is set to $CE^{2+} = 0.44X + 4$ and $CE^{3+} = 0.4X + 5$ (with $X =$
m/z ratio of signal to be fragmented). Ion spray voltage is set to 2.3 kV and the
declustering potential to 60 (Note: If extensive metastable fragmentation is
observable, the declustering potential can be lowered).

3.2.2. Precursor ion scanning

Instead of choosing ions by a common survey scan (compare Section 3.2.1)
for subsequent MS/MS fragmentation, a selection of potential phosphopeptides can be made on basis of specific marker ions for phosphorylation. Thus,
predominantly phosphopeptides can be selected while non-phosphopeptides

are largely excluded from analysis, thereby saving instrument cycle time. Those marker ions can be detected upon fragmentation of precursor ions either in the positive (m/z 216, pY immonium ion) or in the negative mode (m/z −79, PO_3^-). While detection in negative ion mode also accounts for pS and pT containing peptides, scanning for m/z 216 in positive ion mode is rather selective for pY containing peptides. However, false positive assignments may be due to dipeptide ions exhibiting the same m/z ratio such as Glu-Ser, Lys-Ser, or Asn-Thr and/or insufficient mass accuracy.

For precursor ion scanning using a Qtrap4000, the scan range is set to m/z 380–1500, with a scan time of 3.5 s. Although, this results in a rather large cycle time, it is an absolute prerequisite for sensitive detection of phosphopeptides. In this context, it is necessary to note that within this time frame, approximately 1120 fragmentations reactions have to be monitored, so only a fraction of the total time span remains for a single peptide mass. For m/z 216 precursor ion scanning, the spray voltage of a Qtrap4000 system is set to 2.3 kV. It is usually lowered to 1.6–1.8 kV in the negative ion mode for detection of m/z 79 (PO_3^-). However, under the given liquid chromatography conditions, fragmentations for peptide sequencing are more effective in positive ion mode and also the data evaluation is based on spectra acquired in this mode. Therefore, a polarity switch from negative to positive ion mode between precursor ion scan and the following enhanced multiple charge and MS/MS scans is required. To minimize spray fluctuations, a small volume of isopropanol or 80% (v/v) isopropanol, 20% (v/v) acetonitrile can be added to the HPLC flow via a post-column T-split at a ratio of 3:1. The quadrupoles of the Qtrap4000 are operated at unit/unit resolution to obtain high selectivity. The collision energy is adjusted according to the mass of the precursor from 55 to 120. In parallel to the common data dependant survey scan method (compare 2.2), a high resolution scan is used to elucidate exact m/z ratios for subsequent MS/MS spectra (compare also Fig. 7.4).

3.2.3. Multiple reaction monitoring

Assuming, the identity of a mitochondrial protein of interest is already known, but pY-phosphorylation sites remain to be elucidated, a hypothesis driven workflow can be used to screen for potential phosphorylation sites. Therefore, each tyrosine residue of the respective protein is phosphorylated in silico and peptide precursor as well as potential fragment masses are calculated. These masses (or transitions) are then used for multiple reaction monitoring (MRM) in combination with a triple quadrupole mass spectrometer. Data dependent acquisition of MS/MS spectra is only triggered upon detection of the predefined transitions, thereby focusing with high sensitivity on the subset of pY-containing peptides. Usually, no upfront enrichment of phosphopeptides (as described in Sections 3.1.1 and 3.1.2) is performed, as long as the protein of interest itself is accessible in a relatively pure form.

For calculation of peptide/fragment ion pairs, software tools like the MRM-builder (Applied Biosystems) can be used. Since at this step of the workflow potential fragment ions are not yet experimentally validated, several transitions per peptide are recommended. Suitable transitions include the pY-immonium ion (m/z 216) as well as sequence ions with high probability to occur, such as those adjacent to proline or glycine residues. Two further parameters need to be specified for each transition. These are collision energy and dwell time. The collision energy can be calculated depending on the charge of the precursor ion by the values given in 2.1. The dwell time is usually between 15 and 20 ms/transition. This ensures reasonable sensitivity in combination with unit resolution settings of the Q1/Q3 quadrupoles. However, note that the number of possible transitions is limited by the cycle time. The scan time for the summed transitions should not exceed 3 s for a discovery driven workflow; so the total cycle time including subsequent enhanced resolution and MS/MS spectra is between 5 and 6 s.

4. BIOINFORMATIC TOOLS TO ANALYZE THE POTENTIAL ROLE OF TYROSINE PHOSPHORYLATION

The identification of the specific site(s) of phosphorylation is required for the in silico analysis of the potential role of phosphorylation. The first facet to understand the role of phosphorylation is the conservation of the target site throughout evolution. Conservation typically implies an essential role for this site and consequently for its posttranslational modifications in the structure/function of the protein. The alignments of the sequences of different species could be easy performed using ClustalX (ftp://ftp.ebi.ac.uk/pub/software/clustalw2/).

An in silico analysis of the potential role of phosphorylation also required structural informations on the phosphoprotein. PDB databank (http://www.rcsb.org/pdb/) is the single worldwide archive of structural data of biological macromolecules. Protein structure is visualized with Pymol, a molecules visualization system (http://pymol.org/). Other molecular viewers are available such as VMD (http://www.ks.uiuc.edu/Research/vmd/) or Discovery Studio Visualizer (http://accelrys.com/downloads/freeware/).

The position that the phospho-site occupies within the structure of the protein is very informative. Observing the structure, it can be recognized whether the phospho-site is close to a functional part of the molecule such as the active site of an enzyme, or to sites of interactions between subunits or binding sites to the membranes, and so on. One distinctive example is the phospho-sites that we have identified in the adenine nucleotide translocase (ANT). The transmembrane domain of ANT is formed by six α-helices

(H1-H6) mimicking the form of a basket; it is mainly hydrophilic, closed toward the matrix and opened widely toward the outer membrane. The transport substrates bind to the bottom of the cavity and the translocation results from a transient transition from a pit to a channel conformation. Figure 7.5 shows the structure of ANT displaying the two identified phospho-tyrosines (Y190 and Y194). Both tyrosines are found inside the cavity, immediately suggesting that tyrosine phosphorylation should be involved in the regulation of the transport. Moreover, aromatic residues are not uniformly distributed within the cavity, but are grouped along H4. Y194 is the first of three highly conserved tyrosines, Y194/Y190/Y186, with side chains oriented toward the cavity. This arrangement forms a ladder along H4, entering the cavity (see Fig. 7.5C). It has been suggested that a stacking interaction with tyrosine rings could guide the ADP translocation along the tyrosine ladder (Pebay-Peyroula *et al.*, 2003).

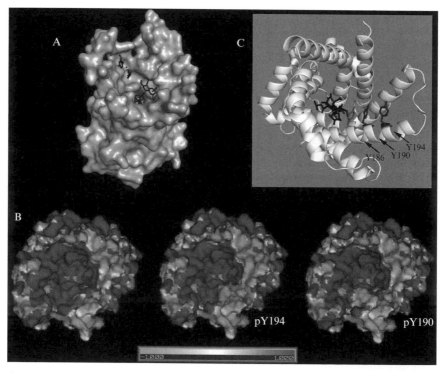

Figure 7.5 Structure of ADP/ATP translocase. (A) Surface representation of ANT (Protein Data Bank entry 1ock, bovine ANT) where the two tyrosines identified as phosphorylated are marked in black sticks. Also the ADP/ATP inhibitor carboxyatractyloside is black. (B) Electrostatic surface before and after addition of phosphate groups to Y194 or Y190. Positive and negative surfaces are shown in blue and red, respectively. (C) Cartoon model where the tyrosine ladder (black sticks) is shown. Carboxyatractiloside is marked in black. Reprinted from FEBS letters (Lewandrowski *et al.*, 2008).

A second relevant facet to take into consideration is the electrostatic surface of the molecule. The gain of a phosphate group means the addition of negative charges that could be responsible for a significant modification of its electrostatic surface. Electrostatic surface of ANT before and after the addition of phosphate group to Y190 and Y194 is calculated with Adaptive Poisson–Boltzmann Solver (http://apbs.sourceforge.net/) (Fig. 7.5). Also the same Pymol could be used for calculate the electrostatic surface but with some approximation. Figure 7.5 shows that before phosphorylation the internal surface of the cavity is positively charged (the nucleotides transported are negatively charged). In contrast, the phosphorylation of Y190 or Y194 clearly changes the electrostatic surface of the cavity (Fig. 7.5).

The position of the sites inside the cavity and the alteration in the electrostatic surface strongly support the above-mentioned hypothesis that tyrosine phosphorylation of Y190 and Y194 alters the rate and/or the specificity of the transport.

Phosphorylation could also affect the functional part of the molecule far from the target site by inducing a structural rearrangement of the overall architecture of the molecule. This issue is analyzed using modeling programs. The Y461 phospho-site that we have identified in Type I hexokinase (HK-I) is localized in the helix that connects the N- and C-terminal halves which show about 30% sequence identity between (Fig. 7.6). The C-terminal half is catalytically active, whereas the N-terminal is not. However, interactions of effectors such as glucose-6-phosphate and Pi to the N-terminal half affect the functionality of the catalytic C-terminal half.

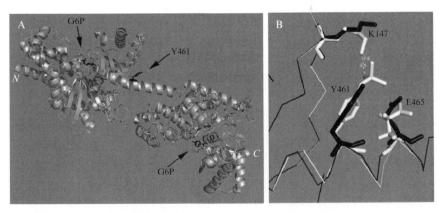

Figure 7.6 Structure of hexokinase. (A) Overview of hexokinase structure (Protein Data Bank entry 1cza, human hexokinase I). Tyrosine residue identified as phosphorylated (Y461) is marked in black and indicated by an arrow. Two molecules of G-6-P, in black, are in the two catalytic sites, shown by arrows. The N-terminus (N) and the C-terminus (C) are shown. (B) *In silico* model of effects of Y461 phosphorylation on protein structure. The structures with (in white) and without (in black) phosphate added to the tyrosine are superimposed. Reprinted from FEBS letters (Lewandrowski *et al.*, 2008).

Binding of the effectors at the N-terminal half determines two different conformations of the C-terminal half, one permits the association of ATP, whereas the other antagonizes ATP binding. These conformational changes are transmitted from the N-terminal to the C-terminal half by a rotation of 6° of the helix that connects the two halves influencing the position of a loop near the catalytic site (Aleshin et al., 2000). Y461 is far from the active and the regulatory half and apparently not involved in the enzymatic regulation. However, inspecting the amino acids within 2–3 Å of Y461, we also find K147 and E465. To rule out the possibility of interaction between added phosphate and K147 and E465 we generated a model adding a phosphate group to the tyrosine using the Insight II Software Inc. (Accelrys Inc., San Diego) and minimizing until the energy reached a minimum using the Discover module. After the addition of phosphate (white sticks) the lysine approaches the added phosphate while the gluta-mate is turned away in the opposite direction due to electrostatic repulsion. These electrostatic interactions may be responsible for a conformational change in the connecting helix (Fig. 7.6). This conformational change could be transmitted to the catalytic site thereby modulating its activity.

ACKNOWLEDGMENT

We are very grateful to Prof. Sir P. Cohen for having supported the project of the identification of src kinase substrates in rat brain mitochondria.

REFERENCES

Aleshin, E. A., Kirby, C., Liu, X., Bourenkov, G. P., Bartunik, H. D., Fromm, H. J., and Honzatko, R. B. (2000). Crystal structure of mutant monomeric hexokinase I reveal multiple ADP binding sites and conformational changes relevant to allosteric regulation. J. Mol. Biol. **296,** 1001–1015.
Beausoleil, S. A., Jedrychowski, M., Scwartz, D., Elias, J. E., Villen, J., Li, J., Cohn, M. A., Cantley, L. C., and Gygi, S. P. (2004). Large-scale characterization of HeLa cell nuclear phosphoproteins. Proc. Natl. Acad. Sci. USA **101,** 12130–12135.
Bodenmiller, B., Mueller, L. N., Mueller, M., Domon, B., and Aebersold, R. (2007). Reproducible isolation of distinct, overlapping segments of the phosphoproteome. Nat. Methods **4,** 231–237.
Cohen, P., and Knebel, A. (2006). KESTREL: A powerful method for identifying the physiological substrates of protein kinases. Biochem. J. **393,** 1–6.
Ficarro, S. B., McCleland, M. L., Stukenberg, P. T., Burke, D. J., Ross, M. M., Shabanowitz, J., Hunt, D. F., and White, F. M. (2002). Phosphoproteome analysis by mass spectrometry and its application to Saccharomyces cerevisiae. Nat. Biotechnol. **20,** 301–305.
Geer, L. Y., Markey, S. P., Kowalak, J. A., Wagner, L., Xu, M., Maynard, D. M., Yang, X., Shi, W., and Bryant, S. H. (2004). Open mass spectrometry search algorithm. J. Proteome Res. **3,** 958–964.

Hunter, T. (1998). The Croonian lecture 1997. The phosphorylation of proteins on tyrosine: Its role in cell growth and disease. *Philos. Trans. R. Soc. Lond. B Biol. Sci.* **353,** 583–605.

Itoh, S., Lemay, S., Osawa, M., Che, W., Duan, Y., Tompkins, A., Brookes, P. S., Sheu, S. S., and Abe, J. (2005). Mitochondrial Dok-4 recruits Src kinase and regulates NF-kappaB activation in endothelial cells. *J. Biol. Chem.* **280,** 26383–26396.

Knebel, A., Morrice, N., and Cohen, P. (2001). A novel method to identify protein kinase substrates: eEF2 kinase is phosphorylated and inhibited by SAPK4/p38delta. *EMBO J.* **20,** 4360–4369.

Larsen, M. R., Jensen, S. S., Jakobsen, L. A., and Heegaard, N. H. (2007). Exploring the sialiome using titanium dioxide chromatography and mass spectrometry. *Mol. Cell. Proteomics* **6,** 1778–1787.

Larsen, M. R., Thingholm, T. E., Jensen, O. N., Roepstorff, P., and Jorgensen, T. J. (2005). Highly selective enrichment of phosphorylated peptides from peptide mixtures using titanium dioxide microcolumns. *Mol. Cell. Proteomics* **4,** 873–886.

Lee, L., Salomon, A. R., Ficarro, S., Mathes, I., Lottspeich, F., Grossman, L. I., and Huttemann, M. (2005). cAMP-dependent tyrosine phosphorylation of subunit I inhibits cytochrome c oxidase activity. *J. Biol. Chem.* **280,** 6094–6100.

Lewandrowski, U., Sickmann, A., Cesaro, L., Brunati, A. M., Toninello, A., and Salvi, M. (2008). Identification of new tyrosine phosphorylated proteins in rat brain mitochondria. *FEBS Lett.* **582,** 1104–1110.

Livigni, A., Scorziello, A., Agnese, S., Adornetto, A., Carlucci, A., Garbi, C., Castaldo, L., Annunziato, L., Avvedimento, E. V., and Feliciello, A. (2006). Mitochondrial AKAP121 links cAMP and src signaling to oxidative metabolism. *Mol. Biol. Cell.* **17,** 263–271.

Miyazaki, T., Neff, L., Tanaka, S., Horne, W. C., and Baron, R. (2003). Regulation of cytochrome c oxidase activity by c-Src in osteoclasts. *J. Cell. Biol.* **160,** 709–718.

Olsen, J. V., Blagoev, B., Gnad, F., Macek, B., Kumar, C., Mortensen, P., and Mann, M. (2006). Global, *in vivo,* and site-specific phosphorylation dynamics in signaling networks. *Cell* **127,** 635–648.

Pagliarini, D. J., and Dixon, J. E. (2006). Mitochondrial modulation: Reversible phosphorylation takes center stage? *Trends Biochem. Sci.* **31,** 26–34.

Pagliarini, D. J., Wiley, S. E., Kimple, M. E., Dixon, J. R., Kelly, P., Worby, C. A., Casey, P. J., and Dixon, J. E. (2005). Involvement of a mitochondrial phosphatase in the regulation of ATP production and insulin secretion in pancreatic beta cells. *Mol. Cell* **19,** 197–207.

Pebay-Peyroula, E., Dahout-Gonzalez, C., Kah, R., Trezeguet, V., Lauquin, G. J. M., and Brandolin, G. (2003). Structure of mitochondrial ADP/ATP carrier in complex with carboxyactractyloside. *Nature* **426,** 39–44.

Perkins, D. N., Pappin, D. J., Creasy, D. M., and Cottrell, J. S. (1999). Probability-based protein identification by searching sequence databases using mass spectrometry data. *Electrophoresis* **20,** 3551–3567.

Pinkse, M. W., Uitto, P. M., Hilhorst, M. J., Ooms, B., and Heck, A. J. (2004). Selective isolation at the femtomole level of phosphopeptides from proteolytic digests using 2D-NanoLC-ESI-MS/MS and titanium oxide precolumns. *Anal. Chem.* **76,** 3935–3943.

Rush, J., Moritz, A., Lee, K. A., Guo, A., Goss, V. L., Spek, E. J., Zhang, H., Zha, X. M., Polakiewicz, R. D., and Comb, M. J. (2005). Immunoaffinity profiling of tyrosine phosphorylation in cancer cells. *Nat. Biotechnol.* **23,** 94–101.

Salvi, M., Brunati, A. M., La Rocca, N., Bordin, L., Clari, G., and Toninello, A. (2002). Characterization and location of Src-dependent Tyrosine phosphorylation in rat brain mitochondria. *Biochim. Biophys. Acta* **1589,** 181–195.

Salvi, M., Brunati, A. M., and Toninello, A. (2005). Tyrosine phosphorylation in mito-chondria: A new frontier in mitochondrial signaling. *Free Radic. Biol. Med.* **38,** 1267–1277.

Salvi, M., Morrice, N., Brunati, A. M., and Toninello, A. (2007). Identification of the flavoprotein of succinate dehydrogenase and aconitase as *in vitro* mitochondrial substrates of Fgr tyrosine kinase. *FEBS Lett.* **581,** 5579–5585.

Salvi, M., Stringaro, A., Brunati, A. M., Agostinelli, E., Arancia, G., Clari, G., and Toninello, A. (2004). Tyrosine phosphatase activity in mitochondria: Presence of Shp-2 phosphatase in mitochondria. *Cell. Mol. Life Sci.* **61,** 2393–2404.

Schindler, J., Lewandrowski, U., Sickmann, A., and Friauf, E. (2008). Aqueous polymer two-phase systems for the proteomic analysis of plasma membranes from minute brain samples. *J. Proteome Res.* **7,** 432–442.

Zahedi, R. P., Lewandrowski, U., Wiesner, J., Wortelkamp, S., Moebius, J., Schutz, C., Walter, U., Gambaryan, S., and Sickmann, A. (2008). Phosphoproteome of resting human platelets. *J. Proteome Res.* **7,** 526–534.

CHAPTER EIGHT

ACETYLATION OF MITOCHONDRIAL PROTEINS

Matthew D. Hirschey,* Tadahiro Shimazu,* Jing-Yi Huang,*
and Eric Verdin

Contents

1. Introduction	138
2. Purification of Enzymatically Active SIRT3	138
2.1. Purification of SIRT3 from mammalian cell culture systems	139
2.2. Purification of SIRT3 from bacterial expression systems	140
3. SIRT3 Enzymatic Deacetylation Assay	141
3.1. Peptide–substrate *in vitro* SIRT3 deacetylation assay	141
3.2. Protein–substrate *in vitro* SIRT3 deacetylation assay	141
3.3. *In vivo* SIRT3 deacetylation assay	143
4. Detection of Acetylated Proteins in Mitochondria	144
4.1. Isolation and purification of the heavy mitochondrial fraction from murine liver	144
4.2. Detection of acetylated mitochondrial proteins by immunoprecipitation	145
4.3. Detection of multiple acetylated mitochondrial proteins by reverse immunoprecipitation	146
5. Conclusion	146
References	147

Abstract

Sirtuins (SIRT1–SIRT7) are a family of NAD^+-dependent protein deacetylases that regulate cell survival, metabolism, and longevity. SIRT3 is localized to the mitochondria where it deacetylates several key metabolic enzymes: acetylcoenzyme A synthetase, glutamate dehydrogenase, and subunits of complex I and thereby regulates their enzymatic activity. SIRT3 is therefore emerging as a metabolic sensor that responds to change in the energy status of the cell via NAD^+ and that modulates the activity of key metabolic enzymes via protein

Gladstone Institute of Virology and Immunology, University of California, San Francisco, California, USA
* These authors contributed equally

Methods in Enzymology, Volume 457   © 2009 Elsevier Inc.
ISSN 0076-6879, DOI: 10.1016/S0076-6879(09)05008-3   All rights reserved.

deacetylation. Here we review experimental approaches that can be used *in vitro* and *in vivo* to study the role of acetylation in mitochondrial cell biology.

1. INTRODUCTION

Mitochondria are double-membrane bound organelles that play a central role in energy production. Increasing evidence suggests a key role for mitochondrial dysfunction in diabetes, aging, various neurodegenerative disorders, and cancer (Lambert and Brand, 2007; Michan and Sinclair, 2007; Reeve *et al.*, 2008). Acetylation has recently emerged as an important posttranslational modification to regulate mitochondrial proteins, and the recent identification of sirtuins as mitochondrial deacetylases has brought intense interest to the role of protein acetylation in mitochondrial biology. Sirtuins in mammals are a family of seven protein deacetylases/ADP ribo-syltransferases (SIRT1–SIRT7) that target a wide range of cellular proteins for deacetylation. Sirtuins have recently been shown to be involved in longevity pathways (Gan and Mucke, 2008), DNA repair (Lombard *et al.*, 2008), and the control of metabolic enzymes (Schwer and Verdin, 2008).

Among the seven sirtuins, SIRT3, SIRT4, and SIRT5 are located in mitochondria (Michishita *et al.*, 2005). SIRT3 exhibits robust deacetylase activity on chemically acetylated histone H4 peptides, whereas SIRT5 has weak but detectable activity, and SIRT4 has no detectable activity on the same substrate (Verdin *et al.*, 2004). Results from SIRT3, SIRT4, and SIRT5 knock out mice suggest that SIRT3 is the major regulator of mitochondria acetylation regulation (Lombard *et al.*, 2007). SIRT3 is also the most studied mitochondrial sirtuin and has been shown to deacetylate acetyl CoA synthetase 2 (Schwer *et al.*, 2006), glutamate dehydrogenase (Lombard *et al.*, 2007; Schlicker *et al.*, 2008), and respiratory complex I proteins (Ahn *et al.*, 2008). Less is known about SIRT4 and SIRT5, with each only having one known substrate (Schlicker *et al.*, 2008). The key to understanding the functional roles of sirtuins in mitochondrial pathways is to study the control sirtuins have over their various substrates in mitochondria. The methods provided herein are experimental protocols routinely used to study SIRT3 and mitochondrial acetylated proteins.

2. PURIFICATION OF ENZYMATICALLY ACTIVE SIRT3

SIRT3 protein with high enzymatic activity can be purified by immunoprecipitation after transient or stable transfection in mammalian cells. Recombinant proteins can also be expressed in *Escherichia coli* in an enzymatically active form.

2.1. Purification of SIRT3 from mammalian cell culture systems

To generate the SIRT3 plasmid, SIRT3 with a C-terminal FLAG tag is first cloned into the pcDNA3.1 vector. After entering the mitochondrial matrix, the N-terminal mitochondrial targeting sequence of SIRT3 is cleaved by the matrix processing peptidase (Schwer et al., 2002). For this reason, fusion proteins incorporating a C-terminal epitope should be used.

Constructs based on pcDNA3.1 are used to transiently express SIRT3 in a variety of mammalian cell lines, including 293, HeLa, Jurkat, NIH3T3, COS7, and CHO cells. Depending on the cell type, 5×10^6–1×10^7 cells (corresponding to a 10 cm culture dish) are transfected according to standard procedures either by the $CaPO_4$ co-precipitation method or by lipofection. Cells are maintained in growth medium and are harvested 48 h later. When the transfection efficiency ranges from 50 to 90%, 5×10^6 -1×10^7 cells are sufficient for expression and immunoprecipitation of enough SIRT3 protein for up to four radioactive enzymatic HDAC assays.

To generate stably transfected SIRT3–FLAG cell lines, pcDNA3.1 vectors encoding human SIRT3 fused to the FLAG-epitope are linearized by restriction enzyme digestion and purified by agarose gel electrophoresis. HEK293 cells of a low passage number are transfected with the linearized pcDNA3.1 SIRT3–FLAG expression plasmid by lipofection according to common procedures. After 48 h, the cells are split into 15-cm tissue culture dishes at 3×10^4, 1×10^5, and 3×10^5 cells/dish. For selection of stable expressing cell clones, cultures are incubated in standard culture medium containing 700 μg/ml G418 sulfate until visible colonies appear. This takes approximately 3 weeks, and the medium should be changed every 3–4 days. Nontransfected cells (1×10^6 cells/dish as a control) die over this selection period. Plates are rinsed with phosphate-buffered saline (PBS), and single colonies are isolated by trypsinization using cloning cylinders. For further culturing of the stable transfected cell clones, the G418 sulfate concentration is reduced to 450 μg/ml. As soon as the clones reach more than 1×10^6 cells, they are frozen in liquid nitrogen. Expression of SIRT3–FLAG is analyzed by western blotting of total cellular extracts with anti-FLAG antibodies, and small-scale immunoprecipitations are tested for SIRT3 activity in enzymatic assays. Indirect immunofluorescence can be used to verify protein expression in the majority of cells. Single clones are expanded for further use if they test positive for expression of exogenous proteins, have high expression levels of SIRT3–FLAG, and have high deacetylase activity in immunoprecipitation–mediated HDAC activity assays.

To purify active SIRT3 enzyme, 293T cells are transiently transfected with the expression vector by the calcium phosphate DNA precipitation method. After 48 h of transfection, cells are washed twice in PBS by centrifugation and the final cell pellet is lysed in five packed cell pellet volumes of lysis buffer (buffer A) in the presence of protease inhibitor cocktail (Roche), for 30 min at 4 °C. After lysis, cellular debris is cleared by centrifugation at 14,000g for 10 min at 4 °C, and the supernatant is transferred to a new tube.

The resulting lysates are subjected to immunoprecipitation with an anti-FLAG M2 agarose affinity gel (Sigma). Protein concentration is measured in cell lysates with a detergent-compatible protein assay kit (BCA protein assay kit, PIERCE). Anti-FLAG M2 affinity gel should be added at 10 μl/ml of lysate and incubated for at least 1 h at 4 °C with constant agitation. After incubation, immune complexes are pelleted with the agarose beads by centrifugation at 6,000g for 1 min at 4 °C. The immune complexes are washed three times in buffer A for 15 min each at 4 °C with agitation. After each wash, the beads are pelleted and the supernatant is replaced with buffer A. The immune complexes are then washed twice for 15 min each in the buffer used for the enzymatic assay. The beads containing the immunoprecipitated enzyme are now ready to use in the assay.

2.2. Purification of SIRT3 from bacterial expression systems

Human SIRT3 cDNA corresponding to the mature protein sequence (amino acid 102–399; Schwer *et al.*, 2002) is cloned into pTrcHis2C (Invitrogen). The resulting proteins are C-terminal myc and 6×His tagged fusion proteins. To generate catalytically inactive mutant of SIRT3 (pTrcHis2C–SIRT3–H248Y), histidine248 is substituted to tyrosine using site-directed mutagenesis kit (Stratagene). These plasmids are transformed in DH5α bacteria (Gibco) for protein expression, because SIRT3 is well expressed in DH5α, but expression and purification of other proteins might be difficult. For protein expression, 1 l of transformed bacterial culture is grown in LB–Amp to an optical density of 0.4 (A_{600}), induced with 0.5 mM IPTG at 30 °C overnight. For preparation of acetylated substrates, 50 mM nicotinamide is added during protein induction period to inhibit bacterial NAD$^+$-dependent deacetylase. The bacteria are pelleted by centrifugation and re-suspended in lysis buffer (1×PBS, 1% Tween-20, 0.2 mM PMSF; 2 ml/g wet weight). This mixture is sonicated on ice (four 20–30 s bursts at 40–60% power) and centrifuged at 4 °C at 14,000g for 10 min. The supernatant (cleared lysate) is bound to Ni–NTA resin [1 ml of 50% Ni–NTA slurry for 4 ml of cleared lysate, batch method (Qiagen)] on a rotary mixer at 4 °C for 60 min. Then, the batch mixture is passed through a commercial column (Polyprep, BioRad), and the flow-through is saved. The resin bed is washed five times with 4 ml of wash buffer (1×PBS, 20 mM imidazole). Bound proteins are eluted three times with 0.5 ml of elution buffer (1×PBS, 250 mM imidazole). All eluted fractions are collected and dialyzed with dialysis buffer (50 mM Tris–HCl, pH 8.0, 50 mM NaCl, 0.2 mM DTT, 10% Glycerol) using Slide-A-Lyzer Dialysis Cassettes, 10K MWCO (Pierce) at 4 °C overnight. Finally, dialyzed proteins are pooled and concentrated using centrifuge concentrating spin columns (Centricon, MWCO 10KD), and recombinant protein is aliquoted and stored at $-80\,^{\circ}C$.

3. SIRT3 ENZYMATIC DEACETYLATION ASSAY

Recombinant SIRT3 protein has been shown to possess deacetylase activity against histone peptide, as well as recombinant AceCS2 (Schwer et al., 2006). Furthermore, we have confirmed many mitochondrial proteins are deacetylated by recombinant SIRT3 (Lombard et al., 2007). Incubation with recombinant SIRT3 and substrates might be applied to show a protein of interest is deacetylated by SIRT3.

3.1. Peptide–substrate *in vitro* SIRT3 deacetylation assay

Deacetylase assays were performed in 100 μl of deacetylase buffer (4 mM MgCl$_2$, 0.2 mM dithiothreitol, 50 mM Tris–HCl, pH 9.0) containing immunoprecipitated proteins or mitochondrial lysates and a peptide corresponding to the first 23 amino acids of histone 4 chemically acetylated *in vitro* (Emiliani et al., 1998). The histone peptide was acetylated *in vitro* by overnight incubation with [3H]-acetate (5 mCi, 5.3 Ci/mmol; NEN) in 500 μl of EtOH in the presence of 0.24 M benzotriazole 1-yloxy-tris (dimethylamino) phosphonium hexafluorophosphate (Sigma–Aldrich) and 0.2 M triethylamine followed by reverse-phase HPLC purification. Where indicated, 1 mM NAD$^+$, 5 mM nicotinamide, or 400 nM TSA (WAKO) was added. Deacetylation reactions were stopped after 2 h of incubation at room temperature by adding 25 μl of stop solution (0.1 M HCl, 0.16 M acetic acid). Released acetate was extracted into 500 μl ethyl acetate, and samples were vigorously shaken for 15 min. After centrifugation for 5 min, 400 μl of the ethyl acetate fraction was mixed with 5 ml of scintillation fluid (Packard), and the released radioactivity was measured with a liquid scintillation counter.

3.2. Protein–substrate *in vitro* SIRT3 deacetylation assay

3.2.1. Preparation of recombinant substrate

To prepare the protein–substrate-based SIRT3 *in vitro* deacetylation assay, human cDNA corresponding to mature AceCS2 (38–689; Schwer et al., 2006) is cloned into pTrcHis2C (Invitrogen). The resulting proteins are C-terminal myc and 6×His tagged fusion proteins. This plasmid is transformed in DH5α bacteria (Gibco) for protein expression. AceCS2 is well expressed in DH5α, but expression and purification of other proteins might be difficult, so it is important to optimize the bacterial expression system for each protein. To express recombinant AceCS2 for the *in vitro* deacetylation assay, 1 l of transformed bacterial culture is grown in LB-Amp to an optical density of 0.4 (A_{600}), induced with 0.5 mM IPTG at 30 °C for overnight. For preparation of substrates, 50 mM nicotinamide is added during protein induction period to inhibit bacterial NAD^+-dependent deacetylase. The bacterial culture is pelleted by centrifugation and re-suspended in lysis buffer (1×PBS, 1% Tween-20, 0.2 mM PMSF; 2 ml/g wet weight). This mixture is sonicated on ice (four 20–30s bursts at 40–60% power) and centrifuged at 4 °C at 14,000g for 10 min. The supernatant (cleared lysate) is bound to Ni–NTA resin [1 ml of 50% Ni–NTA slurry for 4 ml of cleared lysate, batch method (Qiagen)] on a rotary mixer at 4 °C for 60 min. The batch mixture is passed through a commercial column (Polyprep, BioRad), and the flow-through is saved. The resin bed is washed five times with 4 ml of wash buffer (1×PBS, 20 mM imidazole). Bound proteins are eluted three times with 0.5 ml of elution buffer (1×PBS, 250 mM imidazole). All eluted fractions are collected and dialyzed with dialysis buffer (50 mM Tris–HCl, pH 8.0, 50 mM NaCl, 0.2 mM DTT, 10% Glycerol) using Slide-A-Lyzer Dialysis Cassettes, 10K MWCO (Pierce) at 4 °C for overnight. Dialyzed proteins are pooled and concentrated using centrifuge concentrating spin columns (Centricon, MWCO 10KD). Recombinant protein is aliquoted and stored at −80 °C.

It should be noted that expression and purification in DH5α might be difficult for some substrates, although SIRT3 and AceCS2 are well expressed in this strain. BL21(DE3) or any other strain that can grow in the presence of 50 mM nicotinamide can also be used to obtain recombinant substrates.

3.2.2. Enzymatic assay

To perform the recombinant protein–substrate *in vitro* SIRT3 deacetylation assay, 10 μg of purified recombinant AceCS2 and 2 μg of purified recombinant SIRT3 are incubated in 20 μl of SDAC buffer (50 mM Tris–HCl pH 9.0, 4 mM MgCl$_2$, 50 mM NaCl, 0.5 mM DTT, 1 mM NAD^+, 0.5 μM TSA) at 37 °C for 3 h with gentle agitation. The reaction is then stopped by adding 5 μl of 5× SDS sample buffer (250 mM Tris pH 6.8, 500 mM DTT, 10% SDS, 0.1% Bromophenol Blue, 50% glycerol). Deacetylation of substrate is detected by western blotting with anti-acetyllysine specific antibody

It should be noted that (1) SDAC buffer should be filtered (0.4 μm) and stored at −20 °C., and NAD^+ and DTT are added immediately before use, (2) proper incubation time and amount of SIRT3 may vary depending on your protein of interest; and (3) anti-acetyllysine specific antibodies are commercially available from several companies. We recommend either Cell Signaling Technology monoclonal or polyclonal antibody for the first trial. These antibodies can recognize many mitochondrial proteins (Lombard *et al.*, 2007).

3.3. *In vivo* SIRT3 deacetylation assay

Overexpression of SIRT3 in mammalian cells is shown to decrease acetylation of its substrates, whereas siRNA knockdown of SIRT3 increases acetylation (Schwer *et al.*, 2006). We have adopted this *in vivo* assay to identify SIRT3 substrates as well as *in vitro* SIRT3 deacetylation assay described above.

3.3.1. Plasmids and siRNA oligonucleotides

Human SIRT3 cDNA is cloned into a modified pcDNA3.1+ vector (Invitrogen) to yield hSIRT3 with a C-terminal HA- or FLAG-tag (Schwer *et al.*, 2002). Site-directed mutagenesis (QuikChange Mutagenesis Kit; Stratagene) is used to construct human SIRT3 H248Y mutant. Human AceCS2 cDNA is cloned into the pcDNA3.1+ vectors to yield C-terminal FLAG- or HA-tagged AceCS2. Double-stranded siRNAs directed against human SIRT3 are commercially available (Dharmacon).

It should be noted that substrates must *not* have the same tag as SIRT3, because SIRT3 can interact with many endogenously acetylated proteins (unpublished data). This interaction causes high-back ground of western blotting with anti-acetylated lysine antibody.

3.3.2. Cell culture, transfection and RNA interference

Constructs based on pcDNA3.1 are used to transiently co-express SIRT3 and its substrates in a variety of mammalian cell lines, including 293, HeLa, Jurkat, NIH3T3, COS7, and CHO cells. Depending on the cell type, 5×10^6–1×10^7 cells (corresponding to a 10 cm culture dish) are transfected according to standard procedures either by the $CaPO_4$ co-precipitation method or by lipofection. Cells are maintained in growth medium for 24–48 h before harvesting to allow for accumulation of proteins. For the RNA interference experiment, 2×10^6 293T cells are plated on 10 cm culture dish, and transfected with 200 pmol of siRNA using 10 μl of Dharmafect1 (Dharmacon). After 24–48 h of transfection, cells are transfected with the plasmids for AceCS2–FLAG by using the $CaPO_4$ co-precipitation method, and cultured additional 24–48 h.

3.3.3. Immunoprecipitation and immunoblotting

The following buffers were used. (1) IP buffer: 50 mM Tris–HCl, pH 7.5, 150 mM NaCl, 5 mM EDTA, 0.5% NP-40, 1 μM TSA, 5 mM nicotinamide, protease inhibitor cocktail (Sigma); and (2) Wash buffer: 50 mM Tris–HCl, pH 7.5, 500 mM NaCl, 5 mM EDTA, 0.5% NP-40, 1 μM TSA, 5 mM nicotinamide, protease inhibitor cocktail (Sigma). Buffers are filtered (0.4 μm) and stored at 4 °C. Protease inhibitors are added immediately before use.

Fresh or frozen cell pellets are re-suspended in IP buffer containing protease inhibitors at a ratio of five volumes of buffer to one volume of wet cell mass. Cell lysates are sonicated twice on ice for 10–15 s. Lysates are cleared by centrifugation at 14,000g for 10 min at 4 °C. Whole cell lysates are incubated with M2 agarose beads (Sigma) or HA agarose beads (Roche) at 20 μl/ml for 1 h with rocking at 4 °C. Immunocomplexes bound to the agarose beads are washed three times with 10 volumes of wash buffer each. Then the bound proteins are eluted by adding equal volume of 2× SDS sample buffer (100 mM Tris pH 6.8, 200 mM DTT, 4% SDS, 0.04% Bromophenol Blue, 20% glycerol) and heating at 95 °C for 5 min. Acetylation of HA or FLAG tagged proteins are detected by western blotting with anti-acetylated lysine specific antibody (Cell Signaling Technology).

It should be noted that anti-acetyllysine specific antibodies are commercially available from several companies. As described above, either Cell Signaling Technology monoclonal or polyclonal antibodies will be the best for the first trial.

4. DETECTION OF ACETYLATED PROTEINS IN MITOCHONDRIA

Previous studies have demonstrated that multiple proteins are acetylated in mammalian mitochondria (Kim *et al.*, 2006). Acetylation of mitochondrial proteins typically occurs on a single or multiple lysine residues, which can be immunoprecipitated from and detected in purified mitochondria with specific anti-acetyllysine antiserum.

This basic protocol describes the isolation of the heavy mitochondrial fraction from a mitochondrial-rich tissue, such as murine liver. Other sources of mitochondria can be used for detection of acetylated proteins, (i.e., heart, brown adipose tissue, skeletal muscle, cultured cells, yeast). The isolation protocol is a rapid method that only requires low- and high-speed centrifugation and should be accessible to many laboratories. While the resulting mitochondrial fractions will be contaminated to varying degrees by smaller particles, such as lysosomes and peroxisomes, these preparations can be used directly for detection of acetylated mitochondrial proteins or as starting material for further purification by density-gradient separations.

4.1. Isolation and purification of the heavy mitochondrial fraction from murine liver

Adult male mice (~25 g) are sacrificed according to Institutional Animal Care and Use Committee (IACUC) protocols. After opening the abdominal cavity, the gall bladder must be carefully removed without puncture or bile leakage. The liver is then transferred to chilled liver homogenization medium (LHM), containing, 0.2 M mannitol, 50 mM sucrose, 10 mM KCl, 1 mM EDTA, and 10 mM HEPES (pH 7.4). After briefly rinsing, the liver is minced on ice and homogenized in LHM containing a protease inhibitor cocktail (Roche), 500 nm trichostatin A (TSA), and 10 mM nicotinamide (LHM+) using a chilled glass Dounce or Potter–Elvehjem homogenizer. Nicotinamide and TSA are class III and class I histone deacetylase inhibitors, respectively, and are added to prevent the deacetylation of proteins *in vitro*. The liver tissue homogenate is then decanted and stored on ice, and the procedure is repeated until all liver tissue samples have been homogenized. All homogenates are then centrifuged for 10 min at 1000g, 4 °C, using a low-speed centrifuge. The supernatant is decanted and centrifuged for 10 min at 3000g, 4 °C, using a high-speed centrifuge. The resulting supernatant is then aspirated, leaving a brown mitochondrial pellet. Often times, a lipid layer can be seen at the surface after high-speed centrifugation, which should be removed. The mitochondrial pellet is then re-suspended in LHM and recentrifuged for 10 min at 3000g, 4 °C, and repeated for a total of three washes. After the final wash, the mitochondria are re-suspended in a suitable buffer, such as LHM, for further experimentation.

4.2. Detection of acetylated mitochondrial proteins by immunoprecipitation

For the detection of acetylated intra–mitochondrial proteins, mitochondria isolated as described in Section 4.1 are lysed in ice-cold mitochondrial IP (MIP) buffer [1% n-dodecyl-β-D-maltoside (Sigma), 0.5 mM EDTA, 150 mM NaCl, 50 mM Tris–HCl, (pH 7.4), 10 mM nicotinamide, 500nM trichostatin A] containing a protease inhibitor cocktail (Roche). n-Dodecyl-β-D-maltoside is a common detergent added to lyse the mitochondria and solubilize the mitochondrial proteins. The lysates are centrifuged for 10 min at 10,000g, 4 °C, using a high-speed centrifuge. The lysates are then incubated with primary antibodies for 1 h at 4 °C with gentle agitation. Primary antibody is typically added in an amount 10× more concentrated than is used for western blotting; 1 μg/100 μl is a good starting point. The mitochondrial lysate is then incubated further with the mixture of equal amounts of Protein A agarose and Protein G agarose (Santa Cruz) for 1 h. In most cases, 20 μl Protein A/G is sufficient for IP, but it may depend on the primary antibody and sample volume.

The agarose pellets are then washed three times with MIP buffer. To minimize non-specific binding or decrease the stringency, NaCl concentration in the IP buffer can be changed from 50 to 500 mM according to the primary antibody; or, the NaCl concentration in wash buffer can be changed. The bound proteins are extracted with the SDS–PAGE loading buffer by heating at 95 °C for 5 min. Immunoblotting with anti–acetyllysine antibodies (Cell Signaling Technology) will demonstrate whether the immunoprecipitated protein is acetylated.

4.3. Detection of multiple acetylated mitochondrial proteins by reverse immunoprecipitation

Another immunoprecipitation technique, called a reverse-IP, can also be used to detect acetylated proteins. A reverse-IP has multiple advantages, including the ability to immunoprecipitate many acetylated proteins in a single experiment, probe for novel acetylated proteins, and the ability to detect acetylation in proteins whose primary antibodies are not suitable for immunoprecipitation. A major disadvantage of this technique is the inability to directly show a protein is acetylated, which could be mistaken for an associated protein of similar molecular weight. As such, the reverse-IP technique is an important tool to identify and detect acetylated mitochondrial proteins.

Mitochondria isolated as described in Section 4.1 are lysed in ice-cold MIP buffer [1% n-dodecyl-β-D-maltoside (Sigma), 0.5 mM EDTA, 150 mM NaCl, 50 mM Tris–HCl, (pH 7.4), 10 mM nicotinamide, 500 nM trichostatin A] containing a protease inhibitor cocktail (Roche). The lysates are centrifuged for 10 min at 10,000g, 4 °C, using a high-speed centrifuge. Anti-acetyllysine antibodies (Cell Signaling Technologies) are added to the mitochondrial lysate and incubated overnight at 4 °C with gentle agitation. Often times, probing with multiple anti-acetyllysine antibodies in parallel, such as antibodies raised against mouse and against rabbit, will immunoprecipitate different acetylated proteins and increase protein detection coverage. The mitochondrial lysate is then incubated with the mixture of agarose for 1 h.

These complexes were washed four times in MIP buffer. The bound proteins are extracted with the SDS-PAGE loading buffer by heating at 95 °C for 5 min. Immunoblotting with antibodies against the protein of interest will demonstrate whether the acetylated protein is present in greater amounts in one of the samples.

5. CONCLUSION

Mitochondrial acetylation is becoming recognized as an important posttranslational modification to regulate enzymatic activity and global nutrient homeostasis. SIRT3 is the primary mitochondrial deacetylase and can be used in the laboratory to further define the role of acetylation in mitochondria.

REFERENCES

Ahn, B. H., Kim, H. S., Song, S., Lee, I. H., Liu, J., Vassilopoulos, A., Deng, C. X., and Finkel, T. (2008). A role for the mitochondrial deacetylase Sirt3 in regulating energy homeostasis. *Proc. Natl. Acad. Sci. USA* **105,** 14447–14452.

Emiliani, S., Fischle, W., Van Lint, C., Al-Abed, Y., and Verdin, E. (1998). Characterization of a human RPD3 ortholog, HDAC3. *Proc. Natl. Acad. Sci. USA* **95,** 2795–2800.

Gan, L., and Mucke, L. (2008). Paths of convergence: Sirtuins in aging and neurodegeneration. *Neuron* **58,** 10–14.

Kim, S. C., Sprung, R., Chen, Y., Xu, Y., Ball, H., Pei, J., Cheng, T., Kho, Y., Xiao, H., Xiao, L., Grishin, N. V., White, M., *et al.* (2006). Substrate and functional diversity of lysine acetylation revealed by a proteomics survey. *Mol. Cell* **23,** 607–618.

Lambert, A. J., and Brand, M. D. (2007). Research on mitochondria and aging, 2006–2007. *Aging Cell* **6,** 417–420.

Lombard, D. B., Alt, F. W., Cheng, H. L., Bunkenborg, J., Streeper, R. S., Mostoslavsky, R., Kim, J., Yancopoulos, G., Valenzuela, D., Murphy, A., Yang, Y., Chen, Y., *et al.* (2007). Mammalian Sir2 homolog SIRT3 regulates global mitochondrial lysine acetylation. *Mol. Cell Biol.* **27,** 8807–8814.

Lombard, D. B., Schwer, B., Alt, F. W., and Mostoslavsky, R. (2008). SIRT6 in DNA repair, metabolism and ageing. *J. Intern. Med.* **263,** 128–141.

Michan, S., and Sinclair, D. (2007). Sirtuins in mammals: Insights into their biological function. *Biochem. J.* **404,** 1–13.

Michishita, E., Park, J. Y., Burneskis, J. M., Barrett, J. C., and Horikawa, I. (2005). Evolutionarily conserved and nonconserved cellular localizations and functions of human SIRT proteins. *Mol. Biol. Cell* **16,** 4623–4635.

Reeve, A. K., Krishnan, K. J., and Turnbull, D. M. (2008). Age related mitochondrial degenerative disorders in humans. *Biotechnol. J.* **3,** 750–756.

Schlicker, C., Gertz, M., Papatheodorou, P., Kachholz, B., Becker, C. F., and Steegborn, C. (2008). Substrates and regulation mechanisms for the human mitochondrial sirtuins Sirt3 and Sirt5. *J. Mol. Biol.* **382,** 790–801.

Schwer, B., Bunkenborg, J., Verdin, R. O., Andersen, J. S., and Verdin, E. (2006). Reversible lysine acetylation controls the activity of the mitochondrial enzyme acetyl-CoA synthetase 2. *Proc. Natl. Acad. Sci. USA* **103,** 10224–10229.

Schwer, B., North, B. J., Frye, R. A., Ott, M., and Verdin, E. (2002). The human silent information regulator (Sir)2 homologue hSIRT3 is a mitochondrial nicotinamide adenine dinucleotide-dependent deacetylase. *J. Cell Biol.* **158,** 647–657.

Schwer, B., and Verdin, E. (2008). Conserved metabolic regulatory functions of sirtuins. *Cell Metab.* **7,** 104–112.

Verdin, E., Dequiedt, F., Fischle, W., Frye, R., Marshall, B., and North, B. (2004). Measurement of mammalian histone deacetylase activity. *Methods Enzymol.* **377,** 180–196.

CHAPTER NINE

Non-radioactive Detection of Palmitoylated Mitochondrial Proteins Using an Azido-Palmitate Analogue

Morris A. Kostiuk,* Bernd O. Keller,[†] *and* Luc G. Berthiaume*

Contents

1. Introduction	150
2. Synthesis of 12-Azidododecanoate and 14-Azidotetradecanoate	153
2.1. Methyl 12-hydroxydodecanoate	153
2.2. Methyl 12-methanesulfonyloxydodecanoate	154
2.3. Methyl 12-azidododecanoate	154
2.4. 12-Azidododecanoic acid	154
2.5. Methyl 14-hydroxytetradecanoate	154
2.6. 14-Azidotetradecanoic acid	155
3. Synthesis of Tagged Phosphine Probes	155
4. Preparation of Azido-Fatty Acids and Tagged Phosphine Stock Solutions	155
5. Production of 14-Azidotetradecanoyl-CoA Derivatives	156
6. *In Vitro* Labeling of Proteins with 14-Azidotetradecanoyl-CoA and Tagged Phosphines	156
6.1. Protein acylation reaction	156
6.2. Chemoselective ligation of tagged phosphines	157
7. Isolation of Intact Mitochondria from Livers of Sprague–Dawley Rats	157
8. Chromatography of Soluble Mitochondrial Proteins	158
9. Labeling of Anion Exchange Eluates with 14-Azidotetradecanoyl-CoA and Phosphine–Biotin	158

* Department of Cell Biology, Faculty of Medicine and Dentistry, University of Alberta, Edmonton, Alberta, Canada

† Department of Pathology and Laboratory Medicine, Child & Family Research Institute, University of British Columbia, Vancouver, British Columbia, Canada

Methods in Enzymology, Volume 457   © 2009 Elsevier Inc.

ISSN 0076-6879, DOI: 10.1016/S0076-6879(09)05009-5

10. Identification of Labeled Proteins by Mass Spectrometry   159
    10.1. In-gel digestion and peptide extraction   159
    10.2. Mass spectrometric analysis of peptide extracts and
          database searching   159
11. Demonstration that Acylation of Cysteine Residues Occurs via a
    Thioester Bond   160
    11.1. Cysteine alkylation using N-ethylmaleimide   161
    11.2. Treatments of protein samples with neutral hydroxylamine
          and alkali   161
12. Concluding Remarks   162
References   163

Abstract

While palmitoylation is typically thought of as a cytosolic process resulting in membrane attachment of the palmitoylated proteins, numerous mitochondrial proteins have been shown to be palmitoylated following *in vitro* labeling of mitochondria with radioactive or bioorthogonal analogues of fatty acids. The fatty acylation of two liver mitochondrial enzymes, methylmalonyl semialdehyde dehydrogenase and carbamoyl phosphate synthetase 1, has been studied in great detail. In both cases palmitoylation of an active site cysteine residue occurred spontaneously and resulted in inhibition of enzymatic activity, thus, suggesting that palmitoylation may be a direct means to regulate the activity of metabolic enzymes within the mitochondria.

The progress of investigators working on protein fatty acylation has long been impeded by the long exposure time required to detect the incorporation of [^{3}H]-fatty acids into protein by fluorography (often 1–3 months or more). Significant reduction in exposure times has been achieved by the use of [^{125}I]-iodofatty acids but these analogues are also hazardous and not commercially available. Herein, we describe a sensitive chemical labeling method for the detection of palmitoylated mitochondrial proteins. The method uses azidofatty acid analogues that can be attached to proteins and reacted with tagged phosphines via a modified Staudinger ligation. Recently, we used this labeling method, combined with mass spectrometry analysis of the labeled proteins, to identify 21 palmitoylated proteins from rat liver mitochondria.

1. INTRODUCTION

The covalent attachment of fatty acids to proteins affects a wide range of functions including membrane localization and attachment, protein sorting and secretion, protein–protein interactions, enzyme regulation, and viral budding (reviewed in Bhatnagar and Gordon (1997), Bijlmakers and Marsh (2003), Casey (1995), Dunphy and Linder (1998), Linder and Deschenes (2003), Linder and Deschenes (2007), Resh (1999, 2006b, and

Smotrys and Linder (2004)). Myristoylation and palmitoylation (also termed S-acylation) are the two most common types of fatty acylation. The two modifications differ in many characteristics including specificity of acyl chain length, type of bond formation, modified amino acid, enzymology, and functionality. In myristoylation, the saturated 14-carbon fatty acid myristate is added to the N-terminal glycine residue of proteins containing a distinct consensus sequence via a stable amide bond by one of two N-myristoyl transferases (NMT1 and NMT2). In palmitoylation, the saturated 16-carbon fatty acid palmitate is typically added to variably located cysteine residues within a protein via a thioester bond by one of two types of protein fatty acyltransferases (PATs). The two types of enzymes involved in catalyzing the palmitoylation have been identified as transmembrane proteins either containing an aspartate–histidine–histidine–cysteine (DHHC) signature motif in combination with a cysteine rich domain (CRD) (DHHC-CRD) or belonging to the family of membrane-bound O-acyltransferases (MBOATs) (reviewed in Mitchell *et al.* (2006) Miura and Treisman (2006), Resh (2006a), and Tsutsumi *et al.* (2008)). Palmitoylation has also been shown to occur spontaneously on many mitochondrial proteins (Berthiaume *et al.*, 1994; Corvi *et al.*, 2001; Kostiuk *et al.*, 2008; Stucki *et al.*, 1989). The labile nature of the thioester bond allows protein palmitoylation to be a dynamic modification suited for a regulatory role.

While palmitoylation is typically thought of as a cytosolic process resulting in membrane attachment of the palmitoylated protein (Casey, 1995; Dunphy and Linder, 1998; Linder and Deschenes, 2007; Mumby, 1997; Resh, 1999, 2006b), we and others have shown that the mitochondrial compartment contains several palmitoylated proteins and that palmitoylation of mitochondrial proteins could play a role in the regulation of intermediary metabolism (Berthiaume *et al.*, 1994; Corvi *et al.*, 2001; Kostiuk *et al.*, 2008; Stucki *et al.*, 1989). Of note, we previously demonstrated the existence of approximately 40 palmitoylated proteins in rat liver mitochondria (Corvi *et al.*, 2001).

The fatty acylation of two mitochondrial enzymes, methylmalonyl semialdehyde dehydrogenase (MMSDH) (Berthiaume *et al.*, 1994) and carbamoyl phosphate synthetase 1 (CPS 1) (Corvi *et al.*, 2001), has been studied in great detail. Since acylation of MMSDH occurred at its active site cysteine residue, varied with the mitochondrial energy level, was specific for long chain fatty acyl-CoAs and was reversible, it was proposed that active site fatty acylation of MMSDH could act as a novel mode for the regulation of its enzymatic activity (Berthiaume *et al.*, 1994). Extensive characterization of the palmitoylation of CPS 1 has also been carried out (Corvi *et al.*, 2001), and revealed that CPS 1 is as well specifically palmitoylated on at least one of its active site cysteine residues in an apparent spontaneous fashion. Gross and co-workers recently demonstrated that the glycolytic enzyme, glyceraldehydes-3-phosphate dehydrogenase (GAPDH), is also inhibited by palmitoylation of a cysteine residue (Yang *et al.*, 2005). Altogether, these observations further support the idea that protein palmitoylation could regulate the activity of enzymes involved in intermediary metabolism.

Progress in the study of fatty acylation has been hampered by the lack of rapid detection methods. Until recently, detection of fatty acylated proteins relied on the tedious and hazardous incorporation of radioactive [³H]-fatty acids into proteins followed by lengthy fluorographic exposures (often months). Subsequently Resh and co-workers developed the use of [¹²⁵I]-iodofatty acids as labels, which reduced exposure time significantly but these analogues are also hazardous and not commercially available (Berthiaume *et al.*, 1995). Two very recent breakthroughs led to major improvements in detection and identification of fatty acylated proteins: use of the acyl-biotinyl exchange reaction and use of azido-myristate and azido-palmitate analogues.

The acyl-biotin exchange reaction is an indirect labeling method for the detection and purification of palmitoylated proteins from complex protein extracts (Drisdel and Green, 2004; Roth *et al.*, 2006; Wan *et al.*, 2007). It substitutes biotinyl moieties for protein palmitoyl-modifications via a sequence of three chemical treatment steps. The use of azido-fatty acids results in direct labeling of acylated proteins and can also lead to their biotinylation (Kostiuk *et al.*, 2008; Martin *et al.*, 2008).

To gain a better understanding of the role played by palmitoylation of mitochondrial proteins in metabolism, we sought to identify the complement of palmitoylated mitochondrial proteins (mitochondrial palmitoylome). To do so, we developed a sensitive chemical labeling method to detect palmitoylated proteins that is compatible with mass spectrometry analysis to identify the labeled proteins. The three step labeling method described herein employs the isosteric alkyl-azide bioorthogonal (synthetic analogues with functional handles which are tolerated by enzymes) palmitate analogue, 14-azidotetradecanoate. First the analogue is converted to its CoA derivative using acyl-CoA synthetase (Devadas *et al.*, 1992) (Fig. 9.1A). Second, it is spontaneously incorporated into proteins *in vitro* (Kostiuk *et al.*, 2008) (Fig. 9.1B). Third, the 14-azidotetradecanoylated proteins are reacted with tagged phosphines in a highly specific manner (Kostiuk *et al.*, 2008) using a modified Staudinger ligation (Saxon and Bertozzi, 2000) (Fig. 9.1C).

Unlike aryl-azides, alkyl-azides are stable at physiological temperature and under ambient or UV light (Prescher *et al.*, 2004). Furthermore, alkyl-azides are nontoxic to cells or animals (Devadas *et al.*, 1992). To detect palmitoylated proteins our laboratory has successfully used 14-azidotetradecanoic acid to label proteins followed by ligation of the 14-azidotetradecanoylated proteins to phosphine–myc, –fluorescein, and – biotin probes. Detection is carried out by protein blotting or flatbed scanner analysis. Typically the film exposure time required to obtain a

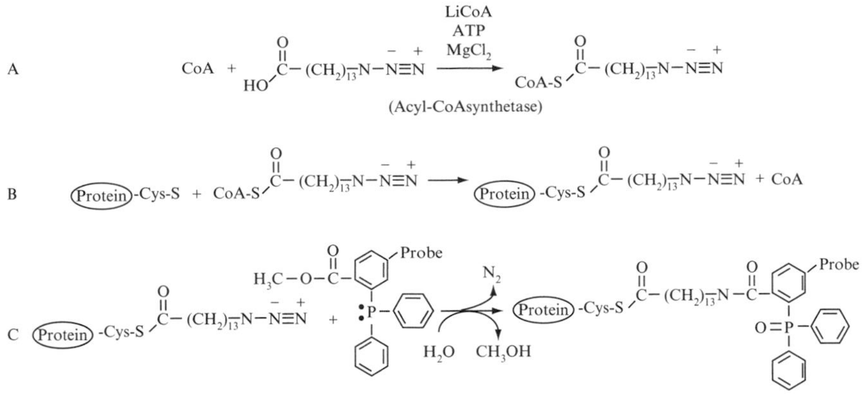

A

$$CoA + \underset{HO}{\overset{O}{\underset{\|}{C}}}-(CH_2)_{\overline{13}}N-N\equiv N \xrightarrow[\text{(Acyl-CoAsynthetase)}]{\overset{\text{LiCoA}}{\overset{\text{ATP}}{\text{MgCl}_2}}} CoA-S\overset{O}{\overset{\|}{C}}-(CH_2)_{\overline{13}}N-N\equiv N$$

B

$$\text{Protein}\text{-Cys-S} + CoA-S\overset{O}{\overset{\|}{C}}-(CH_2)_{\overline{13}}N-N\equiv N \longrightarrow \text{Protein}\text{-Cys-S}\overset{O}{\overset{\|}{C}}-(CH_2)_{\overline{13}}N-N\equiv N + CoA$$

C

Figure 9.1 *In vitro* labeling of palmitoylated mitochondrial proteins. (A) Azidotetra-decanoic acid is readily esterified to its active CoA derivative azidotetradecanoyl-CoA. (B) The azidotetradecanoic acid moiety is non-enzymatically transferred and cova-lently linked to cysteine residues of the target protein(s) via a thioester bond. (C) The azide moiety of azidotetradecanoic acid is coupled to a tagged phosphine forming an amide bond via the Staudinger ligation. Probe = myc epitope, fluorescein, or biotin.

quantifiable signal was in the range of a few seconds to minutes. In addition to improving the detection of palmitoylated proteins, we also ameliorated mass spectrometry techniques allowing identification of the labeled palmitoylated proteins. Recently, we used these improved meth-odologies to identify 21 palmitoylated proteins from rat liver mitochondria (Kostiuk *et al.*, 2008).

2. SYNTHESIS OF 12-AZIDODODECANOATE AND 14-AZIDOTETRADECANOATE

2.1. Methyl 12-hydroxydodecanoate

Concentrated H_2SO_4 (22 mg) is added dropwise with stirring to a 0 °C solution of 12-hydroxydodecanoic acid (Aldrich Chem. Co., 500 mg, 2.31 mmol) in MeOH (25 ml). After stirring at room temperature for 12 h, the solvent is evaporated *in vacuo* to near dryness and the residue is partitioned between EtOAc and water. The organic extract is washed with saturated aqueous $NaHCO_3$ solution, dried over Na_2SO_4, and concentrated *in vacuo*. The residue is purified via SiO_2 column chromatography eluting with 50% EtOAc/hexanes to give methyl 12-hydroxydodecanoate (501 mg, 94%) as a colorless oil. NMR (300 MHz, $CDCl_3$) δ 1.23–1.42 (m, 12 H), 1.51–1.69 (m, 4 H), 2.30 (t, $J = 7.2$ Hz, 2 H), 3.64 (t, $J = 6.9$ Hz, 2 H), 3.66 (s, 3 H).

2.2. Methyl 12-methanesulfonyloxydodecanoate

Methanesulfonyl chloride (498 mg, 4.34 mmol) is added dropwise to a stirring 0 °C solution of the above methyl 12-hydroxydodecanoate (500 mg, 2.17 mmol) and triethylamine (449 mg, 5.49 mmol) in dry CH_2Cl_2 (8 ml). After 20 min at the same temperature, the reaction mixture is diluted with more CH_2Cl_2 (10 ml), then washed sequentially with water and brine, dried over Na_2SO_4, and concentrated *in vacuo*. The pale yellow, oily residue is used in the next reaction without further purification. NMR (300 MHz, CDC_{l3}) δ 1.22–1.48 (m, 14 H), 1.56–1.68 (m, 2 H), 1.69–1.79 (m, 2 H), 2.30 (t, J = 7.2 Hz, 2 H), 3.00 (s, 3 H), 3.67 (s, 3 H).

2.3. Methyl 12-azidododecanoate

A solution of the above methyl 12-methanesulfonyloxydodecanoate (665 mg, 2.06 mmol) and NaN_3 (402 mg, 6.195 mmol) in dry DMF (8 ml) is stirred at 70 °C for 14 h, then cooled to room temperature, diluted with water (20 ml), and extracted with Et_2O (3 × 20 ml). The combined ethereal extracts are washed with water, dried over Na_2SO_4, and concentrated *in vacuo*. The residue is purified via SiO_2 column chromatography eluting with 10% EtOAc/hexanes to give methyl 12-azidododecanoate (473 mg, 90% over two steps) as a colorless oil. NMR (300 MHz, $CDCl_3$) δ 1.22–1.41 (m, 14 H), 1.54–1.68 (m, 4 H), 2.30 (t, J = 7.2 Hz, 2 H), 3.25 (t, J = 6.6 Hz, 2 H), 3.6 (s, 3 H).

2.4. 12-Azidododecanoic acid

LiOH (82 mg, 7.68 mmol) is added to a room temperature solution of the above methyl 12-azidododecanoate (490 mg, 1.92 mmol) in THF/H_2O (7.5 ml, 4:1). After stirring for 12 h, the reaction mixture is acidified with 1 N HCl to pH 4.5 and extracted with EtOAc (3 × 20 ml). The combined organic extracts are washed with brine, dried over Na_2SO_4, and concentrated *in vacuo*. The residue is purified via SiO_2 column chromatography eluting with 20% EtOAc/hexanes to give the known (Nagarajan *et al.*, 1997) 12-azidododecanoic acid (443 mg, 95%) as a colorless oil. NMR (300 MHz, $CDCl_3$) δ 1.22–1.42 (m, 14 H), 1.54–1.69 (m, 4 H), 2.35 (t, J = 7.2 Hz, 2 H), 3.25 (t, J = 6.9 Hz, 2 H).

2.5. Methyl 14-hydroxytetradecanoate

A stream of ozone is passed through a −78 °C solution of methyl 13(E)-docosenoate (350 mg, 0.819 mmol; NuChek Prep, Inc.) in CH_2Cl_2 (8 ml) until TLC analysis indicates complete consumption of the olefin (~30 min).

Excess Ph$_3$P (214 mg, 1.24 mmol) is added in one portion and the mixture is allowed to stir at room temperature for 12 h, at which time all volatiles are removed *in vacuo*. The residue is dissolved in MeOH (4 ml), cooled to 0 °C and NaBH$_4$ (15.5 mg, 0.62 mmol) is added in portions. After 1 h the reaction mixture is concentrated *in vacuo* and the residue is diluted with saturated aqueous NH$_4$Cl (10 ml) and extracted with EtOAc (3 × 20 ml). The combined organic extracts are washed with brine, dried over Na$_2$SO$_4$, and concentrated *in vacuo*. The residue is purified via SiO$_2$ column chromatography eluting with 30% EtOAc/hexanes to give methyl 14-hydroxytetradecanoate (182 mg, 78%) as a colorless oil. NMR (300 MHz, CDCl$_3$) δ 1.23–1.39 (m, 8H), 1.52–1.68 (m, 4H), 2.30 (t, J = 7.2 Hz, 2 H), 3.64 (t, J = 6.9 Hz, 2 H), 3.66 (s, 3 H).

2.6. 14-Azidotetradecanoic acid

The above methyl 14-hydroxytetradecanoate is converted to the corresponding mesylate (94%), azido methyl ester (92%), and finally to the known (Davis *et al.*, 1996) azido free acid (94%) exactly as described above for its C$_{12}$-analogues. Chromatographic and spectral properties are also virtually identical except for the presence of two additional methylene units.

3. SYNTHESIS OF TAGGED PHOSPHINE PROBES

Phosphine–biotin, phosphine–fluorescein, and phosphine–myc are synthesized exactly as previously described (Chang *et al.*, 2007; Kostiuk *et al.*, 2008; Luchansky *et al.*, 2003; Saxon and Bertozzi, 2000).

4. PREPARATION OF AZIDO-FATTY ACIDS AND TAGGED PHOSPHINE STOCK SOLUTIONS

The tagged phosphines are dissolved in 100 mM PBS, pH 7.4 containing 30% *N*,*N*-dimethylformamide (DMF) at a final concentration of 5 mM and stored under argon at −80 °C. Argon storage prevents the oxidation of the tagged phosphines.

12-Azidododecanoic acid and 14-azidotetradecanoic acid are dissolved in dimethyl sulphoxide (DMSO) at a final concentration of 100 mM and stored under argon at −80 °C.

5. PRODUCTION OF 14-AZIDOTETRADECANOYL-CoA DERIVATIVES

A 1 m*M* stock solution of 14-azidotetradecanoyl-CoA is produced through the enzymatic coupling of 14-azidotetradecanoic acid and coenzyme A by *Pseudomonas* acyl-CoA synthetase (Sigma cat. # A 3352) as originally described (Towler and Glaser, 1986).

For convenience we recommend preparation of the following stock solutions:

2× reaction buffer, 20 m*M* Tris–HCl pH 7.4, 2 m*M* DTT, and 0.2 m*M* EGTA (prepared fresh)

Acyl-CoA synthetase in 50 m*M* HEPES pH 7.4 (1 unit/ml), aliquot and store at −80 °C

200 m*M* ATP in water. Adjust to pH 7.0 with NaOH, aliquot and store at −20 °C

20 m*M* LiCoA in water. Adjust to pH 5.0 with NaOH, aliquot and store at −20 °C

10% Triton X-100. Stored at 4 °C in the dark

100 m*M* azido-fatty acids in DMSO. Stored under argon at −80 °C

To make a 500 μl stock solution of 1 m*M* azido-fatty acyl-CoA, the reaction solution is prepared to contain final concentrations of 5 m*M* ATP, 1 m*M* LiCoA, 0.15 units/ml of acyl-CoA synthetase, 0.25% Triton X-100 and 1× buffer in a glass vial at room temperature. A glass vial is used to minimize adhesion of the fatty acid analogues to the walls of the reaction container as is often seen with plastic tubes. To avoid the formation of a precipitant, add the 5 μl of the 100 m*M* stock solution of azido-fatty acid slowly to the reaction solution while vortexing gently. Allow the reaction to proceed at room temperature for 1 h in the dark. Use immediately or store at −20 °C in aliquots. Avoid multiple freeze thaw cycles.

6. *IN VITRO* LABELING OF PROTEINS WITH 14-AZIDOTETRADECANOYL-CoA AND TAGGED PHOSPHINES

6.1. Protein acylation reaction

Proteins are acylated in a final volume of 30 μl containing 1–3 μg purified proteins (10 μg crude mitochondria extracts) in the presence of 100 μ*M* 14-azidotetradecanoyl-CoA in 50 m*M* HEPES pH 7.4 for 30 min at room temperature in the dark.

6.2. Chemoselective ligation of tagged phosphines

Following protein acylation, the reaction solution is adjusted to 1% SDS, 5 mM DTT, 50 mM HEPES pH 7.4, and 200 μM of the appropriate tagged phosphine. The ligation reaction is carried out in the dark for 2 h at room temperature. For SDS–PAGE separation, one fourth volume 5× SDS–PAGE loading buffer containing 200 mM DTT is added and the samples are heated to 100 °C for 2 min prior to performing SDS–PAGE.

7. Isolation of Intact Mitochondria from Livers of Sprague–Dawley Rats

Mitochondria are purified from rat liver following a previously published protocol (Susin *et al.*, 2000) with modifications. *Note*: all buffers should be chilled to 4 °C and contain freshly added Complete™ (Roche) protease inhibitor (1 tablet/50ml). All isolation steps should be carried out on ice whenever possible. To obtain 50–60 mg of liver mitochondria, typically one 10 g rat liver is used. The following protocol is described for one rat liver and can be scaled up accordingly. The liver is promptly removed, rinsed with H buffer (300 mM saccharose, 5 mM HEPES, 200 μM EGTA, pH 7.2) and minced with a razor blade. The tissue is suspended in 3:1 (v/w) H buffer/tissue and homogenized with 20 strokes using a Potter-Thomas homogenizer at 500 rpm with cooling on ice after each fifth stroke. The homogenate is centrifuged for 10 min at 800g at 4 °C. The supernatant is recovered and the pellet is re-suspended in 10 ml of H buffer and centrifuged as described above. The two supernatants are combined and centrifuged for 10 min at 8750g at 4 °C. The pellets from the centrifugation are re-suspended in 1 ml H buffer and loaded on top of a discontinuous Percoll gradient (60%/30%/18%) in P buffer (300 mM saccharose, 10 mM HEPES, 200 μM EGTA pH 6.9) and centrifuged for 30 min at 8750g at 4 °C. Purified mitochondria are recovered from the 30%/60% Percoll gradient interface, diluted 10-fold in H buffer and centrifuged for 10 min at 6800g at 4 °C to remove the Percoll, which is toxic to mitochondria. The supernatants are discarded and the mitochondria are re-suspended in 1 ml IM buffer (250 mM mannitol, 5 mM HEPES pH 7.4, 0.5 mM EDTA, 0.1% essentially fatty acid free BSA, 1 mM DTT). The mitochondrial protein concentration is then determined using the Bradford protein assay following lysis of the mitochondria with 1% Triton X-100. The mitochondrial sample used for the protein assay should contain no more than 0.1% final Triton X-100 concentration to prevent interference with the assay. The mitochondria are either immediately lysed in hypotonic lysis buffer for use in purification as described below or stored in aliquots at −80 °C.

8. Chromatography of Soluble Mitochondrial Proteins

To isolate gel bands originating from a single protein for mass spectrometry analysis, we first subject the soluble mitochondrial proteins to anion exchange chromatography to separate proteins by their charge content followed by electrophoresis to separate the proteins by size.

Fifty milligrams of fresh or thawed mitochondria are centrifuged for 5 min at 10,000*g* at 4 °C to pellet the mitochondria. The IM buffer is discarded and the mitochondrial pellets are re-suspended in 5 ml hypotonic lysis buffer. The hypotonic lysis buffer is made by dissolving one tablet of Complete™ (Roche) protease inhibitor in 50 ml ddH$_2$O. The suspension is incubated on ice for 30 min to allow the mitochondria to swell. The mitochondria are then lysed by sonication using four rounds of 40 s each, at an output setting of five on a Branson Sonifier 450 with 2 min intervals on ice between sonications. The lysed mitochondria are centrifuged for 12 min at 16,000*g* at 4 °C. The supernatant is added to 45 ml of degassed and filtered TD buffer (20 m*M* Tris–HCl pH 8.5, 1 m*M* DTT) and applied to a Source Q (Pharmacia Biotech) anion exchange column containing 8 ml of beads. Bound proteins are washed with two column volumes of TD buffer and eluted with increasing NaCl concentrations in TD buffer. Two milliliter elution fractions are collected from a 30 ml gradient of 0–300 m*M* NaCl followed by a 30 ml gradient of 300 m*M*–1 M NaCl with a flow rate of 1 ml/min. Aliquots of the fractions are used immediately for labeling reactions. The unused portions are adjusted to 20% glycerol and stored at −80 °C.

9. Labeling of Anion Exchange Eluates with 14-Azidotetradecanoyl-CoA and Phosphine–Biotin

Aliquots of 160 μl from each fraction of the ion exchange column are incubated for 30 min at room temperature with 100 μ*M* 14-azidotetradecanoyl-CoA, 50 m*M* HEPES pH 7.4 in a final volume of 200 μl. At the end of the incubation period, the reactions are adjusted to 1% SDS, 5 m*M* DTT, 50 m*M* HEPES pH 7.4, and 200 μ*M* of phosphine–biotin in a final volume of 240 μl. The reaction is allowed to proceed for another 2 h at room temperature in the dark, stopped by addition of one fourth volume 5× SDS–PAGE loading buffer containing 200 m*M* DTT and heating to 100 °C for 2 min. For electrophoresis the samples are vortexed briefly and half the sample is loaded onto each of two 10% SDS gels. One gel is subjected to protein blotting analysis for the detection of the biotin moiety attached to proteins and one gel is stained with Bio-safe Coomassie blue (Bio-Rad) to accommodate coring of the protein bands for mass spectrometry analysis. Detection of the biotinylated–azido-fatty acylated proteins is performed using NeutrAvidin[TM]-HRP, (Pierce Biotech. Rockford, IL) at a dilution of 1:20,000 (1 μg/20 ml) in PBS containing 0.1% Tween-20 and 3.5% BSA, and bands that are 14-azidotetradecanoylated and sufficiently separated from each other are excised from the corresponding Bio-safe Coomassie stained gel and analyzed by mass spectrometry as detailed below.

10. IDENTIFICATION OF LABELED PROTEINS BY MASS SPECTROMETRY

10.1. In-gel digestion and peptide extraction

The sample preparation protocol follows closely the one published online for Coomassie-stained 1D gel bands by the University of the Missouri Proteomics Center (http://proteomics.missouri.edu) and as recently described (Kostiuk et al., 2008). Excised protein bands in 1.5 ml polypropylene vials are minced to ~1 mm^3 pieces, washed three times with 1:1 (v/v) 100 mM NH$_4$HCO$_3$/acetonitrile (ACN), dehydrated with ACN and air dried. Reduction is performed with 10 mM DTT in 100 mM NH$_4$HCO$_3$ for 30 min at 56 °C in a volume sufficient to cover all gel pieces. After cooling to room temperature, excess DTT solution is removed and discarded and the samples are alkylated with sufficient 50 mM iodoacetamide (IAA) in 100 mM NH$_4$HCO$_3$ for 30 min at room temperature in the dark. Excess IAA solution is removed; the gel pieces are washed again with 1:1 (v/v) 100 mM NH$_4$HCO$_3$/ACN and dehydrated for 20 min with ACN. Excess ACN is removed. Gel pieces are rehydrated with sufficient 0.03 μg/μl bovine trypsin in 450:50:500 (v/v/v) 100 mM NH$_4$HCO$_3$/ACN/H$_2$O for 2 h on ice. Excess trypsin solution is removed and sufficient 450:50:500 (v/v/v) 100 mM NH$_4$HCO$_3$/ACN/H$_2$O is added. Incubation is allowed to proceed for 16 h at 37 °C. After centrifugation, excess liquid is transferred into a new 1.5 ml vial, and gel pieces are extracted three times for 10 min each with 25–50 μl of 1% TFA in 60:40 (v/v) ACN/H$_2$O. The combined extracts are then concentrated to approximately 20 μl volumes in a vacuum centrifuge.

10.2. Mass spectrometric analysis of peptide extracts and database searching

For peptide mapping with matrix–assisted laser desorption/ionization time-of-flight mass spectrometry (MALDI TOF MS), 3 μl of the concentrated peptide extract is mixed 1:1 with a saturated α-cyano-4-hydroxy-cinnamic acid (4–HCCA) solution in 30:70 (v/v) methanol/H_2O. For sample spotting, a two–layer method is employed as previously described (Dai *et al.*, 1999). Briefly, ~1 μl of a 12 mg/ml 4–HCCA solution in 80:20 (v/v) acetone/methanol is spotted onto the MALDI stainless steel target to form a thin first layer. As a second layer 0.5 μl of the peptide extract matrix mixture is spotted onto the dried first layer and air dried. Two washes with 0.5 μl H_2O each are performed. For this, the water droplets are blown off with pressurized air 20 s after spotting. For analysis of the MALDI targets, we use a Voyager DE–STR MALDI TOF MS (Applied Biosystems, Framingham, MA), located at the UBC Proteome Core Facility. The instrument is operated in positive reflectron mode. Calibration is done first externally with known peptide standards, and before data analysis, internal calibration is performed with known matrix cluster signals and tryptic autolysis peptide signals, to achieve a mass accuracy of 100 ppm. About 100–200 single laser shot spectra are accumulated per sample. Peptide masses are then tabulated and submitted to the Mascot peptide mapping program (http://www.matrixscience.com) for protein identification retrieval. Only probability scores, p, <0.05 are considered. For proteins for which identity is based on peptide mapping with five or fewer peptides, hypothetical proteins, and proteins identified in gel bands with multiple proteins present, further confirmation of the proteins is performed with ESI–MS/MS analysis of selected tryptic peptides with protein-specific sequences. The instrument we use is a micro–QTOF mass spectrometer (Waters/Micromass, Manchester, UK) equipped with a nanospray source. Protein digests are directly infused with a flow of ~250 nl/min in 20% ACN in H_2O (measured with a 10 μl syringe connected to the nano–LC outlet). Fragment masses are submitted to the Mascot sequence query program (http://www.matrixscience.com).

11. DEMONSTRATION THAT ACYLATION OF CYSTEINE RESIDUES OCCURS VIA A THIOESTER BOND

Although palmitoylation is a term typically used to describe the covalent attachment of long chain fatty acids to cysteine residues via a thioester bond, palmitate and other long chain fatty acids have been shown to be covalently attached to proteins on N-terminal cysteine residues via amide bonds (Pepinsky *et al.*, 1998) and internal serine residues via oxyester bonds (Takada *et al.*, 2006). Since reversible protein palmitoylation (*S*-acylation) appears to be the sole type of fatty acylation occurring in mitochondria labeled *in vitro* with palmitoyl-CoA, we use two assays to confirm that the attachment of 14-azidotetradecanoic acid occurs on cysteine residues of proteins via a thioester bond. To confirm the involvement of cysteine residues in the covalent protein acylation, cysteine residues are alkylated using N-ethylmaleimide (NEM) or other cysteine alkylating compounds prior to the acylation reaction. To determine whether the fatty acids are linked to the proteins via an amide bond or an ester bond, labeled proteins are treated with NaOH, KOH–methanol or hydroxyl-amine (Bizzozero, 1995). While amide bonds are resistant to alkali hydrolysis or hydroxylaminolysis, ester bonds are sensitive to these treatments. Furthermore, hydroxylamine treatment at neutral pH cleaves thioester bonds but not oxyesters. Cleavage of oxyester bonds is carried out using hydroxylamine at pH 10. Therefore, hydroxylamine treatment can be used to differentiate between thioester and oxyester fatty acid-protein linkages (Bizzozero, 1995; Magee *et al.*, 1984; Schmidt and Lambrecht, 1985). To qualify as an *S*-acylated protein *in vitro*, such proteins must satisfy two conditions: the palmitoylation of the protein must be blocked by cysteine alkylation *and* the labeled fatty acid attached to the protein must be removed by hydroxylamine or alkali treatments.

Note: the azido-fatty acids are coupled to the tagged phosphines through an amide linkage, resistant to the alkali or hydroxylamine treatments (Fig. 9.1C).

11.1. Cysteine alkylation using N-ethylmaleimide

For cysteine alkylation, proteins or protein solutions are incubated with 10 mM NEM in 50 mM HEPES pH 7.4 for 30 min to 1 h, to allow the NEM to react with the cysteine residues of the proteins, prior to the addition of the 14-azidotetradecanoic acid for the acylation reaction. Since NEM also reacts with primary amines, buffers containing primary amines such as *tris* (hydroxymethyl) aminomethane (Tris) should not be used for this reaction.

11.2. Treatments of protein samples with neutral hydroxylamine and alkali

11.2.1. In-gel neutral hydroxylamine treatment

For in-gel removal of thioester linked fatty acids or fatty acid analogues, following the labeling reactions and SDS–PAGE the gels are fixed with a 45:10:55 (v/v/v) methanol/acetic acid/water solution for 30 min, rinsed in water, then incubated with rocking for 24 h in 1 M hydroxylamine pH 7.0 in 1:1 (v/v) isopropanol/water, or 1 M Tris–HCl pH 7.0 in 1:1 (v/v) isopropanol/water as a control. The gels are then washed four times for 12 h in 1:1 (v/v) isopropanol/water. When the proteins are labeled with the combination of 14-azidotetradecanoic acid/phosphine–fluorescein, following the washes the gels can be directly scanned on a phosphorimager in the fluorescence mode. Scanning the gels immediately after electrophoresis and again following hydroxylamine treatment allows a direct comparison of the pre- and post-treatment fluorescence. The gels are then stained with Coomassie blue to visualize protein content.

11.2.2. Alkali treatment of PVDF membrane

Since the acylation detection is carried out by protein blotting when ligating the azido–fatty acids to phosphine–biotin or phosphine–myc probes, it is necessary to remove the labeled palmitate analogue from the palmitoylated proteins that are bound to the PVDF (Millipore) membrane. To do so, following transfer of the proteins to the membrane, but prior to probing the membranes with antibodies or the NeutrAvidin^{TM}-HRP, the membranes are incubated at room temperature with rocking for 1 h in 0.1 M KOH in 9:1 (v/v) methanol/water (10 ml 1 M KOH, 90 ml MeOH) or 0.1 M Tris–HCl pH 7.0 in 9:1 (v/v) methanol/water (10 ml 1 M Tris–HCl pH 7.0, 90 ml MeOH) as a control. The membranes are then rinsed once in methanol, once with water, and six times with PBS prior to standard membrane blocking and probing protocols. Since milk may contain biotin or endogenously biotinylated proteins, to avoid any background or quenching of the avidin-HRP conjugates we recommend the use of 3.5% BSA (Fraction V, Sigma A3294) in PBS containing 0.1% Tween-20 for the membrane blocking and probing steps.

11.2.3. Hydroxylamine and alkali treatment of proteins in solution

To remove the labeled fatty acid analogues from proteins in solution, following *in vitro* acylation of the proteins and ligation of the azide moiety to the tagged phosphines, the reaction is adjusted to 1 M hydroxylamine pH 7.0 or 0.2 M NaOH and incubated at room temperature for 16 h or 1 h, respectively. One and 0.2 M Tris–HCl pH 7.0 are used as a control for the hydroxylamine and alkali treatments, respectively. After the incubation period, one fourth volume 5× SDS–PAGE loading buffer containing 200 mM DTT added and the samples are heating to 100 °C and SDS–PAGE is performed. Depending on the tagged phosphine used, the detection of labeled palmitoylated proteins is carried out by either scanning of the gel (for phosphine–fluorescein-tagged proteins) or protein blotting analysis (for phosphine–biotin and phosphine–myc-tagged proteins). Of note, boiling labeled proteins with excessive amounts of reducing agent can lead to hydrolysis of thioester bonds.

12. CONCLUDING REMARKS

The method described herein has proved to be a highly specific, rapid and sensitive means to detect *in vitro* labeled palmitoylated mitochondrial proteins. Using these protocols, we have successfully identified 21

palmitoylated proteins from rat liver mitochondria including the confirmation of two previously characterized palmitoylated proteins, MMSDH and
CPS 1. This thereby further validated our methodology (Kostiuk et al.,
2008). By performing anion exchange chromatography and SDS–PAGE we
were able to sufficiently separate the soluble mitochondrial proteins for
isolation and mass spectrometry identification. Importantly, this protocol
was proved to efficiently separate labeled proteins for identification purposes
since 19 of the 21 bands analyzed contained peptides originating from a single
protein (Kostiuk et al., 2008). We further characterized the palmitoylation
of 3-hydroxy-3-methylglutaryl-CoA synthase (HMGCS2), which like
MMSDH and CPS 1 occurred spontaneously on a cysteine residue (did not
require an enzymatic transfer by a palmitoyl acyltransferase). Spontaneous
palmitoylation of the mitochondrial enzymes MMSDH, CPS 1, and now
HMGCS2 (Berthiaume et al., 1994; Corvi et al., 2001; Kostiuk et al., 2008)
suggests that this labeling method could readily be used to obtain activity-
based protein profiles (Speers and Cravatt, 2004) of proteins in mitochondria
isolated from the tissues of normal and pathophysiological animals.

As a reversible spontaneous reaction dependent on mitochondrial
palmitoyl-CoA concentrations, palmitoylation is emerging as a novel
means to regulate the activity of certain mitochondrial enzymes involved
in intermediary metabolism. The use of our bioorthogonal azido–fatty acid
analogues will actively enable further characterization of the 19 newly
discovered palmitoylated rat liver mitochondrial proteins and might lead
to the identification and characterization of novel roles for protein
palmitoylation.

REFERENCES

Berthiaume, L., Deichaite, I., Peseckis, S., and Resh, M. D. (1994). Regulation of enzymatic
    activity by active site fatty acylation. A new role for long chain fatty acid acylation of
    proteins. J. Biol. Chem. **269,** 6498–6505.
Berthiaume, L., Peseckis, S. M., and Resh, M. D. (1995). Synthesis and use of iodo-fatty acid
    analogs. Methods Enzymol. **250,** 454–466.
Bhatnagar, R., and Gordon, J. I. (1997). Understanding covalent modifications of proteins
    by lipids: Where cell biology and biophysics mingle. Trends cell boil. **7,** 14–20.
Bijlmakers, M. J., and Marsh, M. (2003). The on-off story of protein palmitoylation. Trends
    Cell. Biol. **13,** 32–42.
Bizzozero, O. A. (1995). Chemical analysis of acylation sites and species. Methods Enzymol.
    250, 361–379.
Casey, P. J. (1995). Protein lipidation in cell signaling. Science **268,** 221–225.
Chang, P. V., Prescher, J. A., Hangauer, M. J., and Bertozzi, C. R. (2007). Imaging cell
    surface glycans with bioorthogonal chemical reporters. J. Am. Chem. Soc. **129,**
    8400–8401.
Corvi, M. M., Soltys, C. L., and Berthiaume, L. G. (2001). Regulation of mitochondrial
    carbamoyl-phosphate synthetase 1 activity by active site fatty acylation. J. Biol. Chem.
    276, 45704–45712.

Dai, Y., Whittal, R. M., and Li, L. (1999). Two-layer sample preparation: A method for MALDI-MS analysis of complex peptide and protein mixtures. *Anal. Chem.* **71**, 1087–1091.

Davis, S. C., Sui, Z., Peterson, J. A., and Ortiz de Montellano, P. R. (1996). Oxidation of omega-oxo fatty acids by cytochrome P450BM-3 (CYP102). *Arch. Biochem. Biophys.* **328**, 35–42.

Devadas, B., Lu, T., Katoh, A., Kishore, N. S., Wade, A. C., Mehta, P. P., Rudnick, D. A., Bryant, M. L., Adams, S. P., and Li, Q. (1992). Substrate specificity of *Saccharomyces cerevisiae* myristoyl-CoA: Protein N-myristoyltransferase. Analysis of fatty acid analogs containing carbonyl groups, nitrogen heteroatoms, and nitrogen heterocycles in an *in vitro* enzyme assay and subsequent identification of inhibitors of human immunodeficiency virus I. replication. *J. Biol. Chem.* **267**, 7224–7239.

Drisdel, R. C., and Green, W. N. (2004). Labeling and quantifying sites of protein palmitoylation. *Biotechniques* **36**, 276–285.

Dunphy, J. T., and Linder, M. E. (1998). Signalling functions of protein palmitoylation. *Biochim. Biophys. Acta* **1436**, 245–261.

Kostiuk, M. A., Corvi, M. M., Keller, B. O., Plummer, G., Prescher, J. A., Hangauer, M. J., Bertozzi, C. R., Rajaiah, G., Falck, J. R., and Berthiaume, L. G. (2008). Identification of palmitoylated mitochondrial proteins using a bio-orthogonal azido-palmitate analogue. *FASEB J.* **22**, 721–732.

Linder, M. E., and Deschenes, R. J. (2003). New insights into the mechanisms of protein palmitoylation. *Biochemistry* **42**, 4311–4320.

Linder, M. E., and Deschenes, R. J. (2007). Palmitoylation: Policing protein stability and traffic. *Nat. Rev. Mol. Cell. Biol.* **8**, 74–84.

Luchansky, S. J., Hang, H. C., Saxon, E., Grunwell, J. R., Yu, C., Dube, D. H., and Bertozzi, C. R. (2003). Constructing azide-labeled cell surfaces using polysaccharide biosynthetic pathways. *Methods Enzymol.* **362**, 249–272.

Magee, A. I., Koyama, A. H., Malfer, C., Wen, D., and Schlesinger, M. J. (1984). Release of fatty acids from virus glycoproteins by hydroxylamine. *Biochim. Biophys. Acta* **798**, 156–166.

Martin, D. D., Vilas, G. L., Prescher, J. A., Rajaiah, G., Falck, J. R., Bertozzi, C. R., and Berthiaume, L. G. (2008). Rapid detection, discovery, and identification of post-translationally myristoylated proteins during apoptosis using a bio-orthogonal azidomyristate analog. *FASEB J.* **22**, 797–806.

Mitchell, D. A., Vasudevan, A., Linder, M. E., and Deschenes, R. J. (2006). Protein palmitoylation by a family of DHHC protein S-acyltransferases. *J. Lipid Res.* **47**, 1118–1127.

Miura, G. I., and Treisman, J. E. (2006). Lipid modification of secreted signaling proteins. *Cell. Cycle* **5**, 1184–1188.

Mumby, S. M. (1997). Reversible palmitoylation of signaling proteins. *Curr. Opin. Cell. Biol.* **9**, 148–154.

Nagarajan, S. R., Devadas, B., Zupec, M. E., Freeman, S. K., Brown, D. L., Lu, H. F., Mehta, P. P., Kishore, N. S., McWherter, C. A., Getman, D. P., Gordon, J. I., and Sikorski, J. A. (1997). Conformationally constrained [p-(omega-aminoalkyl)phenacetyl]-L-seryl-L-lysyl dipeptide amides as potent peptidomimetic inhibitors of Candida albicans and human myristoyl-CoA:protein N-myristoyl transferase. *J. Med. Chem.* **40**, 1422–1438.

Pepinsky, R. B., Zeng, C., Wen, D., Rayhorn, P., Baker, D. P., Williams, K. P., Bixler, S. A., Ambrose, C. M., Garber, E. A., Miatkowski, K., Taylor, F. R., Wang, E. A., *et al.* (1998). Identification of a palmitic acid-modified form of human Sonic hedgehog. *J. Biol. Chem.* **273**, 14037–14045.

Prescher, J. A., Dube, D. H., and Bertozzi, C. R. (2004). Chemical remodelling of cell surfaces in living animals. *Nature* **430,** 873–877.

Resh, M. D. (1999). Fatty acylation of proteins: New insights into membrane targeting of myristoylated and palmitoylated proteins. *Biochim. Biophys. Acta* **1451,** 1–16.

Resh, M. D. (2006). Trafficking and signaling by fatty-acylated and prenylated proteins. *Nat. Chem. Biol.* **2,** 584–590.

Resh, M. D. (2006a). Palmitoylation of ligands, receptors, and intracellular signaling molecules. *Sci. STKE* **2006,** re14.

Roth, A. F., Wan, J., Bailey, A. O., Sun, B., Kuchar, J. A., Green, W. N., Phinney, B. S., Yates, J. R., 3rd and Davis, N. G. (2006). Global analysis of protein palmitoylation in yeast. *Cell* **125,** 1003–1013.

Saxon, E., and Bertozzi, C. R. (2000). Cell. surface engineering by a modified Staudinger reaction. *Science* **287,** 2007–2010.

Schmidt, M. F., and Lambrecht, B. (1985). On the structure of the acyl linkage and the function of fatty acyl chains in the influenza virus haemagglutinin and the glycoproteins of Semliki Forest virus. *J. Gen. Virol.* **66,** 2635–2647.

Smotrys, J. E., and Linder, M. E. (2004). Palmitoylation of intracellular signaling proteins: Regulation and function. *Annu. Rev. Biochem.* **73,** 559–587.

Speers, A. E., and Cravatt, B. F. (2004). Chemical strategies for activity-based proteomics. *Chembiochem.* **5,** 41–47.

Stucki, J. W., Lehmann, L. H., and Siegel, E. (1989). Acylation of proteins by myristic acid in isolated mitochondria. *J. Biol. Chem.* **264,** 6376–6380.

Susin, S. A., Larochette, N., Geuskens, M., and Kroemer, G. (2000). Purification of mitochondria for apoptosis assays. *Methods Enzymol.* **322,** 205–208.

Takada, R., Satomi, Y., Kurata, T., Ueno, N., Norioka, S., Kondoh, H., Takao, T., and Takada, S. (2006). Monounsaturated fatty acid modification of Wnt protein: Its role in Wnt secretion. *Dev. Cell.* **11,** 791–801.

Towler, D., and Glaser, L. (1986). Protein fatty acid acylation: Enzymatic synthesis of an N-myristoylglycyl peptide. *Proc. Natl. Acad. Sci. USA* **83,** 2812–2816.

Tsutsumi, R., Fukata, Y., and Fukata, M. (2008). Discovery of protein-palmitoylating enzymes. *Pflugers Arch.* **456,** 1199–1206.

Wan, J., Roth, A. F., Bailey, A. O., and Davis, N. G. (2007). Palmitoylated proteins: Purification and identification. *Nat. Protoc.* **2,** 1573–1584.

Yang, J., Gibson, B., Snider, J., Jenkins, C. M., Han, X., and Gross, R. W. (2005). Submicromolar concentrations of palmitoyl-CoA specifically thioesterify cysteine 244 in glyceraldehyde-3-phosphate dehydrogenase inhibiting enzyme activity: A novel mechanism potentially underlying fatty acid induced insulin resistance. *Biochemistry* **44,** 11903–11912.

CHARACTERIZATION OF MITOCHONDRIAL KINASES, PHOSPHATASES, DYNAMICS, AND RESPIRATORY FUNCTION

CHAPTER TEN

Detection of a Mitochondrial Kinase Complex That Mediates PKA–MEK–ERK-Dependent Phosphorylation of Mitochondrial Proteins Involved in the Regulation of Steroid Biosynthesis

Cristina Paz,* Cecilia Poderoso,* Paula Maloberti,* Fabiana Cornejo Maciel,* Carlos Mendez,* Juan J. Poderoso,† *and* Ernesto J. Podestá*

Contents

1. Introduction	170
2. Steroidogenic Cells	171
3. Analysis of a Mitochondrial Kinase Complex	172
3.1. Involvement of PKA, MEK1/2, and ERK1/2 in mitochondrial function	172
3.2. Localization and activation of MEK and ERK at the mitochondria by western blot and confocal microscopy	174
3.3. Functional analysis of kinase interaction at the mitochondria	178
3.4. StAR as substrate of a kinase complex	181
4. Analysis of a Mitochondrial Acyl-CoA Thioesterase, Acot2 as a Phosphoprotein	186
4.1. Purification of Acot2 to homogeneity preserving its biological activity	187
4.2. Determination of a thioesterase activity	188
4.3. Identification of Acot2 as a phosphoprotein	188
References	190

* IIMHNO, Department of Biochemistry, School of Medicine, University of Buenos Aires, Paraguay, Buenos Aires, Argentina
† Laboratory of Oxygen Metabolism, University Hospital, University of Buenos Aires, Buenos Aires, Argentina

Methods in Enzymology, Volume 457
ISSN 0076-6879, DOI: 10.1016/S0076-6879(09)05010-1

169

Abstract

In order to achieve the goal of this article, as an example we will describe the strategies followed to analyze the presence of the multi-kinase complex at the mitochondria and the posttranslational modification of two key mitochondrial proteins, which participate in the regulation of cholesterol transport across the mitochondrial membranes and in the regulation of steroid biosynthesis. Hormones, ions or growth factors modulate steroid biosynthesis by the posttranslational phosphorylation of proteins. The question still remains on how phosphorylation events transmit a specific signal to its mitochondrial site of action.

Cholesterol transport requires specific interactions in mitochondria between several proteins including a multi-kinase complex. The presence of this multi-kinase complex at the mitochondria reveals the importance of the posttranslational modification of mitochondrial proteins for its activity and functions. The activation of PKA triggers the posttranslational modification of the mitochondrial acyl-CoA thioesterase (Acot2), which releases arachidonic acid (AA) in the mitochondria, and the activation of a kinase cascade that leads to the phoshorylation of the steroidogenic acute regulatory (StAR) protein. The function of StAR is to facilitate the access of cholesterol to the first enzyme of the biosynthesis process and its induction is dependent on Acot2 and intramitochondrial AA release. Truncation of the StAR protein is associated with the steroid deficiency disease, congenital lipoid adrenal hyperplasia.

1. INTRODUCTION

The number of mitochondria present in a cell corresponds closely to the functional status or to the energy requirements of a given tissue. For example, the heart, which contracts constantly, has the highest mitochondrial content of all tissues, whereas type Hx muscle fibers, which are recruited only periodically, have a low mitochondrial volume per fiber. Although the primary role of mitochondria is ATP provision, these organelles are also involved in other processes that are important for cellular function. These include calcium homeostasis, intracellular signal transduction and regulation of apoptosis. For instance, steroid biosynthesis is a complex processes that proceeds after a first step of enzymatic conversion of cholesterol that occurs in the mitochondria once cholesterol is transported across mitochondrial membranes. Thus, steroidogenic cells *contain* a large amount of mitochondria, part of which is dedicated to steroid production.

We describe here the strategies used to analyze the presence of a multi-kinase complex at mitochondria and the posttranslational modifications of two key mitochondrial proteins that participate in the regulation of cholesterol transport across the mitochondrial membranes and the regulation of steroid biosynthesis. In all steroidogenic tissues (testis, adrenal gland, ovary, placenta, and brain), the acute stimulation of steroid synthesis is under the control of a number of phosphorylation-dependent events that implicate the activation of numerous protein kinases. The cyclic-AMP (cAMP)-dependent protein kinase (PKA), the protein kinase C (PKC), the calcium/calmodulin-dependent protein kinase (CamK) and the mitogen activated protein kinases (MAPKs) are among the kinases involved in steroidogenesis. Thus, hormones, ions or growth factors modulate steroid biosynthesis by posttranslational phosphorylation of proteins. The question still remains *on* how phosphorylation events relay a specific signal to its mitochondrial site of action.

Cholesterol transport requires specific interactions in mitochondria between several proteins including the voltage-dependent anion channel (VDAC), the peripheral benzodiazepine receptor (PBR, currently named translocator protein or TSPO), the PBR/TSPO-associated protein (PAP7), an acyl-CoA-thioesterase (Acot2), PKA, the extracellular signal-regulated kinase (ERK), its upstream activator (MEK), and the steroidogenic acute regulatory (StAR) protein. The presence of a multi-kinase complex *in mitochondria* shows the importance of the posttranslational modification of mitochondrial proteins for its activity and functions. One of the mitochondrial proteins phosphorylated throughout the activation of the kinase cascade is StAR. The function of StAR is to facilitate the access of cholesterol to the first enzymatic step of the biosynthetic process. Truncation of StAR protein associates with a human steroid deficiency called congenital lipoid adrenal hyperplasia (Lin *et al.*, 1995). The second protein phosphorylated through a PKA-dependent pathway is the mitochondrial acyl-CoA thioesterase Acot2, which releases arachidonic acid (AA) in the mitochondria, and it is essential for StAR induction (Maloberti *et al.*, 2002, 2005).

2. STEROIDOGENIC CELLS

Steroid biosynthesis is a process finely regulated by hormones, ions, and growth factors that trigger a number of signal transduction pathways involving the activation of different kinases and phosphatases. Thus, steroidogenic cells constitute an appropriate system for the study of protein phosphorylation. Apart from animal models, cell lines are also available as study material. Some of the most used and accepted cell lines are MA-10, Y1, H295R, mLTC and a series of modifications. For example, Y1 cells were used to generate mutants defective in PKA or adenylylcyclase activity, Y1(kin) and Y1(cyc) cells, respectively (Al-Hakim *et al.*, 2004; Rae *et al.*, 1979; Schimmer *et al.*, 1977). There are several advantages in using cell lines.

First, the model is based upon a unique type of cells, thus facilitating the interpretation of results. Second, a large amount of material is easily and rapidly obtained. Third and most important, steroid production is easily evaluated in cell culture media by radioimmunoassay. All of our studies were performed in adrenocortical or interstitial testicular primary cell cultures or using MA-10 Leydig or Y1 adrenocortical cells.

The MA-10 cell line is a clonal strain of mouse Leydig tumor cells that produce progesterone rather than testosterone as the major steroid (Ascoli, 1981). The growth medium consists of Waymouth MB752/1, 20 mM HEPES, 50 μg/ml gentamycin and 15% horse serum. Cells are grown in a humidified atmosphere at 36.5 °C containing 5% CO_2. Stimulation of the cells is performed with chorionic gonadotropin (CG) or 8Br-cAMP in serum-free Waymouth medium, and steroid production is evaluated by determination of progesterone concentrations in the incubation media.

3. Analysis of a Mitochondrial Kinase Complex

3.1. Involvement of PKA, MEK1/2, and ERK1/2 in mitochondrial function

Adrenocorticotropin (ACTH) and Luteinizing Hormone (LH) stimulate steroid production in adrenocortical and testicular Leydig cells, respectively, through a mechanism that involves the activation of PKA. One of the effectors of PKA in those systems is StAR protein. Murine StAR is phosphorylated by PKA in Ser194 while Ser197 is the phosphorylated residue in human StAR, a posttranslational modification that regulates its cholesterol transport capacity.

A-kinase anchor proteins (AKAPs) enhance cAMP-dependent pathways by anchoring PKA near its cellular substrates (Feliciello *et al.*, 2001; Rubin, 1994). It has been demonstrated that murine AKAP121 tethers PKA to the mitochondrial outer surface (Angelo and Rubin, 2000; Diviani and Scott, 2001). In view of this finding, and taking into account the mitochondrial localization of StAR, it seems possible that AKAPs could provide spatial integration between PKA and StAR.

Other kinases involved in the activation of steroid production are ERK1/2 and its upstream activator MEK1/2 which act through genomic and non-genomic effects. The site of action of MEK1/2–ERK1/2 is located between PKA activation and cholesterol transport. Evidence for this type of regulation was obtained by studying steroid production in cells incubated with inhibitors of MEK activation such as U0126 or PD98095. The activation of the MEK1/2–ERK1/2 cascade is dependent on PKA activation, as demonstrated by siRNA experiments described below. The role of ERK in the regulation of steroid biosynthesis was also confirmed by studies using an overexpression approach of wild–type ERK1 and different mutated forms of the kinase (also described below).

3.1.1. Transient transfection of siRNA for PKA using cationic lipids reagents

Small interference RNA is one of the most important research tools for gene silencing since it is target-specific and the knock down efficiency of the specific protein can be easily assessed by immunoblot using antibodies against the targeted protein. The gene silencing strategy is of higher specificity than that achieved with inhibitors such as H89 or PKI commonly used in the study of PKA. Additionally, the siRNA may be directed to different isozymes of PKA making it possible to study the effect of a particular isoform within a system.

Transfection with cationic lipid reagents is considered the most appropriate and less harmful procedure for the introduction of siRNAs into cells. Our siRNA studies were carried out using MA-10 cells. The lipofection mixture consisted of 100 nM of siRNA of the α isoform of the catalytic subunit of PKA and Lipofectamine 2000 Reagent in Opti-MEM medium (Invitrogen) with 15% horse serum and without antibiotics according to Maloberti *et al.* (2005). Cells were transfected for approximately 48 h by adding the mixture to Waymouth medium without antibiotics with 15% horse serum. Transfection efficiency was monitored by transfection of the cells with a plasmid containing the coding sequence of the enhanced green fluorescent protein (pRc/CMVi-EGFP). Following transfection, the medium was replaced with complete Waymouth and cells were allowed to grow for additional 12 h, after which horse serum was removed and the culture remained serum-deprived for an extra 24 h period. ERK1/2 activity was evaluated in the mitochondrial fraction of MA-10 cells (obtained as described below) by means of an enzyme-linked immunosorbent assay (ELISA) that detects the phosphorylated form of the kinase (Sigma, St. Louis, MO) following stimulation of the cells with 0.5 mM 8Br-cAMP. Full activation of ERK1/2 in stimulated MA-10 cells was strictly dependent on the activation of α isoform of the catalytic subunit of PKA in mitochondria.

3.1.2. Constructs of wild-type ERK and ERK H230R for transfection and expression in MA-10 cells

Although inhibitors of MEK1/2 activity are commonly used during functional studies, these compounds are not entirely selective and can affect other kinases. An alternative approach to study the participation of ERK1/2 in a given pathway is the use of molecular biology techniques that modify the levels of the protein in question. That procedure allows for the overexpression of wild–type ERK 1/2 or of negative dominants of the kinase. In our case, we overexpressed wild–type ERK 1/2 and an inactive form of ERK2, the H230R variant, which fails to interact with MEK1 but retains the ability to interact with MEK2 in a weakened fashion (Robinson *et al.*, 2002) in MA-10 cells.

3.1.3. Transient transfection of ERK and ERK H230R by electroporation

MA-10 cell cultures growing in the logarithmic phase are harvested by trypsin digestion, centrifuged and resuspended at a final concentration of 1.6×10^7 cells/ml. Approximately 8×10^6 cells were placed in an electroporation cuvette in 400 μl of antibiotic-free Waymouth MB752/1 medium containing 15% horse serum in the presence of 30 μg of a midipreparation of empty 3xflag-CMV7 plasmid as mock control, 3xflag-CMV7-ERK2 wild type or 3xflag-CMV7-ERK2 H230R mutated form of ERK2. The cell suspension was gently mixed and electropored with a pulse of 0.3 V and 950 μF in the Gene Pulser® II system (Bio-Rad). After the pulse, the cuvettes containing the transfected cells were kept on ice for 10 min and then seeded in 75 cm^2 flasks. After 12 h, the transfection medium was replaced with complete medium and allowed to grow for additional 36 h before the experiment. Again, evaluation of transfection efficiency was monitored by means of a plasmid containing the cDNA for EGFP (pRc/CMVi-EGFP). Usually, 30–35% transfection efficiency was achieved using the described electroporation protocol.

Overexpression of a wild-type form of ERK2 in MA-10 cells resulted in an increment of 8Br-cAMP-stimulated steroid production, while the inactive form of ERK2, H230R variant, failed to stimulate steroidogenesis (Poderoso *et al.*, 2008), thus confirming the role of active ERK1/2 in the regulation of steroid synthesis.

3.2. Localization and activation of MEK and ERK at the mitochondria by western blot and confocal microscopy

The mitochondrion is the key organelle in the steroid production process. Given the proposed role of MEK/ERK in steroidogenesis, it is of interest to explore the possible mitochondrial localization of MEK1/2, ERK1/2, phospho-MEK, and phospho-ERK. For that purpose, we used a combination of western blot and confocal microscopy.

3.2.1. Subcellular fractionation and assessment of the purity of each fraction

For subcellular fractionation, cells were rinsed with PBS, scrapped with 1 ml ice-cold PBS, and collected by centrifugation at 800g for 10 min. Cells were then resuspended in MSHE buffer (250 mM Mannitol, 219 mM Sacarose, 0.02% EGTA, 0.1% BSA, 1.8 mM Hepes pH 7.4) containing protease and phosphatase inhibitors (10 μM leupeptine, 2 μg/ml aprotinine, 1 μM

pepstatine, 1 mM PMSF, 2 mM sodium orthovanadate, 25 mM sodium fluoride, and 40 mM β-Glycerophosphate) and mechanically disrupted by 60 strokes of a 1 ml syringe. Following homogenization, cells were centrifuged for 10 min at 5000g to collect nuclei. The resulting supernatant was centrifuged for 20 min at 15,000g to obtain the mitochondrial (pellet) and cytosolic (supernatant) fractions. The mitochondria-enriched fraction was washed once in MSHE buffer and resuspended in the same buffer before use. The nuclear fraction was rinsed once with washing solution (10 mM Tris pH 7.4, 1.5 mM EDTA, 10% Glycerol, 0.01% NP-40), resuspended in a hyperosmolar solution containing 10 mM Tris pH 7.4, 1.5 mM EDTA, 10% Glycerol, 0.4 M KCl, and homogenized by 15 strokes of a 1 ml syringe. Nuclear soluble proteins were finally obtained by centrifugation for 30 min at 25,000g.

Purity of the different subcellular fractions was determined by measurement of the activity of specific enzymes or by immunoblot detection of specific marker proteins.

Lactate dehydrogenase activity (cytosolic index) was assayed spectrophotometrically by following NADH oxidation at 340 nm. Twenty-five micrograms of protein were added to a 1 ml reaction buffer containing 100 mM KPi, pH 7.0; 30 mM piruvic acid; 1% Triton X-100; and 220 μM NADH. Activity is expressed as micromoles of NADH consumed per milligram of protein per minute.

NADH-cytochrome-c reductase (Complexes I-III) activity (mitochondrial index) was assayed spectrophotometrically by determining cytochrome c reduction at 550 nm in the presence of 30 μM cytochrome c, 1 mM potassium cyanide and 150 μM NADH as electron donor. The velocity of the reaction was determined as a pseudo-first-order constant and expressed as k per minute per milligram of protein.

For immunoblot control of subcellular fractionation, DNA polymerase II was used as nuclear marker and the 39 kDa subunit of the NADH-cytochrome c reductase (complex I) as mitochondrial marker. Primary antibodies were incubated for 16 h at 4 °C and specific binding was detected by horseradish peroxidase-coupled secondary antibodies and the ECL chemiluminescence or the advanced ECL detection system (Amersham). For quantitative analysis, densitometry was performed using a Storm Phosphorimager scanner (Amersham Biosciences, Sweden) and band intensities were analyzed using ImageQuant 5.2 software.

Each subcellular fraction was assumed at least 90% pure by enzymatic and western blot determinations.

3.2.2. Western blot analysis of ERK1/2 and MEK1/2

Samples containing equal amounts of proteins from the different subcellular fractions were subjected to SDS–PAGE in 12% polyacrilamide gels. Separated proteins were electrotransferred to PVDF membranes that were incubated with polyclonal anti–phospho–ERK1/2 or polyclonal anti–phospho–MEK1/2 for the detection of the phosphorylated form of the kinases. Protein loading was evaluated by re-incubation of the membranes with anti–ERK1/2 and anti–MEK1/2, that detect the non-phosphorylated form of the kinases, after striping of the membranes in a stripping buffer containing 62.5 mM Tris–HCl pH 6.8; 2% SDS; 100 mM β-mercaptoethanol.

3.2.3. Confocal analysis of ERK1/2

MA-10 cells were grown to approximately 60–80% confluence on poly-L-lysine glass coverslips, stimulated and stained with a specific mito-chondrial marker, MitoTracker Red 580 (Molecular probes) (300 nM, 45 min at 37 °C) following fixation in 4% paraformaldehyde in PBS for 10 min at room temperature and blocking in 1% BSA; 0.3% Triton X-100; PBS, pH 7.4, in a humidified chamber for 1 h. Cells were also incubated overnight at 4 °C with anti-phosphoERK1/2. Cells were further incubated for 1 h at room temperature with a Cy2-conjugated goat anti-rabbit antibody and coverslips mounted onto slides using Fluorsave mounting media (Calbiochem) for confocal laser scanning microscopy using a Nikon C1 microscope 60×/1.40 (IByME, UBA, Argentina), objective Plan-Apo 40×/0.95. The excitation lines are 488 (blue) and 544 nm (green), emission filters are 515–530 nm (green) and 570-LP (red).

Using immunoblot and confocal microscopy, we located pERK1/2 to the mitochondria of MA-10 cells after stimulation with hCG or 8Br-cAMP (Fig. 10.1A). ERK1/2 phosphorylation was rapid, with a marked increase already 5 min after 8Br-cAMP stimulation, and a slow and progressive decrease of the signal during the first hour of 8Br-cAMP action. hCG induced the activation of ERK1/2 activation with a similar kinetic of that produced by 8Br-cAMP.

Confocal microscopy corroborated 8Br-cAMP-induced activation of mitochondrial ERK1/2 (Fig. 10.1B). As observed with immunoblots, pERK1/2 co-localized with mitochondria after 8Br-cAMP action, an effect observable as early as 5 min and still evident 30 min later.

Two different pools of MEK1/2, containing the non-phosphorylated and the phosphorylated forms of the kinase were detected in the cytosol and mitochondria of MA-10 cells. 8Br-cAMP treatment of the cells induced prolonged MEK1/2 phosphorylation in mitochondria, but had a less pronounced effect on the cytosolic form of the kinase.

cAMP-induced increase in mitochondrial pERK1/2 and pMEK1/2 was abolished by treatment of the cells with the PKA inhibitor H89 and by PKA knockdown experiments. Accordingly, mitochondrial PKA activity increased after 5 min of 8Br-cAMP action (data not shown) in parallel with the appearance of the phosphorylated forms of MEK1/2 and ERK1/2 in the organelle.

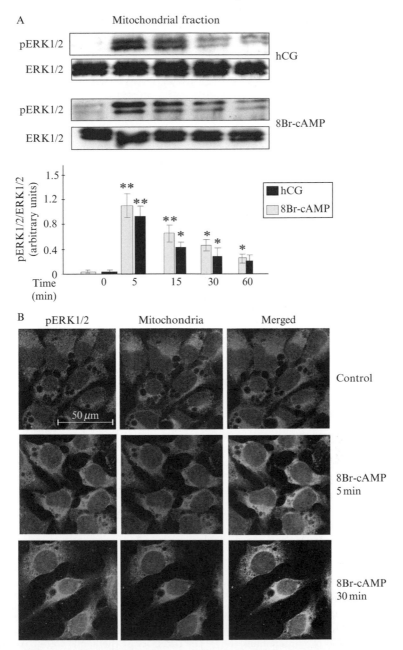

Figure 10.1 Subcellular distribution of hCG/8Br–cAMP-stimulated pERK1/2 activation. MA-10 cells were incubated with or without 20 ng/ml of hCG or 0.5 mM 8Br–cAMP (A) for the indicated times. Next, subcellular fractions were obtained and 40 μg of the total protein of each fraction were subjected to SDS–PAGE and Western blot to

3.3. Functional analysis of kinase interaction at the mitochondria

It is important to localize protein kinases in mitochondria, but it is even more important to analyze the action of a given protein on a given mitochondrial function. For this purpose, one approach is to add the purified protein to a fresh preparation of isolated mitochondria and to evaluate the resulting function. In our case, we studied the action of exogenously added ERK1 and PKA to mitochondria enriched in StAR protein and evaluated the effect on steroid production.

3.3.1. StAR construct and MA-10 cell transfection to enrich the mitochondria of non-stimulated cells with StAR

StAR cDNA (accession N° BC082283) is obtained by PCR from MA-10 cells using primers designed according to the published sequence of mouse StAR. Forward (5′-GGACCTTGAAAGGCTCAGGAAGAACAACCC-3′) and reverse (5′-GGATTAGTAGGGAAGTCGGCACAATGATGG-3′) primers amplify a 1440-bp fragment, corresponding to the 37 kDa protein. The amplified sequence is cloned into pGEM-T easy plasmid (Promega), a basic cloning vector. T vectors are a specific type of cloning vectors that get their name from the T overhangs added to a linearized plasmid. These vectors take advantage of the A overhangs on PCR products after amplification with Taq DNA polymerase by providing compatible ends for ligation.

Following ligation, the fragment containing the double-stranded sequence of StAR cDNA was removed from the vector by EcoRI digestion. The digestion product was subjected to agarose gel electrophoresis and the 1.4 kpb StAR cDNA was subcloned into the eukaryotic expression plasmid, pRc/CMVi. The ligation provided both sense and antisense orientations that were used for transformation of XL-1 *E. coli*. Several clones were then screened, and sense and antisense plasmids obtained by midipreparations. The resulting material was transfected, as described above, in MA-10 cells, and mitochondria obtained from the cells were used for studies in a cell-free assay.

---

detect phospho-ERK1/2 (indicated with pERK1/2). After stripping, total ERK1/2 (indicated with ERK1/2) was detected in the same membrane. The Western blots show the results of a representative experiment performed three times. Graphic bar below shows the quantification of immunoblots above; the intensity of the bands was quantitated using total ERK1/2 as loading control. Bars denote relative levels of pERK1/2 presence in arbitrary units. $\star\star p = 0.01$, $\star p = 0.05$ vs. time 0. (B) Immunofluorescent staining for pERK1/2 (green) and mitochondria (red) in MA-10 cells after treatment with or without 0.5 mM of 8Br-cAMP for the indicated times. Merged images are shown in the right panel. (See Color Insert.)

3.3.2. Cell-free assay as a tool to evaluate mitochondrial function

To test the effect of active ERK1/2 on cholesterol import and progesterone synthesis, we made use of a cell-free assay that allows the determination of progesterone production by isolated mitochondria. This cell-free assay relies on the ability of isolated mitochondria to respond to exogenously added factors if derived from unstimulated cells. Thus, our mitochondrial source consists of non-stimulated MA-10 cells previously enriched in StAR protein by transfection of the corresponding expression construct.

Isolated mitochondria were then incubated (30 °C for 20 min with gentle agitation) in the absence or presence of 50 μM cholesterol as substrate. To analyze the effect of purified kinases on isolated mitochondria, commercially available ERK1 and PKA were used. Mitochondria were incubated in the presence or absence of 300 ng of human recombinant wild-type active ERK1-GST alone or in combination with active PKA catalytic subunit (1 IU). A mutated inactive form of ERK1 (K71A ERK-GST, 500 ng) was also used as control of ERK1 activity. A single amino acid change in ERK1 generates the K71A mutant that lacks autophosphorylation ability and myelin basic protein phosphotransferase activities. K71A is less efficiently phosphorylated by MEK1/2 than its wild-type counterpart (approximately 80%) and that results in partial activation of the myelin basic protein phosphotransferase activity (20%). After that, mitochondria are pelleted by centrifugation at 15,000g for 20 min; media is collected to assess progesterone production by radioimmunoassay. Mitochondrial protein content is analyzed by SDS–PAGE and Western blot to visualize the resulting localization of endogenous and exogenous proteins after the recombination of all these components.

Immunoblot profiles indicate that at the end of the assay, StAR protein is still present in the mitochondria of cells transfected with a sense cDNA plasmid (Fig. 10.2A, western blot of StAR) and that the exogenously added proteins are effectively incorporated and phosphorylated in the organelle (Fig. 10.2A, western blot of PKA catalytic subunit, pERK1/2, and ERK1/2). In this regard, it is worth mentioning that an example of the interaction between the exogenously added proteins and the endogenous mitochondrial protein is the fact that mitochondrial MEK is phosphorylated by the addition of PKA catalytic subunit (Fig. 10.2A, Western blot pMEK1/2). After supplementation of isolated mitochondria with recombinant wild-type ERK1, progesterone yield increased (Fig. 10.2B, bar b vs. a) while addition of the inactive ERK1 mutant (K71A) had no effect on steroid production (Fig. 10.2B, bar c vs. a). The increment in progesterone production was more pronounced when isolated mitochondria were supplemented with the catalytic subunit of recombinant PKA and wild-type ERK1 than when they were challenged with wild-type ERK1 alone (Fig. 10.2B, bar d vs. b). Expectedly, addition of K71A mutant of ERK1 to mitochondria markedly reduced progesterone production (Fig. 10.2B, bar e vs. d).

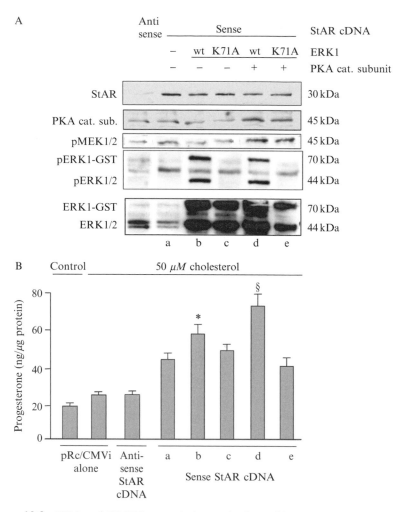

Figure 10.2 PKA and ERK1/2 are strictly required to achieve maximal progesterone production by isolated mitochondria. MA-10 cells were transiently transfected by electroporation with an empty vector (pRc/CMVi) or with a vector containing StAR cDNA (sense or antisense orientations). Mitochondria were incubated in the presence of cholesterol as substrate (a) in the presence of wild-type ERK1-GST protein alone (b) or together with PKA catalytic subunit (d). The mutated inactive form of ERK1-GST (K71A) was also used (c and e). After the indicated incubations, mitochondria were pelleted and subjected to SDS–PAGE and Western blot (A). Specific antibodies that recognize StAR protein, the catalytic subunit of PKA, pMEK1/2, pERK1/2, and total ERK1/2 were used. The panel shows representative Western blots of three independently performed experiments. P4 production is shown in (B). Data are expressed as means ±SD ($n = 3$). $\star p = 0.05$ bar b vs. bar a and $\S p = 0.05$ bar d vs. bar b.

3.4. StAR as substrate of a kinase complex

3.4.1. Sequence analysis

Two important motifs should be present in a protein to be considered as *an* ERK substrate: a docking site for ERK and a phosphorylation consensus site. The typical docking site known as the D domain in ERK substrates shows the sequence **KTKLTWLLSI** (Fantz *et al.*, 2001) while the phosphorylation consensus site is an SP motif.

A potential docking site in StAR sequence was located between amino acids 235 and 244, with a highly conserved profile between species. Although a search for ERK consensus sites in StAR using the Expasy Prosite database (http://expasy.org/prosite/) resulted negative, a detailed inspection of StAR's sequence indicated two Ser-Pro motifs, Ser232 and Ser277 as putative phosphorylatable residues in murine StAR mature protein. Interspecies analysis of StAR sequence indicated that Ser277 is less conserved than Ser232. In addition, Ser232 (PLAGS^{232}PS) lies adjacent to the docking D domain (-2) in a similar fashion as in the classical ERK1/2 substrate Elk-1 (Bardwell *et al.*, 2003) except for the inverted position with respect to the docking domain. In addition, Elk-1's SP motif contains a proline residue in position-1 or -2, whereas StAR has a conserved proline at position-3. As in other substrates, proline is followed by leucine and there is no acidic residue in the motif.

Our findings regarding ERK's docking and phosphorylation sites in StAR protein emphasizes the need for a careful sequence revision to find putative sites for any kinase. Apart from open access databases, it may be necessary to conduct manual alignments of the protein sequence under study.

3.4.2. Interaction of StAR with the kinase complex: Pull-down assay

One possible approach to the study of protein–protein interactions consists of pull-down assays that are based on the precipitation of a known cellular protein and its possible partner(s). Often, pull-down assays are performed after overexpression of one of the proteins suspected to take part in the interaction to maximize the chances of detection. The overexpressed protein is usually tagged in some way, which simplifies the separation of the recombinant protein itself and any other cellular components that interact with it. Finally, a western blot analysis is performed using antibodies against the proteins under study including a loading control.

For our assays using StAR-enriched mitochondria, we isolated the mitochondrial and cytosolic fractions from cells previously stimulated with 8Br-cAMP for 3 h. Cytosolic or mitochondrial proteins (500 or 300 μg, respectively) were then incubated in the presence of human recombinant ERK1-GST bound to agarose beads (Stressgen) previously activated as described by the manufacturer. The pull-down was performed during

16 h at 4 °C in a buffer containing: 50 mM Tris pH 7.4; 150 mM NaCl; 1 mM EDTA; 1 mM EGTA; 10% glycerol; 0.5% Nonidet P-40; 1 mM MgCl$_2$; 1 mM PMSF; 5 μg/ml leupeptin; 5 μg/ml pepstatin; 5 μg/ml aprotinin; 25 mM NaF, and 1mM sodium orthovanadate. After incubation, agarose beads were precipitated by low-speed centrifugation, obtaining an agarose-protein pellet which was washed in the same buffer and boiled for 5 min in SDS–PAGE loading buffer. The sample was then subjected to SDS–PAGE and immunoblot analysis with antibodies against StAR and phospho-ERK1/2 as loading control.

Treatment of the two subcellular fractions (cytosol and mitochondria) with pERK1-GST showed that StAR interacts with pERK1 only in the mitochondrial fraction (Fig. 10.3).

3.4.3. Mutation of the putative ERK1/2 phosphorylation site: Obtention of recombinant wild type and S232A mutated StAR protein

cDNA for the mature form of StAR (30 kDa) is obtained by RT-PCR from MA-10 cells with the following primers: the forward 5′-GGATCCG-CAGGGTGGATGGGTCAA- 3′ and the reverse 5′-GGATTAGTAGG-GAAGTCGGCACAATGATGG- 3′ that amplify a 1200-pb fragment. The obtained fragment is then cloned in the pGEM-T easy plasmid, and further subcloned after purification in the pGEX-4T-3 plasmid (Promega) to add a GST tail to the recombinant protein. The pGEX Vector is included in the GST Gene Fusion System with a Taq promoter for chemically inducible, high-level expression in any *E. coli* host. Very mild elution conditions are needed for release of fusion proteins from the affinity matrix,

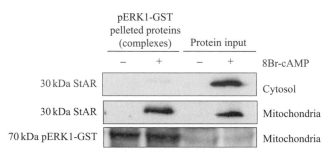

Figure 10.3   StAR interacts with pERK1/2 specifically in mitochondria. MA-10 cells were treated with or without 0.5 mM of 8Br-cAMP for 3 h; cytosolic and mitochondrial subcellular fractions were obtained and incubated in the presence or absence of human pERK1-GST bound to agarose beads. Input and pelleted proteins bound to pERK1-GST (complexes) were analyzed by SDS–PAGE and Western blot. The immunoblots show the bands corresponding to StAR and pERK1/2, as loading control. Data are representative of three independently performed experiments.

thus minimizing effects on antigenicity and functional activity of the protein of interest.

StAR cDNA in the sense orientation was subjected to site-directed mutagenesis obtaining a mutant cDNA, StAR S232A, where serine 232 is changed to alanine. Both, wild-type StAR and the S232A mutant were expressed in BL-21 *E. coli*, recombinant proteins purified by glutathione affinity chromatography (GST Purification Module, Amersham Biosciences) and subjected to thrombin proteolytic cleavage. Obtained proteins were used for *in vitro* phosphorylation experiments.

3.4.4. *In vitro* phosphorylation of recombinant wild type and S232A StAR by wild type and inactive ERK1

As stated earlier, *in vitro* phosphorylation assays are appropriate to demonstrate the viability of a phosphorylation of a protein by a given kinase. Variants carrying a mutation that changes a phosphorylatable amino acid for a non-phosphorylatable are widely used to demonstrate the location of the phospho-acceptor amino acid in the substrate.

Thus, to study the phosphorylation of StAR by ERK1/2, we performed an *in vitro* phosphorylation assay using a purified preparation of mature recombinant StAR protein (30 kDa) and its S232 variant as substrates and two forms of ERK1, the wild-type active and an inactive form of the kinase, and [γ-^{32}P]ATP. Approximately 30 μg of recombinant wild type or S232A StAR were used as substrate for the *in vitro* phosphorylation assay. The reaction was performed in 1.5 ml tubes (Eppendorf) in a reaction buffer containing 20 mM MOPS pH 7.5; 10 mM Mg$_2$Cl$_2$; 5 mM EGTA, 2 $\times$ 10^{-4}% Tween-20; 10 μCi [^{32}P]γ-ATP; 100 μM ATP; 1mM sodium orthovanadate; 1 mM DTT). Wild-type StAR was incubated in the presence or absence of 150 ng of human His-tagged ERK1 catalytically active (Calbiochem) or 300 ng of the mutated and inactive form of ERK1 (K71A ERK-GST) or 1 IU of PKA catalytic subunit (Cell Signaling). PKA was included in our phosphorylation assay by two reasons. First, as a control, since it is known that StAR is phosphorylated by PKA and second, to investigate whether PKA is necessary for the subsequent phosphorylation of StAR by ERK. The phosphorylation status of StAR was analyzed in the presence or absence or 10 μM cholesterol. The reaction was performed for 30 min at 30 °C at a final volume of 30 μl and stopped with 10 μl of SDS–PAGE loading buffer (4$\times$), heated at 80 °C for 5 min and subjected to SDS–PAGE. Polyacrylamide gels were stained, dried and then exposed to autoradiography and scanned using a PhosphorImager.

Wild-type StAR was indeed phosphorylated by wild-type ERK1. Expectedly, the inactive mutant of the kinase (K71A) failed to phosphorylate both wild-type StAR and the S232A variant (Fig. 10.4A and B). Interestingly, phosphorylation of StAR by wild-type ERK1 is enhanced in the presence of cholesterol (Fig. 10.4A), while prior phosphorylation of

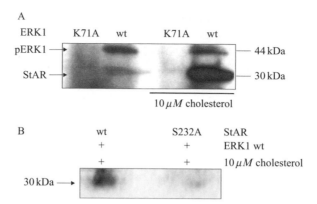

Figure 10.4 StAR is phosphorylated in Ser232 by ERK1 in a cholesterol-dependent manner. Thirty micrograms of recombinant purified 30 kDa wild-type StAR were incubated in the absence or in the presence of 10 μM of cholesterol (A) together with constitutively active His-tagged wild-type ERK1 (wt) or the mutated inactive form of ERK1 (K71A) (A). After the phosphorylation assays, phosphorylated protein levels were analyzed by SDS–PAGE and autoradiography. (B) The wild type and a mutated form (S232A) of StAR were used in the *in vitro* phosphorylation assay described in (A). The phosphorylated products were analyzed by SDS–PAGE and autoradiography. The results in (A) and (B) are representative of three independently performed experiments.

StAR by PKA has no effect on ERK phosphorylation. The conditions in which the phosphorylation reaction takes place within the cells are critical. As for the case presented here, cholesterol is not only a component of the mitochondrial membrane where the reaction occurs, but it is also an endogenous ligand of StAR protein. Thus, our results indicate for the first time that StAR phosphorylation by a kinase different from PKA is modulated by the endogenous ligand of the protein.

3.4.5. Analysis of Ser232 as the phosphorylation site in phospho-StAR: Limited digestion by V8 endoprotease and separation of peptides in gel

An alternative approach to the use of proteins mutants is to analyze the phosphorylated peptides obtained by a limited protease digestion of a radioactively phosphorylated wild-type protein. The sequence of the protein has to be analyzed to choose a protease that renders putative phosphopeptides of different molecular sizes. This allows identification of the phosphorylated amino acid according to the molecular size of the radioactive peptide.

StAR includes within its sequence, cleavage sites for V8 endoprotease that are appropriate for this type of analysis. This protease hydrolyzes peptide bonds containing the α-carboxylic group of glutamic acid residues. StAR is an already known substrate of this protease. Sequence analysis of

StAR digestion using V8 endoprotease predicts the generation of several low molecular peptides with Ser232 as the only serine residue in an identifiable peptide of 6 kDa. Thus, we carried out an *in vitro* phosphorylation assay as described, followed by V8 proteolysis of phospho-StAR and analyzed the resulting peptides by electrophoresis and autoradiography. StAR (30 μg) phosphorylated in the presence of active His-tagged ERK1 and cholesterol was precipitated using chloroform:methanol. The resulting pellet was resuspended and proteolysis done in 100 mM phosphate buffer, pH 7.8 with a rate of enzyme:StAR of 1:30 at 37 °C overnight. The resulting peptides were resolved in polyacrylamide gels following the technique of Schagger and von Jagow (1987). This technique is suited for the identification of low molecular weight peptides and it uses 16.5% polyacrylamide mini gels with two different running buffers. The anode running buffer contains 200 mM Tris–HCl, pH 8.9 and the cathode running buffer contains 100 mM Tris–HCl; 100 mM Tricine, pH 8.25; 0.1% SDS. The electrophoresis was performed at room temperature for 1 h at 30 V in the concentration gel and for 4 h at 105 V in the resolution gel with the following low molecular weight markers: 1.06; 3.5; 6.5; 14.2; 17; 26.6 kDa (Sigma-Aldrich). Gels were then stained, dried, and exposed to autoradiography.

A unique radioactive band with a mass of 6 kDa, matching that of the expected fragment, was detected. This confirms that Ser232 is the target of ERK1/2 phosphorylation as anticipated by the *in vitro* phosphorylation assay already described (Fig. 10.4B).

To further test the identity of Ser232, 30 μg of phosphorylated and unphosphorylated StAR were subjected to SDS–PAGE. The gels were stained with Coomassie Brillant blue and the corresponding bands *were* excised from the gel for high pressure liquid chromatography (HPLC) analysis. Protein samples were partially digested with trypsin, and approximately 50 fractions were detected by peptide ionization. Due to technical issues and low ionization level in the peptides, the Ser232-containing peptide could not be detected in the samples. This situation is not unusual for several proteins of substantial molecular weight (the assay was performed in the Unity of Biochemistry and Analytic Proteomic, Institute Pasteur of Montevideo, Uruguay).

3.4.6. Effect of StAR phosphorylation by ERK1 on residue Ser232 on StAR cellular function

The physiological consequence of protein phosphorylation can be efficiently evaluated by transfection of wild type and different mutants of a given protein, in cells where it is possible to study a certain biological process.

In our case, we made use of two mutants of StAR protein, S232A and S232E, both generated by site directed mutagenesis of serine 232 to alanine or glutamic acid, respectively. The structure and charges of glutamic acid resembles the phosphate residue present in the phosphorylated form of wild-type StAR.

Wild-type StAR and these two variants were transfected to MA-10 cells and steroid production evaluated. Steroid production enhanced by cAMP was inhibited in cells transfected with the non-phosphorylatable S232A mutant. The steroidogenic capacity was less affected by the replacement of Ser232 with Glu (S232E).

It follows from our results that a multi-kinase complex operates in the mitochondria for the regulation of steroid biosynthesis. The fine regulation of kinase activities results in the phosphorylation of a mitochondrial protein, affecting its function and finally, activating steroid biosynthesis.

4. ANALYSIS OF A MITOCHONDRIAL ACYL-CoA THIOESTERASE, ACOT2 AS A PHOSPHOPROTEIN

Acot2, a mitochondrial protein, was identified in experiments aimed at identifying a protein involved in steroid synthesis through the release of AA. Those experiments led us to the isolation of a 43 kDa phosphoprotein (Dada *et al.*, 1991; Mele *et al.*, 1997; Neher *et al.*, 1982; Paz *et al.*, 1994). Further cloning and sequencing of the cDNA encoding the 43 kDa phosphoprotein revealed its primary structure (Finkielstein *et al.*, 1998) and the protein resulted 100% homologous to a mitochondrial-acyl-CoA thioesterase (Acot2) and 92.5% homologous to a cytosolic thioesterase (Acot1) (Lindquist *et al.*, 1998; Svensson *et al.*, 1998). Acot2 and Acot1 are members of a new acyl-CoA thioesterase family with very long chain and long chain acyl-CoA thioesterase activity (Lindquist *et al.*, 1998; Svensson *et al.*, 1998). The family includes four isoforms with different subcellular locations and a high degree of homology.

The sequence of Acot2 includes a mitochondrial leader peptide that targets this enzyme to the inner mitochondrial membrane (Finkielstein *et al.*, 1998; Stocco and Clark, 1996; Svensson *et al.*, 1998). In accordance with the postulated role of Acot2 in steroidogenesis, we detected the protein and its mRNA in all steroidogenic tissues including placenta and brain (Finkielstein *et al.*, 1998). Acot2 has an important role in the generation and export of arachidonic acid in the mitochondria of steroidogenic tissues (Castillo *et al.*, 2006). Interestingly, Acot2 takes part in a similar transduction pathway that regulates fatty acid export in heart mitochondria of diabetic rats (Gerber *et al.*, 2006).

Acyl-CoA thioesterases (EC 3.1.2.1. and EC 3.1.2.2.) are enzymes that catalyze the hydrolysis of CoA esters to free fatty acids and coenzyme A (CoA) (Hunt and Alexson, 2002; Yamada *et al.*, 2005). In the literature, these enzymes have also been referred to as acyl–CoA hydrolases, acyl–CoA thioester hydrolases, and palmitoyl–CoA hydrolases. Although the roles of many acyl–CoA thioesterases in this protein family are not fully understood, they are considered to regulate intracellular levels of CoA esters, the corresponding free fatty acid and CoA and, consequently, cellular processes involving these compounds.

The Acot2 sequence contains a lipase serine motif as well as a Gly–Xaa–His motif close to the C-terminal region which has been shown to be required for the hydrolytic activity of the thioesterase. Antibodies raised against a synthetic peptide that includes the lipase serine motif blocked the activity of the enzyme (Finkielstein et al., 1998).

The Acot2 sequence also contains consensus sites for different protein kinases such as PKA, PKC, CamK, and MAPKs. In agreement with this observation, Acot2 activity and its state of phosphorylation are dependent on hormone action. ACTH treatment of rat adrenal glands resulted in the appearance of multiple phosphorylated forms of the protein which were sensitive to acidic phosphatase treatment (Maloberti et al., 2000). Acot2 activity is dependent on cAMP and PKA but its expression appears to be controlled by EGF (Castilla et al., 2008) and not by cAMP (Cornejo Maciel et al., 2005).

4.1. Purification of Acot2 to homogeneity preserving its biological activity

Acot2 was purified following standard procedures of ion exchange and gel filtration chromatography using homogenates of rat adrenal glands as starting material. Although the result was a 300-fold purification after several chromatography steps, SDS–PAGE analysis of the eluates still revealed the presence of several protein bands. Thus, purification to homogeneity of the protein was achieved by elution of Acot2 from an SDS–PAGE, a procedure that increased the purification to 13,000-fold. In general, protein elution from polyacrylamide gels is recommended in case other purification methods do not achieve further purification of the protein of interest.

The starting material in our protocol was a bioactive pool collected from a Superose-HPLC, but as a general rule, the samples should come from other purification steps and show as high purity as possible by conventional methods. The sample is first subjected to preparative electrophoresis, loading several lanes of an SDS–PAGE gel and running the electrophoresis under ordinary conditions. The lanes are cut into 2–4 mm slices and the slices extracted with 0.4 ml of 0.5% SDS at 4 °C for 16 h. The extracts are then saturated with urea in a final volume of 0.6 ml with the further addition of 40 μg of BSA. Sephadex G-10 equilibrated with 100 mM Tris–HCl, pH 8, was used as gel filtration resin to remove SDS. Aqueous urea (8 M, 0.1 ml) is applied to the column before sample loading. The protein fractions from each column are collected and the content of the protein of interest evaluated in each fraction. The evaluation method is selected according with the known properties of the protein, that is, determination of biological or enzymatic activity or antibody recognition using dot blot or SDS–PAGE and western blot analysis. Regardless of the selected methodology, the eluates with higher content of the protein of interest are pooled and used for any further test, analysis or characterization. In our case, we made use of the known protein biological activity, that is, progesterone production in a cell-free assay, and of peptide mapping. Determination of the enzymatic activity is the most usual approach in case the function of protein of interest is known. After cloning, it became apparent that our protein was a thioesterase, therefore, we measured this activity and how it was regulated by hormones in our experimental model. Peptide mapping resulted also very convenient as it served for the generation of anti-peptide antibodies useful during cloning and identification of the protein by immunoblot or immunoprecipitation throughout the purification process (Paz *et al.*, 1994).

4.2. Determination of a thioesterase activity

In our case, analysis of enzymatic activity implies the determination of hydrolysis of the thioesther bond present in arachidonoyl-CoA. Acyl-CoA thioesterase activity is measured using $1-^{14}C$-arachidonoyl-CoA (specific activity 51.6 mCi/mmol, concentration 0.02 mCi/ml) as substrate. The reaction was carried out using 0.1 μg of protein in 10 mM Hepes, 50 mM KCl, pH 7.4, and 15 μM of the substrate at 37 °C under vigorous shaking. Arachidonic acid released during the reaction was extracted from the aqueous phase with *n*-hexane and quantified by scintillation counting.

At this point, several facts pointed to the phosphoprotein nature of Acot2 (Fig. 10.5): (i) the purified protein evidenced as a double band after SDS–PAGE and silver staining; (ii) immunoblot analysis using an antibody raised against the N-terminal region of the thioesterase detected also a double band, indicating immunoreactivity of both components of the duplet; (iii) 2D electrophoresis resolved the protein in several spots. Taking together, these findings suggest that Acot2 is, in fact, a phosphoprotein.

4.3. Identification of Acot2 as a phosphoprotein

Metabolic labeling of cells with ^{32}P is the standard method for monitoring the level of protein phosphorylation. The combination of metabolic labeling with specific immunoprecipitation and one- or two-dimensional gel electrophoresis is a powerful tool for the identification of phosphoproteins. Other methods useful for the investigation of phosphoproteins include

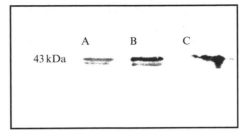

Figure 10.5   Analysis of Acot2. (A) SDS–PAGE and silver staining of elutes of Acot2 from SDS–polyacrilamide gels. (B) Western blot analysis using anti-peptide antibodies. The double band suggests that there is only one protein in different phosphorylation states. (C) 2D-electrophoresis and western blot. The development of several spots confirms the multiple phosphorylation of the protein.

in vitro phosphorylation and dephosphorylation of purified proteins or from recombinant proteins expressed in bacteria.

In vitro phosphorylation of a given protein is useful for determining its behavior as substrate for different kinases. This can be done by using a particular kinase, the purified protein and $[\gamma\text{-}^{32}P]$-ATP as donor. Since the protein has to be in its dephosphorylated form, purification from the tissue or cells under basal (non-activated) conditions or production in bacteria by recombinant cDNA procedures is a prerequisite. The incorporation of the radioactive phosphate into the protein is evaluated in an aliquot of the reaction mixture by scintillation counting of TCA or acetone-precipitated proteins, autoradiography of SDS–polyacrilamide gels or after trapping the sample in phosphocellulose paper and thorough washing before counting. In the case of Acot2 phosphorylation by PKA, the reaction is performed in a total volume of 40 μl of a reaction buffer that contains 50 mM phosphate buffer, pH 6.5; 20 mM MgCl; 5 μg of Acot2 and 2 units of the catalytic subunit of PKA during 10 min at 30 °C. Once the reaction is completed, the mixture is subjected to SDS–PAGE, and the resulting gel stained with Coomassie Brilliant Blue, dried and exposed to autoradiography films or to phosphorimager screens.

The dephosphorylation assay relies on phosphatases that act on the phosphorylated form of the protein. This assay is usually performed in combination with 2D-electrophoresis. In steroidogenic cells, the phosphorylation of Acot2 is under the control of hormone action. Thus, confirmation of the existence of different phoshorylation states of the thioesterase was performed using Acot2 partially purified from adrenal glands of ACTH-treated animals as substrate and potato acid phosphatase as enzyme. The reaction mixture contains 120 mIU of the enzyme and 5–20 μg protein in 10 mM Tris–HCl pH 6.5 buffer (total volume 9 μl) and the reaction proceeds during 30 min at 37 °C. The sample is then subjected to two-dimensional gel electrophoresis and immunoblot analysis. Potato acid phosphatase treatment of Acot2 was effective in reducing the number of spots detected after 2D electrophoresis. While the starting material (the phosphorylated form of the enzyme) shows about seven different spots, phosphatase treatment renders the thioesterase as a double spotted protein, thus confirming protein phosphorylation as an important posttranslational modification of Acot2.

Hormone action on steroidogenic tissues does not affect the expression of Acot2. Therefore, Acot2 constitutes a good example of the importance of posttranslational modifications by protein phosphorylation in mitochondria that trigger the protein's full activity. Finally, Acot2 activation depends on substrate availability and its covalent modification.

REFERENCES

Al-Hakim, A., Rui, X., Tsao, J., Albert, P. R., and Schimmer, B. P. (2004). Forskolin-resistant Y1 adrenal cell mutants are deficient in adenylyl cyclase type 4. *Mol. Cell. Endocrinol.* **214,** 155–165.

Angelo, R. G., and Rubin, C. S. (2000). Characterization of structural features that mediate the tethering of *Caenorhabditis elegans* protein kinase A to a novel a kinase anchor protein. Insights into the anchoring of PKAI isoforms. *J. Biol. Chem.* **275,** 4351–4362.

Ascoli, M. (1981). Characterization of several clonal lines of cultured Leydig tumor cells: Gonadotropin receptors and steroidogenic responses. *Endocrinology* **108,** 88–95.

Bardwell, A. J., Abdollahi, M., and Bardwell, L. (2003). Docking sites on mitogen-activated protein kinase (MAPK) kinases, MAPK phosphatases and the Elk-1 transcription factor compete for MAPK binding and are crucial for enzymic activity. *Biochem. J.* **370,** 1077–1085.

Castilla, R., Gadaleta, M., Castillo, A. F., Duarte, A., Neuman, I., Paz, C., Cornejo Maciel, F., and Podesta, E. J. (2008). New enzymes involved in the mechanism of action of epidermal growth factor in a clonal strain of Leydig tumor cells. *Endocrinology* **149,** 3743–3752.

Castillo, F., Cornejo Maciel, F., Castilla, R., Duarte, A., Maloberti, P., Paz, C., and Podesta, E. J. (2006). cAMP increases mitochondrial cholesterol transport through the induction of arachidonic acid release inside this organelle in MA-10 Leydig cells. *FEBS J.* **273,** 5011–5021.

Cornejo Maciel, F., Maloberti, P., Neuman, I., Cano, F., Castilla, R., Castillo, F., Paz, C., and Podesta, E. J. (2005). An arachidonic acid-preferring acyl-CoA synthetase is a hormone-dependent and obligatory protein in the signal transduction pathway of steroidogenic hormones. *J. Mol. Endocrinol.* **34,** 655–666.

Dada, L. A., Paz, C., Mele, P., Solano, A. R., Cornejo Maciel, F., and Podesta, E. J. (1991). The cytosol as site of phosphorylation of the cyclic AMP-dependent protein kinase in adrenal steroidogenesis. *J. Steroid Biochem. Mol. Biol.* **39,** 889–896.

Diviani, D., and Scott, J. D. (2001). AKAP signaling complexes at the cytoskeleton. *J. Cell Sci.* **114,** 1431–1437.

Fantz, D. A., Jacobs, D., Glossip, D., and Kornfeld, K. (2001). Docking sites on substrate proteins direct extracellular signal-regulated kinase to phosphorylate specific residues. *J. Biol. Chem.* **276,** 27256–27265.

Feliciello, A., Gottesman, M. E., and Avvedimento, E. V. (2001). The biological functions of A-kinase anchor proteins. *J. Mol. Biol.* **308,** 99–114.

Finkielstein, C., Maloberti, P., Mendez, C. F., Paz, C., Cornejo Maciel, F., Cymeryng, C., Neuman, I., Dada, L., Mele, P. G., Solano, A., and Podesta, E. J. (1998). An adrenocorticotropin-regulated phosphoprotein intermediary in steroid synthesis is similar to an acyl-CoA thioesterase enzyme. *Eur. J. Biochem.* **256,** 60–66.

Gerber, L. K., Aronow, B. J., and Matlib, M. A. (2006). Activation of a novel long-chain free fatty acid generation and export system in mitochondria of diabetic rat hearts. *Am. J. Physiol. Cell Physiol.* **291,** C1198–C1207.

Hunt, M. C., and Alexson, S. E. (2002). The role Acyl-CoA thioesterases play in mediating intracellular lipid metabolism. *Prog. Lipid Res.* **41,** 99–130.

Lin, D., Sugawara, T., Strauss, J. F. 3rd, Clark, B. J., Stocco, D. M., Saenger, P., Rogol, A., and Miller, W. L. (1995). Role of steroidogenic acute regulatory protein in adrenal and gonadal steroidogenesis. *Science* **267,** 1828–1831.

Lindquist, P. J., Svensson, L. T., and Alexson, S. E. (1998). Molecular cloning of the peroxisome proliferator-induced 46-kDa cytosolic acyl-CoA thioesterase from mouse and rat liver—Recombinant expression in Escherichia coli, tissue expression, and nutritional regulation. *Eur. J. Biochem.* **251,** 631–640.

Maloberti, P., Castilla, R., Castillo, F., Maciel, F. C., Mendez, C. F., Paz, C., and Podesta, E. J. (2005). Silencing the expression of mitochondrial acyl-CoA thioesterase I and acyl-CoA synthetase 4 inhibits hormone-induced steroidogenesis. *FEBS J.* **272,** 1804–1814.

Maloberti, P., Lozano, R. C., Mele, P. G., Cano, F., Colonna, C., Mendez, C. F., Paz, C., and Podesta, E. J. (2002). Concerted regulation of free arachidonic acid and hormone-induced steroid synthesis by acyl-CoA thioesterases and acyl-CoA synthetases in adrenal cells. *Eur. J. Biochem.* **269,** 5599–5607.

Maloberti, P., Mele, P. G., Neuman, I., Cornejo Maciel, F., Cano, F., Bey, P., Paz, C., and Podesta, E. J. (2000). Regulation of arachidonic acid release in steroidogenesis: Role of a new acyl-CoA thioestrase (ARTISt). *Endocr. Res.* **26,** 653–662.

Mele, P. G., Dada, L. A., Paz, C., Neuman, I., Cymeryng, C. B., Mendez, C. F., Finkielstein, C. V., Cornejo Maciel, F., and Podesta, E. J. (1997). Involvement of arachidonic acid and the lipoxygenase pathway in mediating luteinizing hormone-induced testosterone synthesis in rat Leydig cells. *Endocr. Res.* **23,** 15–26.

Neher, R., Milani, A., Solano, A. R., and Podesta, E. J. (1982). Compartmentalization of corticotropin-dependent steroidogenic factors in adrenal cortex: Evidence for a post-translational cascade in stimulation of the cholesterol side-chain split. *Proc. Natl. Acad. Sci. USA* **79,** 1727–1731.

Paz, C., Dada, L., Cornejo Maciel, F., Mele, P., Cymeryng, C., Neuman, I., Mendez, C., Finkielstein, C., Solano, A., Minkiyu, P., Fisher, W. H., Towbin, H., *et al.* (1994). Purification of a novel 43-kDa protein (p43) intermediary in the activation of steroidogenesis from rat adrenal gland. *Eur. J. Biochem.* **224,** 709–716.

Poderoso, C., Converso, D. P., Maloberti, P., Duarte, A., Neuman, I., Galli, S., Maciel, F. C., Paz, C., Carreras, M. A. C., Poderoso, J. J., and Podestá, E. J. (2008). A mitochondrial kinase complex is essential to mediate an ERK1/2-dependent phosphorylation of a key regulatory protein in steroid biosynthesis. *PLoS ONE* **3,** e1443.

Rae, P. A., Gutmann, N. S., Tsao, J., and Schimmer, B. P. (1979). Mutations in cyclic AMP-dependent protein kinase and corticotropin (ACTH)-sensitive adenylate cyclase affect adrenal steroidogenesis. *Proc. Natl. Acad. Sci. USA* **76,** 1896–1900.

Robinson, F. L., Whitehurst, A. W., Raman, M., and Cobb, M. H. (2002). Identification of novel point mutations in ERK2 that selectively disrupt binding to MEK1. *J. Biol. Chem.* **277,** 14844–14852.

Rubin, C. S. (1994). A kinase anchor proteins and the intracellular targeting of signals carried by cyclic AMP. *Biochim. Biophys. Acta* **1224,** 467–479.

Schagger, H., and von Jagow, G. (1987). Tricine-sodium dodecyl sulfate-polyacrylamide gel electrophoresis for the separation of proteins in the range from 1 to 100 kDa. *Anal. Biochem.* **166,** 368–379.

Schimmer, B. P., Tsao, J., and Knapp, M. (1977). Isolation of mutant adrenocortical tumor cells resistant to cyclic nucleotides. *Mol. Cell. Endocrinol.* **8,** 135–145.

Stocco, D. M., and Clark, B. J. (1996). Regulation of the acute production of steroids in steroidogenic cells. *Endocr. Rev.* **17,** 221–244.

Svensson, L. T., Engberg, S. T., Aoyama, T., Usuda, N., Alexson, S. E., and Hashimoto, T. (1998). Molecular cloning and characterization of a mitochondrial peroxisome proliferator-induced acyl-CoA thioesterase from rat liver. *Biochem. J.* **329**(Pt 3), 601–608.

Yamada, J., Hamuro, J., Sano, Y., Maruyama, K., and Kinoshita, S. (2005). Allogeneic corneal tolerance in rodents with long-term graft survival. *Transplantation* **79,** 1362–1369.

CHAPTER ELEVEN

ISOLATION OF REGULATORY-COMPETENT, PHOSPHORYLATED CYTOCHROME C OXIDASE

Icksoo Lee,* Arthur R. Salomon,[†] Kebing Yu,[†] Lobelia Samavati,[‡] Petr Pecina,* Alena Pecinova,* *and* Maik Hüttemann*

Contents

1. Introduction	194
2. Purification of Mitochondria Maintaining Protein Phosphorylation	195
2.1. General considerations	196
2.2. Preparation of activated vanadate	197
2.3. Animal tissue handling and pretreatment	197
2.4. Incubation of ground tissue with compounds that modify signaling pathways (optional)	197
2.5. Isolation of mitochondria	198
3. Isolation of Cytochrome c Oxidase from Mitochondria	199
3.1. Determination of cytochrome c oxidase concentration and purity	199
3.2. Cytochrome c oxidase isolation	201
4. Analysis of Cytochrome c Oxidase Phosphorylation	204
4.1. Western blot analysis	205
4.2. Phosphoepitope determination via immobilized metal affinity chromatography/nano-liquid chromatography/electrospray ionization mass spectrometry	206
5. Concluding Remarks	208
References	209

* Center for Molecular Medicine and Genetics, Wayne State University School of Medicine, Detroit, Michigan, USA
[†] Department of Molecular Biology, Cell Biology, and Biochemistry, Brown University, Providence, Rhode Island, USA
[‡] Department of Medicine, Division of Pulmonary/Critical Care and Sleep Medicine, Wayne State University School of Medicine, Detroit, Michigan, USA

Methods in Enzymology, Volume 457
ISSN 0076-6879, DOI: 10.1016/S0076-6879(09)05011-3

Abstract

The role of posttranslational modifications, specifically reversible phosphorylation as a regulatory mechanism operating in the mitochondria, is a novel research direction. The mitochondrial oxidative phosphorylation system is a particularly interesting unit because it is responsible for the production of the vast majority of cellular energy in addition to free radicals, two factors that are aberrant in numerous human diseases and that may be influenced by reversible phosphorylation of the oxidative phosphorylation complexes. We here describe a detailed protocol for the isolation of mammalian liver and heart mitochondria and subsequently cytochrome *c* oxidase (CcO) under conditions maintaining the physiological phosphorylation state. The protocol employs the use of activated vanadate, an unspecific tyrosine phosphatase inhibitor, fluoride, an unspecific serine/threonine phosphatase inhibitor, and EGTA, a calcium chelator to prevent the activation of calcium-dependent protein phosphatases. CcO purified without manipulation of signaling pathways shows strong tyrosine phosphorylation on subunits II and IV, whereas tyrosine phosphorylation of subunit I can be induced by the cAMP- and TNFα-dependent pathways in liver. Using our protocol on cow liver tissue we further show the identification of a new phosphorylation site on CcO subunit IV tyrosine 11 of the mature protein (corresponding to tyrosine 33 of the precursor peptide) via immobilized metal affinity chromatography/nano-liquid chromatography/electrospray ionization mass spectrometry (IMAC/nano-LC/ESI-MS). This phosphorylation site is located close to the ATP and ADP binding site, which adjusts CcO activity to cellular energy demand, and we propose that phosphorylation of tyrosine 11 enables allosteric regulation.

1. INTRODUCTION

The mitochondrial oxidative phosphorylation (OxPhos) process provides more than 15 times the energy in the form of ATP compared to anaerobic glycolysis and is thus essential to life in all higher organisms. OxPhos consists of two parts, the electron transport chain (ETC) and ATP synthase (complex V). The ETC is an assembly line of protein complexes that transfer electrons mainly derived from food or energy stores to oxygen, which is reduced to water. It consists of NADH dehydrogenase (complex I), succinate dehydrogenase (complex II), ubiquinone, bc_1–complex (complex III), cytochrome *c* (Cyt *c*), and cytochrome *c* oxidase (CcO; complex IV). Complexes I, II, and IV pump protons across the inner mitochondrial membrane generating the mitochondrial proton motive force Δp_m, which is finally used by ATP synthase to generate ATP out of ADP and phosphate.

Since cellular energy requirements vary considerably during diverse conditions such as rest, exercise, or hibernation, several means of regulation exist to adjust energy production to demand. In the traditional view mainly based on studies in bacteria, in addition to the availability of substrates, regulation of the activity of the ETC complexes is mediated through Δp_m, leading to an inhibition of the proton pumps at high Δp_m levels. In higher organisms several additional mechanisms evolved, including allosteric regulation of CcO and Cyt *c* by ATP and ADP, and expression of tissue-specific and developmentally regulated isoforms of several CcO subunits and Cyt *c*. Importantly, it is now evident that all OxPhos complexes are targets of signaling pathways and can be phosphorylated (Hüttemann *et al.*, 2007), including tyrosine phosphorylation, which is mainly found in higher organisms but hardly encountered in bacteria or yeast.

The necessity of additional regulatory mechanisms found in higher organisms may be explained by a side product of OxPhos, free radicals, which are produced at high mitochondrial membrane potentials (Liu, 1999). These free radicals can serve as a signal, e.g., during the induction of apoptosis, triggered by a hyperpolarization of the membrane potential (Kadenbach *et al.*, 2004). Thus, we recently proposed that a main function of phosphorylation is a structural/functional change in the proton pumping complexes I, III, and IV, which already results in inhibition of proton pumping activity at physiologically lower membrane potentials, avoiding the production of free radicals under healthy conditions (Hüttemann *et al.*, 2008). In contrast, mitochondria isolated using traditional protocols that do not consider protein phosphorylation contain dephosphorylated OxPhos complexes that are highly active and generate high mitochondrial membrane potentials leading to excessive free radical production.

We here describe a protocol for the isolation of mitochondria and subsequently CcO maintaining its phosphorylation state. Using this method, we have discovered a new phosphorylation site as identified by mass spectrometry, tyrosine 11 of the mature CcO subunit IV isoform 1 from cow liver. The protocol described here can be easily adapted for the purification of other mitochondrial proteins in their phosphorylated state.

2. PURIFICATION OF MITOCHONDRIA MAINTAINING PROTEIN PHOSPHORYLATION

Traditional protocols for the isolation of mitochondria omit phosphatase inhibitors and sometimes even include calcium. Calcium is the most potent activator of mitochondrial function (Robb-Gaspers *et al.*, 1998) and induces dephosphorylation of most mitochondrial proteins (Hopper *et al.*, 2006), likely mediated via calcium activated phosphatases (Hüttemann *et al.*, 2007). As a consequence, our buffers include EGTA (2 mM), a calcium chelating agent that binds calcium during tissue homogenization and subsequent steps.

2.1. General considerations

The protocol described herein allows the maintenance of protein phosphorylation during enzyme purification. However, if a particular signaling pathway is to be studied several aspects have to be considered, including incubation temperature and time, and effector concentration. For example, we have reported tyrosine 304 phosphorylation on CcO subunit I followed by strong enzyme inhibition in cow liver tissue in the presence of high cAMP levels (Lee *et al.*, 2005) and after TNFα treatment (Samavati *et al.*, 2008). cAMP-dependent phosphorylation can be achieved at lower incubation temperatures of 10–25 °C, whereas the incubation with TNFα requires higher incubation temperatures of about 30 °C, while hardly any effect can be observed at 4 °C.

The incubation time is of equal importance. For example, the activation of the epidermal growth factor receptor (EGFR) pathway leads to a translocation of EGFR to the mitochondria where the receptor directly interacts with CcO (Boerner *et al.*, 2004). EGFR translocation to the mitochondria is dynamic and maximal after 20 min, whereas no translocation is detectable after 60 min, similar to the non-stimulated condition.

To account for these variables when studying a novel pathway where optimal conditions have yet to be determined, we recommend a range finding study with variable concentrations of the test compound using cell culture cells or tissue samples. Cell signaling usually operates quickly and incubation times of 2–30 min should be assessed next in order to determine an optimal incubation period. For studies of CcO the effect on enzyme activity is an excellent indicator and can be quickly determined using total cell homogenates via oxygen electrode as described (Lee *et al.*, 2005; Samavati *et al.*, 2008).

The use of cultured cells has the advantage that experimental conditions can be well controlled, but preparative purification of mitochondria and subsequently mitochondrial proteins is not recommended due to the large amount of starting material needed. Therefore, for protein purifications the use of animal tissues is much more economical but harbors several complicating factors. For example, the time period from sacrificing the animal until the organ is processed or stored frozen should be kept as short as possible, because ischemia and stress signaling may affect the phosphorylation status of mitochondrial proteins. In addition, for some studies it is essential to know the starved/fed state of the animal, which can be easily controlled when working with laboratory animals, but which is difficult to control when using slaughterhouse tissues. For instance, we have shown that activation of the cAMP-dependent pathway leads to phosphorylation of CcO subunit I Tyr304 in cow liver tissue followed by strong CcO inhibition (Lee *et al.*, 2005). cAMP is a starvation signal in liver. Thus, the starved/fed state of the animal is an important aspect for cAMP-signaling in liver. Accordingly, we have observed high basal levels of Tyr304 phosphorylation in one cow liver sample, which did not further increase significantly after the addition of reagents that increase cellular cAMP levels, presumably because the animal was starved.

2.2. Preparation of activated vanadate

During mitochondria and subsequent CcO isolation unspecific phosphatase inhibitors fluoride and vanadate are added to the isolation buffers. Vanadate, an unspecific tyrosine phosphatase inhibitor, forms polymers in aqueous solutions and has to be depolymerized ("activated") for maximal inhibition of protein phosphotyrosine–phosphatases (Gordon, 1991). A yellowish color of vanadate-containing solutions is indicative of polymers whereas colorless solutions contain monomers and shorter oligomers.

A 500 mM sodium orthovanadate stock solution is adjusted to pH 8 with HCl and boiled until it turns colorless (approximately 10 min) and cooled to room temperature. pH adjustment and boiling is repeated two more times. Aliquots may be stored at -80 °C for convenience. However, at least the final boiling step should be repeated prior to each use.

2.3. Animal tissue handling and pretreatment

All steps are performed on ice or at 4 °C except for the optional incubation step to modify a signaling pathway. If it is the goal to modulate a signaling pathway, the starting tissue amount should be doubled to allow side-by-side isolation of the treated and control samples: per preparation, 300 g of frozen cow liver, stored at -80 °C, is partially thawed. Blood vessels, connective tissue, and membranes are removed. Animal tissue samples should be cut into smaller pieces and briefly but thoroughly rinsed with ice-cold water to remove blood. Serum proteins have a high binding capacity for various compounds and it may be difficult to control the effective concentration of the test compound that modifies a signal transduction pathway if the majority of the blood has not been removed. The tissue, still partially frozen, is then ground using a household meat grinder and collected in 3 L of precooled phosphate buffered saline (pH 7.4), briefly stirred, and collected using a fine metal sieve. This second washing step removes residual blood.

If the manipulation of a signaling pathway is not the goal and mitochondria and CcO are to be isolated as they are, partially thawed tissue can be directly subjected to homogenization using a commercial blender as described in Section 2.5.

2.4. Incubation of ground tissue with compounds that modify signaling pathways (optional)

We recently showed that both the cAMP-dependent and the TNFα signaling pathways lead to tyrosine 304 phosphorylation on CcO subunit I in cow liver tissue (Lee *et al.*, 2005; Samavati *et al.*, 2008). Those treatments will serve as examples.

Before starting, a stock solution of serine protease inhibitor phenyl-methylsulphonyl fluoride (PMSF; 200 mM) is freshly prepared in warm isopropanol and is added to the mitochondria incubation/isolation buffer (250 mM sucrose, 20 mM Tris–Cl (pH 7.4), 2 mM EDTA, 10 mM KF, 2 mM EGTA, 1 mM PMSF) immediately prior to use. For activation of the cAMP-dependent pathway the buffer is supplemented with phosphodiesterase inhibitor theophylline (10 mM), adenylyl cyclase activator forskolin (75 nM), or a combination of both. To trigger the inflammatory pathway the incubation buffer is supplemented with TNFα (20 ng/mL).

Ground tissue is supplemented with five volumes of the mitochondria isolation buffer. The incubation step is performed in a container with a large surface, such as a big beaker or bucket, under continuous manual stirring using a large plastic spoon to allow adequate aeration for mitochondrial respiration. Incubation is performed for 20 min at 20 °C for cAMP and for 5 min at 30 °C for TNFα signaling. Temperature and pH are monitored and readjusted if necessary.

2.5. Isolation of mitochondria

Immediately after incubation and prior to homogenization the crude tissue suspension is supplemented with activated vanadate solution (1 mM final concentration) to preserve tyrosine phosphorylation. Vanadate must not be added earlier during the incubation because it can interfere with signaling pathways, such as the AMP-kinase pathway.

The tissue is homogenized using a precooled commercial blender at the lowest speed four times for 6–8 s with 20 s intervals to prevent heating of the sample. If only a powerful blender is available, operate it with a dimmer to reduce speed. Higher speeds during blending should be avoided in order to prevent fragmentation of cellular organelles including the mitochondria, which would result in decreased yield and purity. The pH might drop down below 7.0 due to lactic acid production and thus has to be monitored and readjusted to pH 7.4 (pH indicator paper) using 2 M Tris base. The suspension is centrifuged at 650g for 10 min to sediment nuclei and unbroken cells. The supernatant is collected through cheesecloth, and the pellet is resuspended in 1 L mitochondria isolation buffer (= mitochondria incubation buffer supplemented with 1 mM activated vanadate; this buffer does not contain theophylline, forskolin, or TNFα) and homogenized and centrifuged as above to increase the yield of the mitochondria. The combined supernatants are then centrifuged at 14,000g for 20 min. The supernatant is discarded and the mitochondrial pellet is washed by resuspension in a small volume (400 mL) of isolation buffer using a glass-teflon homogenizer at 150 rpm for 4–5 strokes (until homogeneous). The suspension is diluted with isolation buffer to ≥2.5 L and centrifuged at 300g for 8 min to remove further precipitating contaminants in the pellet.

The supernatant is centrifuged at 14,000*g* for 20 min to collect mitochondria. The mitochondria are then washed two more times by resuspension using a glass-teflon homogenizer and centrifugation at 14,000*g* for 20 min. Finally, the mitochondrial pellet is resuspended in a small volume of the isolation buffer with a glass-teflon homogenizer, protein concentration is determined and adjusted to 20–30 mg/mL, and mitochondria can be stored at −80 °C.

3. ISOLATION OF CYTOCHROME *c* OXIDASE FROM MITOCHONDRIA

3.1. Determination of cytochrome *c* oxidase concentration and purity

During CcO isolation from mitochondria the presence or absence of CcO in certain fractions is monitored. In addition the total amount and purity of CcO can be determined during and at the end of the enzyme isolation. This is done using a spectrophotometer (double beam, if available).

The concentration of CcO (heme aa_3) in total mitochondria and during CcO isolation from mitochondria is estimated from the difference spectrum of air-oxidized (reference cell or baseline) and sodium dithionite-reduced samples (after the addition of a small amount, i.e., a tip of a spatula of solid sodium dithionite) measured in 200 mM KPi (pH 7.4), 5% TX-100 from 500 to 650 nm. $A_{603-630\ nm}$ is calculated and CcO concentration is calculated using a millimolar extinction coefficient for cytochrome aa_3 of $\varepsilon = 24\ cm^{-1}$ (Büge and Kadenbach, 1986; Von Jagow and Klingenberg, 1972). A typical spectrum of isolated liver mitochondria is shown in Fig. 11.1A.

Purity and concentration of isolated CcO (210 kDa per CcO monomer) are determined from the absolute spectra of the air-oxidized and sodium dithionite-reduced enzyme (Fig. 11.1C). The spectrophotometer baseline is run with 200 mM KPi (7.4), 2% Na-cholate (w/v), 0.5% Na-deoxycholate. The ratio A280/A420 of the oxidized enzyme serves as a measure of its purity (Kadenbach *et al.*, 1986). Lower values indicate higher purity and values should fall below 3. Usually heart CcO is purer compared to liver CcO with an A280/A420 ratio of close to 2.6. The air-oxidized (measured first) and sodium dithionite-reduced (see above) spectra of CcO are measured from 260 to 670 nm against buffer in the reference cuvette. $A_{[reduced\ 605-650\ nm]}$ and $A_{[reduced\ 443-490\ nm]}$ are calculated and CcO cytochrome aa_3 concentration is calculated with a millimolar extinction coefficients of $\varepsilon_{(605-650\ nm)} = 40\ cm^{-1}$ and $\varepsilon_{(443-490\ nm)} = 204\ cm^{-1}$, respectively (Wikström *et al.*, 1981). The two obtained values are averaged.

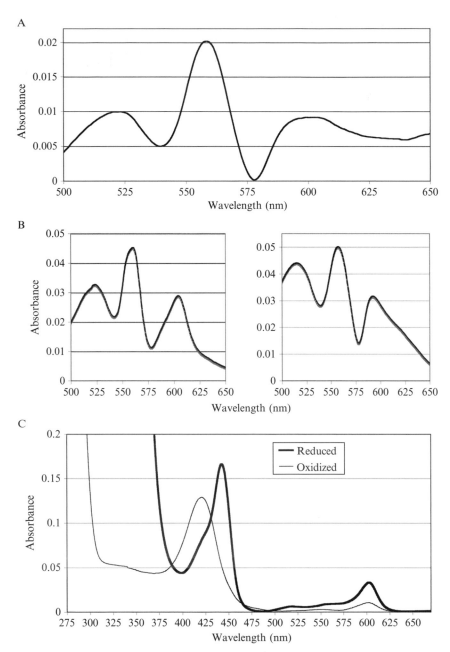

Figure 11.1 *Spectra of isolated liver mitochondria and isolated CcO.* (A) Cow liver mitochondria were analyzed in 200 m*M* KPi (pH 7.4), 5% TX-100 using a double-beam spectrophotometer. Shown is the difference spectrum of air-oxidized and sodium-dithionite reduced mitochondria obtained after putting the oxidized sample in one cuvette and the reduced sample in the other cuvette. The peaks at about 520 nm

3.2. Cytochrome *c* oxidase isolation

All steps are performed on ice or at 4 °C, and all buffers are supplemented with 10 mM KF and 2 mM EGTA. Activated sodium vanadate (1 mM final concentration) is added to all buffers up to the CcO extraction step and excluded from the diethylaminoethyl (DEAE) sepharose column step until the end of the purification. Homogenization/resuspension steps are performed using a glass-teflon homogenizer.

Frozen mitochondria as prepared in the previous section (100 mL of 30 mg/mL total protein) are thawed and poured into a beaker, and 25 mL of 1M KPi (pH 7.4) supplemented with KF, EGTA, and vanadate is added to the mitochondria under stirring to a final concentration of 200 mM (1/4 (v/v) mitochondria) to switch the buffer system from Tris to phosphate. To break the mitochondrial membranes 20% Triton X-114 is then added drop by drop under constant stirring. One milliliter of 20% Triton X (TX)-114 is applied to 1 g of total mitochondrial protein, i.e., 3 mL is added for liver mitochondria; if heart mitochondria are used twice the amount of TX-114 has to be added. The sample will become darker and more transparent due to the solubilization process. It is important not to add the detergent too quickly and not to use more than the recommended amount to avoid partial extraction of CcO at this step. After the addition of the detergent the mixture is stirred continuously for 15 min. The sample is subjected to ultracentrifugation at 195,000g for 30 min (250,000g, if available) to remove soluble components. It should be noted that the pellet is sometimes loose, and care must be taken for the separation of the supernatant. To increase yield it is advisable to discard less of the supernatant, since the pellet will be subsequently washed. The supernatant is discarded after confirming that the fraction does not contain CcO via spectrophotometer (see Fig. 11.1B, right spectrum: the sample does not contain CcO, whereas the left spectrum indicates presence of CcO due to the strong peak at 603 nm). The pellet is washed by homogenization with a glass-teflon homogenizer (until homogeneous; 5–10 strokes) using ≥100 mL of 200 mM KPi (pH 7.4) supplemented with KF, EGTA, and vanadate, and centrifuged for 25 min at 195,000g to remove residual TX-114 and other

---

correspond to the β band of hemes b and c, the peak at about 560 nm corresponds to heme b of bc_1 complex, and the peak at 603 nm corresponds to the α band of heme aa_3. Note that in contrast to liver mitochondria, intensities of the peaks at 560 and 603 nm are similar in heart mitochondria. (B) Difference spectrum of air oxidized and sodium dithionite-reduced supernatant during CcO extraction in 200 mM KPi (pH 7.4), 5% TX-100. Note that the left spectrum shows a strong absorption peak at 603 nm indicating the presence of CcO. In contrast, the right spectrum lacks that peak indicating the absence of CcO. (C) Air oxidized and sodium dithionite-reduced CcO spectra as indicated, showing the characteristic heme aa_3 gamma (between 410 and 460 nm) and alpha ($\sim$603 nm) absorption bands.

soluble components. A loose pellet can also be obtained during the washing steps and one proceeds as above; if necessary the washing step can be repeated. Again, the supernatant is analyzed by spectrophotometer to confirm absence of CcO and discarded. The spectral assessment of fractions that normally do not contain CcO is performed to confirm that extraction conditions are not too harsh. If CcO is found in any of those fractions, the fractions cannot be used and have to be discarded. In this case detergent concentrations should be checked and decreased where necessary.

CcO is extracted from the pellet, which contains the mitochondrial membrane fraction, using ≥ 150 mL of extraction buffer (200 mM KPi (pH 7.4), 5% TX-100, supplemented with KF, EGTA and vanadate). After homogenization (5–10 strokes) the sample is subjected to ultracentrifugation at 195,000g (250,000g, if available) for 20 min. The supernatant is collected and analyzed for CcO content via spectrophotometer. The supernatant containing CcO is collected and kept for the subsequent purification through DEAE ion exchange chromatography. The remaining pellet is re-extracted with extraction buffer repeatedly as above until no CcO is left in the pellet after ultracentrifugation (usually two to three re-extractions are performed). In the beginning of the CcO extraction the supernatants have a dark red-brown color due to high content of hemes b and c. At later extraction steps the color of the supernatant becomes more greenish, which is the color of CcO, i.e., bc_1 complex is extracted more easily and thus more concentrated in the early fractions.

CcO-containing supernatants are pooled and three volumes of dH$_2$O supplemented with fluoride and EGTA are added to dilute the TX-100 concentration to 1.25%. At higher detergent concentrations CcO would not bind to the weak anion exchange DEAE column. In addition, vanadate is omitted from here on, because it forms poly-anions that may affect protein binding to the column. The phosphate concentration of the sample is increased to 100 mM by addition of 1 M KPi (pH 7.4). At this concentration CcO will bind to the DEAE column, but binding of bc_1 complex is blocked. Similar to CcO, bc_1 complex also has a negatively charged area for the interaction with cytochrome c and can bind to the DEAE column. However, this binding is more sensitive to the phosphate concentration compared to CcO and can be suppressed. Therefore, the determination of the optimal concentration of phosphate allowing binding of CcO but not bc_1 complex is very important and can range from 50 to 150 mM. For example, 50 mM KPi is sufficient for cow heart CcO, but 100–150 mM KPi is required for cow liver CcO. In addition, the optimal KPi concentration may depend on cow subspecies. Furthermore, the phosphorylation state of CcO and bc_1 complex may affect their binding to the DEAE column.

After the addition of phosphate, the sample is loaded onto the DEAE column. A 300 mL column bed volume is sufficient, and the column is equilibrated with 100 mM KPi (pH 7.4), 0.5% TX-100, 10 mM KF, 2 mM

EGTA. The flow-through is collected from time to time to check whether CcO is bound on the column. If the flow-through contains CcO, the concentration of KPi of the loading sample and that of the column equilibration buffer should be decreased. In addition, the applied amount of TX-100 may have been too high and should be decreased in future CcO isolations. As CcO binds to the DEAE column, the binding area turns dark green. If significant amounts of bc_1 complex bind together with CcO, the column color will turn red. In this case it is necessary to increase the phosphate concentration of the sample and the column equilibration buffer by 20 mM KPi increments. After loading the sample, the column is washed with 1 L equilibration buffer, 50–150 mM KPi (pH 7.4) depending on tissue type, 0.5% TX-100, supplemented with KF and EGTA. Again, for cow liver CcO 100 mM KPi is a good starting point. To decrease the detergent concentration for the subsequent AmSO$_4$ fractionation, the column is washed with 1 L of 50–150 mM KPi (pH 7.4), 0.1% TX-100, supplemented with KF and EGTA. CcO is eluted by a applying a linear gradient from 50 to 600 mM KPi (pH 7.4), 0.1% TX-100, with a total volume of 600 mL elution buffer. As the concentration of KPi increases, the CcO band moves down and is eluted from the column (CcO from cow heart elutes at about 200 mM KPi, whereas CcO from liver elutes at >200 mM). The fractions (each 8 mL) are analyzed for the presence of CcO via spectrophotometer. The CcO-containing fractions are bright green whereas the bc_1 complex-containing fractions are yellow. The fractions mainly containing bc_1 complex are discarded. CcO-containing fractions are pooled in a beaker.

Sodium cholate is added to 1.5% (w/v) and the pH is readjusted to 7.4. If cholate is not added, CcO will not form a pellet during AmSO$_2$ precipitation and instead will form a floating oil-like layer. Saturated AmSO$_4$ solution (pH 7.4) is added to a final concentration of 28% under stirring and the solution is stirred for 8–14 h. AmSO$_4$ should be added dropwise; otherwise local precipitations occur. If the volume of the sample is too high or the protein concentration of the sample is too low, using powdered AmSO$_4$ instead of an AmSO$_4$ solution is another option. In this case, AmSO$_4$ is first pulverized with a mortar and pestle or, more conveniently, a coffee grinder before addition, allowing fast dissolving. It is slowly added to the sample, and the pH of the sample has to be monitored and adjusted to 7.4. The sample becomes turbid during this precipitation step and is centrifuged at 27,000g for 20 min. The clear supernatant is collected in a beaker and saturated AmSO$_4$ (pH 7.4) solution is added dropwise to a final concentration of 37%. The sample is stirred for 30 min to 1 h and centrifuged as before. The supernatant is subjected to further fractionations at 42–45% and subsequently at 48% AmSO$_4$. At these higher AmSO$_4$ concentrations the sample is centrifuged immediately after the addition of AmSO$_4$. Each fractionated pellet is dissolved in a very small volume of mitochondria isolation buffer. Most of the time the 28% AmSO$_4$ pellet is large and grey, and dissolves in 1–2 mL buffer. As assessed by spectrophotometer this fraction contains more impurities than CcO, so it is rarely used for analytical or functional experiments. With higher AmSO$_4$ concentrations green precipitates are obtained and dissolved in 200–500 μL buffer. Usually the CcO concentration in the buffer is in the range of 10–30 mg/mL. Sometimes at high AmSO$_4$ concentrations (>42%), sodium cholate precipitates with CcO producing a big white pellet. In this case, the whole pellet is dissolved in buffer and transferred to a 2 mL centrifuge tube. Cholate can be removed as a white pellet via brief centrifugation for 20 s using a tabletop mini centrifuge, whereas CcO remains in the supernatant. After removal of excessive cholate by repeated centrifugation steps, the sample is desalted and concentrated using Ultra-15 centrifugal filter units with Ultracel-50 membrane (Millipore), and its spectrum is measured. The absolute spectrum of each fraction is recorded to determine purity and the concentration of CcO (Fig. 11.1C). The purest CcO fractions are obtained in the 37–48% AmSO$_4$ fractions. Finally, CcO aliquots are stored at $-80\,^\circ$C. Repeated freezing and thawing damages CcO. Thus, CcO should be stored in small aliquots. Using 300 g of liver tissue as starting material usually produces a yield of about 20–30 mg pure CcO (not including the 28% AmSO$_4$ fraction).

4. ANALYSIS OF CYTOCHROME *c* OXIDASE PHOSPHORYLATION

The effect of CcO phosphorylation can be functionally studied using an oxygen electrode with isolated CcO, or with total tissue homogenates and isolated mitochondria after solubilization of the sample, as we have described in detail (Lee *et al.*, 2005; Samavati *et al.*, 2008). Particularly when testing a new compound that may lead to changes in CcO phosphorylation, this method allows quick optimization of variable parameters, such as effector concentration, and incubation time and temperature using tissue homogenates. We have shown that cAMP- and TNFα-dependent phosphorylation of CcO subunit 1 Tyr304 results in up to 70% inhibition of CcO activity at maximal turnover and a shift from more hyperbolic (Tyr304 non–phosphorylated) to more sigmoidal kinetics (Lee *et al.*, 2005; Samavati *et al.*, 2008). The inhibitory effect on CcO activity is even more pronounced at lower cytochrome *c* substrate concentration. To directly assess the phosphorylation state, Western blot analysis with commercially available anti–phosphoserine/threonine/tyrosine antibodies can be performed, and tandem mass spectrometry can later be utilized to identify phosphorylated amino acids as outlined below.

4.1. Western blot analysis

Using CcO isolated from cow liver tissue after treatment with TNFα as an example, we briefly describe Western blot analysis with anti-phosphoSer/Thr/Tyr antibodies, followed by mass spectrometry, revealing a novel phosphorylation site on CcO subunit IV isoform 1.

SDS–PAGE of liver CcO after TNFα treatment was carried out using a gel system with a high urea concentration and a long separation gel as described (Kadenbach *et al.*, 1983). This gel system allows the separation of most (and sometimes all) of the 13 CcO subunits (Fig. 11.2, left lane). If separation of the smaller subunits is not of interest other gel systems including 10% acrylamide mini gels can be used as an alternative. Protein transfer time on a nitrocellulose or PVDF membrane was 60 min to facilitate efficient transfer of larger CcO subunits. For studies of the smaller CcO subunits transfer time can be reduced. For blocking and antibody dilution use 2% BSA instead of non-fat dry milk in Tris buffered saline (TBS, pH 7.4), 0.1% Tween 20. Anti-phosphoserine and -threonine antibodies were sets of four (1C8, 4A3, 4A9, and 16B4; Calbiochem) and three (1E11, 4D11,

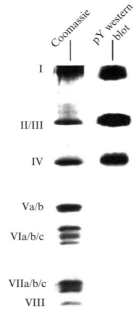

Figure 11.2 *SDS–PAGE and Western blot analysis of isolated liver CcO.* Cow liver CcO was isolated after incubation of tissue with TNFα. CcO subunits were separated by SDS–PAGE and protein bands were stained with Coomassie blue (left lane; subunits as indicated). Western analysis with a phosphotyrosine-specific antibody revealed a strong signal for subunits I, II or III, and IV (right lane).

and 14B3; Calbiochem) individual monoclonal antibodies, respectively, while a single anti-phosphotyrosine antibody was used (4G10; Upstate Bio-technology, now Millipore). The anti-phosphotyrosine antibody is highly specific, whereas the anti-phosphoserine and -threonine antibodies are less specific due to the smaller size of the phosphorylated amino acid, and their binding can vary depending on the phosphoepitope, which is influenced by amino acids adjacent to the phosphorylation site. They may also cross-react with phosphotyrosine epitopes under less stringent conditions. It is thus advisable to search for newly developed anti-phosphoserine and -threonine antibodies with improved specificity.

Western analysis was performed with a 1:1000 (for anti-phosphoSer/Thr) or a 1:5000 dilution (for anti-phosphoTyr) of primary antibodies and horseradish peroxidase-conjugated secondary antibodies (1:10,000 dilution, GE Healthcare), and signals were detected using the ECL plus western blotting detection kit (GE Healthcare). The anti-phosphotyrosine specific antibody produced a strong signal for CcO subunits I, II or III, and IV (Fig. 11.2, right lane). The anti-phosphoserine and -threonine antibodies did not produce strong signals (not shown).

4.2. Phosphoepitope determination via immobilized metal affinity chromatography/nano-liquid chromatography/electrospray ionization mass spectrometry

The definite assignment of phosphorylation sites to a protein is very desirable and a key step, together with functional experiments, to determine structure–function relationships. Tandem mass spectrometry, as performed here, is the method of choice and often allows the unambiguous assignment of a phosphate group to a particular amino acid. We here present data on CcO isolated after TNFα treatment revealing a new phosphorylation site on Tyr11 of the mature CcO subunit IV isoform 1. CcO was used as without gel-electrophoretic separation into its 13 subunits and processed as follows.

CcO isolated after TNFα treatment (100 μL, 2.2 mg/mL) was diluted with 280 μL of 95% formic acid. A small crystal of CNBr was added and the mixture was incubated for 12 h at room temperature in the dark. On the following day, digests were diluted fivefold with dH$_2$O and dried under vacuum. Pellets were reconstituted in 100 μL of 50 mM NH$_4$HCO$_3$ (pH 8.9) and digested with sequence-modified trypsin at a trypsin:protein ratio of 1:100 (w/w) overnight at 37 °C. Peptides were dried under vacuum thereafter. For LC-MS, dried peptides were reconstituted in 40 μL of 0.1 M acetic acid in dH$_2$O and subjected to automated desalt/IMAC/nano-RPLC/ESI-MS analysis as described previously (Ficarro *et al.*, 2005). Briefly, CNBr/trypsin double digested peptides were loaded onto a desalting reversed-phase column, fitted with an inline microfilter, packed with 12 cm Self Pack POROS 10 R2 resin. Peptides were rinsed with

0.1 M acetic acid/water (solvent A) at a flow rate of 10 μL/min and then eluted to an Fe^{3+}-activated IMAC column packed with 15 cm Self Pack POROS 20 MC resin with a HPLC gradient of 0–70% solvent B (0.1 M acetic acid/acetonitrile) in 17 min at a flow rate of 1.8 μL/min. The column was washed with 40 μL of a 25:74:1 acetonitrile/water/acetic acid mixture containing 100 mM NaCl followed by 4 min solvent A at a flow rate of 35 μL/min. Enriched phosphopeptides were eluted to a pre-column containing 2 cm of 5 μm Monitor C18 resin with 40 μL of 25 mM potassium phosphate (pH 9.0) and rinsed with solvent A for 30 min. Peptides were eluted into the mass spectrometer (LTQ-FT, Thermo Electron, San Jose, CA) through an analytical column (360 μm OD×75 μm ID fused silica with 12 cm of 5 μm Monitor C18 particles with an integrated ~4 μm ESI emitter tip fritted with 3 μm silica) with an HPLC gradient (0–70% solvent B in 30 min). Static peak parking was performed via flow rate reduction from the initial 200 to ~20 nL/min when peptides began to elute as judged from a BSA peptide scouting run as described previously (Ficarro *et al.*, 2005). The electrospray voltage of 2.0 kV was applied in a split flow configuration as described (Ficarro *et al.*, 2005). Spectra were collected in positive ion mode and in cycles of one full MS scan in the FT (*m/z*: 400–1800; ~1 s each) followed by MS/MS scans in the LTQ (~0.3 s each) sequentially of the five most abundant ions in each MS scan with charge state screening for +1, +2, and +3 ions and dynamic exclusion time of 30 s. MS/MS spectra were assigned to peptide sequences from the NCBI non-redundant protein database sliced in Bioworks 3.1 for bovine proteins and searched with the SEQUEST algorithm. SEQUEST search parameters designated variable modifications of +79.9663 Da on Ser, Thr, and Tyr for phosphorylation. Assigned MS/MS spectra were manually verified. Applying this procedure we unambiguously identified a new phosphorylation site on Tyr11 of the mature CcO subunit IV isoform 1 (Fig. 11.3; Tyr11 of the mature peptide corresponds to Tyr33 of the precursor protein). Peptide sequence was confirmed by fragment ions y2, y3, y4, y5, y6, y7, y8, y9, y10 and b3, b4, b5, b7, b9, b10, b11. The phosphorylation site was assigned by fragment ions y8, y9, y10 and b3, b4, b5.

Tyrosine phosphorylation of subunit IV tyrosine 11 is probably not triggered by TNFα, because subunit IV also gives a strong signal with the anti-phosphotyrosine antibody after isolation of CcO without the manipulation of a signaling pathway, whereas subunit I phosphorylation is triggered by TNFα or cAMP (Lee *et al.*, 2005; Samavati *et al.*, 2008).

CcO subunit IV-1 is ubiquitously expressed, has a single transmembrane helix and a dumbbell-like structure with a matrix domain and another domain on the intermembrane space side of the inner mitochondrial membrane. Tyr11 is part of the matrix domain, which contains the binding site for ADP and ATP. This site mediates allosteric regulation of CcO depending on the energetic state of the cell: inhibition at high and activation at low

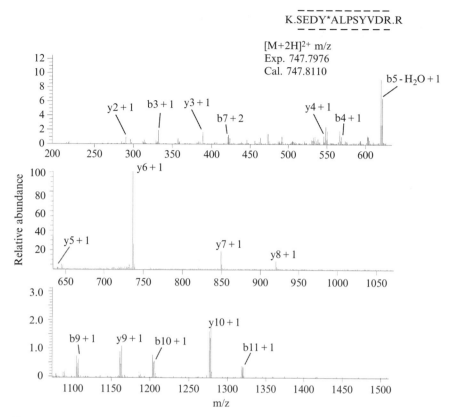

Figure 11.3 Metal affinity chromatography/nano-liquid chromatography/electro-spray ionization mass spectrometry (IMAC/nano-LC/ESI-MS) reveals phosphorylation of tyrosine 11 of the mature CcO subunit IV (tyrosine 33 of the precursor peptide). Peptide sequence was confirmed by fragment ions y2, y3, y4, y5, y6, y7, y8, y9, y10 and b3, b4, b5, b7, b9, b10, b11. The phosphorylation site was unambiguously assigned to tyrosine 11 by fragment ions y8, y9, y10 and b3, b4, b5. Tyrosine 11 is located on the matrix side.

ATP/ADP ratios. We have recently shown that dephosphorylation of CcO abolishes allosteric regulation (Hüttemann *et al.*, 2008), and Tyr11 may be the residue that enables allosteric regulation when phosphorylated due to its proximity to the proposed binding site based on computer modeling (Hüttemann *et al.*, 2001).

5. Concluding Remarks

Analyzing the effect of reversible phosphorylation on mitochondrial proteins is possible by inclusion of unspecific protein phosphatase inhibitors and calcium chelating agents during enzyme isolation. It is important to note that vanadate interferes with signaling pathways and must therefore be added after tissue or cells are broken up for mitochondria isolation but not during the optional initial incubation step, in which signaling pathways can be manipulated.

Vanadate also directly interferes with mitochondrial function (Soares *et al.*, 2007) and may therefore not be the reagent of choice for studies on isolated mitochondria. For example, we have observed that activated vanadate can directly affect the membrane potential of liver mitochondria, which can be accompanied by a gradual decrease in the rate of respiration in state 3 after the addition of ADP (which normally plateaus until all ADP is converted into ATP). However, if vanadate is omitted, for example in the last washing step during mitochondria isolation, tyrosine phosphorylation decreases within minutes, presumably due to the presence of mitochondrial tyrosine phosphatases. Therefore, the inclusion of activated vanadate for protein isolation is desirable and efficient for protecting the physiological phosphorylation state during enzyme isolation, but functional studies using isolated mitochondria may require the substitution of vanadate with a yet-to-be identified tyrosine phosphatase inhibitor that does not interfere with mitochondrial function.

REFERENCES

Boerner, J. L., Demory, M. L., Silva, C., and Parsons, S. J. (2004). Phosphorylation of Y845 on the epidermal growth factor receptor mediates binding to the mitochondrial protein cytochrome c oxidase subunit II. *Mol. Cell Biol.* **24**, 7059–7071.

Büge, U., and Kadenbach, B. (1986). Influence of buffer composition, membrane lipids and proteases on the kinetics of reconstituted cytochrome-c oxidase from bovine liver and heart. *Eur. J. Biochem.* **161**, 383–390.

Ficarro, S. B., Salomon, A. R., Brill, L. M., Mason, D. E., Stettler-Gill, M., Brock, A., and Peters, E. C. (2005). Automated immobilized metal affinity chromatography/nano-liquid chromatography/electrospray ionization mass spectrometry platform for profiling protein phosphorylation sites. *Rapid Commun. Mass Spectrom.* **19**, 57–71.

Gordon, J. A. (1991). Use of vanadate as protein-phosphotyrosine phosphatase inhibitor. *Methods Enzymol.* **201**, 477–482.

Hopper, R. K., Carroll, S., Aponte, A. M., Johnson, D. T., French, S., Shen, R. F., Witzmann, F. A., Harris, R. A., and Balaban, R. S. (2006). Mitochondrial matrix phosphoproteome: Effect of extra mitochondrial calcium. *Biochemistry* **45**, 2524–2536.

Hüttemann, M., Kadenbach, B., and Grossman, L. I. (2001). Mammalian subunit IV isoforms of cytochrome c oxidase. *Gene* **267**, 111–123.

Hüttemann, M., Lee, I., Pecinova, A., Pecina, P., Przyklenk, K., and Doan, J. W. (2008). Regulation of oxidative phosphorylation, the mitochondrial membrane potential, and their role in human disease. *J. Bioenerg. Biomembr.* **40**, 445–456.

Hüttemann, M., Lee, I., Samavati, L., Yu, H., and Doan, J. W. (2007). Regulation of mitochondrial oxidative phosphorylation through cell signaling. *Biochim. Biophys. Acta* **1773**, 1701–1720.

Kadenbach, B., Arnold, S., Lee, I., and Hüttemann, M. (2004). The possible role of cytochrome *c* oxidase in stress-induced apoptosis and degenerative diseases. *Biochim. Biophys. Acta* **1655,** 400–408.

Kadenbach, B., Jarausch, J., Hartmann, R., and Merle, P. (1983). Separation of mammalian cytochrome *c* oxidase into 13 polypeptides by a sodium dodecyl sulfate-gel electrophoretic procedure. *Anal. Biochem.* **129,** 517–521.

Kadenbach, B., Stroh, A., Ungibauer, M., Kuhn-Nentwig, L., Büge, U., and Jarausch, J. (1986). Isozymes of cytochrome *c* oxidase: Characterization and isolation from different tissues. *Methods Enzymol.* **126,** 32–45.

Lee, I., Salomon, A. R., Ficarro, S., Mathes, I., Lottspeich, F., Grossman, L. I., and Hüttemann, M. (2005). cAMP-dependent tyrosine phosphorylation of subunit I inhibits cytochrome *c* oxidase activity. *J. Biol. Chem.* **280,** 6094–6100.

Liu, S. S. (1999). Cooperation of a "reactive oxygen cycle" with the Q cycle and the proton cycle in the respiratory chain—superoxide generating and cycling mechanisms in mitochondria. *J. Bioenerg. Biomembr.* **31,** 367–376.

Robb-Gaspers, L. D., Burnett, P., Rutter, G. A., Denton, R. M., Rizzuto, R., and Thomas, A. P. (1998). Integrating cytosolic calcium signals into mitochondrial metabolic responses. *EMBO. J.* **17,** 4987–5000.

Samavati, L., Lee, I., Mathes, I., Lottspeich, F., and Hüttemann, M. (2008). Tumor necrosis factor α inhibits oxidative phosphorylation through tyrosine phosphorylation at subunit I of cytochrome *c* oxidase. *J. Biol. Chem.* **283,** 21134–21144.

Soares, S. S., Gutierrez-Merino, C., and Aureliano, M. (2007). Decavanadate induces mitochondrial membrane depolarization and inhibits oxygen consumption. *J. Inorg. Biochem.* **101,** 789–796.

Von Jagow, G., and Klingenberg, M. (1972). Close correlation between antimycin titer and cytochrome *b* (T) content in mitochondria of chloramphenicol treated *Neurospora crassa. FEBS Lett.* **24,** 278–282.

Wikström, M., Krab, K., and Saraste, M. (1981). "Cytochrome Oxidase: A Synthesis." Academic Press, Ltd, London, England.

CHAPTER TWELVE

Using Functional Genomics to Study PINK1 and Metabolic Physiology

Camilla Scheele,* Ola Larsson,† *and* James A. Timmons‡

Contents

1. Introduction	212
2. Studying Mitochondrial Modulation in Humans *In Vivo*	214
2.1. Increase in mitochondrial activity	214
2.2. Decrease in mitochondrial activity	215
2.3. Mitochondrial dysfunction	215
3. Cellular Mitochondrial Models	216
3.1. Studies in human muscle primary myotubes	216
3.2. Knockdown of target genes in brown adipocytes	217
4. Considerations When Using Short Interfering RNAs	218
4.1. Off-target effects of siRNA	219
4.2. siRNA and apoptosis	220
5. *Ex Vivo* Gene Expression Analysis	222
5.1. RNA isolation	222
5.2. Global gene expression analysis—microarray	223
5.3. Targeted gene expression—quantitative real-time PCR	224
5.4. Studying candidate loci/genes in detail	225
References	226

Abstract

Genome sequencing projects have provided the substrate for an unimaginable number of biological experiments. Further, genomic technologies such as micro-arrays and quantitative and exquisitely sensitive techniques such as real-time quantitative polymerase chain reaction have made it possible to reliably generate millions of data points per experiment. The data can be high quality and yield entirely new insights into how gene expression is coordinated under complex physiological situations. It can also be that the data and interpretation

* The Centre of Inflammation and Metabolism, Department of Infectious Diseases and CMRC, Rigshospitalet, The Faculty of Health Sciences, University of Copenhagen, Denmark
† Department of Biochemistry, McGill University, Montreal, Canada
‡ The Wenner-Gren Institute, Arrhenius Laboratories, Stockholm University, Stockholm, Sweden

Methods in Enzymology, Volume 457
ISSN 0076-6879, DOI: 10.1016/S0076-6879(09)05012-5

are meaningless because of a lack of physiological context or experimental control. Thus, functional genomics is now being applied to study metabolic physiology with varying degrees of success. From the genome sequencing projects we also have the information needed to design chemical tools that can knock down a gene transcript, even distinguishing between splice variants in mammalian cells. Use of such technologies, inspired by nature's endogenous RNAi mechanism–microRNA targeting, comes with significant caveats. While the discipline of Pharmacology taught us last century that inhibitor action specificity is dependent on the concentration used, these experiences have been ignored by users of siRNA technologies. What we provide in this chapter is some considerations and observations from functional genomic studies. We are largely concerned with the phase that follows a microarray study, where a candidate gene is selected for manipulation in a system that is considered to be simpler than the *in vivo* mammalian tissue and thus the methods discussed largely apply to this cell biology phase. We apologize for not referring to all relevant publications and for any technical considerations we have *also* failed to factor into our discussion.

1. INTRODUCTION

Numerous methods have been developed during the endeavor to understand gene function. As a consequence of the often complex mechanisms which govern gene–network interactions, caution is needed when applying these methods, both when selecting the model system and during data analysis. When assessing the function of a gene in cell culture systems, it is common to use both overexpression and knockdown to subsequently evaluate if the cellular responses are coherent with the hypothesized mechanism(s). However, these commonly used approaches can be associated with major shortcomings, which limit the conclusions one can reach. For example, experiments involving overexpression very often achieve nonphysiological expression levels and usually no consideration is given to the multiplicity of functions that protein structure may have on cell biochemistry. Such issues are very problematic in the context of biochemical reactions that are regulated by concentration gradients and feedback loops, where the newly overexpressed protein could act as a chemical sink for intermediates involved in many pathways. The applications of functional genomics methods to study mitochondrial related genes and metabolic physiology is highly plausible and represent an alternative tool to discover new regulatory control processes. It is notable that murine knock-out technology has, after detailed analysis, rarely yielded convincing new understanding of human metabolism. Only rarely do investigators (Cannon and Nedergaard, 2004) study their gene on interest on sufficient genetic backgrounds to make robust conclusions. For approaches employing knockdown, using RNA interference, there are numerous reports on non-specific effects, which have unfortunately entirely been ignored by most. There is a great need for standards in the application of short-interfering RNAs to study gene function, to be established.

In this article we describe some functional genomics methods utilized to study the function of PINK1, a kinase localized mainly to the mitochondrial inner membrane. The physiological function of PINK1 is still unclear, but several studies have suggested a protective role against oxidative stress (Pridgeon et al., 2007; Wang et al., 2006) and recent results from Drosophila experiments further suggest that PINK1 promotes mitochondrial fission (Yang et al., 2008). The hypotheses and the studies performed on PINK1 to develop these ideas must be seen in the light of the major discovery in 2004, where the PINK1 locus was linked to hereditary Parkinson's disease (Valente et al., 2004). Prior to this discovery, the function of PINK1 was unknown. Since then there has been a clear bias in the design of studies and data interpretation favoring phenotypes which can be related back to Parkinson's disease. By analogy, the biological research programs have established model systems that are similar to lead-development strategies applied in industrial drug discovery, which has been focused on the use of assays that provide reliable biochemical endpoints rather than on in vivo physiological relevance (the exception being the knockdown studies, which we will return to later).

Favored models to study PINK1's function include Drosophila, mice and neuroblastoma cell lines. Whereas these studies have resulted in important biochemical observations, they are unlikely to fully reflect the metabolic interactions occurring in humans (Fredriksson et al., 2008). Here we describe human in vivo models for mitochondrial modulation which can be used in screens for global gene expression changes: a human primary muscle cell model which can be used as a tool to examine how these differences in gene expression come about; and a murine RNA interference model which can be used to elucidate the genes involvement in mitochondrial biogenesis. In relation to RNA interference, we discuss limitations when using short interfering RNAs; in particular when investigating target genes involvement in apoptosis. In relation to studies of gene expression, we provide examples where consideration of splice variants and natural antisense RNAs during primer design will improve study design. Overall, the approach we are advocating is multidisciplinary and thus requires the integration of skills and methods developed in many different biological fields.

2. Studying Mitochondrial Modulation in Humans *In Vivo*

The first example we provide is analysis of PINK1 regulation in human skeletal muscle, a very suitable tissue for studying mitochondrial dynamics. For instance, it allows manipulation of metabolism and direct tissue sampling *in vivo* (Greenhaff and Timmons, 1998; Timmons *et al.*, 1998b). Through the measurement of hormones and metabolites (e.g., insulin-, glucose- and secreted protein-levels) in blood samples and gene expression in muscle biopsies, we provide a framework for using *in vivo* physiological inference to design cell biology studies. Muscle biopsies are obtained using the percutaneous needle method with suction (Bergstrom and Hultman, 1966). Prior to each biopsy, local anesthesia (lidocaine, 20 mg/ml; SAD, Denmark) is applied to the skin and superficial fascia of the biopsy site. The muscle biopsies are quickly cleaned from blood and connective tissue; flash-frozen in liquid nitrogen to prevent RNA degradation; and stored at $-150\ ^{\circ}$C until RNA isolation. The process of RNA degradation from tissue samples can be fast, particularly in partially frozen tissue. Therefore, working quickly is crucial for preservation of RNA quality (Thompson *et al.*, 2007). Measuring the gene expression (mRNA or non-coding RNA abundance) in samples from the various physiological states enables the assessment of whether a correlation exists between gene expression of our specific gene of interest and any physiological parameter, such as VO_2max, or insulin and glucose levels in plasma. The strength of this simple approach is that the physiological manipulation is known (e.g., muscle contraction) to incrementally increase oxidative metabolism. In contrast, the cause and effect relationship obtained from expression profiling of clinical patient material alone is unknown.

2.1. Increase in mitochondrial activity

During aerobic exercise, mitochondrial activity increases in skeletal muscle to support oxidative ATP regeneration and prevents muscle fatigue (Timmons *et al.*, 1997). If carried out frequently, periods of aerobic exercise ("exercise training") result in mitochondrial biogenesis (Kiessling *et al.*, 1974) and altered mitochondrial control. Importantly, sample material can be obtained before, during, and after the training period so that each individual serves as one's own control, reducing subject-to-subject variation and providing a temporal parameter for cause–effect inference. More infrequently, a non-training control group is used to observe differences occurring over time or associated with the schedule for sample collection. More typically, subjects are studied at the same time of day, under similar dietary conditions for at least 24 h following the last previous exercise study.

In our studies of PINK1, we utilized material from an endurance training study of 24 young sedentary but healthy men (Timmons *et al.*, 2005b). The use of sedentary subjects is ideal because training will have a more dramatic effect on these individuals compared to athletes for instance. Thus differences in gene expression and physiological improvements will be more pronounced. Subjects performed a fully supervised 6-week training program consisting of four 45-min cycling sessions per week at an intensity corresponding to 70% of pre-training VO_2max ($\sim$100% compliance). It is essential to supervise training to minimize variation in training dose caused to differences in compliance (e.g., motivation) between subjects. Cycling VO_2max was determined during an exhaustive incremental cycling test (Rodby, Sweden), with continuous analysis of respiratory gases using the SensorMedics ventilator (SensorMedics, USA). Physiological measurements and muscle biopsies in *vastus lateralis* were taken prior to the training program and 24 h after the last training session, as previously described (Timmons *et al.*, 1998a, 2005b).

2.2. Decrease in mitochondrial activity

Forced bed rest results in many changes in skeletal muscle, involving not only a lack of sustained mitochondrial biogenesis and activity, but also a degeneration of the muscle, with potential severe consequences for metabolic homeostasis (Fredriksson *et al.*, 2008). Studying target gene expression during inactivity in healthy subjects can be considered a model for decrease in mitochondrial activity. We utilized a model of skeletal muscle unloading (Berg *et al.*, 1991, 1993; Haddad *et al.*, 2005) to impose 5 weeks of muscle inactivity in the lower leg muscle of healthy volunteers. Six healthy adult subjects, four men, and two women volunteered for this study. Muscle biopsies were obtained from the Soleus before and after this intervention period and analyzed for alterations in mitochondrial gene expression (Timmons *et al.*, 2006). During these studies we discovered that PINK1 gene expression is downregulated *in vivo* in humans during a period when mitochondrial loss can be expected. This coincided with the discovery that it is gene responsible for the PARK6 loci association with Parkinson's disease.

2.3. Mitochondrial dysfunction

Obesity and type 2 diabetes represent disturbances in the metabolic homeostasis typically characterized by an elevated fasting glucose concentration or impaired insulin action, or both. Mitochondrial dysfunction has been suggested to play a central role in the pathophysiology of these diseases and to do so across a variety of organs (Befroy *et al.*, 2007; Lin *et al.*, 2004; Petersen *et al.*, 2004). Functional genomics approaches have also been presented in a manner to support the idea that oxidative phosphorylation genes are downregulated in the muscle of humans with Type II diabetes or insulin resistance (Mootha *et al.*, 2003; Patti *et al.*, 2003). Yet each of these studies *has* serious caveats, including a complete lack of age-matching (Patti *et al.*, 2003), sampling tissue after a non-relevant pharmacological insulin infusion protocol (Mootha *et al.*, 2003) or generally blurring the lines between genetic inheritance and familial inheritance combined with shared environmental influences (Petersen *et al.*, 2004). We were particularly concerned about the value of the functional genomics (RNA profiling studies) so we therefore isolated RNA from muscle samples from a much larger cross-sectional cohort and assessed gene expression, along with associated physiological parameters, to re-examine the mitochondrial-insulin resistance hypothesis.

3. CELLULAR MITOCHONDRIAL MODELS

Cell lines offer a *relatively* homogenous biological model in which gene expression can be easily manipulated. It is common to choose a cell line established from the tissue of interest. However, in many cases it is uncertain how much of the original tissue specific features remain conserved in cell cultures (Sandberg and Ernberg, 2005). As cell lines most commonly are derived from tumors, many of the primary phenotype features have most likely been altered prior to cell culture isolation. Furthermore, it has been established that many adaptations occur as a direct consequence of the *in vitro* conditions impacting on many aspects of cell physiology. Thus, caution is needed when inferring physiological conclusions based on results from cell lines. Indeed, these models are more suitable, in our opinion, for exploring mechanistic gene–gene interactions. Primary cultures or cultures of primary cells which have been immortalized with telomerase (Bodnar *et al.*, 1998) could be considered as physiologically closer to the parent tissue compared to tumor derived cell lines. Limitations of a non-immortalized primary cell model include that the cultures can be heterogeneous and that primary cells cannot divide indefinitely, entering senescence after a certain number of cell divisions (Hayflick, 1965; Hayflick and Moorhead, 1961). Thus, the appropriate cell model to use varies depending on scientific question, time, and availability of samples for establishing primary cell cultures.

3.1. Studies in human muscle primary myotubes

Skeletal muscle tissue can be easily obtained to create primary cultures of human myotubes. This system can be used to follow-up functional genomic relationships derived from *in vivo* human studies. Muscle biopsies are taken in the same manner used to analyze gene expression (Scheele *et al.*, 2007a), but instead of being flash-frozen the biopsies are harvested in cell culture media.

Following removal of pieces of fat and connective tissue, the muscle is first minced and then digested in 5 ml HamF10 media containing 50% trypsin EDTA (1×) and 5 mg collagenase IV and BSA at 37 °C for 5 min. The digestion solution now containing satellite cells is transferred to a tube with 2 ml serum on ice, to inactivate the digestion process. Then 5 ml digestion solution is added to the remaining tissue and the digestion is allowed to proceed for 10 more minutes. The digestion solution is filtered and centrifuged at 800g for 7 min. Following a wash with HamF10, cells are pre-plated in a 60 mm culture plate for 3 h. This step will allow fibroblasts to bind to the plate, whereas the satellite cells, still in their quiescent state, will not immediately attach to the plate. This diminishes contamination of fibroblasts in the muscle cell preparation. After the 3 h incubation, the media containing the satellite cells are transferred to a 25 cm culture flask, pre-coated with matrigel (BD Science). Cell culture media, HamF10, 20% fetal bovine serum (FBS), 1% Penicillin/Streptomycin (PS), and 1% Fungizone, are changed on day 4 after isolation and thereafter every other day. Once activated, satellite cells can proliferate for several passages and can be frozen during this process. They can also be differentiated into myotubes for further studies.

Myocytes are differentiated into myotubes by first changing the culture media to DMEM low glucose, 10% FBS, 1% PS. When cells are 100% confluent (they develop a slim, pine–leaf–like morphology and start to line-up side by side) the media is changed to DMEM, high glucose, 2% horse serum, 1% PS. Both confluence and reducing the serum content of the media removes FGF receptor signaling, thus allowing the cells to progress down the myogenesis pathway (Berkes and Tapscott, 2005). From this point, the myocytes will be fully differentiated within 5–7 days and will survive for at least 2 weeks as long as the culture media is changed every second day. Once differentiated, myotubes can be used to investigate results obtained by statistical inference from human *in vivo* functional genomic studies (Timmons *et al.*, 2005b, 2007; Fredriksson *et al.*, 2008). Interestingly, there are studies indicating that muscle cells derived from biopsies tend to conserve the phenotype of their donor in culture. For example, insulin resistance (Henry *et al.*, 1995, 1996; McIntyre *et al.*, 2004), possibly due to genetic variation and methylation patterns previously established *in vivo*. This may enable comparisons between myotubes derived from healthy, versus insulin resistant subjects, and evaluation of their responses to various treatments providing an additional experimental variable to facilitate the cause–effect nature of gene expression changes under common experimental conditions.

3.2. Knockdown of target genes in brown adipocytes

To investigate if PINK1 was involved in mitochondrial biogenesis, we utilized a well-established mitochondrial model (Cannon and Nedergaard, 2008). For each experiment, 10 Male NMRI mice (age 3–4 weeks; B&K,

Stockholm, Sweden) were euthanized using CO_2, and brown adipose tissue (BAT) (from the interscapular, cervical and axillary depots) was dissected and the brown pre-adipocytes are isolated as described previously (Nechad *et al.*, 1987). Briefly, minced tissues are digested in buffer containing collagenase (type II; SigmaAldrich) for 30 min at 37 °C. The cell suspension is filtered and kept on ice for 20 min. After discarding the top layer (mature adipocytes) the suspension is filtered and washed in Dulbecco's modified Eagle's media (DMEM) and re-suspended in 0.5 ml culture media per animal. Cells are cultured in 6-well plates (Falcon) in DMEM supplemented with 10% (v/v) newborn-calf serum (Invitrogen), 2.4 nM insulin, 25 μg/ml sodium ascorbate, 10 mM HEPES, 4 mM glutamine, 50 units/ml penicillin, and 50 μg/ml streptomycin. 1.8 ml of culture media is added to each well followed by 0.2 ml of cell suspension. Following 24 h in culture, cells are transfected using 0.32% Lipofectamine 2000 (Invitrogen) and 20 nM short interfering RNA (siRNA) in 2.5 ml newborn-calf serum-containing cell culture media (without penicillin/streptavidin) according to the manufacturer's protocol and as described previously (Scheele *et al.*, 2007b). At day 3 of culture (following 48 h of transfection), cells are harvested for RNA isolation in TRIzol (Invitrogen). Subsequent RNA isolation is performed as described below in Section 5.1.

4. Considerations When Using Short Interfering RNAs

Administration of short double-stranded RNAs (~20 nucleotides), called short interfering RNAs (siRNAs), to cells results in efficient gene silencing in mammalian cells (Elbashir *et al.*, 2001). SiRNA molecules complementary to particular mRNA sequence will bind the mRNA through a complex process mediated by the RNA induced silencing complex (RISC), leading to cleavage and degradation of the target mRNA. The efficiency of a particular siRNA is thought to depend on its sequence and structure, as well as the sequence context of the target sequence. Prediction software is available to increase the chance of designing a potent siRNA (Chalk *et al.*, 2004). The advantages of utilizing siRNA in mammalian cell models include ease of use, moderate cost, high success rate in knockdown and that the mechanisms is posttranscriptional, thus no modification of the genomic DNA is necessary. However, this has led to the indiscriminant use of siRNA in functional genomic studies, with little concern paid to their specificity; a trend which is not helping biomedical research. Many researchers take an experimental tool that may be relatively selective at 10–50 nM and overwhelm the cell with micromolar levels of the nucleotide and do not consider, in a statistically robust manner, the specificity of their experimental intervention. Note that this form of "high throughput" biology (e.g., large scale RNAi screens), using single function nonphysiological read-outs may well fail as spectacularly as such approach have done (so far) in the pharmaceutical industry and thus at this stage we do not see them as a superior strategy to the *in vivo* functional genomics microarray approach (see below).

4.1. Off-target effects of siRNA

Approaches for avoiding non-specific effects of siRNA (quaintly referred to as "off-target" effects) remain to be either standardized or optimized. There are thought to be two broad types of off-target effects; motif-dependent toxic effects, and silencing of genes other than the target genes. Both of these are dependent on the sequence of the siRNA and thus are "sequence dependant" non-specific actions. A specific motif, UGUGU in siRNA sequences, has been shown to induce interferon response *in vivo* in mice and *in vitro* in human peripheral blood mononuclear cells (Judge *et al.*, 2005). Another motif, UGGC, was identified to cause a reduction in cell survival when present in the antisense strand of the siRNA (Fedorov *et al.*, 2006). Expression profiling using microarrays, revealed that in some cases, 11 base complementarities between siRNA and mRNA was enough to cause a target specific off-target effect (Jackson *et al.*, 2003). This was further investigated in another study, demonstrating that off-target effects are associated with the hexamer or heptamer seed region (positions 2–7 or 2–8) of the antisense strand of the siRNA when perfectly complementary to the 3'UTR of several target mRNAs (Birmingham *et al.*, 2006). This manner of silencing, similar to regulation of gene expression by endogenously expressed miRNAs, was confirmed in another study (Jackson *et al.*, 2006b). Thus there should be major concerns regarding the specificity of any siRNA used and an example of gene-knockdown carried out with a singular siRNA should not be accepted for publication.

The potential for off-target effects can be appreciated by the similarity between siRNA and endogenous miRNA actions, whereby a single miRNA can potentially regulate large sets of genes (targeting a complimentary sequence in the 3'UTR) involved in a variety of physiological processes. A chemical modification of the guide (antisense) strand, a 2'-O-methyl ribosyl substitution at position 2, has been shown to reduced the silencing of 3'UTR targeted transcripts (miRNA like mechanism) while the silencing of perfectly matched mRNA transcripts were unaffected (Jackson *et al.*, 2006a). Thus, siRNAs as a tool to study gene function requires adherence to a rigorous experimental protocol and most current data in the cell biology literature should be interpreted with some caution. Indeed, given manner in which commercial companies market siRNA technology, it has become possible for laboratories to use this approach without significant knowledge of the potential problems with siRNA methodology. There are still too many examples of studies drawing conclusions based on only one siRNA. We suggest that at least two siRNAs targeting different sites of the target gene should be used (but still with caution when describing the outcome), and perhaps up to 5 siRNA could be used to make rigorous claims about cell phenotype changes. Indeed, depending on the condition studied, two siRNAs may not be enough. Below we illustrate such an example from one of our new PINK1 knockdown experiments in neuroblastoma cells.

4.2. siRNA and apoptosis

Given the role of mitochondria in apoptosis, control and the potential role for apoptosis in Parkinson's disease, one of the prevailing hypotheses regarding PINK1's function has been that it protects against oxidative stress and the subsequent apoptosis. Indeed, it was not long before several laboratories produced evidence for this affect, typically using high concentrations of siRNA against 1 or 2 locations in the gene (Deng *et al.*, 2005; MacKeigan *et al.*, 2005). We also investigated this hypothesis by knockdown of PINK1 mRNA levels in neuroblastoma cells using siRNAs and determined the level of apoptosis. The neuroblastoma cell line SK-N-MC was sub-cultured in minimal essential media (MEM) plus Earle's supplemented with 10% FBS, 2 mM L-glutamine, 1 mM sodium pyruvate, 1× nonessential amino acids, 100 U/ml penicillin and 100 μg/ml streptomycin and maintained at 37 °C in 5% CO_2. For maximal reproducibility, a fixed number of cells were used in replicate experiments. Cells were therefore counted using a hemacytometer and 50,000 SK-N-MC cells were seeded per well into 6-well plates 48 h before transfection. SK-N-MC cells attach slowly to the plate surface after splitting and are sensitive to transfection if confluence is less than 75%. Transfection was performed using 0.25% Lipofectamine 2000 (Invitrogen, Sweden) and 20 nM siRNA in 2.5 ml serum-containing cell culture media (without antibiotics). Briefly, Lipofectamine 2000 is pre-mixed with Optimem, incubated for 5 min and then mixed with siRNA. Following 20 min of incubation, the Lipofectamine 2000–siRNA mix is applied to cells. Media are changed 4 h later and cells are harvested 48 h after transfection for RNA isolation or 63 h after transfection for apoptosis assays using annexin/propidium iodide staining and quantified using Flow cytometry.

The siRNAs we used for targeting PINK1 were described as being "pre-validated" (siRNAs PinkA and PinkB) or "pre-designed" (siRNAs PinkC and PinkD) siRNAs from Ambion which refers to whether or not their "activity" against the target gene had been established or not. Notably, this gives no assurance that the (expensive) reagents you are buying are selectively targeting your gene of interest. The control siRNA was pre-designed by Ambion not to target any gene in the genome

(Silencer®Negative Control#1, Ambion) and thus represents a tool that should have no effect on your gene. As demonstrated in Fig. 12.1, despite efficient knockdown by all four siRNAs, targeting different parts of the PINK1 gene, the percentage of apoptosis varied between siRNAs, indicating that the induced apoptosis may well depend on "off-target" effects rather than gene-specific effects. Interestingly, one of the siRNAs targeting PINK1 actually demonstrated a significantly lower apoptosis than the control siRNA (note that both C and D target the long and short variant

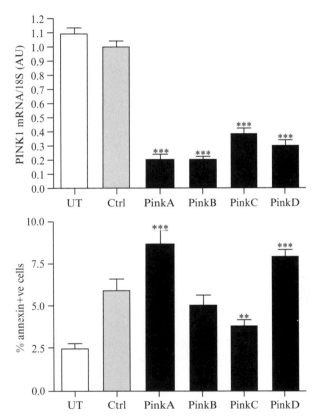

Figure 12.1 Apoptosis in SK-N-MC cells following knockdown of PINK1 (A and B). Cells were transfected with siRNAs targeting four different sites of the PINK1 mRNA. The different siRNAs were assigned PinkA, PinkB, PinkC, PinkD. As controls, untreated cells (UT) and cells treated with a nontargeting siRNA control (Ctrl) were included. (A) Knockdown efficiency was evaluated by using qPCR analysis. (B) Apoptosis were assessed by PI and annexin staining with subsequent flow cytometry analysis. Differences in gene expression and annexin staining were tested using one-way ANOVA. Tukey post-tests provided the p-values and represents comparisons to the siRNA control samples (Ctrl). $N = 9$–13, data are mean ±se. $\star P < 0.05$, $\star\star P < 0.01$, $\star\star\star P < 0.001$.

of PINK1 (Scheele *et al.*, 2007c)). At the same time, articles published in *high profile* biomedical journals, using only 1 or 2 siRNA molecules at high concentration or nonphysiological overexpression approaches, have "established" that PINK1 is anti-apoptotic (Deng *et al.*, 2005; MacKeigan *et al.*, 2005; Wood-Kaczmar *et al.*, 2008). We believe this to be a premature conclusion.

5. *Ex Vivo* Gene Expression Analysis

To study gene expression in human and cell models, total RNA is isolated and used either for targeted (quantitative real-time PCR) or global (microarray) gene expression analysis.

5.1. RNA isolation

Chloroform–phenol extraction for isolation of total RNA is a well established and widely used method. However, a shift in downstream detection methods from Northern blots to quantitative real-time PCR and microarrays have increased the demands of purity and quality of the isolated RNA. We use TRIzol, a phenol-based reagent from Invitrogen (but available from several manufacturers) for RNA isolation principally to keep our studies as compatible as possible across time. For homogenization, we use as Tissuelyser (Qiagen) a bead beater or a motor-driven homogenizer (Polytron, Kinematica, Newark, USA). It is important to keep everything on dry ice prior to homogenization to ensure inactivation of RNases. Following homogenization, 200 μl chloroform/ml TRIzol is added, tubes are shaken vigorously for 15 s, and centrifuged at 12,000g for 15 min and the aqueous phase is transferred into a new tube. RNA is precipitated by adding 1:1 isopropanol followed by thoroughly vortexing and incubation for 10 min in room temperature. Precipitated RNA is pelleted by centrifuging for 10 min at 12,000g. The RNA pellet is washed with ice-cold 75% EtOH and finally air dried and dissolved in nuclease free water.

We typically use a low volume spectrophotometer (Nanodrop) to measure the RNA concentration, which in addition to 260/280 ratio also calculates the 260/230 ratio. We further evaluate the RNA quality using a micro-capillary gel system (Bioanalyzer, Agilent). Small quantities of RNA are loaded onto chips and an RNA integrity number (RIN) score is estimated from the RNA electrophoresis profile. The RIN score represents a standardized RNA quality control. The algorithm for the RIN score was extracted from a large collection of microcapillary electrophoretic RNA measurements by constructing regression models based on a Bayesian learning technique (Schroeder *et al.*, 2006). RIN scores seem comparable from each tissue isolation device, for example, in general RIN scores from frozen human muscle rarely exceed 8.5 (average ~8), while control RNA isolated at the same time is greater than 9.

5.2. Global gene expression analysis—microarray

Comparing gene expression between healthy controls and people with a disease is often done to find candidate genes that may contribute to a disease. The power of this approach is that rather than studying one gene, where variations within the control group and disease group, respectively, could mask involvement in the disease, the entire canonical pathway can be examined, and a high statistical rigor applied. Thus, unlike research initiated from the perspective of an *in vitro* observation, which is often out of context of the biological system *that* the gene is supposed to be involved with, an initial *ex vivo* analysis (e.g., gene network analysis) can provide a robust statistical framework for the identification of the gene networks involved in a complex physiological process for subsequent follow-up. This approach reduces the dimensionality of all possible single gene studies that could be done using isolated cell systems. However, it is currently underutilized. Indeed, few major laboratories in metabolic physiology can generate the initial *ex vivo* human tissue analysis. This bias must change and can rapidly change with greater collaboration between leaders in molecular metabolism and human physiology. When comparing patient with a control at a single time point, the analysis cannot determine whether a change in gene expression is *causing* the disease or if it rather reflects a *secondary effect*. Various approaches can be utilized to enhance the likelihood that a change is important. For example, most disease indicators (e.g., fasting glucose) are represented by a continuum within the patient population (albeit a non-normally distributed one in this case) and if a gene is not only differentially expressed between control and patient, but is so in proportion to disease status, then this implies that it is more central to the biological process.

As with all studies, functional genomics studies with microarray analyses must make efforts to control known physiological variables before making new hypotheses based on differential gene expression. One example of this type of problem recently occurred in the type 2 diabetes field (Mootha *et al.*, 2003). The study involved a diabetes and a control group. Closer inspection reveals that a set of non–obese normal glucose tolerant people was compared with a blend of non–obese and obese diabetics following the pharmacological infusion of insulin. Thus, the difference between the categories arises from the confounding factor (obesity) and/or differential responses to pharmacological insulin levels rather than the diabetes disease process. In contrast to the problems of adjusting for confounding factors in the case-control design, which requires large sample sizes for correct stratification or alternatively use of rodent out-bred models (Wisloff *et al.*, 2005), *in vivo* intervention studies (Timmons *et al.*, 2005b) offer a cleaner system for functional genomics studies. In these studies, materials (i.e., blood samples and biopsies) can be collected before, during and after the intervention or at various generations. In the human studies, the material collected before the intervention will serve as control for each individual and thus changes in the endogenous gene expression profile can be directly linked to the intervention. Obviously, differences in study effects (and gene expression) can arise as a consequence of individual subject characteristics thus leading to variation in the extent of expression in the muscle tissue (Timmons *et al.*, 2005a).

5.3. Targeted gene expression—quantitative real-time PCR

Threshold (Ct) values. The Ct value reflects the PCR cycle when the signal from a specific sample reaches an arbitrary threshold above baseline noise, set for the particular gene (selected to be in the log–linear phase of amplification for all measured samples). Quantitative real-time PCR (qPCR) is a commonly used method to compare relative gene expression levels between samples. The method utilizes fluorescent dyes that are activated during amplification so that each PCR cycle will result in an increase in fluorescence. The cycle at which a given amount of signal is detected is directly related to the amount of starting gene specific cDNA in the sample and can thus be used to accurately determine the total expression level in the tissue or cell. Prior to the qPCR reaction, first-strand cDNA synthesis is performed using total RNA as template. The same amount of total RNA is used for all samples in a study, but to adjust for variations in reverse transcription efficiency, the gene expression of an endogenous control is measured in the same samples. Commonly used controls include 18S and so-called house-keeping genes such as GAPDH and β-actin, but these should be assessed to ensure absence of systematic changes between the samples categories (Thellin *et al.*, 1999). We normally use 18S in our studies and this approximates to accurate measurement of high purity total RNA input using a Nanodrop. The raw data obtained from qPCR are cycle threshold (Ct) values. As 18S constitutes a major part of total RNA, Ct values that vary largely between the samples may indicate problems with RNA integrity or purity (reflected in different RNA inputs into the cDNA reaction) rather than real differences in target gene expression. To detect target genes, gene specific primers are designed based on the reference sequence (RefSeq), which is the most curated version of the mRNA sequence. There are often several different RefSeq for a gene, reflecting occurrence of splice-variants. Prior to design, careful assessment of the genomic locus is required using a genome browser so that one is very clear which transcript is being measured.

5.4. Studying candidate loci/genes in detail

5.4.1. Splice variants, natural antisense, and primer design

Because many genes are regulated through alternative splicing, a closer look at the genomic locus is necessary when studying a novel gene and in particular when designing gene specific qPCR primers. Some splice variants are translated into proteins and may have a different and even an opposite function compared to the target splice-variant. Even if not translated, a higher abundance of alternative splice variants can lead to a lower abundance of the target gene if the gene is regulated at the posttranscriptional level. Splice variants are commonly treated with ambivalence during primer design, a procedure that can result in biased results and lead to faulty conclusions. This can include detection of massive increases in gene expression from baseline, because a rare transcript is the main target for the primers and it is only really expressed after the intervention. Always be concerned if you see induction of gene expression changes *in vivo*, in human tissue that is claimed to be many 1000-fold increases as this usually means a new cell type has moved into the tissue (e.g., inflammation) or that baseline expression was close to zero. In addition to splice variants, a thorough assessment of the genomic locus to assess if any natural antisense transcripts (NATs) are predicted to be transcribed from the opposite DNA strand is necessary. NATs are a subset of the non protein-coding transcripts, which show modest or major sequence complementarities with coding sequences in the genome. These NATs can be transcribed from the antisense DNA strand (*cis*-NATs) or from a separate loci (*trans*-NATs) and have been suggested to influence the transcriptional output of the genome by regulating protein-coding transcripts (Lapidot and Pilpel, 2006; Li *et al.*, 2006). Some NATs have been characterized in more detail at a functional level and *cis*-NATs have been suggested to regulate the expression of their target genes in several biological conditions including development (Coudert *et al.*, 2005), circadian rhythms (Kramer *et al.*, 2003), hypoxia (Uchida *et al.*, 2004), and imprinting (Sleutels *et al.*, 2002). Although the regulatory mechanisms are still largely unknown, *cis*-NATs have been suggested to act through transcriptional interference, RNA masking, double-stranded RNA-dependent mechanisms and chromatin remodeling (Lapidot and Pilpel, 2006; Lavorgna *et al.*, 2004). In conclusion, there are strong arguments for avoiding overlap with splice variants and natural antisense molecules when designing gene specific primers for qPCR gene expression analysis. For this purpose, we utilized Vega Genome browser http://vega.sanger.ac.uk/index.html.

5.4.2. Verification of novel transcripts

While a transcript can be viewed on a Genome browser, often it is good practice to verify that the expression of the novel transcript identified in genomic sequencing projects are not sequencing artifacts. For this purpose

5′ Rapid Amplification of cDNA ends (RACE) can be applied using RNA from the tissue in question (i.e., to show that variant is expressed in your target organ). Total RNA is used as template for the reaction. A 45 base RNA adaptor oligonucleotide is ligated to the 5′ end of the de-capped RNA population. The RNA is reverse transcribed in a Thermal Cycler at 25 °C for 10 min, 48 °C for 30 min and 95 °C for 5 min. To ensure specificity, we perform nested PCR using the cDNA from the reverse transcribed RNA as template. The first PCR reaction is performed with a primer complementary to the ligated adaptor sequence (provided in the RLM-RACE kit, Ambion) and a primer complementary to the region downstreams of the predicted 3′UTR of the novel transcript. Using the PCR product from this reaction, a second PCR analysis is performed using an inner primer toward the ligated adaptor sequence (provided in RLM-RACE kit, Ambion) and an inner gene-specific primer also designed downstream of the predicted 3′UTR of the novel transcript. The sequence can finally be confirmed by a third PCR analysis using internal primers toward the previously amplified PCR product. The 5′ internal primer targets and thus the product include the start of the novel transcript sequence. The product obtained from 5′ RACE can then be cloned and finally sequenced to confirm correct sequence identity.

REFERENCES

Befroy, D. E., Petersen, K. F., Dufour, S., Mason, G. F., de Graaf, R. A., Rothman, D. L., and Shulman, G. I. (2007). Impaired mitochondrial substrate oxidation in muscle of insulin-resistant offspring of type 2 diabetic patients. *Diabetes* **56,** 1376–1381.

Berg, H. E., Dudley, G. A., Haggmark, T., Ohlsen, H., and Tesch, P. A. (1991). Effects of lower limb unloading on skeletal muscle mass and function in humans. *J. Appl. Physiol.* **70,** 1882–1885.

Berg, H. E., Dudley, G. A., Hather, B., and Tesch, P. A. (1993). Work capacity and metabolic and morphologic characteristics of the human quadriceps muscle in response to unloading. *Clin. Physiol.* **13,** 337–347.

Bergstrom, J., and Hultman, E. (1966). Muscle glycogen synthesis after exercise: An enhancing factor localized to the muscle cells in man. *Nature* **210,** 309–310.

Berkes, C. A., and Tapscott, S. J. (2005). MyoD and the transcriptional control of myogenesis. *Semin. Cell Dev. Biol.* **16,** 585–595.

Birmingham, A., Anderson, E., Reynolds, A., Ilsley-Tyree, D., Leake, D., Fedorov, Y., Baskerville, S., Maksimova, E., Robinson, K., Karpilow, J., Marshall, W., and Khvorova, A. (2006). 3[prime] UTR seed matches, but not overall identity, are associated with RNAi off-targets. *Nat. Methods* **3,** 199–204.

Bodnar, A. G., Ouellette, M., Frolkis, M., Holt, S. E., Chiu, C. P., Morin, G. B., Harley, C. B., Shay, J. W., Lichtsteiner, S., and Wright, W. E. (1998). Extension of life-span by introduction of telomerase into normal human cells. *Science* **279,** 349–352.

Cannon, B., and Nedergaard, J. (2004). Brown adipose tissue: Function and physiological significance. *Physiol. Rev.* **84,** 277–359.

Cannon, B., and Nedergaard, J. (2008). Studies of thermogenesis and mitochondrial function in adipose tissues. *Methods Mol. Biol.* **456,** 109–121.

Chalk, A. M., Wahlestedt, C., and Sonnhammer, E. L. (2004). Improved and automated prediction of effective siRNA. *Biochem. Biophys. Res. Commun.* **319,** 264–274.

Coudert, A., Pibouin, L., Vi-Fane, B., Thomas, B., MacDougall, M., Choudhury, A., Robert, B., Sharpe, P., Berdal, A., and Lezot, F. (2005). Expression and regulation of the Msx1 natural antisense transcript during development. *Nucleic Acids Res.* **33,** 5208–5218.

Deng, H., Jankovic, J., Guo, Y., Xie, W., and Le, W. (2005). Small interfering RNA targeting the PINK1 induces apoptosis in dopaminergic cells SH-SY5Y. *Biochem. Biophys. Res. Commun.* **337,** 1133–1138.

Elbashir, S., Harborth, J., Lendeckel, W., Yalcin, A., Weber, K., and Tuschl, T. (2001). Duplexes of 21-nucleotide RNAs mediate RNA interference in cultured mammalian cells. *Nature* **411,** 494–498.

Fedorov, Y., Anderson, E., Birmingham, A., Reynolds, A., Karpilow, J., Robinson, K., Leake, D., Marshall, W., and Khvorova, A. (2006). Off-target effects by siRNA can induce toxic phenotype. *RNA* **12,** 1188–1196.

Fredriksson, K., Tjader, I., Keller, P., Petrovic, N., Ahlman, B., Scheele, C., Wernerman, J., Timmons, J. A., and Rooyackers, O. (2008). Dysregulation of mitochondrial dynamics and the muscle transcriptome in ICU patients suffering from sepsis induced multiple organ failure. *PLoS ONE* **3,** e3686.

Greenhaff, P. L., and Timmons, J. A. (1998). Interaction between aerobic and anaerobic metabolism during intense muscle contraction. *Exerc Sport Sci. Rev.* **26,** 1–30.

Haddad, F., Baldwin, K. M., and Tesch, P. A. (2005). Pretranslational markers of contractile protein expression in human skeletal muscle: Effect of limb unloading plus resistance exercise. *J. Appl. Physiol.* **98,** 46–52.

Hayflick, L. (1965). The limited *in vitro* lifetime of human diploid cell strains. *Exp. Cell Res.* **37,** 614–636.

Hayflick, L., and Moorhead, P. S. (1961). The serial cultivation of human diploid cell strains. *Exp. Cell Res.* **25,** 585–621.

Henry, R. R., Abrams, L., Nikoulina, S., and Ciaraldi, T. P. (1995). Insulin action and glucose metabolism in nondiabetic control and NIDDM subjects. Comparison using human skeletal muscle cell cultures. *Diabetes* **44,** 936–946.

Henry, R. R., Ciaraldi, T. P., Abrams-Carter, L., Mudaliar, S., Park, K. S., and Nikoulina, S. E. (1996). Glycogen synthase activity is reduced in cultured skeletal muscle cells of non-insulin-dependent diabetes mellitus subjects. Biochemical and molecular mechanisms. *J. Clin. Invest.* **98,** 1231–1236.

Jackson, A., Bartz, S., Schelter, J., Kobayashi, S., Burchard, J., Mao, M., Li, B., Cavet, G., and Linsley, P. (2003). Expression profiling reveals off-target gene regulation by RNAi. *Nat. Biotechnol.* **21,** 635–637.

Jackson, A., Burchard, J., Leake, D., Reynolds, A., Schelter, J., Guo, J., Johnson, J., Lim, L., Karpilow, J., Nichols, K., Marshall, W., Khvorova, A., *et al.* (2006a). Position-specific chemical modification of siRNAs reduces "off-target" transcript silencing. *RNA* **12,** 1197–1205.

Jackson, A., Burchard, J., Schelter, J., Chau, B., Cleary, M., Lim, L., and Linsley, P. (2006b). Widespread siRNA "off-target" transcript silencing mediated by seed region sequence complementarity. *RNA* **12,** 1179–1187.

Judge, A. D., Sood, V., Shaw, J. R., Fang, D., McClintock, K., and MacLachlan, I. (2005). Sequence-dependent stimulation of the mammalian innate immune response by synthetic siRNA. *Nat .Biotechnol.* **23,** 457–462.

Kiessling, K. H., Pilstrom, L., Bylund, A. C., Saltin, B., and Piehl, K. (1974). Enzyme activities and morphometry in skeletal muscle of middle-aged men after training. *Scand. J. Clin. Lab. Invest.* **33,** 63–69.

Kramer, C., Loros, J., Dunlap, J., and Crosthwaite, S. (2003). Role for antisense RNA in regulating circadian clock function in *Neurospora crassa. Nature* **421,** 948–952.

Lapidot, M., and Pilpel, Y. (2006). Genome-wide natural antisense transcription: Coupling its regulation to its different regulatory mechanisms. *EMBO Rep.* **7,** 1216–1222.

Lavorgna, G., Dahary, D., Lehner, B., Sorek, R., Sanderson, C. M., and Casari, G. (2004). In search of antisense. *Trends Biochem. Sci.* **29,** 88–94.

Li, Y.-.Y, Qin, L., Guo, Z.-M, Liu, L., Xu, H., Hao, P., Su, J., Shi, Y., He, W.-Z, and Li, Y.-X (2006). In silico discovery of human natural antisense transcripts. *BMC Bioinformatics* **7,** 18.

Lin, J., Wu, P. H., Tarr, P. T., Lindenberg, K. S., St-Pierre, J., Zhang, C. Y., Mootha, V. K., Jager, S., Vianna, C. R., Reznick, R. M., Cui, L., Manieri, M., *et al.* (2004). Defects in adaptive energy metabolism with CNS-linked hyperactivity in PGC-1alpha null mice. *Cell* **119,** 121–135.

MacKeigan, J. P., Murphy, L. O., and Blenis, J. (2005). Sensitized RNAi screen of human kinases and phosphatases identifies new regulators of apoptosis and chemoresistance. *Nat. Cell Biol.* **7,** 591–600.

McIntyre, E. A., Halse, R., Yeaman, S. J., and Walker, M. (2004). Cultured muscle cells from insulin-resistant type 2 diabetes patients have impaired insulin, but normal 5-amino-4-imidazolecarboxamide riboside-stimulated, glucose uptake. *J. Clin. Endocrinol. Metab.* **89,** 3440–3448.

Mootha, V. K., Lindgren, C. M., Eriksson, K. F., Subramanian, A., Sihag, S., Lehar, J., Puigserver, P., Carlsson, E., Ridderstrale, M., Laurila, E., Houstis, N., Daly, M. J., *et al.* (2003). PGC-1alpha-responsive genes involved in oxidative phosphorylation are coordinately downregulated in human diabetes. *Nat. Genet.* **34,** 267–273.

Nechad, M., Nedergaard, J., and Cannon, B. (1987). Noradrenergic stimulation of mitochondriogenesis in brown adipocytes differentiating in culture. *Am. J. Physiol.* **253,** C889–C894.

Patti, M. E., Butte, A. J., Crunkhorn, S., Cusi, K., Berria, R., Kashyap, S., Miyazaki, Y., Kohane, I., Costello, M., Saccone, R., Landaker, E. J., Goldfine, A. B., *et al.* (2003). Coordinated reduction of genes of oxidative metabolism in humans with insulin resistance and diabetes: Potential role of PGC1 and NRF1. *Proc. Natl. Acad. Sci. USA* **100,** 8466–8471.

Petersen, K. F., Dufour, S., Befroy, D., Garcia, R., and Shulman, G. I. (2004). Impaired mitochondrial activity in the insulin-resistant offspring of patients with type 2 diabetes. *N. Engl. J. Med.* **350,** 664–671.

Pridgeon, J. W., Olzmann, J. A., and Chin, L. S. L. L. (2007). PINK1 protects against oxidative stress by phosphorylating mitochondrial chaperone TRAP1. *PLoS Biol.* **5.**

Sandberg, R., and Ernberg, I. (2005). Assessment of tumor characteristic gene expression in cell lines using a tissue similarity index (TSI). *Proc. Natl. Acad. Sci. USA* **102,** 2052–2057.

Scheele, C., Nielsen, A. R., Walden, T. B., Sewell, D. A., Fischer, C. P., Brogan, R. J., Petrovic, N., Larsson, O., Tesch, P. A., Wennmalm, K., Hutchinson, D. S., Cannon, B., *et al.* (2007a). Altered regulation of the PINK1 locus: A link between type 2 diabetes and neurodegeneration? *FASEB J.* **21,** 3653–3665.

Scheele, C., Petrovic, N., Faghihi, M. A., Lassmann, T., Fredriksson, K., Rooyackers, O., Wahlestedt, C., Good, L., and Timmons, J. A. (2007b). The human PINK1 locus is regulated *in vivo* by a non-coding natural antisense RNA during modulation of mitochondrial function. *BMC Genomics* **8,** 74.

Schroeder, A., Mueller, O., Stocker, S., Salowsky, R., Leiber, M., Gassmann, M., Lightfoot, S., Menzel, W., Granzow, M., and Ragg, T. (2006). The RIN: An RNA integrity number for assigning integrity values to RNA measurements. *BMC Mol. Biol.* **7,** 3.

Sleutels, F., Zwart, R., and Barlow, D. P. (2002). The non-coding Air RNA is required for silencing autosomal imprinted genes. *Nature* **415,** 810–813.

Thellin, O., Zorzi, W., Lakaye, B., De Borman, B., Coumans, B., Hennen, G., Grisar, T., Igout, A., and Heinen, E. (1999). Housekeeping genes as internal standards: Use and limits. *J. Biotechnol.* **8,** 291–295.

Thompson, K. L., Pine, P. S., Rosenzweig, B. A., Turpaz, Y., and Retief, J. (2007). Characterization of the effect of sample quality on high density oligonucleotide microarray data using progressively degraded rat liver RNA. *BMC Biotechnol.* **7,** 57.

Timmons, J. A., Gustafsson, T., Sundberg, C. J., Jansson, E., and Greenhaff, P. L. (1998a). Muscle acetyl group availability is a major determinant of oxygen deficit in humans during submaximal exercise. *Am. J. Physiol.* **274,** E377–380.

Timmons, J. A., Gustafsson, T., Sundberg, C. J., Jansson, E., Hultman, E., Kaijser, L., Chwalbinska-Moneta, J., Constantin-Teodosiu, D., Macdonald, I. A., and Greenhaff, P. L. (1998b). Substrate availability limits human skeletal muscle oxidative ATP regeneration at the onset of ischemic exercise. *J. Clin. Invest.* **101,** 79–85.

Timmons, J. A., Jansson, E., Fischer, H., Gustafsson, T., Greenhaff, P. L., Ridden, J., Rachman, J., and Sundberg, C. J. (2005a). Modulation of extracellular matrix genes reflects the magnitude of physiological adaptation to aerobic exercise training in humans. *BMC Biol.* **3,** 19.

Timmons, J. A., Larsson, O., Jansson, E., Fischer, H., Gustafsson, T., Greenhaff, P. L., Ridden, J., Rachman, J., Peyrard-Janvid, M., Wahlestedt, C., and Sundberg, C. J. (2005b). Human muscle gene expression responses to endurance training provide a novel perspective on Duchenne muscular dystrophy. *FASEB J.* **19,** 750–760.

Timmons, J. A., Norrbom, J., Scheele, C., Thonberg, H., Wahlestedt, C., and Tesch, P. (2006). Expression profiling following local muscle inactivity in humans provides new perspective on diabetes-related genes. *Genomics* **87,** 165–172.

Timmons, J. A., Poucher, S. M., Constantin-Teodosiu, D., Macdonald, I. A., and Greenhaff, P. L. (1997). Metabolic responses from rest to steady state determine contractile function in ischemic skeletal muscle. *Am. J. Physiol.* **273,** E233–E238.

Timmons, J. A., Wennmalm, K., Larsson, O., Walden, T. B., Lassmann, T., Petrovic, N., Hamilton, D. L., Gimeno, R. E., Wahlestedt, C., Baar, K., Nedergaard, J., and Cannon, B. (2007). Myogenic gene expression signature establishes that brown and white adipocytes originate from distinct cell lineages. *Proc. Natl. Acad. Sci. USA* **104,** 4401–4406.

Uchida, T., Rossignol, F., Matthay, M., Mounier, R., Couette, S., Clottes, E., and Clerici, C. (2004). Prolonged hypoxia differentially regulates hypoxia-inducible factor (HIF)-1alpha and HIF-2alpha expression in lung epithelial cells: Implication of natural antisense HIF-1alpha. *J. Biol. Chem.* **279,** 14871–14878.

Valente, E. M., Abou-Sleiman, P. M., Caputo, V., Muqit, M. M., Harvey, K., Gispert, S., Ali, Z., Del Turco, D., Bentivoglio, A. R., Healy, D. G., Albanese, A., Nussbaum, R., et al. (2004). Hereditary early-onset Parkinson's disease caused by mutations in PINK1. *Science* **304,** 1158–1160.

Wang, D., Qian, L., Xiong, H., Liu, J., Neckameyer, W. S., Oldham, S., Xia, K., Wang, J., Bodmer, R., and Zhang, Z. (2006). Antioxidants protect PINK1-dependent dopaminergic neurons in Drosophila. *Proc. Natl. Acad. Sci. USA* **103,** 13520–13525.

Wisloff, U., Najjar, S. M., Ellingsen, O., Haram, P. M., Swoap, S., Al-Share, Q., Fernstrom, M., Rezaei, K., Lee, S. J., Koch, L. G., and Britton, S. L. (2005). Cardiovascular risk factors emerge after artificial selection for low aerobic capacity. *Science* **307,** 418–420.

Wood-Kaczmar, A., Gandhi, S., Yao, Z., Abramov, A. S., Miljan, E. A., Keen, G., Stanyer, L., Hargreaves, I., Klupsch, K., Deas, E., Downward, J., Mansfield, L., et al. (2008). PINK1 is necessary for long term survival and mitochondrial function in human dopaminergic neurons. *PLoS ONE* **3,** e2455.

Yang, Y., Ouyang, Y., Yang, L., Beal, M. F., McQuibban, A., and Vogel, H. (2008). PINK1 regulates mitochondrial dynamics through interaction with the fission/fusion machinery. *Proc. Natl. Acad. Sci. USA* **105,** 7070–7075.

CHAPTER THIRTEEN

FUNCTIONAL CHARACTERIZATION OF PHOSPHORYLATION SITES IN DYNAMIN-RELATED PROTEIN 1

J. Thomas Cribbs *and* Stefan Strack

Contents

1. Introduction	232
2. Replacement of Endogenous with Phosphorylation-Site Mutant Drp1	234
2.1. Selection and characterization of Drp1-specific shRNAs	235
2.2. Site-directed mutagenesis	236
2.3. Transient transfection and generation of stable cell lines	237
3. Analysis of Drp1 Phosphorylation	237
3.1. Metabolic labeling with ^{32}P and immunoprecipitation of Drp1	238
3.2. Generation of phosphorylation-specific antibodies	239
4. Production and Assays of Recombinant Drp1	240
4.1. Expression and purification of recombinant Drp1	241
4.2. *In vitro* phosphorylation	242
4.3. GTPase activity assays	242
5. Cell Death Assays	243
5.1. Viability assays in stable PC12 cell lines	243
5.2. Apoptosis/necrosis assays in primary astrocytes	244
6. Mitochondrial Shape Analysis with ImageJ	245
6.1. Visualization of mitochondria	246
6.2. Image enhancement	246
6.3. Morphometry	249
7. Appendix: Morphometry Macro	250
References	251

Abstract

Dynamin-related protein 1 is member of the dynamin-family of large GTPases. Similar to the endocytosis motor dynamin, Drp1 uses GTP hydrolysis to power constriction of the outer mitochondrial membrane and ultimately mitochondrial

Department of Pharmacology, University of Iowa Carver College of Medicine, Iowa City, Iowa, USA

Methods in Enzymology, Volume 457
ISSN 0076-6879, DOI: 10.1016/S0076-6879(09)05013-7

division. Dynamin phosphorylation in its unique C-terminal proline-rich domain interferes with binding of accessory proteins that induce membrane curvature and inhibits clathrin-mediated endocytosis. Evidence within the last few years indicates that Drp1 is also regulated by the phosphorylation/dephosphorylation cycle. Drp1 regulation is complex, in that both inhibitory and activating phosphorylations have been described that lead to, respectively, mitochondrial elongation and shortening. In this chapter, we describe methods for the identification and functional characterization of Drp1 phosphorylation sites. Among these methods is replacement of the endogenous protein by phosphorylation-site mutant Drp1 via combined shRNA and RNAi-resistant cDNA expression from the same plasmid. We also discuss primary astrocyte cultures as a model for regulation of cell death and mitochondrial morphology by Drp1 and present ImageJ macro source code for unbiased quantification of mitochondrial shape changes.

1. INTRODUCTION

Highly conserved from yeast to man, Drp1 belongs to the dynamin-family of large GTPases. Drp1 most likely functions as a mechanoenzyme that forms ring- or spiral-shaped superstructures to constrict and eventually sever mitochondria (Hoppins *et al.*, 2007). In addition to catalyzing mito-chondrial division, Drp1 participates in cristae remodeling and cytochrome *c* mobilization during apoptosis (Suen *et al.*, 2008). In mammalian cells, Drp1 is recruited from the cytosol to the mitochondrial membrane by a multi-protein complex that includes the membrane-anchored adaptor protein Fis1. Fis1/Drp1 is also responsible for fission of peroxisomes, much smaller organelles delimited by a single membrane (Koch *et al.*, 2003, 2005). Even though Drp1 knockout mice have not been reported, it is likely that Drp1 is an essential enzyme in mammals, because a dominant-negative mutation in the fission GTPase causes severe birth defects in humans (Waterham *et al.*, 2007).

Drp1 is a ~80 kDa protein with a domain structure similar to the dynamins (Fig. 13.1). Shared with the dynamins are the N-terminal GTPase, the middle (MID) domain, and the GTPase effector domain (GED). Drp1 lacks the poly-proline domain that follows the GED in the dynamins. The poly-proline domain is the major regulatory region of the dynamins, containing PKC and CDK5 phosphorylation sites as well as docking sites for SH3-domain containing proteins including amphiphysin (Smillie and Cousin, 2005; Takei *et al.*, 2005). Whereas a lipid-binding PH domain separates the MID and GED in the dynamins, a poorly conserved variable domain (VD) of unknown function takes its place in Drp1 (Fig. 13.1). Mammalian Drp1 is diversified by alternative splicing of three short (11–26 amino acid) exons, the most N-terminal of which separates

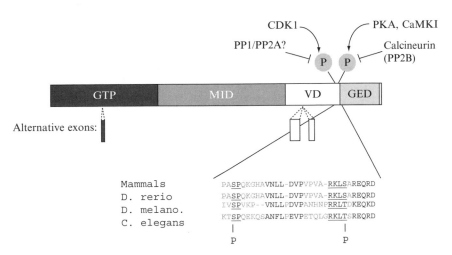

Figure 13.1 Drp1 domain diagram with phosphorylation sites. Drp1 consists of an N-terminal GTPase domain, followed by a middle (MID) and variable (VD) domain, as well as a C-terminal GTPase effector domain (GED). Alternative splicing of three exons generates eight splice variants and changes the numbering of the two highly conserved phosphorylation sites that have been characterized thus far. Kinases (↓) and phosphatases (⊥) that target the two phosphorylation sites are discussed in Section 1.

subdomains A and B of the GTPase domain, whereas the other two alternative exons are in the VD. While the functional consequences of alternative splicing are unknown, all eight possible splice variants are represented as expressed sequence tags (ESTs). The existence of multiple Drp1 variants has confused the phosphorylation site nomenclature (Jahani-Asl and Slack, 2007), and care should be taken to refer to sequence numbers only in conjunction with the species origin and splice variant.

Posttranslational regulation of Drp1 is an active area of research, with ubiquitylation, sumoylation, and phosphorylation having been documented (Chang and Blackstone, 2007; Cribbs and Strack, 2007; Han *et al.*, 2008; Harder *et al.*, 2004; Nakamura *et al.*, 2006; Wasiak *et al.*, 2007). Phosphorylation at two sites has been shown to have functional consequence (Fig. 13.1). The first report came from Mihara's group, who reported that CDK1/cyclin B phosphorylates Drp1 at a Ser residue in the VD drives the breakdown of the mitochondrial reticulum during early mitosis (Taguchi *et al.*, 2007). Only 21 residues upstream of this activating phosphorylation site at the border of the GED is a Ser residue, phosphorylation of which by PKA inhibits the mitochondria-severing activity of Drp1 in HeLa and PC12 cells (Chang and Blackstone, 2007; Cribbs and Strack, 2007). This residue is dephosphorylated by the calcium/calmodulin-dependent phosphatase calcineurin (a.k.a. protein phosphatase 2B, PP2B), contributing to the calcium-dependent fragmentation that ensues following mitochondrial depolarization (Cereghetti *et al.*, 2008; Cribbs and Strack, 2007). However, the opposite result was also reported, in that calcium/calmodulin-dependent kinase I (CaMKI)–mediated phosphorylation of the same Drp1 site was found to participate in high potassium-induced mitochondrial scission in neurons (Han *et al.*, 2008). While different Drp1 splice variants were examined in these conflicting studies, our results so far suggest that alternative exons in the VD do not influence mitochondrial shape changes induced by Drp1 phosphorylation-site mutants (Fig. 13.5). Clearly, more work is required to understand the interplay between different phosphorylation events and the mechanism by which phosphorylation of Drp1 leads to mitochondrial shape changes and altered apoptotic sensitivity. This chapter describes some of the methods that our laboratory uses to characterize Drp1 phosphorylation sites.

2. Replacement of Endogenous with Phosphorylation-Site Mutant Drp1

A common way to infer the effect of site-specific phosphorylation on protein function is to replace the phospho-acceptor with an amino acid that cannot be phosphorylated, or with a negatively charged amino acid that mimics the phosphate group. To create non-phosphorylatable mutants, Ser/Thr residues are most often replaced by Ala, while Tyr is best replaced with Phe. Constitutive phosphorylation may be achieved by introduction of an Asp or Glu. We usually replace Ser with Asp and Thr with Glu to match for size. As there are no naturally occurring amino acids with a negatively charged aromatic side chain, phosphomimetic substitution of Tyr is difficult. Even for Ser and Thr, there is no guarantee that an acidic amino acid substitution will faithfully replicate the effects of phosphorylation, since a phosphate group is considerably larger and carries twice the negative charge of a carboxylate. For instance, autophosphorylation of calcium/calmodulin–dependent protein kinase II (CaMKII) at Thr286 results in activation of the enzyme to ~50% of its maximal activity in the presence of calcium/calmodulin. However, replacing Thr286 with Asp only results in ~10% CaMKII autonomous activity, implying that this supposedly phosphomimetic substitution does not fully remove the auto-inhibitory region from the catalytic cleft of the kinase (Brickey *et al.*, 1994).

In the most easily interpretable scenario, the cellular activity of the wild type, phosphorylatable protein will fall between that of the Ala and that of the Asp/Glu substitution. If the Ala substitution alters activity, but the acidic substitution does not, then either Glu/Asp is not a faithful phosphomimetic or basal phosphorylation levels at the site in question are high. An essential structural role of the Ser/Thr is suggested in the case that both types of mutations are found to impair activity of the protein.

Overexpression of proteins can lead to artifactual results. Especially in the case of homo-oligomeric proteins like Drp1, the presence of the endogenous protein may obscure effects of mutating specific phosphorylation sites. For these reasons, we routinely combine shRNA-mediated knockdown with expression of the RNAi-resistant cDNAs to replace wild type with mutant Drp1 in the cell. To ensure co-expression of shRNA and cDNA, we express both from the same plasmid backbone.

2.1. Selection and characterization of Drp1-specific shRNAs

To select RNAi target sites, we use criteria derived from an empirical study of siRNA efficiency by Reynolds and colleagues at Dharmacon (Reynolds *et al.*, 2004). Criteria include low G/C content, low internal stability at the sense strand 3′-end, lack of inverted repeats, and base preferences at specific positions. This target selection algorithm has been implemented as an Excel macro and as a Web tool accessible at the Dharmacon Web site. We also use other Web tools (siRNA scales: http://gesteland.genetics.utah.edu/ siRNA_scales/; DSIR: http://cbio.ensmp.fr/dsir/) that scored favorably in a recent evaluation of available siRNA selection algorithms (Matveeva *et al.*, 2007). A point to consider is that these tools may not optimally predict the efficiency of vector-encoded shRNAs, which generally do not reach cellular concentrations of chemically synthesized siRNAs.

We generate between 3 and 5 shRNA encoding plasmid for each gene target and find that generally one or two silences the expression of the cotransfected target's cDNA by 80% or better. Target sites are selected that fulfill three criteria: (1) an efficiency score in the upper 10% (score 6 or greater with the Reynolds *et al.* algorithm), (2) perfect conservation in human and rat, the origin of most of our cell lines and primary cell cultures, (3) no more than 15 consecutive matches with mRNA from any other vertebrate organism to minimize off-target effects. We use the hairpin design and H1 promoter of the original pSUPER vector described in Brummelkamp *et al.* (2002). The H1 promoter-shRNA cassette is amplified from pSUPER or related vectors by PCR using a forward primer that anneals 5′ of the promoter sequence:

```
cgcacatgtgaacgctgacgtcatcaacccg
```

and a reverse primer that encodes the specific shRNA (in this example, Drp1-specific bases are in capital letters):

```
cgcacatgtccggattccaaaaaGAGTGTAACTGATTCAtctcttgaa
TGAATCAGTTACACTCTTCggggatctgtggtctcatacagaac
```

Both primers include a PciI site (underlined) for ligation; the reverse primer also contains a BspEI site, so the direction of insertion can be easily determined. We use a proofreading polymerase, such as Pfu Ultra

(Stratagene) to minimize PCR errors and an initial annealing temperature of 60 °C. Depending on the stability of the hairpin in the reverse primer, PCR conditions may have to be optimized, for instance by varying the annealing temperature from 55 to 65 °C. The PCR product is gel purified (we use a Qiagen kit), digested with PciI, and ligated into the PciI site of the pEGFP series of vectors from Clontech/Takara. We found that insertion of the H1 promoter:shRNA cassette into the unique PciI site, which is located near the plasmid's origin of replication, does not interfere with cDNA expression from the CMV promoter or expression of the G418/kanamycin resistance gene.

To select the most efficient shRNA for subsequent functional studies, we initially test three to four candidate shRNA plasmids for silencing of cotransfected expression plasmids, transfecting COS or HEK293 cells with increasing shRNA to cDNA plasmid ratios (e.g., 0.2:1 to 3:1). If the target cDNA is available fused to a fluorescent protein, knockdown efficiency can be visually estimated by live cell imaging. More commonly though, we quantify knockdown of epitope-tagged cDNAs by western blotting and subsequent densitometry using the gel analysis macro set of ImageJ.

2.2. Site-directed mutagenesis

To replace endogenous with phosphorylation-site mutant fission/fusion proteins (e.g., Drp1), the cDNA has to be first rendered insensitive to the co-expressed shRNA. To this end, two to five silent (non-coding) base changes are incorporated into the shRNA target site using protocols adapted from the QuikChange mutagenesis kit. Briefly, double-stranded oligonucleotides with 19–25 bp flanking a central region with up to eight base changes are used for linear amplification of the template plasmid with Pfu Ultra. For screening purposes, we additionally incorporate a silent restriction site into the mutagenic oligonucleotide. To render Drp1 resistant to the shRNA described above, we used the following sense primer (shRNA target sequence underlined, base changes capitalized):

```
gccaactggacattaacaataagaaATCAgtGactgattcaatccgtga
  tgagtatgc
```

The plasmid template is destroyed by digestion with *DpnI*, a methylation-specific restriction enzyme, and the remaining mutated DNA is transformed into competent *E.coli*, where the DNA is circularized by single-strand break repair. Through additional subcloning, the H1 promoter:shRNA and cDNA expression cassettes are incorporated into the same plasmid back-bone to ensure co-expression. Mutagenesis of Ser/Thr phosphorylation sites is carried out identically, using the plasmid that expresses both shRNA and shRNA-resistant cDNA as a template.

2.3. Transient transfection and generation of stable cell lines

Vectors for replacing endogenous with phosphorylation-site mutant fission/fusion enzymes can be used for both transient and stable expression of shRNA and cDNA, with stable expression utilizing the neomycin (G418) resistance gene that is part of the pEGFP series of plasmids. We use transient transfection for analysis of mitochondrial morphology (see Section 6), subcellular localization of Drp1, and most biochemical assays, while stable cell lines are desirable for assays that require a homogenous cell population (e.g., 96-well plate growth and viability assays). We most commonly employ chemical transfection using Lipofectamine 2000 for adherent cells according to the manufacturer's instructions (BD Biosciences). Most cell lines we use routinely (COS, HEK293, PC12) are transfected with 2 μg DNA and 4 μl Lipofectamine 2000 per ml of OptiMEM growth media. For primary neuronal cultures, which do not tolerate high concentrations of cationic lipids as well, we decrease the Lipofectamine concentration to 0.8–1.5 μl/ml and shorten the incubation period from 4–5 to 3 h or less. While cDNA expression may peak earlier, we mostly analyze cells around 72 h after transfection to ensure that the endogenous protein is maximally downregulated by the co-expressed shRNA.

To generate stable cell lines, cells are trypsinized 48–72 h after transfection and divided over several larger dishes (e.g., cells transfected in a 60 mM dish are re-seeded into 5–6 100 mM dishes), adding 500 μg/ml G418 to the media to select for stable genomic integration of the transfected plasmid. After 2–3 weeks, individual colonies are isolated and expanded, lowering the G418 concentration to 100 μg/ml for maintenance. We pick individual colonies under the dissecting microscope, using a p200 pipettor to scrape cells from the substrate while releasing the plunger to draw up the cell suspension. If the overexpressed protein carries a fluorescent protein tag (e.g., GFP-Drp1), positive clones can by identified by epifluorescence microscopy prior to isolation, which reduces the number of clones that have to be expanded for characterization. Clones are characterized for expression of the transfected cDNA and knockdown of the endogenous protein by immunoblotting (Fig. 13.2).

3. ANALYSIS OF DRP1 PHOSPHORYLATION

Metabolic labeling with radioactive orthophosphate $\left(^{32}PO_4^{2-}\right)$ is a relatively straightforward method to assess the degree to which a protein is phosphorylated in intact cells under various stimulation conditions. Mass spectrometry has emerged as a powerful tool for the unbiased identification of phosphorylation sites. However, despite recent advances (Macek et al., 2009), full sequence coverage and reliable quantification of phosphorylation

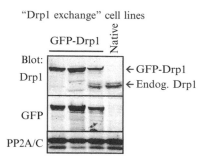

Figure 13.2   Stable replacement of endogenous with GFP-tagged, mutant Drp1. PC12 cells were transfected with a vector that expresses a Drp1-targeting shRNA together with the RNAi-resistant, GFP-tagged cDNA. After selection in G418, clonal cell lines were immunoblotted for Drp1, GFP, and the catalytic subunit of PP2A (PP2A/C) as a loading control. Cell lines analyzed in the first two lanes have replaced endogenous with similar levels of GFP-tagged Drp1, whereas the line in lane three failed to silence endogenous Drp1 expression.

is currently difficult to achieve by mass spectroscopy. For now, metabolic labeling with ^{32}P remains the method of choice for initially establishing a protein as phosphorylated, and for implicating specific kinases and phosphatases in its regulation. Once phosphorylation sites have been confirmed *in vitro* phosphorylation assays and by metabolic labeling of phosphorylation-site mutant proteins, production of phosphorylation-specific antibodies should be initiated to allow for quantitative analysis of site-specific phosphorylation in intact cells and organisms.

3.1. Metabolic labeling with ^{32}P and immunoprecipitation of Drp1

To assess ^{32}P incorporation into your protein of interest in intact cells, it is necessary to separate it from other ^{32}P-labeled proteins and nucleic acids. While this can be achieved by multiple methods, most commonly, proteins are immunoprecipitated with specific antibodies. For Drp1, we either immunoprecipitate the endogenous protein using a monoclonal antibody available from Transduction Laboratories, or we use hemagglutinin (HA) tag-directed antibodies to isolate HA-Drp1 (e.g., carrying mutations of phosphorylation sites) after transient transfection using Lipofectamine 2000 (see above).

Transfected or untransfected cells are grown to subconfluency in 60 mm dishes. To reduce isotope dilution with non-radioactive phosphate, cells are incubated for 30–60 min with phosphate-free RPMI (BD Biosciences) containing 1% dialyzed fetal bovine serum (labeling media) prior to labeling. The media are then replaced with the same media containing 0.5 mCi/ml ^{32}P-orthophosphate, and dishes are placed in a Tupperware or similar container before returning them to the incubator to minimize radioactive contamination. Cells are allowed to incorporate ^{32}P for 5–6 h in the incubator, including kinase and phosphatase activity modulators (e.g., 50 μM forskolin to stimulate PKA, 250 nM okadaic acid to inhibit PP2A) during the last 1–2 h of incubation. After a rinse in phosphate-free RPMI, cells are lysed in 400 μl buffer containing 1% Triton X-100, 20 mM Tris pH 7.5, 1 mM EDTA, 1 mM EGTA and both protease (1 mM PMSF, 1 mM benzamidine, 1 μg/ml leupeptin) and phosphatase inhibitors (1 mM β-glycerolphosphate, 1 mM Na$_3$VO$_4$, 1 mM Na$_4$P$_2$O$_7$, 1 μM microcystin-LR). After precipitating insoluble material (10 min centrifugation in a table top centrifuge at maximum speed at 4 °C), lysates are incubated with antibody and agarose beads (10 μl HA antibody–agarose conjugate for transfected Drp1 or 4 μl Drp1 antibody and 10 μl protein G-agarose for endogenous Drp1) for 3–4 with end-over-end rotation at 4 °C. Immunoprecipitates are recovered by centrifugation (2 min at 2000g) and washed four times in lysis buffer, transferring beads to a fresh tube after the first wash to minimize carry-over of proteins from the lysate. Proteins are eluted from the beads by boiling in SDS sample buffer and resolved on 10% polyacrylamide gels. After transferring proteins to nitrocellulose membranes, blots are first labeled for Drp1 by immunoblotting, and then exposed to PhosphorImager screens to visualize ^{32}P incorporation. Relative phosphorylation is quantified as the ratio of the ^{32}P signal to the Drp1 immunoblot signal in each lane. We use the Gel Analyzer plug-in from ImageJ for densitometry of blots and ^{32}P images.

3.2. Generation of phosphorylation-specific antibodies

Once sufficient evidence has been gathered that a residue is phosphorylated (e.g., by *in vitro* phosphorylation and metabolic labeling of wild type vs. phosphorylation-site mutant proteins), phosphorylation-specific antibodies are useful tools for tracking site-specific phosphorylation in intact cells and tissues. Antibodies are raised against a short peptide into which the phosphorylated residue is incorporated during peptide synthesis. Phosphopeptides are directionally coupled to maleimide-acitvated keyhole limpet hemocyanin (KLH, Pierce) via a cysteine residue added to either the N- or C-terminus of the peptide. Employing the services of an on-campus facility or a company, we then inject the peptide–KLH conjugated into mice or rabbits for mono- or polyclonal antibody production, respectively. Design considerations for the phospho-peptide include: (1) Short peptides ($\leq$12 amino acids) are more likely to yield antibodies that include the phospho-Ser/Thr as part of their epitope. (2) The end of the peptide that is farthest from the phosphorylated residue should be sulfhydryl-coupled to the carrier protein so as to maximize accessibility of the phosphorylation site. (3) The unconjugated, exposed end of the peptide should be modified by acetylation (N-terminus) or amidation (C-terminus) to increase resemblance to the native protein. To generate antibodies against the PKA site in Drp1 (Ser656 in rat splice variant 1), we conjugated the following peptide via the C-terminal Cys to maleimide-activated KLH and raised mouse monoclonal antibodies at the University of Iowa Hybridoma Facility:

```
acetyl-Arg-Lys-Leu-[phospho-Ser]-Ala-Arg-Glu-Gln-
  Arg-Asp-Cys
```

Supernatants of hybridoma clones were screened by immunoblotting GST fusion proteins of the Drp1 GED domain (residues 643–755) with or without prior phosphorylation by PKA (see Section 4.2). Several clones were identified that recognize Drp1 only when phosphorylated at the PKA site.

We also generated polyclonal rabbit antibodies against the CDK1/cyclin B phosphorylation site in Drp1 that was initially identified by Mihara's group (Taguchi *et al.*, 2007) using the following peptide as the immunogen:

```
acetyl-Met-Pro-Ala-[phospho-Ser]-Pro-Gln-Lys-Gly-
  His-Ala-Cys
```

The rabbit's immune system is likely to produce not only the desired phospho-specific antibody, but also non-discriminating antibodies that recognize an epitope adjacent to the phosphorylated residue, or even antibodies specific for the non-phosphorylated peptide (presumably as a result of dephosphorylation in the animal). Therefore, polyclonal antisera generally have to be affinity purified to increase their specificity. Minimally, antibodies are subjected to a positive selection step by passing antisera over a column to which the phosphorylated peptide was covalently coupled (ultralink Iodoacetyl gel, Pierce), and then eluting immunoglobulins with 20 mM glycine–HCl, pH 2.0. If the eluted antibodies display significant reactivity toward the non-phosphorylated form of the protein, they should be depleted by adsorption to a second column containing the dephospho-form of the peptide (synthesized separately or enzymatically dephosphory-lated by alkaline phosphatase). In the case of the phospho-CDK1/cyclin B site antibody, affinity purification against the phospho-peptide was sufficient to produce antibodies that discriminate between phospho- and non-phospho-Drp1.

4. PRODUCTION AND ASSAYS OF RECOMBINANT DRP1

Recombinant proteins are indispensable for showing that a particular kinase or phosphatase directly targets a cellular substrate (as opposed to acting indirectly by modulating the activity of other kinases or phosphatases). When combined with phosphorylation-blocking or -mimicking mutations, they are also useful for demonstrating effects of phosphorylation on *in vitro* enzymatic activities or protein–protein interactions.

4.1. Expression and purification of recombinant Drp1

To express recombinant, full-length Drp1 in *E. coli*, cDNAs are cloned into the pCAL-n-EK vector (Strategene) using PCR-based strategies. The pCAL series of vectors includes an N-terminal calmodulin binding peptide (CBP) tag, which is used for one-step purification of Drp1 on calmodulin-agarose in the presence of calcium. CBP-Drp1 is then eluted by chelating calcium with EGTA. Because calmodulin is a eukaryotic protein and because both binding and elution conditions are gentle, the CBP purification tag results in a relatively pure ($\sim$90%) Drp1 preparation compared to other affinity tags (e.g., His- or GST-Drp1).

BL21 (DE3) Star chemically competent *E. coli* (Invitrogen) are transformed with Drp1 expressing vectors, and 100 ml cultures of SOB media (20 g/l Tryptone, 5 g/l yeast extract, 100 mM NaCl, 25 mM KCl, 1 mM MgCl$_2$, 1 mM MgSO$_4$) containing ampicillin (100 μg/ml) are inoculated with individual colonies and grown over night at 37 °C. One liter cultures are inoculated with starter cultures to an OD$_{600}$ of 0.3 and grown to an OD$_{600}$ of 0.6–0.7 before 0.5 mM isopropyl-thiogalactoside (IPTG) is added to induce expression of CBP-Drp1. After 5–6 h continued growth at 37 °C, bacterial pellets are collected by centrifugation (30 min at 20,000g) and stored at -70 °C until further processing. Bacteria are lysed by 30 min shaking incubation at 4 °C in 25 ml of Lysis Buffer (1% Triton X-100, 1 mg/ml lysozyme, 25 mM HEPES, 25 mM PIPES pH 7.5, 150 mM NaCl, 0.5 mM EDTA, 10 mM β-mercaptoethanol, 1 mM PMSF, 1 mM benzamidine), followed by sonication with a probe-tip sonicator (1 min at maximum power) and clarification by centrifugation (25 min at 30,000g). After addition of 5 mM MgCl$_2$ and 2 mM CaCl$_2$ to the lysate, CBP-Drp1 is allowed to bind to 1 ml calmodulin-Sepharose for 2 h at 4 °C, followed by centrifugation (5,000g, 5 min), aspiration of unbound proteins, and washing of the beads with 50 ml Wash Buffer (25 mM HEPES pH 7.5, 50 mM KCl, 5 mM MgCl$_2$, 2 mM CaCl$_2$, 10 mM β-mercaptoethanol). Calmodulin-bound Drp1 is eluted by incubation of the beads with 15 ml Elution Buffer (25 mM HEPES pH 7.5, 50 mM KCl, 5 mM EGTA) for 1–2 h at 4 °C. Calmodulin eluates are concentrated with Amicon ultra filters (Millipore, 50,000 molecular weight cut-off) to ca. 200 μl, and supplemented to 50% glycerol for storage at -20 °C. Protein concentrations are determined using the Bradford assay (Pierce) or by Coomassie blue staining of SDS–PAGE gels with known amounts of BSA as standards.

To demonstrate site-specific phosphorylation of Drp1 *in vitro*, we express fragments of the protein as GST fusion proteins and use them in *in vitro* phosphorylation reactions with purified kinases. To this end, Drp1 domains are PCR-amplified and subcloned into pGEX-2T or -4T vectors. Bacterial growth and induction protocols are similar to those described above for full-length CBP-Drp1 expression, except that 100 ml cultures are grown over night (16–18 h) at 23 °C in the presence of 0.5 mM IPTG to induce GST-Drp1 expression. Bacterial pellets are lysed in 5 ml 1% Triton X-100, 1 mg/ml lysozyme, 50 mM Tris pH 7.5, 150 mM NaCl, 1 mM EDTA, 2 mM benzamidine, 0.1 mM PMSF (30 min, 4 °C, with shaking), and sonicated and centrifuged as above. Cleared supernatants are incubated with 1 ml of a 50% slurry of glutathione–agarose for 2 h at 4 °C. Beads are washed once with 10 ml 0.05% Triton X-100, 20 mM HEPES pH 7.5, 0.1 mM EDTA and are stored at −20 °C in ~0.5 ml of the same buffer containing 50% glycerol. GST fusion proteins are retained on glutathione–agarose beads, which facilitates removal of unincorporated ^{32}P after *in vitro* phosphorylation. Protein concentrations are determined by SDS–PAGE with BSA standards and Coomassie blue staining.

4.2. *In vitro* phosphorylation

In vitro phosphorylations with PKA are carried out in a 200 μl reaction volume containing 10–20 μl GST-Drp1 (or GST only control) agarose slurry, 25 units of PKA catalytic subunit (2.5 μl 10 U/μl stock of PKA$_c$ from bovine heart, Sigma), 5–15 μl [γ-^{32}P]ATP (5 mCi/ml, Amersham), 200 μM unlabeled ATP, 50 mM Tris pH 7.5, 10 mM MgCl$_2$, 2 mM DTT, 1 mM EGTA, 0.01% Triton X-100. Reactions started by the addition of GST-Drp1 proceed for 15 min at 30 °C on an Eppendorf thermomixer (2 s 1000 rpm mixing every 1 min) and are stopped by the addition of 1 ml Tris-buffered saline containing 10 mM EDTA. After precipitation of the GST-Drp1 agarose (2 min at 3,000g) and removal of the supernatant, SDS sample buffer is added and proteins are separated by 10% polyacrylamide gels, stained with Coomassie blue, and exposed to PhosphorImager cassettes to quantify ^{32}P incorporation. Coomassie blue-stained proteins are imaged using the 700 nm laser line of the Odyssey imager (Licor). Relative phosphorylation levels of, e.g., wild type versus phosphorylation-site mutant GST-Drp1 fusion proteins are determined by digital densitometry utilizing the Gel Analyzer plug-in of ImageJ, dividing ^{32}P signals by protein signals in each lane.

4.3. GTPase activity assays

For assaying the GTPase activity of wild type and mutant Drp1, we use a rapid method based on the release of ^{32}P-labeled orthophosphate from [γ-^{32}P]GTP and adsorption of non-hydrolyzed [γ-^{32}P]GTP by charcoal (Brandt and Ross, 1986). Reactions are started by the addition of 50 μl purified CBP-Drp1 (~2 μg) to 150 μl of 25 mM HEPES pH 7.5, 50 mM

KCl, 10 mM MgCl$_2$, 25 μM unlabeled GTP, 2 μl [γ-^{32}P]GTP (10 mCi/ml, Amersham). Reactions proceed at 30 °C with intermittent mixing (2 s 1000 rpm every min) using an Eppendorf thermomixer, while 10 μl aliquots are removed at 0, 5, 15, 30, 45, 60 min, etc. time points and added to 700 μl of an activated charcoal suspension (10% w/v Charcoal Norit A in PBS, 50–200 mesh, Fisher). For determination of total [γ-^{32}P]GTP, 10 μl reaction aliquots are added to 700 μl PBS. After centrifugation of the charcoal-containing tubes (7 min at 20,000g), triplicate 50 μl aliquots of the supernatants are added to 1.5 ml Biosafe II scintillation fluid and subjected to liquid scintillation counting. Percent hydrolysis is calculated by dividing the cpm in the charcoal supernatants by the cpm in the PBS-diluted aliquots.

5. CELL DEATH ASSAYS

The mitochondrial fission/fusion equilibrium is an important determinant in cellular life and death decisions. In particular, inhibiting mitochondrial fission by RNA interference or expression of dominant-negative of Drp1 (GTPase-defective Lys38 mutant) slows both apoptotic and non-apoptotic cell death (Suen *et al.*, 2008). Our laboratory has shown that PC12 cell lines expressing Drp1 pseudo-phosphorylated at the PKA site (S656D, rat variant 1 numbering) are remarkably resistant to several apoptotic stimuli; conversely, non-phosphorylatable Drp1 (S656A) confers increased sensitivity to the same set of stimuli (Cribbs and Strack, 2007). In the following, we describe two cell death assays currently used in the laboratory.

5.1. Viability assays in stable PC12 cell lines

Cell viability assays in a 96-well plate format allow for rapid testing of multiple toxic compounds at different concentrations. However, these assays require a homogenous cell population (i.e., cell lines expressing phosphorylation-site mutant Drp1 stably, see Section 2.3) and cannot distinguish between apoptotic and non-apoptotic cell death mechanisms. Cell lines expressing proteins of interest from an inducible promoter allow comparison to an uninduced cell population (e.g., Strack *et al.*, 2004). If inducible cell lines are unavailable, multiple clonal cell lines expressing the same mutant protein constitutively should be analyzed to avoid clonal selection artifacts.

Stable PC12 cell lines are seeded at 8000 cells/well in 96-well plates (100 μl media/well). Four wells (half a column) are assayed for each condition; however, we frequently exclude the edge and corner wells from the final analysis, because liquid in these wells is prone to evaporation. After

2 days of growth (to ca. 80% confluency), media are replaced with fresh media containing the desired concentrations of drugs, which are prepared by serial dilution. Cell viability is determined by a colorimetric assays that is based on reduction of 3-(4,5-dimethylthiazol-2-yl)-5-(3-carboxymethoxyphenyl)-2-(4-sulfophenyl)-2H-tetrazolium, inner salt (MTS) to formazan (CellTiter 96^{TM} AQueous non-radioactive cell proliferation assay, Promega). While the MTS assay measures metabolic capacity as opposed to cell density, we find that results closely mirror those obtained by counting viable (Trypan-blue excluding) cells. Two days after drug addition, we replace the old media with fresh media containing MTS and phenazine-methosulfate (PMS) using an 8-channel pipettor. Residual MTS/PMS-containing media is pipetted into an empty 96-well plate to provide blank values. The 96-well plates are returned to the incubator for 2–4 h until sufficient brown formazan reaction product has formed, and the absorbance at 490 nm is determined using a plate reader. After background subtraction, cell viability is expressed relative to control wells incubated without drugs.

5.2. Apoptosis/necrosis assays in primary astrocytes

Astrocytes from neonatal rodent brain provide a convenient and physiologically relevant cell system to investigate the role of Drp1 in mitochondrial restructuring and cell death. Tumor derived or artificially immortalized, established cell lines such as HeLa or COS cells produce much of their ATP through glycolysis. In contrast, primary cells depend on mitochondrial ATP generation, making them a superior *in vitro* model for all but cancer-related studies of mitochondrial processes. Still, primary astrocytes share many of the advantages of established cell lines, in that they proliferate robustly for several weeks in serum-containing media on unmodified tissue culture plastic, and are easily transfectable by standard techniques. Perhaps most importantly, astrocytes are large and flat, which facilitates visualization of nuclear and mitochondrial morphology by epifluorescence microscopy (Figs. 13.4 and 13.5).

To prepare primary astrocyte cultures, 4–8 neonatal rat pups (0–3 days old) are euthanized by cervical transection. Forebrains excluding the cerebellum and hindbrain are dissected and the meninges are removed with fine forceps under sterile conditions. The tissue is minced with a scalpel or razor blade and resuspended in 5–10 volumes of Hank's buffered salt solution (HBSS) containing 0.25% trypsin and 1 mM EDTA. Following 15 min incubation at 37 °C with intermittent mixing to digest the extracellular matrix, the tissue is triturated with a fire-polished glass pipette and passed through a cell strainer (70–100 μm mesh size). The cell suspension is diluted to 0.5–1.0 × 10^5 cells/ml in DMEM with 10% fetal bovine serum and plated onto uncoated 100 mm dishes. After 3–4 h, dishes are tapped and the media are replaced to remove poorly adherent cells (neurons and microglia).

This procedure can be repeated as needed. We culture astrocytes for up to 3 weeks, passaging cells when they reach confluency. Astrocytes can be frozen, stored in liquid N_2, and thawed with plating efficiencies similar to established cell lines. Astrocytes can be transfected using a variety of methods; we use Lipofectamine 2000 at a concentration of 0.2% to achieve about 20–30% transfection efficiency.

Assaying cell death in cells that have been transiently transfected to replace endogenous with phosphorylation-site mutant Drp1 requires microscopy and is therefore more time-consuming than the MTS assay described above. On the other hand, transient transfection-based assays avoid the delays and pitfalls associated with generating stable cell lines and allow differentiation of cell death mechanisms. Because the classification into live, apoptotic, and necrotic cells can be somewhat subjective, the experimenter scoring cell death should be blinded to the identity of the transfected plasmids. For cell death assays in primary astrocytes, cells are seeded in poly-L-lysine-coated 4-chambered coverglasses (#1, Nunc) and transfected at day 4–12 in vitro with GFP-tagged Drp1. Allowing 3 days for knockdown of endogenous Drp1, cultures are subjected to various toxic compounds for 3–10 h. We have employed 1 μM staurosporine to induce apoptosis and 2–5 U/ml glucose oxidase, which generates H_2O_2 from glucose in the medium, to induce necrosis. Five minutes prior to the end of the incubation period, the red fluorescent dye propidium iodide (5 $\mu g/ml$) is added to the cultures to label the nuclei of cells that have lost membrane integrity, one of the hallmarks of necrotic cell death. After a quick wash with PBS, cells are fixed with 4% paraformaldehyde in PBS for 15 min at 37 °C in the presence of 1 $\mu g/ml$ Hoechst 33342, a blue-fluorescent DNA dye. After 4–5 additional washes in PBS, astrocytes are scored for nuclear morphology and dye uptake using the 63× lens of an inverted, fluorescence microscope. Transfected (GFP-positive) cells that have undergone apoptotic death are characterized by propidium iodide negative nuclei with aberrant morphologies (condensed, fragmented Hoechst-stain). Early-stage necrotic astrocytes have propidium iodide-positive nuclei of normal morphology, while late-stage necrotic cell death is associated with both propidium iodide uptake and nuclear fragmentation/condensation.

6. MITOCHONDRIAL SHAPE ANALYSIS WITH IMAGEJ

Drp1 activity appears to be regulated at multiple interrelated levels, e.g., the intrinsic GTPase activity of the enzyme, its distribution between cytosolic and mitochondrial pools, and its assembly into homo- and hetero-oligomeric complexes. Ultimately however, changes in Drp1 activity (through for instance (de)phosphorylation) translate into changes in the rate of mitochondrial fission which in turn alters mitochondrial morphology. Quantifying mitochondrial fission rates by time–lapse imaging is prohibitively labor intensive, leaving steady–state morphology assays as the readout for cellular Drp1 activity. Live cell–based assays of mitochondrial interconnectivity have been developed that are based on fluorescence recovery after photobleaching (FRAP) of matrix–targeted EYFP, or the spread of a photo-activatable, matrix–targeted fluorescent protein (mito-PAGFP (Karbowski et al., 2004)). Because these methods are time–consuming and can be confounded by changes in mitochondrial motility, our laboratory prefers unbiased image analysis of fluorescently labeled and fixed mitochondria as the principal Drp1 activity assay. For image enhancement and quantification, we use the free ImageJ, a Java application that runs on virtually every computer platform (http://rsbweb.nih.gov/ij/), as well as the ImageJ plug-ins and macros described below.

6.1. Visualization of mitochondria

For microscopic analysis, cells (e.g., primary astrocytes or HeLa cells) are cultured on 4-chambered coverglasses (#1, Nunc) and transfected using Lipofectamine 2000. To label mitochondria, either we incubate cells with MitoTracker Red (Invitrogen, 200 nM) for 15 min prior to fixation, we co-transfect with matrix-targeted fluorescent proteins (EYFP, EGFP, dsRed), or we immunofluorescently stain endogenous mitochondrial proteins. EYFP/EGFP signals can be enhanced by GFP-directed primary antibodies and green-fluorescent secondary antibodies (we use the GFP ab290 antibody from Abcam, with Alexa 488-conjugated, rabbit-directed secondary antibodies from Invitrogen). The monoclonal antibody against cytochrome oxidase subunit II (MTCO2, Labvision) gives in our hands by far the brightest and most specific mitochondrial staining, with other antibodies (e.g., HSP60 Ab2 from Neomarkers, COXIV from Cell Signaling) requiring antigen retrieval for optimal results. Unfortunately, the MTCO2 antibody is primate specific, labeling mitochondria from HeLa and COS cells, but not, e.g., rat astrocytes. After fixation, optional immunofluorescence labeling, and staining of nuclei with 1 μg/ml Hoechst 33342 to assess cell health, multicolor digital images are captured on an inverted epifluorescence or confocal microscope, using a 63× oil immersion lens.

6.2. Image enhancement

Depending on the quality of the mitochondria channel in terms of signal to noise ratio and separation of individual mitochondria, images may be subjected to a 2D deconvolution step, which is meant to compensate for optical imperfections of the microscope. To this end, we use the "Interative

Deconvolution" and "Diffraction Limit PSF" plug-ins written by Bob Dougherty for ImageJ (http://www.optinav.com/imagej.html). The latter plug-in computes a point spread function (PSF), which can be used by the deconvolution algorithm in lieu of an image of a fluorescent particle close to the optical resolution limit. Both plug-ins accept several parameters, which should be optimized for each cell type, mitochondria staining method, and microscope. Because deconvolution can take several minutes depending on image dimensions and hardware, we have written a macro that calls the deconvolution plug-in to batch-process a folder-full of images (available by request from the authors). Figure 13.3 shows images of HeLa cells stained with the MTCO2 antibody before and after deconvolution. Note the overall improvement in contrast, but also the appearance of artificial discontinuities in what appears to the eye as interconnected mitochondrial networks. We find that deconvolution of images of marginal quality decreases the cell-to-cell variability in mitochondrial size metrics, thus amplifying experimental effects (of e.g., phosphorylation-site mutant Drp1 expression). Higher quality images, like those of astrocytes expressing

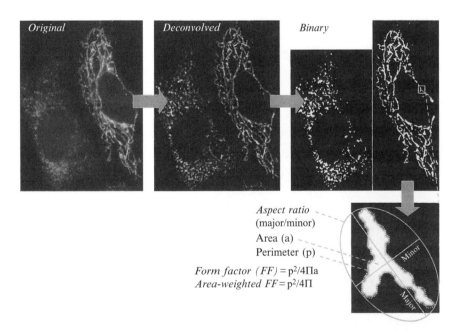

Aspect ratio (major/minor)
Area (a)
Perimeter (p)
Form factor (FF) = p²/4Πa
Area-weighted FF = p²/4Π

Figure 13.3 Flowchart of mitochondrial morphometry with ImageJ. The raw image (HeLa cells fluorescently labeled with mitochondrial MTC02 antibody) on the left is first enhanced (deblurred) by 2D deconvolution and then converted to a binary (black and white) image. Individual particles in the binary image are analyzed for area, perimeter, as well as major and minor axis of a bounding rectangle using the "Analyze Particles" function of ImageJ. These measurements are used to calculate aspect ratio, form factor and area-weighted form factor (product of area and form factor).

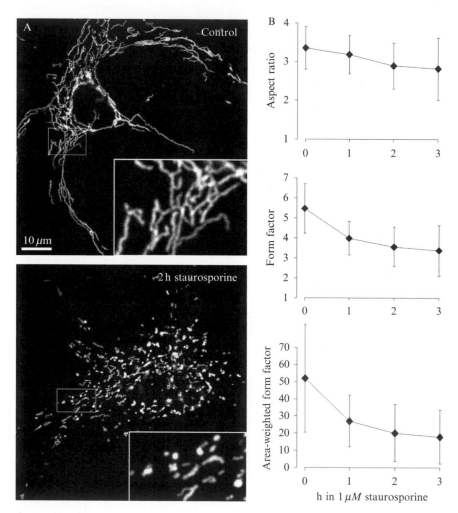

Figure 13.4 Comparison of three mitochondrial shape metrics. Rat primary astro-cytes were transfected with mitochondrial dsRed, treated for 0–3 h with 1 μM staurosporine, fixed, and analyzed for mitochondrial morphology using the ImageJ macro listed in Section 7. Representative epifluorescence images are shown in (A) (with magnified areas in the insets), while three mitochondrial shape metrics are plotted in (B) (means ± SD of ~130 cells per time point). Staurosporine results in mitochondrial fragmentation that is significant by all metrics (values at the 2 and 3 h time point are not significantly different from one another, while pairwise comparisons between all other time points result in p values <0.01).

mitochondria-targeted dsRed shown in Figs. 13.4 and 13.5, may actually suffer from deconvolution, so we subject them to only minor image processing before quantifying mitochondrial shape. A relatively "safe" image manipulation is the "subtract background" function built into

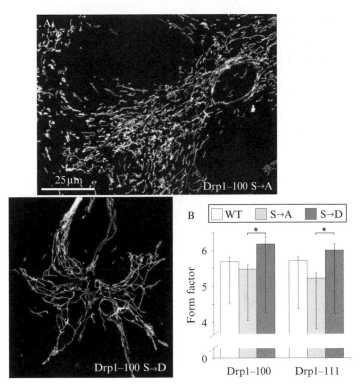

Figure 13.5   Effect of Drp1 phosphorylation on mitochondrial morphology is not influenced by alternative splicing. Rat primary astrocytes were transfected with mitochondrial dsRed and plasmids expressing both Drp1-targeting shRNA and GFP-Drp1 cDNAs. Phosphorylation-blocking (S→A) and -mimicking (S→D) substitutions of the PKA/CaMKI phosphorylation site (Ser656 in rat splice variant 111) were compared to wild type (WT) in the context of Drp1 splice variants with (111) and without (100) the two alternative exons in the variable domain (see Fig. 13.1). Shown are representative epifluorescence images of astrocytes with dsRed-labeled mitochondria (A), as well as mitochondrial shape quantified as form factor (mean ± SD/SEM (lower/upper bars) of ~100 cells/condition). $\star p < 0.005$ by Student's t-test.

ImageJ, which is invoked by the morphometry macro listed in Section 7. ImageJ features several other useful image enhancement functions, including kernel-based convolution and fast Fourier transform bandpass filters.

6.3. Morphometry

After optional image enhancement, mitochondrial shape metrics are reported by the morphometry macro, which at its core calls and processes the output of ImageJ's "Analyze Particles" function. Since this function requires a two-intensity (binary) image, the mitochondria channel is first transformed by binary thresholding (Fig. 13.3). This step invariably degrades the image, making closely spaced mitochondria appear as fused and introducing gaps into clearly contiguous mitochondria. A careful comparison between original and thresholded images is the basis for evaluating different preprocessing options (e.g., deconvolution parameters), and images in which mitochondrial morphology cannot be approximated by the binary transformation should be excluded from the analysis. For each mitochondrial particle, the "Analyze Particles" function tabulates area, perimeter, and major/minor axis of a bounding ellipse (Fig. 13.3). The macro uses these values to compute form factor, area-weighted form factor (AWFF), and aspect ratio, and outputs them to the "Log" window as an average for all particles in the selected region of interest. The aspect ratio (major axis divided by minor axis) is a useful shape metric for simple rod-like mitochondria, but it does not faithfully represent the shape of kinked, branched, or highly interconnected mitochondria. The form factor takes into account perimeter and area and can therefore capture complex mitochondrial shapes. As the inverse of circularity, form factors range from 1 for a perfect circle to infinity as the ratio of particle perimeter to area increases. The AWFF metric is an intuitively sensible modification that affects the way form factors are averaged across all particles in the current selection. In multiplying form factor by area, mitochondria contribute to the population's AWFF in proportion to their mass. The AWFFs fluctuate with greater amplitude than form factors or aspect ratios and tend to better approximate subjective assessments of mitochondrial morphology in cells with mitochondria of broad size distribution. A comparison of aspect ratio, form factor, and AWFF of mitochondria from astrocytes treated for various times with staurosporine is shown in Fig. 13.4.

The benchmark for evaluating mitochondrial shape changes (at least at the light microscopic level) remains the observer's perception. Especially when establishing a new cell model or mitochondria staining protocol, it is therefore important to validate the image analysis algorithms described above by comparison to subjective scores. For these visual scores, coded and randomized images are assigned a value between 0 (completely fragmented, circular mitochondria) and 4 (highly elongated and interconnected) based on comparison to a set of reference images (Dagda et al., 2008).

7. Appendix: Morphometry Macro

```
// This ImageJ macro reports three metrics that describe
// the average particle shape in the current image (selection)
// ImageJ is available at: rsbweb.nih.gov/ij/
```

```
macro "Morphometry [F7]" { // F7 is short-cut
  requires("1.41");
  title = getTitle();
  if (isOpen("binary")) // close previous working image
  {selectWindow("binary"); close(); selectWindow(title); }
  run("Duplicate...", "title=binary");
  run("Subtract Background...","rolling=50");
  run("Make Binary");
  run("Restore Selection");
  run("Set Measurements...", "decimal=5 area perimeter fit");
  run("Analyze Particles...",
      "size=9-Infinity circularity=0.00-1.00 show=Nothing
        clear");
  awff = ff = ar = sum_a = 0;
  for (i = 0; i < nResults; i++) {
    a = getResult('Area', i);
    p = getResult('Perim.', i);
    ar += getResult("Major", i)/getResult("Minor", i);
    // ar = length/width
    sum_a += a;
    awff += b = (p * p)/(4 * 3.14159265358979);
    // awff = ff * (a/sum_area)
    ff += b/a;
    // ff = p^2 / (4 * pi * a)
  }
  awff = awff / sum_a;
  ff = ff / nResults;
  ar = ar / nResults;
  print(title+" -\t area-weighted form factor=\t"+awff+
        "\t, form factor=\t"+ff+"\t, aspect ratio=\t"+ar);
}
```

REFERENCES

Brandt, D. R., and Ross, E. M. (1986). Catecholamine-stimulated GTPase cycle. Multiple sites of regulation by beta-adrenergic receptor and Mg^{2+} studied in reconstituted receptor-Gs vesicles. *J. Biol. Chem.* **261,** 1656–1664.

Brickey, D. A., Bann, J. G., Fong, Y. L., Perrino, L., Brennan, R. G., and Soderling, T. R. (1994). Mutational analysis of the autoinhibitory domain of calmodulin kinase II. *J. Biol. Chem.* **269,** 29047–29054.

Brummelkamp, T. R., Bernards, R., and Agami, R. (2002). A system for stable expression of short interfering RNAs in mammalian cells. *Science* **296,** 550–553.

Cereghetti, G. M., Stangherlin, A., Martins de Brito, O., Chang, C. R., Blackstone, C., Bernardi, P., and Scorrano, L. (2008). Dephosphorylation by calcineurin regulates translocation of Drp1 to mitochondria. *Proc. Natl. Acad. Sci. USA* **105,** 15803–15808.

Chang, C. R., and Blackstone, C. (2007). Cyclic AMP-dependent protein kinase phosphorylation of Drp1 regulates its GTPase activity and mitochondrial morphology. *J. Biol. Chem.* **282,** 21583–21587.

Cribbs, J. T., and Strack, S. (2007). Reversible phosphorylation of Drp1 by cyclic AMP-dependent protein kinase and calcineurin regulates mitochondrial fission and cell death. *EMBO Rep.* **8,** 939–944.

Dagda, R. K., Merrill, R. A., Cribbs, J. T., Chen, Y., Hell, J. W., Usachev, Y. M., and Strack, S. (2008). The spinocerebellar ataxia 12 gene product and protein phosphatase 2A regulatory subunit Bbeta 2 antagonizes neuronal survival by promoting mitochondrial fission. *J Biol Chem.* **283,** 36241–36248.

Han, X. J., Lu, Y. F., Li, S. A., Kaitsuka, T., Sato, Y., Tomizawa, K., Nairn, A. C., Takei, K., Matsui, H., and Matsushita, M. (2008). CaM kinase I alpha-induced phosphorylation of Drp1 regulates mitochondrial morphology. *J. Cell Biol.* **182,** 573–585.

Harder, Z., Zunino, R., and McBride, H. (2004). Sumo1 conjugates mitochondrial substrates and participates in mitochondrial fission. *Curr. Biol.* **14,** 340–345.

Hoppins, S., Lackner, L., and Nunnari, J. (2007). The machines that divide and fuse mitochondria. *Annu. Rev. Biochem.* **76,** 751–780.

Jahani-Asl, A., and Slack, R. S. (2007). The phosphorylation state of Drp1 determines cell fate. *EMBO Rep.* **8,** 912–913.

Karbowski, M., Arnoult, D., Chen, H., Chan, D. C., Smith, C. L., and Youle, R. J. (2004). Quantitation of mitochondrial dynamics by photolabeling of individual organelles shows that mitochondrial fusion is blocked during the Bax activation phase of apoptosis. *J. Cell Biol.* **164,** 493–499.

Koch, A., Thiemann, M., Grabenbauer, M., Yoon, Y., McNiven, M. A., and Schrader, M. (2003). Dynamin-like protein 1 is involved in peroxisomal fission. *J. Biol. Chem.* **278,** 8597–8605.

Koch, A., Yoon, Y., Bonekamp, N. A., McNiven, M. A., and Schrader, M. (2005). A role for Fis1 in both mitochondrial and peroxisomal fission in mammalian cells. *Mol. Biol. Cell* **16,** 5077–5086.

Macek, B., Mann, M., and Olsen, J. V. (2009). Global and site-specific quantitative phosphoproteomics: Principles and applications. *Annu. Rev. Pharmacol. Toxicol.* **49,** 199–221.

Matveeva, O., Nechipurenko, Y., Rossi, L., Moore, B., Saetrom, P., Ogurtsov, A. Y., Atkins, J. F., and Shabalina, S. A. (2007). Comparison of approaches for rational siRNA design leading to a new efficient and transparent method. *Nucleic Acids Res.* **35,** e63.

Nakamura, N., Kimura, Y., Tokuda, M., Honda, S., and Hirose, S. (2006). MARCH-V is a novel mitofusin 2- and Drp1-binding protein able to change mitochondrial morphology. *EMBO Rep.* **7,** 1019–1022.

Reynolds, A., Leake, D., Boese, Q., Scaringe, S., Marshall, W. S., and Khvorova, A. (2004). Rational siRNA design for RNA interference. *Nat. Biotechnol.* **22,** 326–330.

Smillie, K. J., and Cousin, M. A. (2005). Dynamin I phosphorylation and the control of synaptic vesicle endocytosis. *Biochem. Soc. Symp.* **72,** 87–97.

Strack, S., Cribbs, J. T., and Gomez, L. (2004). Critical role for protein phosphatase 2A heterotrimers in mammalian cell survival. *J. Biol. Chem.* **279,** 47732–47739.

Suen, D. F., Norris, K. L., and Youle, R. J. (2008). Mitochondrial dynamics and apoptosis. *Genes Dev.* **22,** 1577–1590.

Taguchi, N., Ishihara, N., Jofuku, A., Oka, T., and Mihara, K. (2007). Mitotic phosphorylation of dynamin-related GTPase Drp1 participates in mitochondrial fission. *J. Biol. Chem.* **282,** 11521–11529.

Takei, K., Yoshida, Y., and Yamada, H. (2005). Regulatory mechanisms of dynamin-dependent endocytosis. *J. Biochem. (Tokyo).* **137,** 243–247.

Wasiak, S., Zunino, R., and McBride, H. M. (2007). Bax/Bak promote sumoylation of DRP1 and its stable association with mitochondria during apoptotic cell death. *J. Cell Biol.* **177,** 439–450.

Waterham, H. R., Koster, J., van Roermund, C. W., Mooyer, P. A., Wanders, R. J., and Leonard, J. V. (2007). A lethal defect of mitochondrial and peroxisomal fission. *N. Engl. J. Med.* **356,** 1736–1741.

CHAPTER FOURTEEN

FUNCTIONAL CHARACTERIZATION OF A MITOCHONDRIAL SER/THR PROTEIN PHOSPHATASE IN CELL DEATH REGULATION

Gang Lu,* Haipeng Sun,* Paavo Korge,[†] Carla M. Koehler,[‡] James N. Weiss,[†,§] and Yibin Wang*,[†,§]

Contents

1. Introduction	256
2. Identification of Protein Phosphatases in Mitochondria	258
2.1. Databases used for identification of mitochondria targeting sequence in PP2C family members	258
2.2. Molecular cloning of mitochondria PP2Cm	258
3. Characterization of Mitochondrial Localization of PP2Cm in Mammalian Cells	260
3.1. Detecting mitochondrial localization of PP2Cm by immunofluorescent microscopy	260
3.2. Determination of mitochondrial targeting of PP2Cm via fractionation	261
3.3. Localization of mitochondrial proteins in different mitochondrial compartments	262
3.4. Solubility assay of PP2Cm in mitochondria	262
3.5. Osmotic shock assay	262
4. Functional Characterization of PP2Cm in Cultured Cells	264
4.1. Characterization of recombinant PP2Cm phosphatase activity	264
4.2. Generation of phosphatase-dead PP2Cm mutants	265
4.3. Knocking down PP2Cm with siRNA expressing vector	265
4.4. Adenovirus-mediated gene deliver in mouse liver	266

* Department of Anesthesiology, Division of Molecular Medicine, UCLA, Los Angeles, California, USA
[†] Department of Medicine, Division of Molecular Medicine, UCLA, Los Angeles, California, USA
[‡] Department of Biological Chemistry, Division of Molecular Medicine, UCLA, Los Angeles, California, USA
[§] David Geffen School of Medicine, Cardiovascular Research Laboratories, UCLA, Los Angeles, California, USA

Methods in Enzymology, Volume 457
ISSN 0076-6879, DOI: 10.1016/S0076-6879(09)05014-9

5. Functional Characterization of PP2Cm in Cell
   Death and Mitochondrial Regulation   267
   5.1. PP2Cm deficiency on cell viability and mitochondria inner
        membrane potential using JC-1 fluorescent assay   267
   5.2. PP2Cm deficiency on mitochondria oxidative
        phosphorylation and respiration   268
   5.3. PP2Cm deficiency on mitochondria permeability
        transition pore regulation (Weiss *et al.*, 2003)   270
6. Mitochondrial Phosphatase in Cell Death Regulation   270
Acknowledgments   271
References   271

Abstract

Protein phosphorylation is a major form of posttranslational modification critical to cell signaling that also occurs in mitochondrial proteome. Yet, only very limited studies have been performed to characterize mitochondrial-targeted protein kinases or phosphatases. Recently, we identified a novel member of PP2C family (PP2Cm) that is a resident mitochondrial protein phosphatase which plays an important role in normal development and cell survival. In this chapter, we will describe the methods applied in the identification of PP2Cm as a resident mitochondrial protein phosphatase based on sequence analysis and biochemical characterization. We will also provide experimental protocols used to establish the intracellular localization of PP2Cm, to achieve loss and gain function of PP2Cm in cultured cells and intact tissue, and to assess the impact of PP2Cm deficiency on cell death, mitochondria oxidative phosphorylation and permeability transition pore opening.

1. INTRODUCTION

Protein phosphorylation/dephosphorylation is regulated by the interplay of protein kinases and phosphatases and plays a pivotal role in cell signaling important for a wide spectrum of cell functions. There are at least 518 protein kinases (human kinome) identified in the human genome and many of them have been extensively investigated (Caenepeel *et al.*, 2004). However, our knowledge on protein phosphatases is still relatively limited. Tyrosine phosphatase has the largest number of family members whereas dual-specific phosphatases and Ser/Thr protein phosphatases represent minorities in human phosphatome. Ser/Thr protein phosphatases are further divided into four sub-families based upon sequence and structural features, namely, PP1, PP2A, PP2B (calcineurin), and PP2C (Cohen, 1994). Different from other phosphatase sub-families, the PP2C family members are monomeric with two characteristic structural domains:

a common N-terminal catalytic domain and a C-terminal region with substantial structure and sequence variations among different isoforms. PP2Cs belong to Mg^{2+}- or Mn^{2+}-dependent protein phosphatases (PPM) family and are highly conserved in both eukaryotes and some prokaryotes (Cohen, 1994). Recent findings have established PP2C isoforms as novel players in cell signaling regulation with exquisite specificity and potency (Lu and Wang, 2008; Stern et al., 2007). Thus, PP2C family members are interesting players in cell signaling that merit detailed investigations.

The mitochondrial network is an essential organelle for cellular metabolism and survival, it is also a convergent and integrative site for cell signaling pathways (Horbinski and Chu, 2005; Ravagnan et al., 2002; Newmeyer and Ferguson-Miller, 2003). Wide-spread and dynamic phosphorylation/dephosphorylation has been reported in mitochondrial proteome (Hopper et al., 2006). A number of Ser/Thr protein kinases, including PKA, PKB/AKT, PKC, and JNK, are also located in mitochondria (Alto et al., 2002; Huang et al., 1999; Nantel et al., 1999; Wang et al., 2003; Wiltshire et al., 2002). These kinases have been implicated in metabolic or apoptotic regulations by targeting matrix, inner, outer membrane components ranging from the ATP synthase to VDAC and BAD/Bax (Baines et al., 2003; Bijur and Jope, 2003; Brichese et al., 2004; Chen et al., 2001; Cohen et al., 2000; Deng et al., 2000; Kang et al., 2003; Scacco et al., 2000; Schroeter et al., 2003; Technikova-Dobrova et al., 2001). Since protein phosphorylation is a dynamic process involving a balancing act of kinases and phosphatases (Shenolikar, 1994), mitochondrial protein phosphorylation will also most likely be regulated by mitochondrial-targeted protein phosphatases. Protein phosphatase 1 (PP1) and PP2A are localized to mitochondria (Brichese et al., 2004; Dagda et al., 2003; Ruvolo et al., 2002; Tamura et al., 2004), although their specific contribution to mitochondrial function remains unclear. Another novel tyrosine protein phosphatase, PTPMT1, has been reported to be located specifically in the mitochondrial matrix and to play an important role in both ATP production and insulin secretion (Pagliarini et al., 2005). In short, protein phosphorylation and dephosphorylation in mitochondria are an important mechanism of cell signaling to regulate metabolic and apoptotic activities.

Recently, we identified a novel Ser/Thr protein phosphatase, named *PP2Cm*, that is targeted exclusively to the mitochondrial matrix (Lu et al., 2007). In this chapter, we will describe the bioinformatics and biochemical methods used in the identification of PP2Cm as a resident mitochondrial protein phosphatase. We will also provide experimental protocols utilized to establish the mitochondrial localization of PP2Cm, to achieve loss and gain function of PP2Cm in cultured cells, and to assess the impact of PP2Cm on cell death. Finally, we will describe the experimental approaches employed to establish gain- and loss-of function of PP2Cm in intact mouse liver as an *in vivo* model system to gain insights into its function in regulating mitochondrial oxidative phosphorylation and the mitochondrial permeability transition pore.

 ## 2. IDENTIFICATION OF PROTEIN PHOSPHATASES IN MITOCHONDRIA

2.1. Databases used for identification of mitochondria targeting sequence in PP2C family members

PP2Cs are Mg^{2+}- or Mn^{2+}-dependent protein phosphatases with a number of isoforms encoded by different homologous genes (Cohen, 1994; Stern *et al.*, 2007). In mammals, at least 17 PP2C family members have been identified. They include PP2Cα, PP2Cβ, PP2Cγ, PP2Cε, PP2Cη, PP2Cm, TA-PP2C, ILKAP, NERRP, Wip1/PPM1D, POPX1, POPX2, PHLPP1, PHLPP2, PDP1, PDP2, and PPM1H (Lu and Wang, 2008). Among them, human PP2Cm (also named PP2Cκ, PPM1K, NM 152542) was listed in Genbank as a putative protein with no known function prior to our publication. Based on sequence alignment, PP2Cm is conserved among human, mouse and zebrafish, but no homolog was identified in *Drosophila* genome. PP2Cm contains a highly conserved catalytic domain in its C-terminal portion as commonly seen in other PP2C family members (Lu *et al.*, 2007). However, the N-terminus of PP2Cm contains a putative mitochondrial targeting sequence as predicted using two different programs: Mitoprot (http://ihg.gsf.de/ihg/mitoprot.html) that gave a probability of 0.9738 for mitochondrial targeting and iPSORT (http://hc.ims.u-tokyo.ac.jp/iPSORT/), see Table 14.1 for the data output.

2.2. Molecular cloning of mitochondria PP2Cm

Full-length human, mouse, and zebrafish PP2Cm cDNA are amplified by PCR from a human EST clone (BG713950), a mouse heart cDNA library and an embryonic cDNA library using primers listed in Table 14.2. The full-length ORF of human and mouse PP2Cm are cloned into pShuttle-CMV vectors (Stratagene) with either GFP or 3XFLAG tag at the C-terminus for adenovirus construction. Human PP2Cm truncation mutant lacking the mitochondrial targeting signal deletion mutant are created by PCR and subcloned into pShuttle-CMV vector tagged with 3XFLAG at the C terminus. Human PP2Cm wild-type and point mutants H129A, R236G, and D298A are generated by site directed mutagenesis using PCR and cloned into pGEX vector (Amersham) for generation of recombinant proteins. All adenoviruses are concentrated with a cesium

Table 14.1 Prediction of mitochondria targeting for PP2Cm

MitoProt II (http://ihg.gsf.de/ihg/mintoprot.html):	
Net charge of query sequence	: −4
Analyzed region	: 29
Number of basic residues in targeting sequence	: 5
Number of acidic residues in targeting sequence	: 0
Cleavage site	: 30
Cleaved sequence	: MSTAALITLVRSGGNQVRRVLLSSRLLQ
Probability of export to mitochondria	: 0.9738

iPSORT Prediction (http://he.ims.u-tokyo.ac.jp/IPSORT/):				
Values used for reasoning				
Node	Answer	View	Substring	Value(s)
1. Signal peptide?	No	Average hydropathy (KYTJ820101)	[6,20]	−0.453333 (≥0.953? No)
2. Mitochondrial?	Yes	Average net charge (KLEP 840101)	[1,30]	0.133333(≥0.083? Yes)
Indexing:All				
Pattern:22121122(ins/del≤3)			[1,30]	MSTAALITLVRSGGN QV‑R‑RRVL LSSRLLQD 22222122212000 22‑1‑1122 22212220 221121122

Table 14.2   Primers and template information for cloning PP2Cm

Gene	Primer sequences for PCR	Template
Human PP2Cm:	Sense: 5′-TAAAAGATCTGCCACC ATGTCAACAGCTGCCTTA-3′ Anti-sense: 5′-TAAACTCGAGTC AGGCCCATCGTCCACTGGA GGC-3′	Human EST clone (BG713950)
Mouse PP2Cm:	Sense:5′- TAAAGCGGCCGCCA CCATGTTATCAGCGGGCTT-3′ Anti-sense: 5′- TAAACTCGAGTC AGGCCCATCTCCCACTGGA-3′	Mouse heart cDNA library

chloride gradient, and further purified on PD10 gel filtration column (Amersham), and stored at −80 °C in PBS plus 10% glycerol before usage as described earlier (Lu *et al.*, 2006).

3. Characterization of Mitochondrial Localization of PP2Cm in Mammalian Cells

3.1. Detecting mitochondrial localization of PP2Cm by immunofluorescent microscopy

Neonatal rat ventricular myocytes or COS1 cells are cultured on 12 mm coverslips coated with laminin. Forty eight hours after adenovirus transfection, cells are incubated with 200 nM Mito-Tracker Red (Molecular Probes) for 45 min at 37 °C. For detection of Flag-tagged PP2Cm, cells are washed twice with PBS, fixed with 10% formalin for 10 min, permeabilized with 0.2% Triton X-100 for 10 min, and blocked in PBS with 3% BSA and 5% donkey serum for 1 h. After incubation with anti-FLAG M2 antibody (1:5000) for 2 h, cells are washed fours times with PBS and then incubated with Alexa488 conjugated Donkey anti-Mouse IgG (Molecular Probes) for 2 h. For GFP-tagged PP2Cm, cells are washed and fixed only without further processing. Coverslips are extensively rinsed with PBS and mounted onto glass slides with Anti-Fade regents (Molecular Probes). Images are captured using a laser scanning confocal microscope (Olympus Fluoview) equipped with an Argon 488 laser for Alexa 488 or GFP signals, a HeNe Green 543 laser for MitoTracker Red signal, respectively. Different fluorescent signals from the same images are recorded separately as digital image files and analyzed using MetaMorph program (Universal Imaging Corp) to generate merged images (Fig. 14.1A). Co-localization of the two signals can be quantified if necessary based on protein proximity index (PPI) calculated using a custom made software program as described elsewhere (Lu *et al.*, 2006).

3.2. Determination of mitochondrial targeting of PP2Cm via fractionation

Liver is a rich source of mitochondria with relatively easy enrichment protocol. In our study, mouse liver is removed and homogenized in STE buffer (250 mM sucrose, 5 mM Tris, 1 mM EGTA, pH 7.4) using a Teflon-glass Dounce homogenizer. One to three mouse livers are broken by three to five passes with the homogenizer. Unbroken cells and cellular debris are removed by centrifugation at 1000g for 3 min. The crude mitochondrial fraction is obtained in the pellet by centrifugation at 10,000g for 10 min. The cytosolic soluble and ER enriched membrane fractions in the resulting supernatant are further separated by centrifugation at 100,000g for 1 h (see Fig. 14.1B). The specificity of each fraction is confirmed by immunoblot

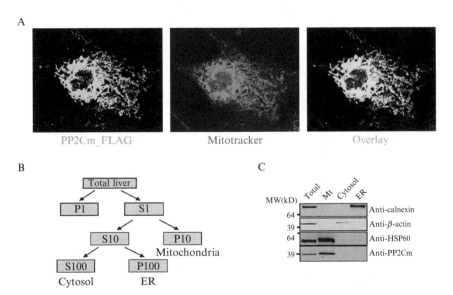

Figure 14.1 Intracellular localization of PP2Cm in mitochondria. (A) Immunofluo-rescent images of PP2Cm-Flag and Mitotracker, as well as their merged image in COS1 cells expressing PP2Cm-Flag fusion gene. (B) Fractionation scheme for cytosol, ER and mitochondria from mouse liver. S, supernatant; P, pellet; 1, 10, 100 represent 1000g, 10,000g, and 100,000g centrifugations, respectively. (C) Immunoblot of cellular fractions using specific antibodies as shown. Mt, mitochondria. Reproduced from (Lu *et al.*, 2007) with permission from CSHL Press.

using specific marker proteins, including Calnexin for ER, β-actin for cytosol and Hsp60 for mitochondria. Based on immunoblot analysis of each subcellular component using anti-PP2Cm antibody, PP2Cm is detected only in mitochondrial fraction (see Fig. 14.1C), further supporting it is a mitochondrial resident protein.

3.3. Localization of mitochondrial proteins in different mitochondrial compartments

Mitochondrial proteins can be divided into four compartments, that is, outer membrane (OM), inner membrane (IM), intermembrane space (IMS), and matrix. The proteins in each compartment of mitochondria are engaged in different functions. Outer membrane and IMS proteins are critical to classic apoptotic activities; matrix and inner membrane proteins are mostly involved in respiratory chain reaction and oxidative phosphory-lation; and both inner/outer membrane proteins are implicated in mito-chondria permeability transition pore opening important to cell death regulation. Therefore, the sub-mitochondrial localization of mitochondrial proteins has major implications with respect to their functions. To date, it is not possible to predict with high confidence to which compartment a mitochondrial protein is located and whether a mitochondrial protein is soluble in its matured form or remains membrane anchored. These ques-tions therefore must be addressed experimentally according to a scheme employing carbonate extraction and osmotic shock assays.

3.4. Solubility assay of PP2Cm in mitochondria

To establish first whether the mitochondrial-targeted PP2Cm is soluble or membrane associated, we perform carbonate extraction (Fujiki *et al.*, 1982) by re-suspending mitochondria pellet in 0.1 M Na_2CO_3 at pH 11.0 and incubating on ice for 20 min followed by $100,000g$ for 10 min. The soluble proteins partition to the supernatant and the integral membrane proteins partition in pellet; fractionation is analyzed by immunoblotting using an integral membrane protein Tom40 as a positive control for the membrane fraction and an IMS soluble protein Timm13 as a positive control for the soluble fraction (Fig. 14.2A).

3.5. Osmotic shock assay

In the osmotic shock assay, mitochondria are diluted into a hypotonic buffer that causes the inner membrane to swell and ruptures the outer membrane (Koehler *et al.*, 1998). Thus, the soluble IMS contents are released and "mitoplasts," which contain the intact inner membrane, matrix, and adherent, ruptured OM fragments, can be recovered by centrifugation.

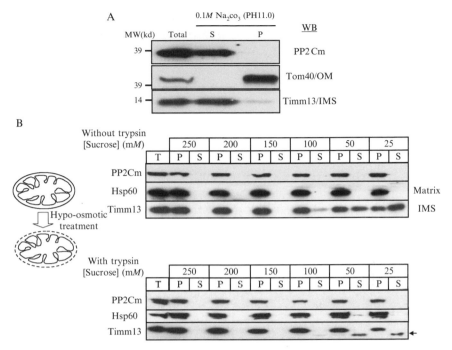

Figure 14.2 Determination of PP2Cm as a soluble protein located in mitochondrial matrix. (A) Immunoblot for PP2Cm, TOM40, and Timm13 following carbonate extraction using 0.1 M Na$_2$CO$_3$ (pH 11.0). S, supernatant; P, pellet from centrifugation after Na$_2$CO$_3$ treatment. (B) Immunoblotting for PP2Cm, Hsp60 (matrix protein marker) and Timm13 (IMS marker) following mitoplasting assay using sucrose concentrations as indicated to produce hypotonic conditions. T, total; P, pellet; and S, supernatant following centrifugation. Reproduced from (Lu *et al.*, 2007) with permission from CSHL Press.

Purified mitochondria are re-suspended in serial hypotonic STE buffers containing 5.0 mM Tris, 1.0 mM EGTA, pH 7.4, plus 250, 200, 150, 100, 50, or 25 mM of sucrose. After 30 min incubation on ice followed by centrifugation at 16,000g for 10 min, the pellets, which contain the mitoplasts, and supernatants are then analyzed by regular SDS–PAGE and immunoblotting. To further demonstrate mitochondria outer membrane rupture, 50 ug/mL soybean trypsin (Sigma) is added to hypotonic STE buffer during the incubation and the mitochondrial preparation is treated with 20 ug/mL trypsin inhibitor for an additional 10 min before the centrifugation. Matrix proteins (such as Hsp60) remain in the pellets under all hypotonic conditions and resistant to trypsin digestion. On the other hand, soluble IMS proteins (such as Timm13) begin to be detected in the supernatant with increasing strength of osmotic shock (lower sucrose concentration) and become sensitive to trypsin digestion (Fig. 14.2B). Based on these assays, PP2Cm is established as a soluble protein in the mitochondrial matrix.

4. Functional Characterization of PP2Cm in Cultured Cells

4.1. Characterization of recombinant PP2Cm phosphatase activity

Protein phospatase 2C family has no known specific inhibitors and the specificities of each isoforms toward their targets are unclear based on earlier studies. Therefore, there is no method available to directly measure the specific activity of the endogenous PP2Cm in tissues or cells. However, quantitative enzymatic activity can be measured *in vitro* using recombinant proteins on generic phosphor–protein or phosphor–peptide substrates.

4.1.1. Radioactive assay

Ser/Thr protein phosphatase activity is determined by using myelin basic protein (MBP) pre-phosphorylated by the catalytic subunit of cyclic AMP-dependent protein kinase (PKA). To prepare phosphorylated MBP as substrate, 500 μg of MBP (M1891, Sigma) is incubated with 25 units of PKA (P2645, Sigma) in kinase buffer (50 mM Tris–HCl (pH 7.5), 10 mM MgCl$_2$, 1 mM DTT, and 3.70 MBq [γ-32P]ATP) at 30 °C overnight. The phosphorylated MBP was then purified by Quick Spin protein desalting column (Cat# 100973, Roche) and further concentrated by Micron Millipore filter unit (Cat#42404, Milllipore). Phosphatase assays are performed using 50–500 ng purified GST fusion protein and 1.0–10.0 μg of ^{32}P-labeled MBP substrate in 50 mM Tris–HCl (pH 7.0), 0.1 mM EDTA, 5 mM DTT, and 0.01% Brij35. The reactions are done at 30 °C and terminated by addition of 100 μl of 20% (w/v) trichloroacetic acid at specific time points. Precipitated proteins are removed by centrifugation at 16,000g for 30 min. Radioactivity of free phosphates (^{32}Pi) released in the supernatants and total protein substrate (^{32}P-MBP) is measured using a scintillation counter and the percent of phosphate release (free phosphate versus total MBP signals) is used to calculate protein phosphatase activities. The dependence of PP2Cm on MnCl$_2$ was determined by adding different concentrations of MnCl$_2$ to the reaction buffer (between 0 and 20 mM). It is important to recognize that MBP is not a natural substrate of PP2Cm and the *in vitro* assay condition is far from optimal comparing with its native environment in mitochondrial matrix. Therefore, the enzyme activity of PP2Cm measured *in vitro* may not represent its optimal physiological activity against its endogenous substrates.

4.1.2. Colorimetric assay

PP2Cm assay can also be conducted in a more high-throughput manner using a colorimetric assay in 96-well plates with a synthetic phosphopeptide instead of ^{32}P-labeled MBP as substrate. Specifically, 50 μl total reaction solution is prepared containing 0.1–0.5 μg of GST-PP2Cm protein and 1 μg of phosphopeptide substrate (H-Arg-Arg-Ala-pThr-Val-Ala-OH, Cat # P152–0001, Biomol) in phosphatase reaction buffer (50 mM Tris–HCl (pH 7.0), 0.1 mM EDTA, 5 mM DTT, and 0.01% Brij35). After incubation at 30 °C for a specific period of time (determined experimentally within linear reaction curve), the reaction is terminated by adding 100 μl of Biomol Green (Cat #AK111–250, Biomol) and further incubation at 30 °C for additional 30 min. The free phosphate is measured based on absorbance at 620 nm. As for any colorimetric assays, a standard curve using specific concentrations of phosphate (0–10 nM, KI-102, Biomol) should be established to ensure the absorbance values are within the linear detection range. It is also important to make sure the recombinant proteins are prepared in phosphate-free buffer (Tris based) and eliminate contamination of other free phosphate throughout the assay. Since there is no known commercial source of small molecule inhibitors with sufficient specificity and potency against PP2C family of protein phosphatases, this assay can be useful for high-throughput screening of novel compounds as PP2C inhibitors or agonists.

4.2. Generation of phosphatase-dead PP2Cm mutants

Loss of function study can be very powerful to reveal novel gene functions. A loss of function can be achieved by generating a phosphatase-dead mutant or knocking down endogenous PP2Cm expression using siRNA strategy. Based on sequence alignment and earlier studies on PP2Cα/β isoforms, H129, R236, and D298 of PP2Cm are predicted to be essential residues necessary for its phosphatase activity. Site-directed mutagenesis is carried out to replace alanine in the three sites and the resulting H129A, R236A, and D298A recombinant proteins are tested for enzymatic activity in $vitro$ as described above (Lu et $al.$, 2007). Indeed, the mutant PP2Cm proteins show little phosphatase activity comparing to wild-type PP2Cm, suggesting further that PP2Cm is a bona fide protein phosphatase, although the endogenous substrate remains to be revealed.

4.3. Knocking down PP2Cm with siRNA expressing vector

PP2Cm protein and mRNA are detected at high levels in adult mouse brain and heart, and its expression in heart is diminished in hypertrophic or failing hearts induced by mechanical stress (Lu et $al.$, 2007). To establish the physiological function of endogenous PP2Cm and to better understand the functional significance of PP2Cm downregulation in response to stress, a loss of function approach is employed to knockdown PP2Cm expression in cultured cardiac myocytes. Due to the low efficiency of transfection, recombinant adenovirus vectors are used to achieve efficient gene transfer. First, synthetic 64-nt oligonucleotide hairpin pairs targeting mouse and rat PP2Cm (same sequence) are annealed and ligated into a modified version of pSUPER vector (Oligoengine), pShuttle-pSUPER, which is constructed by subcloning the H1-RNA expression cassette into pShuttle (Adeasy, stratagene) (kind gift from Dr. J. Han). As a control, recombinant adenoviral shRNA vector targeting firefly luciferase is also constructed. The targeting sequences are listed as follows:

PP2Cm shRNA1, 5′-TCTGGGATAACCGCATTGA-3′;
PP2Cm shRNA2, 5′-GAAGCTGACCACTGACCAT-3′;
PP2Cm shRNA3, 5′-GAAGCTGACCACTGACCAT-3′;
Luciferase shRNA, 5′-CTGACGCGGAATACTTCGA-3′.

The recombinant adenoviruses expressing PP2Cm and Luciferase shRNAs are generated according to an established protocol (Lu *et al.*, 2006). The efficacy of shRNA mediated knockdown is confirmed by immunoblotting using a PP2Cm specific antibody developed in our lab (Fig. 14.3A).

4.4. Adenovirus-mediated gene deliver in mouse liver

To establish the physiological function for PP2Cm, a genetic-targeted mouse model would be ideal. However, in the absence of a readily available PP2Cm−/− mouse model, adenovirus mediated gene transfer can be used to achieve efficient gain and loss of function studies in targeted tissue, including liver (Wang *et al.*, 1996). As mentioned earlier, liver is the rich source of mitochondria and thus ideally suited to adenovirus-mediated gene knockdown. For PP2Cm, the Adv-PP2Cm–shRNA vectors are generated as described earlier. Twenty to twenty-five microliters of viral solution (approximately 5×10^{11} particles/ml) are injected via tail vein for each adult mouse (2 months old, C57BL/5) to achieve an average dose of 5×10^9 particle/g body weight. In addition to PP2Cm–shRNA vectors, an adv–luciferase shRNA vector is required as a negative control. Five to seven days post-viral infection, liver tissues are excised for mitochondria isolation as described above. It is well established that systemic adenovirus adminis-tration will preferentially target liver hepatocytes (Wang *et al.*, 1996). However, one of the drawbacks of this approach is the transient nature of the target gene expression (with peak expression within a week) and viral associated inflammatory response which may complicate the data interpre-tation. Finally, the efficacy of knockdown for PP2Cm is confirmed by immunoblotting as PP2Cm protein is significantly reduced following shRNA expression (Fig. 14.3A).

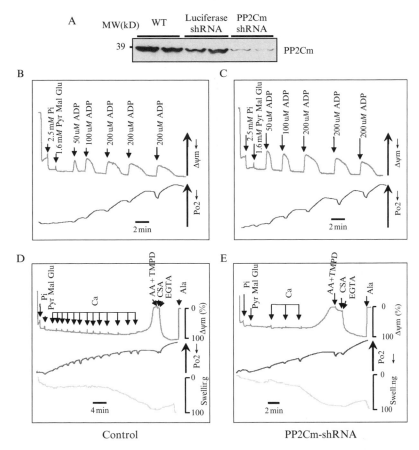

Figure 14.3 PP2Cm deficiency on mitochondrial function. (A) Immunoblot of PP2Cm showing efficient shRNA mediated knockdown in mouse liver following adv-shRNA injection. (B) and (C) Innermembrane potential ($\psi\Delta m$) and oxygen consumption simultaneously recorded in response to ADP pulses in Control and PP2Cm deficient mitochondria. (D) and (E) Innermembrane potential ($\psi\Delta m$), oxygen consumption and mitochondrial swelling simultaneously recorded in response to 2 μM Ca^{2+} stimulations. Reproduced from (Lu *et al.*, 2007) with permission from CSHL Press.

 ## 5. FUNCTIONAL CHARACTERIZATION OF PP2CM IN CELL DEATH AND MITOCHONDRIAL REGULATION

5.1. PP2Cm deficiency on cell viability and mitochondria inner membrane potential using JC-1 fluorescent assay

Neonatal myocytes are cultured on collagen coated glass cover slips in serum free media for 1 day following isolation when they are infected with Adv-PP2Cm–shRNA or control vectors. Under normal conditions, the neonatal myocytes would survive without appreciable level of cell death for more than 5 days. However, PP2Cm–shRNA infected myocytes have a much higher death rate. To directly demonstrate if the cell death is caused by mitochondria permeability transition pore opening, we use JC-1 fluorescent activity assay to measure mitochondrial inner membrane potential ($\Delta\Psi_m$) in living myocytes (Reers *et al.*, 1995). JC-1 is a cationic fluorescent dye (5,5′,6,6′-tetrachloro-1,1′,3,3′-tetraethylbenzimidazolylcarbocyanine iodide, Molecular Probes) and has a unique ability to accumulate in polarized mitochondria when added to the medium, resulting in the formation of mitochondrial specific red fluorescent aggregates (absorbance/emission maxima of 585/590 nm). In apoptotic cells, mitochondrial inner membrane potential $\Delta\Psi$ collapsed and JC-1 fails to accumulate in mitochondria and remains in the green fluorescent monomeric form (absorbance/emission maxima of 510/527 nm). Therefore, the relative intensity of red versus green fluorescence serves as an indirect measurement of D *Y* polarity (Reers *et al.*, 1995). To perform JC-1 assay, cells are stained with 2 μg/mL of JC-1 (Molecular Probes) in growth medium at 37 °C for 20 min following by a double wash with growth media. The fluorescent signals are viewed using a laser scanning confocal microscope (Olympus Fluoview). The red fluorescence JC-1 aggregates in healthy polarized mitochondria is detected at 585–640 nm, and the green fluorescence of JC-1 monomers in apoptotic mitochondria is detected at 510–530 nm. We use a 488 Argon Laser to excite green fluorescent signal and a 543 HeNe laser light to excite red fluorescence. The red and green fluorescence emissions from each view are recorded digitally under the same setting of exposure and light intensity without digital manipulation. Their respective pixel intensities were quantified by MetaMorph (Universal Imaging Corp). Finally, the ratio of total red (aggregated JC-1 from polarized intact mitochondria) versus total green (monomers from cells with collapsed mitochondria $\Delta\Psi_m$) signals from the same image was calculated as an indirect value of mitochondria membrane potential. Our results from this experiment suggests that loss of PP2Cm leads to cell death associated with loss of $\Delta\Psi_m$. Thus, PP2Cm is a critical molecule for mitochondrial permeability transition pore regulation and cell survival.

5.2. PP2Cm deficiency on mitochondria oxidative phosphorylation and respiration

After achieving an efficient knockdown of endogenous PP2Cm (Fig. 14.3A), mitochondria are isolated from control and PP2Cm deficient liver as described in Section 3.2. To measure oxidative phosphorylation (Korge *et al.*, 2005), isolated mitochondria (0.25 mg/mL) are added into 100 KCl buffer containing 10m*M* HEPES pH7.4 and 0.2% BSA in a closed, continuously stirred cuvette at room temperature. Mitochondrial membrane potential $\Delta\Psi_m$, matrix volume, and oxygen consumption are simultaneously recorded. For mitochondria $\Delta\Psi_m$, an optic fiber spectrofluorimeter (Ocean Optics) is used to detect tetramethylrhodamine methyl ester (TMRM) fluorescence at 580 nm. For mitochondrial matrix volume, 90° light scattering is recorded at 520 nm. For oxygen consumption, pO_2 in the buffer is measured via a fiber–optic oxygen sensor (Fig. 14.4). Mitochondria are energized using 2.5 mM Pi and complex I substrates (1.6 mM of pyruvate, malate, and glutamate). After being stabilized for approximately 5 min, ADP is added to cuvette at 50, 100, 200 μM doses with sufficient time interval to allow complete recovery of $\Delta\Psi_m$. The integrity of mitochondrial respiratory activity is determined based on the time and magnitude of $\Delta\Psi_m$ recovery associated with oxygen consumption. If there is a defect observed in experimental samples related to controls, different complex substrates can be used to detect the specific defects in respiration chain. Our studies suggest that PP2Cm deficiency does not affect mitochondria oxidative phosphorylation (Fig. 14.3B and C).

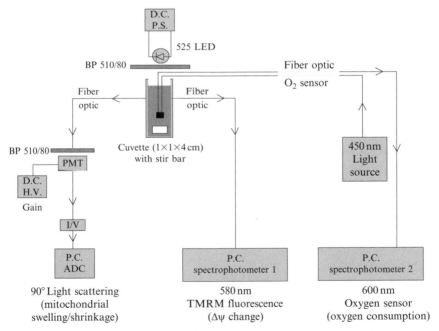

Figure 14.4 Configuration of mitochondrial assay system. Mitochondria preparation is suspended in stirred cuvette. Excitation source is light emitting diode (LED) and BP 510/80 Fluorescence filter. Mitochondrial swelling is detected by 90 degree scattered light at photomultipler tube (PMT). TMRM fluorescence is detected by Spectrophotometer 1, and oxygen is detected using a fiber–optic sensor inside of the mitochondrial solution by Spectrophotometer 2. Signals are logged by analog to digital converter (ADC) in a PC.

5.3. PP2Cm deficiency on mitochondria permeability transition pore regulation (Weiss *et al.*, 2003)

In the same preparation of isolated mitochondria described above, mitochondria (0.25 mg/mL) are added in 150 mM KCl buffer containing 0.2% BSA. Mitochondria are also energized with 2.5 mM Pi and 1.6 mM complex I substrates (pyruvate, malate, and glutamate). Mitochondrial membrane potential $\Delta\Psi_m$, matrix volume, and oxygen consumption are simultaneously recorded. Ca^{2+} pulses (2.0 μM) are added successively with a sufficient time delay (usually 2–3 min) to allow full recovery of $\Delta\Psi_m$ between pulses. After a sufficient number of Ca^{2+} has been added to trigger the permeability transition, as indicated by depolarization of inner membrane potential ($\Delta\Psi_m$ approaching 0) with simultaneous matrix swelling, we add the complex IV substrates ascorbic acid (2 mM) and N,N,N', N'-tetramethyl-p-phenylenediamine (TMPD, 0.2 mM), together with the mitochondrial permeability transition pore inhibitor Cyclosporin A (CsA, 1.5 μM) and EGTA (1 mM). These treatments should recover the mitochondrial membrane potential if the earlier loss of Ψ_m is truly mediated by the opening of permeability transition pore. Finally, to quantify the extent of permeability transition pore opening, 10 μg alamethicin (Ala) is added to induce maximal mitochondrial swelling and Ψ_m dissipation. In our experiments, the number of Ca^{2+} pulses (2 μM) required for 95% mitochondrial matrix swelling relative to alamethicin was greater in control than in PP2Cm knockdown mitochondrial preparations (Fig. 14.3D and E). This result shows PP2Cm deficient mitochondria are much more sensitive to Ca^{2+} induced permeability transition pore opening.

Our studies in isolated mitochondria suggest that PP2Cm deficiency sensitizes mitochondria permeability transition pore opening in response to Ca^{2+} overload and can predispose cells to apoptosis. To further investigate this findings, we carried out direct apoptosis assays on PP2Cm deficient liver tissue using TUNEL immunostaining methods. This is a well-established procedure based on caspase induced DNA fragmentation present in the apoptotic nuclei. In addition to counter staining with DAPI to illustrate total number of nuclei, a positive control is performed using DNase I pretreated slides (Lu *et al.*, 2007).

6. MITOCHONDRIAL PHOSPHATASE IN CELL DEATH REGULATION

Based on this set of comprehensive studies, we establish the intracellular location of a novel PP2C family member as a mitochondrial resident protein. PP2Cm is present in soluble form in mitochondria matrix. It is essential to cell survival and permeability transition pore opening, but has not effect on oxidative phosphorylation and respiratory chain activity. PP2Cm is a novel regulator in cell death regulation and plays an important role in normal development. Further efforts are needed to identify specific downstream targets of PP2Cm and the mechanisms involved in its down-regulation by stress in heart. Protein phosphorylation remains poorly characterized in the mitochondrial proteome. Our study shows the potential importance of protein phosphorylation/dephosphorylation in mitochondria matrix for cellular survival and function. Research in this area holds great promise for uncovering novel mechanisms in cell signal transduction mediated via mitochondrial regulation, toward the ultimate goal of identifying molecular targets for novel therapies to treat human diseases.

ACKNOWLEDGMENTS

Authors wish to thank Dr John Parker for his assistant in technical help and manuscript preparation. This work was partially supported by Laubisch Foundation and grants from National Institutes of Health to PK, CMK, JNW, and YW, Predoctoral fellowship to GL and Postdoctoral fellowship to HS from AHA Western States Affiliate, and YW is an Established Investigator of AHA.

REFERENCES

Alto, N. M., Soderling, J., and Scott, J. D. (2002). Rab32 is an A-kinase anchoring protein and participates in mitochondrial dynamics. *J. Cell Biol.* **158,** 659–668.

Baines, C. P., Song, C. X., Zheng, Y. T., Wang, G. W., Zhang, J., Wang, O. L., Guo, Y., Bolli, R., Cardwell, E. M., and Ping, P. (2003). Protein kinase C epsilon interacts with and inhibits the permeability transition pore in cardiac mitochondria. *Circ. Res.* **92,** 873–880.

Bijur, G. N., and Jope, R. S. (2003). Rapid accumulation of Akt in mitochondria following phosphatidylinositol 3-kinase activation. *J. Neurochem.* **87,** 1427–1435.

Brichese, L., Cazettes, G., and Valette, A. (2004). JNK is associated with Bcl-2 and PP1 in mitochondria: Paclitaxel induces its activation and its association with the phosphorylated form of Bcl-2. *Cell Cycle* **3,** 1312–1319.

Caenepeel, S., Charydczak, G., Sudarsanam, S., Hunter, T., and Manning, G. (2004). The mouse kinome: Discovery and comparative genomics of all mouse protein kinases. *Proc. Natl. Acad. Sci. USA* **101,** 11707–11712.

Chen, L., Hahn, H., Wu, G., Chen, C. H., Liron, T., Schechtman, D., Cavallaro, G., Banci, L., Guo, Y., Bolli, R., Dorn, G. W. 2nd, and Mochly-Rosen, D. (2001). Opposing cardioprotective actions and parallel hypertrophic effects of delta PKC and epsilon PKC. *Proc. Natl. Acad. Sci. USA* **98,** 11114–11119.

Cohen, P. T. W. (1994). Nomenclature and chromosomal localization of human protein serine/threonine phosphatase genes. *Adv. Protein Phosphatases* **8,** 371–376.

Cohen, M. V., Baines, C. P., and Downey, J. M. (2000). Ischemic preconditioning: From adenosine receptor to KATP channel. *Annu. Rev. Physiol.* **62,** 79–109.

Dagda, R. K., Zaucha, J. A., Wadzinski, B. E., and Strack, S. (2003). A developmentally regulated, neuron-specific splice variant of the variable subunit Bbeta targets protein phosphatase 2A to mitochondria and modulates apoptosis. *J. Biol. Chem.* **278,** 24976–24985.

Deng, X., Ruvolo, P., Carr, B., and May, W. S. Jr., (2000). Survival function of ERK1/2 as IL-3-activated, staurosporine-resistant Bcl2 kinases. *Proc. Natl. Acad. Sci. USA* **97,** 1578–1583.

Fujiki, Y., Hubbard, A. L., Fowler, S., and Lazarow, P. B. (1982). Isolation of intracellular membranes by means of sodium carbonate treatment: Application to endoplasmic reticulum. *J. Cell Biol.* **93,** 97–102.

Hopper, R. K., Carroll, S., Aponte, A. M., Johnson, D. T., French, S., Shen, R. F., Witzmann, F. A., Harris, R. A., and Balaban, R. S. (2006). Mitochondrial matrix phosphoproteome: Effect of extra mitochondrial calcium. *Biochemistry* **45,** 2524–2536.

Horbinski, C., and Chu, C. T. (2005). Kinase signaling cascades in the mitochondrion: A matter of life or death. *Free Radic Biol. Med.* **38,** 2–11.

Huang, L. J., Wang, L., Ma, Y., Durick, K., Perkins, G., Deerinck, T. J., Ellisman, M. H., and Taylor, S. S. (1999). NH2-Terminal targeting motifs direct dual specificity A-kinase-anchoring protein 1 (D-AKAP1) to either mitochondria or endoplasmic reticulum. *J. Cell Biol.* **145,** 951–959.

Kang, B. P., Urbonas, A., Baddoo, A., Baskin, S., Malhotra, A., and Meggs, L. G. (2003). IGF-1 inhibits the mitochondrial apoptosis program in mesangial cells exposed to high glucose. *Am. J. Physiol. Renal. Physiol.* **285,** F1013–F1024.

Koehler, C. M., Merchant, S., Oppliger, W., Schmid, K., Jarosch, E., Dolfini, L., Junne, T., Schatz, G., and Tokatlidis, K. (1998). Tim9p, an essential partner subunit of Tim10p for the import of mitochondrial carrier proteins. *EMBO J.* **17,** 6477–6486.

Korge, P., Honda, H. M., and Weiss, J. N. (2005). K+-dependent regulation of matrix volume improves mitochondrial function under conditions mimicking ischemia-reperfusion. *Am. J. Physiol. Heart Circ. Physiol.* **289,** H66–77.

Lu, G., Kang, Y. J., Han, J., Herschman, H. R., Stefani, E., and Wang, Y. (2006). TAB-1 modulates intracellular localization of p38 MAP kinase and downstream signaling. *J. Biol. Chem.* **281,** 6087–6095.

Lu, G., Ren, S., Korge, P., Choi, J., Dong, Y., Weiss, J., Koehler, C., Chen, J. N., and Wang, Y. (2007). A novel mitochondrial matrix serine/threonine protein phosphatase regulates the mitochondria permeability transition pore and is essential for cellular survival and development. *Genes Dev.* **21,** 784–796.

Lu, G., and Wang, Y. (2008). Functional diversity of mammalian type 2C protein phosphatase isoforms: New tales from an old family. *Clin. Exp. Pharmacol. Physiol.* **35,** 107–112.

Nantel, A., Huber, M., and Thomas, D. Y. (1999). Localization of endogenous Grb10 to the mitochondria and its interaction with the mitochondrial-associated Raf-1 pool. *J. Biol. Chem.* **274,** 35719–35724.

Newmeyer, D. D., and Ferguson-Miller, S. (2003). Mitochondria: Releasing power for life and unleashing the machineries of death. *Cell* **112,** 481–490.

Pagliarini, D. J., Wiley, S. E., Kimple, M. E., Dixon, J. R., Kelly, P., Worby, C. A., Casey, P. J., and Dixon, J. E. (2005). Involvement of a mitochondrial phosphatase in the regulation of ATP production and insulin secretion in pancreatic beta cells. *Mol. Cell* **19,** 197–207.

Ravagnan, L., Roumier, T., and Kroemer, G. (2002). Mitochondria, the killer organelles and their weapons. *J. Cell Physiol.* **192,** 131–137.

Reers, M., Smiley, S. T., Mottola-Hartshorn, C., Chen, A., Lin, M., and Chen, L. B. (1995). Mitochondrial membrane potential monitored by JC-1 dye. *Methods Enzymol.* **260,** 406–417.

Ruvolo, P. P., Clark, W., Mumby, M., Gao, F., and May, W. S. (2002). A functional role for the B56 alpha-subunit of protein phosphatase 2A in ceramide-mediated regulation of Bcl2 phosphorylation status and function. *J. Biol. Chem.* **277,** 22847–22852.

Scacco, S., Vergari, R., Scarpulla, R. C., Technikova-Dobrova, Z., Sardanelli, A., Lambo, R., Lorusso, V., and Papa, S. (2000). cAMP-dependent phosphorylation of the nuclear encoded 18-kDa (IP) subunit of respiratory complex I and activation of the complex in serum-starved mouse fibroblast cultures. *J. Biol. Chem.* **275,** 17578–17582.

Schroeter, H., Boyd, C. S., Ahmed, R., Spencer, J. P., Duncan, R. F., Rice-Evans, C., and Cadenas, E. (2003). c-Jun N-terminal kinase (JNK)-mediated modulation of brain mitochondria function: New target proteins for JNK signalling in mitochondrion-dependent apoptosis. *Biochem. J.* **372,** 359–369.

Shenolikar, S. (1994). Protein serine/threonine phosphatases—New avenues for cell regulation. *Annu Rev Cell Biol.* **10,** 55–86.

Stern, A., Privman, E., Rasis, M., Lavi, S., and Pupko, T. (2007). Evolution of the metazoan protein phosphatase 2C superfamily. *J. Mol. Evol.* **64,** 61–70.

Tamura, Y., Simizu, S., and Osada, H. (2004). The phosphorylation status and anti-apoptotic activity of Bcl-2 are regulated by ERK and protein phosphatase 2A on the mitochondria. *FEBS Lett.* **569,** 249–255.

Technikova-Dobrova, Z., Sardanelli, A. M., Speranza, F., Scacco, S., Signorile, A., Lorusso, V., and Papa, S. (2001). Cyclic adenosine monophosphate-dependent phosphorylation of mammalian mitochondrial proteins: Enzyme and substrate characterization and functional role. *Biochemistry* **40,** 13941–13947.

Wang, Y., Krushel, L. A., and Edelman, G. M. (1996). Targeted DNA recombination *in vivo* using an adenovirus carrying the cre recombinase gene. *Proc. Natl. Acad. Sci. USA* **93,** 3932–3936.

Wang, W. L., Yeh, S. F., Chang, Y. I., Hsiao, S. F., Lian, W. N., Lin, C. H., Huang, C. Y., and Lin, W. J. (2003). PICK1, an anchoring protein that specifically targets protein kinase C alpha to mitochondria selectively upon serum stimulation in NIH 3T3 cells. *J. Biol. Chem.* **278,** 37705–37712.

Weiss, J. N., Korge, P., Honda, H. M., and Ping, P. (2003). Role of the mitochondrial permeability transition in myocardial disease. *Circ. Res.* **93,** 292–301.

Wiltshire, C., Matsushita, M., Tsukada, S., Gillespie, D. A., and May, G. H. (2002). A new c-Jun N-terminal kinase (JNK)-interacting protein, Sab (SH3BP5), associates with mitochondria. *Biochem. J.* **367,** 577–585.

CHAPTER FIFTEEN

DISTINGUISHING MITOCHONDRIAL INNER MEMBRANE ORIENTATION OF DUAL SPECIFIC PHOSPHATASE 18 AND 21

Matthew J. Rardin,*,† Gregory S. Taylor,§ and Jack E. Dixon†,‡

Contents

1. Introduction	276
2. Analysis of Phosphatase Activity of DSP18	277
2.1. Bacterial expression constructs and mutagenesis	277
2.2. Expression and purification of recombinant DSP18	278
2.3. Use of the phosphotyrosine analog pNPP for determining phosphatase activity	279
3. Isolation of Highly Purified Rat Kidney Mitochondria and Subfractionation	279
3.1. Differential centrifugation to isolate crude mitochondria	280
3.2. Histodenz gradient purification of mitochondria	281
3.3. Subfractionation of highly purified mitochondria	282
4. Mitochondrial Inner Membrane Association of DSP18 and DSP21	283
4.1. Trypsin digestions of mitochondrial inner membranes	283
4.2. Integral versus peripheral membrane protein	285
Acknowledgments	286
References	287

Abstract

Dual specificity phosphatase (DSP) 18 and 21 are members of a poorly understood subfamily of protein tyrosine phosphatases (PTP) that are unique in their ability to dephosphorylate both phosphotyrosine and phosphoserine/threonine residues *in vitro*. Because of the difficulty in identifying substrate specificity, determining subcellular localization can help to resolve biological function of these phosphatases. DSP18 and DSP21 are targeted to mitochondria by

* Biomedical Sciences Graduate Program, University of California, San Diego, La Jolla, California, USA
† Departments of Pharmacology, Cellular and Molecular Medicine, and Chemistry and Biochemistry, University of California, San Diego, La Jolla, California, USA
‡ The Howard Hughes Medical Institute, University of California, San Diego, La Jolla, California, USA
§ Department of Biochemistry and Molecular Biology, University of Nebraska, Nebraska Medical Center, Omaha, Nebraska, USA

Methods in Enzymology, Volume 457
ISSN 0076-6879, DOI: 10.1016/S0076-6879(09)05015-0

internal localization signals. Surprisingly, DSP18 and DSP21 are both periph-erally associated with the mitochondrial inner membrane, however, DSP18 is oriented toward the intermembrane space while DSP21 is facing the matrix compartment. This chapter describes methodology for purification of recombi-nant protein and demonstration of phosphatase activity, for mitochondrial purification and subfractionation of mitochondria to determine submitochon-drial localization and for determining membrane orientation and strength of membrane association.

1. INTRODUCTION

The PTPs are crucial regulators of signal transduction pathways in the cell that catalyze the removal of a phosphoryl group from proteins, lipids, and complex carbohydrates (Guan *et al.*, 1991; Maehama and Dixon, 1998; Tonks *et al.*, 1988; Worby *et al.*, 2006). Their ability to dephosphorylate both serine/threonine and tyrosine residues is due to a shallower active site cleft which allows these enzymes to accommodate multiple substrates *in vitro*. DSP18 and DSP21 are members of the atypical DSP subgroup of PTPs. This subgroup of phosphatases is typically 150–250 amino acids in length and lack clearly defined motifs beyond their highly conserved PTP catalytic motif Cys–X_5–Arg (CX_5R). Although similar to the MAP kinase phosphatases, the atypical DSPs do not contain Cdc25 homology domains or kinase interaction motifs (Alonso *et al.*, 2004). Because of their minimal sequence identity to other PTPs, it is difficult to predict what role these atypical DSPs play *in vivo*; therefore, localization of these phosphatases to particular compartments within the cell can help to distinguish possible functions.

The regulation of the pyruvate dehydrogenase complex by phosphory-lation was the first example of reversible phosphorylation participating in mitochondrial function (Linn *et al.*, 1969). Mutations within this complex, including the phosphatases that regulate the three phosphorylation sites of the E1α subunit, have been associated with several mitochondrial diseases (Maj *et al.*, 2006). However, since the discoveries of Linn *et al.* (1969) there have been a limited number of examples of protein kinases and phosphatases playing a direct role in mitochondrial biology. Most recently, the mito-chondrially localized kinase PTEN induced kinase 1 (PINK1) was shown to be mutated in a rare form of early-onset Parkinson's disease (Valente *et al.*, 2004) suggesting that the role of phosphorylation in the mitochondria has been underestimated.

We initially discovered the first PTP to be predominantly localized to the matrix compartment of mitochondria (Pagliarini *et al.*, 2005). Therefore, we began to explore the possibility that other PTPs may also be targeted to this organelle. Using green fluorescent protein, we tagged several poorly characterized DSPs and assessed their subcellular localization using immuno-fluorescence (Rardin et al., 2008). This approach allowed us to identify two DSPs that are localized to mitochondria. We demonstrated that the widely expressed DSP18 is peripherally associated with the mitochondrial inner membrane (MIM) facing the intermembrane space (IMS) compartment in highly purified rat kidney mitochondria. In contrast, the testis specific DSP21 was shown to be peripherally associated with the MIM, but facing the matrix compartment. This chapter describes assays and methods for demonstrating phosphatase activity, rat kidney gradient purified mitochondria, membrane orientation, and strength of membrane association.

2. ANALYSIS OF PHOSPHATASE ACTIVITY OF DSP18

The presence of the PTP catalytic motif CX_5R within the primary sequence of DSP18 and DSP21 indicates that these proteins are catalytically active phosphatases. We generated a DSP18–His_6 fusion protein to confirm that it is an active phosphatase. To do this we took advantage of the ability of phosphatases to dephosphorylate the phosphotyrosine analog p-nitrophenyl phosphate (pNPP). pNPP is a chromogenic substrate that is used in an endpoint assay whereby a phosphatase can remove the phosphate group from pNPP forming p-nitrophenol. The reaction is quenched by the addition of NaOH, which removes the phenolic proton and forms p-nitrophenolate. p-Nitrophenolate is yellow in color and can be measured spectrophotometrically at 410 nm.

2.1. Bacterial expression constructs and mutagenesis

To generate recombinant protein to test the phosphatase activity of DSP18, the full-length murine reading frame was first cloned into the pET21a vector (EMD Biosciences), which contains a C-terminal His_6 tag. The coding region of DSP18 (accession #NM_173745) was amplified using PCR with 5'-EcoRI and 3'-XhoI restriction enzyme sites designed into the primers. The PCR product is digested with EcoRI and XhoI (New England BioLabs) and ligated into the pET21a vector that has been linearized by digestion with EcoRI and XhoI at these sites and purified. Note that the 3' stop codon has not been included in the amplification product. This cloning effort led to the generation of pET21a–DSP18 with a carboxy-terminal hexa histidine tag (His_6) for use in Ni^+ affinity chromatography.

Generation of catalytically inactive mutants is a customary strategy for testing in vitro phosphatase activity. We replaced the catalytic cysteine in the CX_5R motif with a serine using the Quickchange site-directed mutagenesis (SDM) kit (Stratagene).

PCR primers for amplification of DSP18 and SDM to produce pET21a–DSP18–His$_6$ and pET21a–DSP18–His$_6$ (C104S) with the codon switched to encode a serine underlined were as follows:

Forward *Eco*RI DSP18: 5′-GGGAATTCGATTCTACGGAAAACCTG TATTTTC
Reverse *Xho*I DSP18: 5′-AACCTCGAGCAGTGGGATCATCAAACG GGT
SDM DSP18: 5′-ACACTGTTGCATTCTGCTGCTGGGGTG
SDM DSP18: 5′-CACCCCAGCAGCAGAATGCAACAGTGT

2.2. Expression and purification of recombinant DSP18

The reagents used were *lysis buffer* (50 mM Tris–HCl pH 8.0, 0.3 M NaCl, 10 mM imidazole, fresh 0.05% β-mercaptoethanol [BME] and fresh EDTA-free complete protease inhibitors [Roche]); *wash buffer* (50 mM Tris–HCl pH 8.0, 0.3 M NaCl, 20 mM imidazole, fresh 0.05% β-mercaptoethanol [BME] and fresh EDTA-free complete protease inhibitors [Roche]), and *elution buffer* (50 mM Tris–HCl pH 8.0, 0.3 M NaCl, 250 mM imidazole, fresh 0.05% β-mercaptoethanol [BME] and fresh EDTA-free complete protease inhibitors [Roche]). Triton X-100 (Sigma), Ni^{2+} agarose slurry (Qiagen), Glycerol (Sigma).

BL21 (DE3) Codon Plus RIL competent cells were transformed with 100 ng of pET21a–DSP18–His$_6$ and plated overnight at 37 °C on agar plates containing antibiotics (100 μg/ml ampicillin and 34 μg/ml chloramphenicol) (Sigma). Roughly 100 bacterial colonies were used to inoculate 1 l of 2×YT medium (1-l flask 16 g Bacto Tryptone, 10 g Bacto Yeast Extract, 5 g NaCl, pH 7.0) containing 100 μg/ml ampicillin and 34 μg/ml chloramphenicol. Cultures are grown at 37 °C with shaking until an OD$_{600\ nm}$ of 0.6–0.7 is reached. Cultures are then incubated on ice for 20 min. Fresh antibiotics are then added and protein expression is induced by addition of isopropyl-β-thiogalactopyranoside (IPTG) (Denville) to a final concentration of 0.5 mM. Cultures are then incubated overnight at room temperature with shaking.

Cells are harvested by centrifugation at 5000g for 10 min at 4 °C and the supernatant is discarded. The bacteria from a 1-l culture are re-suspended in 30 ml of lysis buffer (50 mM Tris–HCl pH 8.0, 0.3 M NaCl, 10 mM imidazole, fresh 0.05% β-mercaptoethanol [BME] and fresh EDTA-free complete protease inhibitors [Roche]) and disrupted by two passes through a Microfluidizer (Avestin) at 12–15 kpsi or alternatively a French pressure cell can be used. Triton X-100 is added to 0.05% and lysates are spun at 18,000g for 30 min at 4 °C to remove insoluble cellular debris. Lysates are then incubated with 0.75 ml Ni^{2+} agarose slurry (Qiagen) at 4 °C for 2 h to bind the His-tagged protein. *Note* wash agarose slurry twice with lysis buffer to remove ethanol and free Ni^{2+} before incubation with bacterial extracts.

The agarose beads are then washed five times with wash buffer (50 mM Tris–HCl pH 8.0, 0.3 M NaCl, 20 mM imidazole, fresh 0.05% BME, fresh protease inhibitors) containing 0.5% Triton X-100 for 5 min at 4 °C, followed by two 5-min washes with wash buffer without detergent. Recombinant DSP18–His$_6$ is then eluted from the Ni^{2+} agarose by washing with elution buffer (50 mM Tris–HCl pH 8.0, 0.3 M NaCl, 250 mM imidazole, fresh 0.05% BME, fresh protease inhibitors). The protein is eluted by two times 5 min washes (0.5 ml each) followed by three times 10 min washes (0.5 ml each). The supernatants containing the purified protein are combined and passed through a 0.2-μm filter. DTT (Fermentas) and glycerol are added to 2 mM and 25% (v/v), respectively. Samples can be used immediately or flash frozen in liquid nitrogen.

2.3. Use of the phosphotyrosine analog pNPP for determining phosphatase activity

To confirm that DSP18 is a catalytically active phosphatase, we tested its ability to dephosphorylate the phosphotyrosine analog pNPP (Sigma). Reactions are performed in the presences of 2 mM DTT, increasing concentrations of pNPP (0, 0.1, 0.5, 1.0, 5.0, 10.0, 25.0, 50.0) in 50 μl volumes of 1×-assay from a 5×-assay buffer stock solution (0.1 M sodium acetate, 0.05 M bis·Tris, 0.05 M Tris pH 6.0). Note you can use this assay to empirically determine the optimum pH for enzymatic activity by varying the pH of the assay buffer. Each sample is performed in triplicate in addition to a sample without enzyme to be used to subtract the background signal. All reagents are mixed except the enzyme and pre-warmed at 30 °C for ~5 min prior to the beginning of the assay. Five hundred nanograms of enzyme [DSP18–His$_6$ or DSP18 (C104S)–His$_6$] are then added and the reaction is allowed to proceed for 15 min at 30 °C. The reaction is terminated by addition of 200 μl of 0.25 M NaOH. Samples are then analyzed by a spectrophotometer at 410 nm and the background signal is subtracted. Using this method allowed us to easily demonstrate that DSP18–His$_6$ is a catalytically active phosphatase against pNPP (Fig. 15.1). However, when we mutate the catalytic cysteine residue, we abolish the activity.

3. ISOLATION OF HIGHLY PURIFIED RAT KIDNEY MITOCHONDRIA AND SUBFRACTIONATION

The observation that DSP18 targets to mitochondria when overexpressed in COS7 cells with a carboxy-terminal GFP tag led us to investigate where in the mitochondria endogenous DSP18 is localized (Rardin *et al.*, 2008). We generated an antibody against full-length recombinant DSP18

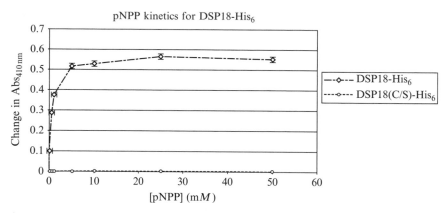

Figure 15.1 DSP18–His$_6$ and the catalytically inactive mutant DSP18(C/S)–His$_6$ were tested against increasing concentrations of pNPP at 30 °C for 15 min. The reaction was stopped with the addition of 0.25 M NaOH and absorbance was measured at 410 nm. Absorbance (Abs) versus substrate concentration was fitted to the graph.

and observed high levels of protein expression in kidney compared to other tissues. To determine DSP18's localization, we isolated crude mitochondria from rat kidney tissue using differential centrifugation (Section 3.1), followed by gradient purification to remove small vesicle contaminants, which enriched for highly purified mitochondria (Section 3.2). The double membrane structure of mitochondria creates multiple subcompartments to which proteins can be localized. At a minimum, these compartments include the mitochondrial outer membrane (MOM), IMS, MIM, and matrix. Therefore, we subfractionated our highly purified mitochondria and separated out individual mitochondrial compartments (Section 3.3) to examine DSP18's localization.

3.1. Differential centrifugation to isolate crude mitochondria

Kidney tissue is excised from 7 to 8 Sprague–Dawley adult rats and immediately placed into ice-cold MSHE + BSA buffer (210 mM mannitol, 70 mM sucrose, 5 mM HEPES [pH 7.4 using KOH], 2 mM EGTA, 0.5% fatty acid-free BSA, and EDTA-free complete protease inhibitor cocktail [Roche]). The tissue is rinsed twice in MSHE + BSA buffer to remove excess blood and hair and should remain on ice for the duration of the procedure. The renal capsule is removed and the tissue is minced and washed three times in MSHE + BSA buffer. After the tissue is sufficiently minced it is then homogenized using a Potter–Elvehjem tissue homogenizer (~10 g tissue/5 volumes of buffer, four passes at rheostat 70%). Remove a small aliquot (0.1–0.5 ml) of the homogenate to be used for Western blot analysis. To remove unbroken cells and large organelles, the homogenate is centrifuged at 600g for 10 min at 4 °C. The post-nuclear supernatant (PNS) is passed through cheesecloth and the pellet is re-suspended in MSHE + BSA buffer, homogenized and spun again at 600g for 10 min at 4 °C. Once again filter the supernatant through the cheese-cloth and discard the pellet. Remove a small aliquot of PNS for later WB analysis. The PNSs are combined and centrifuged at 15,000g for 10 min 4 °C to pellet crude mitochondria. Save a small aliquot of the post-mitochondrial supernatant (PMS) and then discard the PMS.

The crude mitochondrial pellet (dark brown in color) is gently dispersed into a minimal amount of MSHE + BSA buffer using a fine-tipped paintbrush. Following re-suspension of the pellet, additional MSHE + BSA buffer (~25 ml/pellet) is added and the mixture is spun at 15,000g for 10 min at 4 °C. The supernatant is removed, and using a fine-tipped aspirator the white layer containing microsomes and endoplasmic reticulum contaminants is removed. The removal of contaminating vesicles is preferred over the quanti-tative yield of mitochondrial protein. Wash the pellet as above and repeat once in MSHE + BSA buffer, followed by a wash using BSA-free MSHE (all subsequent steps will use MSHE buffer without BSA). Re-suspend the final crude mitochondrial pellet in a minimal amount of MSHE.

3.2. Histodenz gradient purification of mitochondria

To further remove remaining contaminants, we gradient purify our crude mitochondria to minimize the quantity of endoplasmic reticulum, peroxi-somes, and so on. Gradients are prepared in ultra-clear tubes (14 × 95 mm for an SW40 Ti swinging bucket rotor [Beckman]) by layering 3 ml of 17% histodenz on top of 2 ml of 35% histodenz prepared from a 70% wt/v stock of histodenz (Sigma) in MSHE buffer (prepare at least 1 day in advance at 4 °C and allow for considerable swelling of the histodenz). Five to six milliliters of Percoll (Sigma) diluted to 6% in MSHE are then layered on top of the 17% histodenz layer. Layer differential centrifugation purified mitochondria (≤30 mg/gradient) on top of the Percoll gradient that will fall through the top layer. Fill the tube with additional MSHE as needed to prevent collapse of the tube during high-speed centrifugation. Histodenz gradients are then spun at 45,500g for 45 min at 4 °C. Following centrifu-gation, collect the light mitochondrial layer between the 6% Percoll and 17% histodenz layers. Remove intervening material and collect the heavy mitochondrial layer between the 35% and 17% histodenz layers. Wash heavy and light fractions two times with 20–30 ml of MSHE buffer by centrifugation at 15,000g for 10 min 4 °C. Re-suspend the light membrane and heavy mitochondrial fraction in a minimal amount of MSHE. Samples from each step of the purification can be separated out by SDS–PAGE and blotted for markers of mitochondria and endoplasmic reticulum to demonstrate purity of the heavy mitochondrial fraction (Fig. 15.2).

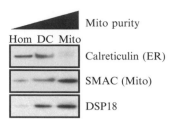

Figure 15.2 Evaluation of mitochondrial purity. Equal amounts of rat kidney homogenate (Hom), differential centrifugation purified mitochondria (DC), and histodenz gradient purified mitochondria (Mito) were separated out by SDS–PAGE and immunoblotted with anti-calreticulin (endoplasmic reticulum marker), anti-voltage-dependent anion channel (VDAC; mitochondrial marker), and anti-DSP18 antibodies. Modification of this figure was reprinted from Rardin *et al.* (2008) with permission.

3.3. Subfractionation of highly purified mitochondria

Mitochondria were placed in a hypotonic solution (10 mM KCl, 2 mM HEPES, pH 7.2) at a concentration of 2 mg/ml for 20 min on ice with gentle agitation. This causes the mitochondria to swell breaking off the MOM because of the greater surface area of the MIM. Addition of a one-third volume of hypertonic solution (1.8 mM sucrose, 2 mM ATP, 2 mM MgSO$_4$, 2 mM HEPES, pH 7.2) then shrinks the matrix and MIM back down. Sonicating the mitochondria for 15 s at 3 A with a small probe sonicator then largely disrupts the MOM/MIM contact sites. Although this step is used to release MOM fragments from the intact matrix and inner membrane, and thus increase yield of MOM, this step should be optimized for duration and strength of sonication to avoid rupture of the MIM. Consequent release of matrix contents can be monitored by Western blotting for matrix proteins in the IMS fraction. The swollen/shrunk mitochondrial solution is then layered on top of a step-wise sucrose gradient containing 0.76, 1, and 1.32 M sucrose (8 ml each) in ultra-clear tubes (25 × 89 mm for an SW28 swinging bucket rotor [Beckman]). Gradients are spun at 75,000g for 3 h at 4 °C. This allows for isolation of the IMS-soluble fraction collected from the uppermost supernatant. The MOM is isolated from the 0.76 to 1 M sucrose interface, washed with MSHE and pelleted by centrifugation at 120,000g. Mitoplasts (MP—intact matrix and MIM) were collected from the pellet, washed with MSHE and pelleted by centrifugation at 15,000g 10 min at 4 °C.

Purified MP can also be used to generate outer membrane and matrix depleted submitochondrial particles (SMPs) instead of using heavy mitochondria to make conventional SMPs (Pedersen and Hullihen, 1978). We found this to be convenient when the yield of gradient purified mitochondria is limited. SMPs in this experiment are inside-out inner membranes. Re-suspend mitochondria at a concentration of ∼10 mg/ml in a small beaker on ice. Mitochondria are sonicated $3\times$ 2 min at 50% power on ice with 1-min intervals in between sonication steps to allow the mixture to cool. The solution should transition to a darker color and become more translucent. This solution is then spun at 15,000g for 10 min at 4 °C to remove intact MP. The supernatant is then spun at 120,000g for 45 min at 4 °C to pellet SMP. The supernatant contains the soluble matrix (SM) fraction. Wash the SMP pellet with MSHE then re-suspend using a microfuge homogenizer to aid in disrupting the compact SMP pellet. This collection of fractions: IMS, MOM, MP, SMP, and SM were separated out by SDS–PAGE and a variety of antibodies were used against known mitochondrial marker proteins. Our results showed that DSP18 is enriched in MP and inner membrane SMP demonstrating that DSP18 is associated with the MIM of kidney mitochondria (Fig. 15.3B). Similar results were obtained for DSP21 from mitochondria isolated from rat testis tissue (Rardin *et al.*, 2008). Note that a similar mitochondrial isolation procedure was used for testis tissue; however, a larger number of animals (10) were required for sufficient quantities to perform the subfractionation procedure.

4. MITOCHONDRIAL INNER MEMBRANE ASSOCIATION OF DSP18 AND DSP21

We observed that in mitochondria isolated from testes. To determine which compartment or membrane of mitochondria DSP18 is being localized to, we further subfractionated mitochondria (Fig. 15.3A). Gradient purified and kidney tissue that DSP18 and DSP21 were both being localized to the MIM. Mitochondrial compartments can vary quite distinctly in their function and signaling processes. Therefore, we sought to distinguish whether DSP18 and DSP21 were oriented on the matrix side or the IMS side of the MIM. Furthermore, because DSP18 and DSP21 are membrane associated, we evaluated the strength of their membrane association.

4.1. Trypsin digestions of mitochondrial inner membranes

Mitochondrial fractions are isolated fresh; do not freeze your samples prior to protease treatment. Freezing of samples will cause rupture of the membranes and allow the protease access to both sides of the membrane. This method will allow you to determine which side of the MIM your protein of interest is facing:

1. Two hundred micrograms of mitoplasts or submitochondrial particles are spun down (mitoplasts at 15,000g for 10 min at 4 °C and SMP at 120,000g for 15 min at 4 °C). We typically use 200 μg of sample; however, this can be altered depending on the availability of starting sample.

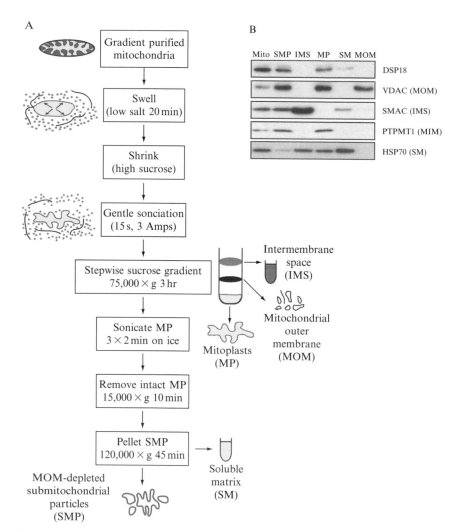

Figure 15.3 (A) Schematic diagram and flowchart of mitochondrial fractionation procedure of histodenz gradient purified rat kidney mitochondria. (B) Equal amounts of each submitochondrial fraction were separated out by SDS–PAGE and immunoblotted with markers to look for enrichment of soluble matrix (SM; anti-Hsp70, heat-shock protein 70), inner membrane (IM; anti-PTPMT1), intermembrane space (IMS; second mitochondria-derived activator of caspase), mitochondrial outer membrane (MOM; VDAC). VDAC is present at contact sites between the MIM and MOM, so it is common to see it in both fractions. Modifications of these figures were reprinted from Rardin *et al.* (2008) with permission.

2. Four separate samples are re-suspended in 100 μl trypsin buffer (0.125 M sucrose, 10 mM Tris–HCl pH 8.0) containing 2.5 μg of trypsin (Sigma) and incubated at 37 °C for 0-, 5-, 10-, and 20-min intervals.

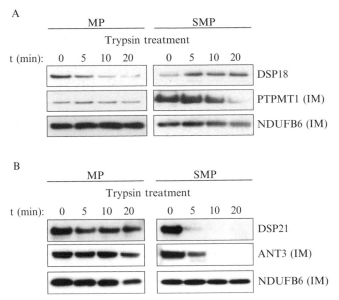

Figure 15.4 (A) MP and SMP isolated from rat kidney mitochondria were treated with trypsin for the indicated amounts of time. Samples were separated by SDS–PAGE and immunoblotted with anti-DSP18 and anti-PTPMT1. The Complex I subunit NDUFB6 was not susceptible to trypsin digestion and was used as a loading control. MP treated with trypsin showed a decrease in signal for DSP18; however, there was no change in the matrix-oriented PTP-MT1. Alternatively SMP treated with trypsin did not show a decrease in DPS18, demonstrating that DSP18 is a MIM protein facing the IMS compartment. (B) MP and SMP isolated from rat testis mitochondria were treated with trypsin for the indicated amounts of time. Surprisingly DSP21 did not show of a loss of signal in MP but did show a decrease in SMP suggesting that DSP21 is a MIM protein facing the matrix compartment in testis mitochondria. Modifications of these figures were reprinted from Rardin *et al.* (2008) with permission.

3. Reactions are quenched with the addition of 10 μg soybean trypsin inhibitor (Sigma) on ice for 5 min.
4. Laemmli buffer is added to the reactions and the samples are separated by SDS–PAGE and analyzed by Western blotting with the appropriate antibodies (Fig. 15.4).

4.2. Integral versus peripheral membrane protein

Classically, membrane proteins are considered to be either peripheral or integral, the latter being defined as permanently attached to the membrane typically via a transmembrane domain. To determine the strength of protein–membrane associations, membranes can be treated with either a high salt wash (200 mM KCl, 2 mM HEPES, pH 7.2) which can disrupt

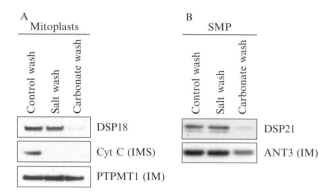

Figure 15.5   (A) To determine the strength of association of DSP18 to the MIM, we washed MP with high salt (200 m*M* KCl, 2 m*M* HEPES, pH 7.2), an alkaline wash (0.1 *M* Na$_2$CO$_3$, pH 11.5), or control buffer. Samples were separated out by SDS–PAGE and immunoblotted with antibodies against DSP18, cyt *c*, and PTPMT1. DSP18 was released following the alkaline wash, unlike the integral membrane protein PTPMT1, suggesting that DSP18 is a peripheral membrane protein. (B) Similarly DSP21 was also released from the MIM when treated with the alkaline wash. Suggesting that DSP21 is also a peripheral membrane protein. Modifications of these figures were reprinted from Rardin *et al.* (2008) with permission.

weak ionic membrane associations or a more stringent alkaline wash (0.1 *M* Na$_2$CO$_3$, pH 11.5) which causes the membranes to flatten and release peripherally associated membrane proteins (Fujiki *et al.*, 1982). Using these two membrane treatments, we were able to determine that DSP18 and DSP21 are peripherally associated membrane proteins of the MIM (Fig. 15.5).

1. One hundred microgram of freshly isolated MP or SMPs are spun down (MP at 15,000*g* for 10 min at 4 °C and SMP at 120,000*g* for 15 min at 4 °C).
2. Re-suspend MP or SMP in control wash (MSHE), salt wash (200 m*M* KCl, 2 m*M* HEPES, pH 7.2), or alkaline wash (0.1 *M* Na$_2$CO$_3$, pH 11.5).
3. Samples are incubated on ice for 20 min with gentle agitation.
4. Samples are pelleted (MP at 15,000*g* for 10 min at 4 °C and SMP at 120,000*g* for 15 min at 4 °C) and washed once in MSHE.
5. Samples are re-suspended in Laemmli buffer and equal amounts of each sample are separated by SDS–PAGE and immunoblotted.

ACKNOWLEDGMENTS

We thank Anne Murphy, Sandra Wiley, and members of the Dixon laboratory for help in preparation of this manuscript. NIH Grant 18849 and the Walther Cancer Institute (to J.E.D). M.J.R was supported by Pharmacology Training Grant NIH 2 T32 GM07752–25.

REFERENCES

Alonso, A., Rojas, A., Godzik, A., and Mustelin, T. (2004). The dual-specific protein tyrosine phosphatase family. *Top. Curr. Genet.* **5,** 333–358.

Fujiki, Y., Hubbard, A. L., Fowler, S., and Lazarow, P. B. (1982). Isolation of intracellular membranes by means of sodium carbonate treatment: Application to endoplasmic reticulum. *J. Cell Biol.* **93,** 97–102.

Guan, K. L., Broyles, S. S., and Dixon, J. E. (1991). A Tyr/Ser protein phosphatase encoded by vaccinia virus. *Nature* **350,** 359–362.

Linn, T. C., Pettit, F. H., and Reed, L. J. (1969). Alpha-keto acid dehydrogenase complexes, X. Regulation of the activity of the pyruvate dehydrogenase complex from beef kidney mitochondria by phosphorylation and dephosphorylation. *Proc. Natl. Acad. Sci. USA* **62,** 234–241.

Maehama, T., and Dixon, J. E. (1998). The tumor suppressor, PTEN/MMAC1, dephosphorylates the lipid second messenger, phosphatidylinositol 3,4,5-trisphosphate. *J. Biol. Chem.* **273,** 13375–13378.

Maj, M. C., Cameron, J. M., and Robinson, B. H. (2006). Pyruvate dehydrogenase phosphatase deficiency: Orphan disease or an under-diagnosed condition? *Mol. Cell Endocrinol.* **249,** 1–9.

Pagliarini, D. J., Wiley, S. E., Kimple, M. E., Dixon, J. R., Kelly, P., Worby, C. A., Casey, P. J., and Dixon, J. E. (2005). Involvement of a mitochondrial phosphatase in the regulation of ATP production and insulin secretion in pancreatic beta cells. *Mol. Cell* **19,** 197–207.

Pedersen, P. L., and Hullihen, J. (1978). Adenosine triphosphatase of rat liver mitochondria. Capacity of the homogeneous F1 component of the enzyme to restore ATP synthesis in urea-treated membranes. *J. Biol. Chem.* **253,** 2176–2183.

Rardin, M. J., Wiley, S. E., Murphy, A. N., Pagliarini, D. J., and Dixon, J. E. (2008). Dual specificity phosphatases 18 and 21 target to opposing sides of the mitochondrial inner membrane. *J. Biol. Chem.* **283,** 15440–15450.

Tonks, N. K., Diltz, C. D., and Fischer, E. H. (1988). Characterization of the major protein-tyrosine-phosphatases of human placenta. *J. Biol. Chem.* **263,** 6731–6737.

Valente, E. M., Abou-Sleiman, P. M., Caputo, V., Muqit, M. M., Harvey, K., Gispert, S., Ali, Z., Del Turco, D., Bentivoglio, A. R., Healy, D. G., Albanese, A., Nussbaum, R., *et al.* (2004). Hereditary early-onset Parkinson's disease caused by mutations in PINK1. *Science* **304,** 1158–1160.

Worby, C. A., Gentry, M. S., and Dixon, J. E. (2006). Laforin, a dual specificity phosphatase that dephosphorylates complex carbohydrates. *J. Biol. Chem.* **281,** 30412–30418.

Monitoring Mitochondrial Dynamics with Photoactivateable Green Fluorescent Protein

Anthony J. A. Molina *and* Orian S. Shirihai

Contents

1. Mitochondrial Dynamics	290
1.1. Monitoring dynamics	291
2. PAGFPmt	292
2.1. Photoconversion	292
2.2. Targeting mitochondrial matrix	292
2.3. Generating lentiviral and adenoviral vectors	293
3. Tracking Individual Fusion and Fission Events	293
3.1. Co-labeling mitochondria with TMRE versus dsRED versus Mitotracker Red	294
3.2. Spatially precise photoconversion with 2-photon laser	295
3.3. Confocal imaging	296
4. Quantifying Networking Activity in Whole Cells: Whole Cell Mitochondrial Dynamics Assay	296
4.1. Photoconversion of mitochondrial subpopulations for the whole cell mitochondrial dynamics assay	297
4.2. Time lapse confocal imaging for cellular mitochondrial dynamics quantification	298
4.3. Analysis	299
5. Potential Artifacts and Important Controls	301
5.1. Phototoxicity generated by 2-photon laser excitation	301
5.2. Photobleaching and saturation	301
5.3. Thresholding	302
6. Comparison with an Alternative Method	302
Acknowledgments	303
References	303

Department of Medicine, Boston University, Massachusetts, USA

Methods in Enzymology, Volume 457
ISSN 0076-6879, DOI: 10.1016/S0076-6879(09)05016-2

Abstract

Mitochondria are dynamic organelles that undergo continuous cycles of fusion and fission. Monitoring and quantification of mitochondrial dynamics has proved to be challenging because these processes are distinctly different from movement and apposition. While the majority of contact events do not lead to fusion, fission can occur without translocation, leaving the two mitochondria juxtaposed. The advent of photoactivateable fluorescent proteins has enabled researchers to distinguish mitochondrial fusion and fission. These genetically encoded fluorophores can be targeted to the mitochondrial compartments of interest to visualize how these intermix and segregate between dynamic mitochondria over time. The PAGFPmt-based mitochondrial dynamics assay has proved to be a powerful technique for revealing the treatments and cellular processes that affect fusion and fission. By using this technique in combination with other parameters, such as measurements of mitochondrial membrane potential, we have begun to understand the processes that control fusion and fission as well as the significance of mitochondrial dynamics.

1. MITOCHONDRIAL DYNAMICS

Mitochondrial dynamics refers to two distinct processes that are both essential for normal cellular function. First, dynamics can refer to the movement and redistribution of mitochondria. For example, mitochondria must redistribute during mitosis to populate both daughter cells. Second, mitochondrial dynamics also refer to processes of fusion and fission that mitochondria continuously undergo. Fusion of both the outer and inner membranes of mitochondria leads to the intermixing of membrane proteins. In addition, matrix solutes and mitochondrial DNA are able to mix upon fusion of two mitochondria. These processes underlie complementation processes that play an important role in the maintenance of the mitochondrial population. On the cellular level, mitochondrial dynamics affects respiration, calcium handling, mitosis, and apoptosis (Detmer and Chan, 2007; Herzig and Martinou, 2008; Jeong and Seol, 2008; Szabadkai et al., 2006). At the level of the organism, mitochondrial dynamics is involved in disease, aging, and development (Chan, 2006; Chen et al., 2003; Zuchner et al., 2004).

The number, shape, and size of mitochondria are mediated by the balance of fusion and fission. Under most physiological conditions, mitochondrial shape and size remain relatively constant (within 1 or 2 μm) because fusion and fission events occur in pairs (Twig et al., 2008). When perturbations in mitochondrial dynamics occur, either favoring fission or fusion, dramatic changes in mitochondrial morphology have been observed (Bereiter-Hahn and Voth, 1994; Hoffmann and Avers, 1973; Mozdy and Shaw, 2003; Nunnari et al., 1997; Scott et al., 2003).

In recent years, much interest has been devoted to studying the proteins, processes, and treatments that affect mitochondrial fusion and fission. To this end, various tools have been developed to observe, monitor, and quantify mitochondrial dynamics. The quantification of fusion in particular can be complicated because it is necessary to differentiate between apposition, partial fusion, and full fusion. This is further complicated by the finding that fusion of the inner and outer mitochondrial membranes is independent events (Malka *et al.*, 2005). For these reasons, looking at morphological changes alone, at the light microscopy resolution, does not allow for the measurement of fusion and fission events. By using a photoactivateable form of GFP (PAGFP), we have been able to differentiate between these events. Furthermore, simultaneous measurements of other biophysical parameters add another level of understanding and can aid in the differentiation of certain fission and fusion events (Twig *et al.*, 2006).

PAGFP can be targeted to different parts of the mitochondrion in order to examine the mixing properties that underlie fusion events between these organelles. Once targeted, spatially precise photoconversion can be used to induce GFP fluorescence. The movement and dilution of photoconverted GFP within an individual mitochondrion or mitochondrial population can then be tracked by time lapse confocal imaging. We have found that targeting PAGFP to the mitochondrial matrix is useful for assessing size and fusion because the fluorophore diffuses freely in this space. When full fusion occurs and the matrix temporarily becomes continuous between mitochondria, the fluorophore is transferred between organelles and provides an indicator for the establishment of matrix continuity and the transfer of matrix solutes. Mitochondrial DNA in the matrix has also been shown to be redistributed between mitochondria upon fusion. If PAGFP is targeted to a membrane bound protein, it is possible to visualize how this protein becomes redistributed when two mitochondria fuse. For example, we have observed that redistribution of the inner membrane bound protein, ABC-me, does not occur at the same time as matrix solute equilibration (Twig *et al.*, 2006).

1.1. Monitoring dynamics

In recent years, imaging based methodologies for the observation of mitochondrial fusion and fission have allowed investigators to begin characterizing the processes that control mitochondrial dynamics. The use of genetically encoded fluorescent proteins has further enabled investigators to observe fusion events between mitochondria. Full fusion is characterized by luminal continuity and is responsible for the passage of DNA, proteins, and matrix solutes between mitochondria. By targeting fluorescent indicators to the mitochondrial matrix, it is possible to observe these complete fusion events by imaging the passage of fluorophores between mitochondria.

2. PAGFPмт

Targeting PAGFP to the mitochondrial matrix delineates the borders of the mitochondrial inner membrane. By photoconverting regions within a mitochondrion with a 2-photon laser, photoconverted GFP molecules in the matrix will trace the extent of luminal continuity (Twig *et al.*, 2006) as GFP molecules move freely through the matrix space. The movement of GFP within this space is not hindered despite protein density and high viscosity of the matrix (Partikian *et al.*, 1998). In addition to quantifying mitochondrion size, the diffusion ability of GFP molecules within the mitochondrial matrix can be used to observe mitochondrial fusion events in real time. PAGFPmt can be used alone or in combination with other probes for a number of different applications that can measure the following parameters; mitochondrial movement, membrane potential of individual mitochondria over time, fusion frequency, fusion site/localization, fusion rate of a cell's mitochondrial population, and the transfer and organization of proteins in fusing mitochondria. Although this review will focus on the quantification of mitochondrial dynamics, the methodologies described can be easily applied to the measurement of all these parameters.

2.1. Photoconversion

The photoactivateable form of GFP increases fluorescence intensity (FI) 100-fold after irradiation with 413 nm light (Patterson and Lippincott-Schwartz, 2002). The development of a photoactivatable GFP that is useful at physiological conditions has opened new doors in the study of temporal and spatial dynamic interactions within a cell. Combined with 2-photon laser stimulation, it is possible to specifically stimulate individual organelles within a living cell and to monitor its interactions with other organelles.

Wild-type GFP is a mixed population of fluorophores with a major and minor absorbance peaks at 397 and 475 nm, respectively. Intense illumination with ultraviolet light causes the fluorophore population to give rise to the anionic form which demonstrates an increase in the minor peak absorbance. This causes an increase in fluorescence with subsequent 488 nm excitation. PAGFPmt is a variant that possesses a minor absorbance peak (475 nm) that is significantly lower than wild type. This further enlarges the increase in fluorescence emission detected following photoconversion if excitation is done with a 488 nm laser.

2.2. Targeting mitochondrial matrix

We added a mitochondrial matrix targeting sequence to PAGFP cDNA (Karbowski *et al.*, 2004; Twig *et al.*, 2006). DNA coding for the mitochondrial targeting sequence of COXVIII was amplified by PCR and inserted 5′

to GFP thereby targeting it to the mitochondrial matrix. Transfection of this construct works well in many systems such as COS7 cells, primary human myocytes, hippocampal neurons, and MEF cells (Karbowski *et al.*, 2004). Expression of PAGFPmt becomes evident after 48 h. PAGFPmt expression can be visualized by eye with blue light excitation and green emission. Alternatively expression can be verified by western blot analysis with GFP antibody. Transfection is a stressful treatment and in some cells may lead to a change in mitochondrial architecture and dynamics. If cells are transfected with the PAGFPmt plasmid using lipofection, it is recommended that mitochondrial architecture of transfected and non-transfected cells be compared. We have observed some level of mitochondrial fragmentation to occur due to the stress of lipofection in the clonal beta cell line, INS1. To prevent lipofection induced stress, we generated lentiviral and adenoviral vectors. Although the initial infection may cause some degree of cell death (1–10%) after 48 h, this becomes less evident over time and is not observed in subsequent passages of the cell line.

2.3. Generating lentiviral and adenoviral vectors

We have packaged PAGFPmt for lentiviral and adenoviral delivery using pWPI (Trono) or pAdEasy (Adenoeasy), respectively. We find lentiviral transduction highly efficient in cell lines such as INS1 while adenoviral transduction exhibits better efficiency in primary preparations such as beta cells from the islets of Langerhans. In addition, lentiviral transduction allows the PAGFPmt to integrate into the host genome. We have found expression to be stable for as many as 10 passages. Freezing the cells and storing in liquid nitrogen leads to noticeably lower expression when the cells are thawed for use. This may be due to selection influences during the freeze thaw cycle.

3. TRACKING INDIVIDUAL FUSION AND FISSION EVENTS

By tagging individual mitochondria with photoconverted PAGFPmt, we have observed individual fusion events (Fig. 16.1). These events occur under normal conditions and without stimulation or stress. By generating time lapse z-stacks we find that we are able to capture these events and document and quantify their occurrence. A fusion event is characterized by the transfer of photoconverted PAGFPmt molecules from the tagged mitochondrion to another previously unlabeled unit (Fig. 16.1). Fission events typically follow fusion events and are characterized by the loss of PAGFPmt continuity. We have observed that the average duration of a fusion event is ~1 min. It is notable that fission can occur without a change in the

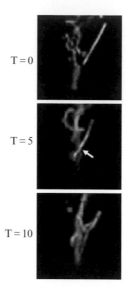

Figure 16.1 Mitochondrial fusion results in the transfer of PAGFPmt. At time zero, a photolabeled PAGFPmt positive mitochondrion is depicted as an isolated unit within an INS1 cell. Five min later, this mitochondrion has moved and undergone fusion with an adjacent mitochondrion. The arrow points to the presumptive fusion site. PAGFPmt has been transferred to another mitochondrion which now appears yellow due to the mixing of the green and red fluorophore. At $T = 10$ min, more fusion events have occurred and the average PAGFPmt fluorescence intensity has gone down due to dilution of photoconverted molecules. (See Color Insert.)

apposition of the two daughter mitochondria, a process we refer to as "hidden fission" (Twig *et al.*, 2006). We find that fission events often generate daughter mitochondria with disparate membrane potential that can be appreciated when using a potential sensitive dye such as TMRE. Daughter mitochondria resulting from a fission event will appear more red when hyperpolarized and stained with TMRE or more green when depolarized due to the presence of PAGFPmt. Therefore, some "hidden fission" events can be identified by the two daughter mitochondria having disparate changes in membrane potential.

3.1. Co-labeling mitochondria with TMRE versus dsRED versus Mitotracker Red

Mitochondria were labeled with the mitochondrion-specific dye tetramethylrhodamine ethyl ester perchlorate (TMRE; Invitrogen). TMRE concentration should be adjusted for the cell type with a lower concentration being preferred. Keep in mind that laser toxicity is proportional to the dye concentration in the mitochondria. Typically, for freshly isolated primary cells, 3–5 nM should be sufficient; immortalized cell lines may require higher concentrations, 7–15 nM. Freshly prepared TMRE was added to culture in DMSO to give a final concentration and incubated for 45 min in a 37 °C incubator before imaging. Cells loaded with TMRE should be kept in dark to avoid phototoxicity. At the end of the loading period, the dye is not removed from the media. TMRE can be used to dynamically monitor membrane potential in mitochondria. Increases in TMRE fluorescence indicate hyperpolarization while decreases report depolarization. Since membrane potential influences mitochondrial fusion, it is expected that mitochondria with reduced TMRE intensity will have reduced probability for a fusion event within the duration of the experiment. During a fission event, the concentration of matrix targeted PAGFPmt in the two daughters is identical. It is therefore possible to use the ratio (R) of TMRE/PAGFPmt for ratio imaging and comparison of membrane potential between the two daughter mitochondria generated during the fission event. The membrane potential difference between daughter (a) and daughter (b) can be calculated in millivolts ($\Delta\psi = 61.5$ $\text{Log}(R_a/R_b)$) in experiments performed at 37 °C.

Other fluorophores such as the dsRED protein and Mitotracker Red dye (MTR, Invitrogen) can be used to identify and characterize non-fusing mitochondria. In mitochondria, we have observed that slight increase in the intensity of 2-photon laser (750 nm) will result in dsRED bleaching during the photoconversion of PAGFPmt. This characteristic can be used to identify non-fusing mitochondria, because these will have very high dsRED fluorescence. In addition, cells expressing mitochondrial dsRED can be fixed with 4% paraformaldehyde for 15 min while preserving fluorescence and mitochondrial architecture. This allows the user to further characterize the non-fusing mitochondrial subpopulation. For example, using an antibody to probe for the mitochondrial fusion protein OPA1 in fixed cells, we have found that OPA1 expression is decreased in the non-fusing population (Twig *et al.*, 2008). MTR loading into mitochondria is dependent on $\Delta\psi$. Therefore, short pulses of MTR exposure can be used to identify polarized mitochondria versus those that are depolarized. Once the dye is loaded, it does not leave the mitochondria during fixation allowing further characterization of MTR stained mitochondria by immunofluorescence.

3.2. Spatially precise photoconversion with 2-photon laser

Cells transfected or virally transduced with PAGFPmt should be allowed to accumulate the protein in the mitochondrial matrix for 48 h. A transition to its active (fluorescent) form is achieved by photoisomerization with a 2-photon laser (750 nm) to give a 375-nm photon equivalence at the focal plane. This allows for selective photoconversion of areas as small as 0.5 μm^2 with a thickness of less than 0.5 μm. In the absence of photoconversion, PAGFPmt protein molecules remained stable in their pre-converted form. The presence of pre-converted PAGFPmt was detected with high-intensity excitation at 488 nm (25-mW laser set at 1%) in combination with a fully opened pinhole. Spatially precise laser excitation can be used to label individual segments of the mitochondrial network at a time. The extent that photoconverted PAGFPmt is able to travel within a mitochondrion can be measured in order to quantify the size distribution of mitochondrial populations.

3.3. Confocal imaging

Confocal microscopy was performed on live cells in glass bottomed dishes (MatTek, Ashland, MA) with a Zeiss LSM 510 Meta microscope with a plan apochromat 100× (numerical aperture 1.4) oil immersion objective. Three configurations were set using the multitrack mode: one for detection of the pre-converted PAGFPmt (higher 488 nm intensity), a second for photoconversion (750 nm with 2P laser), and a third for recording photoconverted PAGFPmt (low intensity 488 nm). Red-emitting TMRE was excited with a 1-mW, 543 nm helium/neon laser set at 0.3%, and emission was recorded through a BP 650–710 nm filter. Photoconverted PAGFPmt protein was excited with a 25-mW, 488 nm argon laser set between 0.2 and 0.5%. Emission was recorded through a BP 500–550 nm filter.

4. Quantifying Networking Activity in Whole Cells: Whole Cell Mitochondrial Dynamics Assay

PAGFPmt can be similarly used to monitor and quantify networking activity in a whole cell (Karbowski et al., 2004). By photoconverting PAGFPmt in a subpopulation of mitochondria, we have been able to observe the spread of photoconverted PAGFPmt signal throughout a cell via fusion and fission events and by mitochondrial movement as well (Fig. 16.2). Fusion events not only lead to the spread of the photoconverted PAGFPmt across the networking population, but it also leads to a dilution in the concentration of photoconverted molecules. This is translated into a reduction in the average GFP fluorescent intensity in the mitochondria that carry the photoconverted form. Therefore, by monitoring the decrease in PAGFPmt fluorescence intensity over time, we can distinguish fusion events that result in the transfer of PAGFPmt between mitochondria from the spread of PAGFPmt due to mitochondrial movement alone. This type of analysis can be used to compare the rate of mitochondrial dynamics between cells and due to various treatments. For example, we have reported

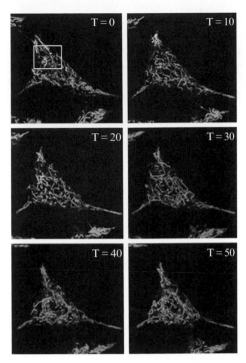

Figure 16.2 Monitoring mitochondrial fusion within the entire population. At time zero, the white box designates the area targeted with 2-photon laser for photoconversion of PAGFPmt to an active form. Notice that by the time the imaging acquisition has started, noticeable movement of mitochondria has already occurred. Subsequent images are obtained every 10 min for 50 min. Images are a single slice through the middle of an INS1 cell. The PAGFPmt signal is diluted over time leading to the appearance of yellow mitochondria. The yellow mitochondria appear more red over time due to further dilution of the green fluorophore. (See Color Insert.)

that mitochondrial fusion is halted in pancreatic beta cells with exposure to toxic nutrient levels.

4.1. Photoconversion of mitochondrial subpopulations for the whole cell mitochondrial dynamics assay

The size of the mitochondrial subpopulation to be photoconverted should be kept constant if the user wishes to compare the rate of fusion between different conditions or cells. Photoconverting larger subpopulations will lead to shorter equilibration times. Two numerical values can be used to quantify the rate of mitochondrial dynamics:

A. The extent of dilution after a specified period of time (30 min or 1 h)
B. Time to steady state (equilibration time), defined by time after which no further dilution is measured

Although the size of the area of photoconversion can be kept constant by using the same zoom value for activation, the number of mitochondria and size of the photoconverted population can still vary. This is due to the ability of matrix targeted GFP molecules to diffuse freely through any mitochondria with interconnected lumen and variations in the density of mitochondria. We find that with INS1 cells, activating an area that is 20% of the total cell area with 2P laser will provide an average equilibration time of around 45 min.

The same laser settings used for the monitoring of single mitochondria can be used for activating subpopulations. However, it is important to ensure that the 2-photon laser intensity is sufficient to photoconvert GFP while leaving the TMRE signal intact. The loss of TMRE fluorescence is indicative of phototoxicity and mitochondrial depolarization.

For our experiments, we used a Coherent Mira 900 fs laser (Santa Clara, CA). We determined the minimum intensity and duration of laser exposure that initiated changes in $\Delta\psi_m$ and/or mitochondrial morphology in cells treated with TMRE. The parameters utilized in the reported experiments were well below these thresholds. To determine the safety limits of 2-photon laser stimulation in INS1 cells, excitation was delivered over a wide range of intensities and durations. We found that excitation for 600 ms/μm^2 at 1 mW laser intensity at the objective is the threshold dosage for INS1 and COS7 cells above which a reduction in mitochondrial membrane potential can be observed. All subsequent experiments using 2-photon illumination were conducted with duration of 150 ms/μm^2 and an intensity of 1 mW. Due to variability in laser output, it is suggested that the user determine these values for the particular system being used. These intensity values can be used as a starting point and fine tuned.

4.2. Time lapse confocal imaging for cellular mitochondrial dynamics quantification

It is sufficient to collect six images from different focal planes at each time point (this is compared to 20 images or more that would be required for 3D reconstruction) because the extent of fusion activity is derived from the dilution of the photoconverted PAGFPmt. After photoconversion, a z-stack of six images is collected every 5 min for 50 min. This can be adjusted to ensure that photobleaching or phototoxicity does not reduce the cellular PAGFPmt or TMRE fluorescent intensity. It is conceivable that PAGFPmt bleaching may contribute to a decrease in PAGFPmt signal over time. This would present an artifact in the analysis and quantification of PAGFPmt dilution. When fusion is inhibited, we find that the PAGFPmt intensity/(pixel area) remains stable over 50 min. For this measurement pixel area is defined as the total area of photoconverted PAGFP. Figure 16.3 shows that without fusion and dilution of PAGFPmt, there is no bleaching due to repeated excitation and no loss of fluorescence intensity over a period of 50 min.

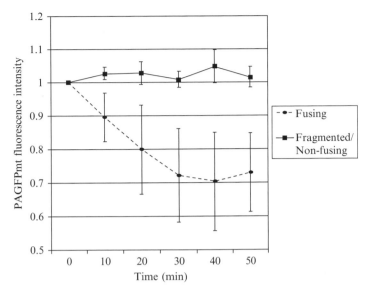

Figure 16.3   The average intensity of PAGFPmt fluorescence is monitored in order to assess mitochondrial fusion activity. An area encompassing 20% of the total cell is targeted with 2-photon laser activation in order to convert PAGFPmt into its active form. Z-stacks consisting of six slices each are taken every 10 min for 50 min. The average PAGFPmt fluorescence intensity is measured and normalized to time zero. The decrease in fluorescence indicates that under normal conditions, when INS1 cell mitochondria form elongated elaborate structures, fusion occurs between mitochondria ($n = 6$). When we induce apoptosis and mitochondria become fragmented, there is no observed dilution of the PAGFPmt signal ($n = 6$). Notice that the fluorescence intensity remains stable throughout the recording time indicating that photobleaching does not occur using the image acquisition parameters.

4.3. Analysis

Monitoring the dilution of photoconverted PAGFPmt is an efficient way of quantification the sharing of GFP between mitochondria (Fig. 16.3). Theoretically, when one mitochondrion carrying a matrix targeted photoconverted PAGFPmt fuses with another, the number of photoconverted molecules equilibrates between the two units and each ends up with half, causing a decrease in fluorescence intensity.

4.3.1. Monitoring GFP dilution

Quantification of fusion was performed using Metamorph (Molecular Devices CA) by measuring the average FI of the mitochondria that became PAGFPmt positive. The procedure involved first the elimination of non-mitochondrial pixels from the green (PAGFPmt) image followed by the measurement of green FI from mitochondria that were PAGFPmt and TMRE positive.

Prior to measuring FI, we used an "Integrated Morphometry Analysis" function designed for these experiments in order to extract TMRE (or dsRED) positive structures that were larger than 10 pixels. These areas were interpreted as mitochondria, and their PAGFPmt was recorded. This procedure enabled the selection of mitochondrial structures from which PAGFPmt was measured using very low threshold levels in the green channel (approximately 10% of the image average intensity) assuring that over 90% of the mitochondrial pixels were included for analysis. We verified that all intensity measurements were below saturation, see Section 5.2.

A low threshold ($\sim$10%) was applied to the green channel to identify the PAGFPmt positive mitochondria. Average FI (PAGFPmt) was measured from thresholded areas using Region Measurement. To set the threshold level, a test-threshold function first measured the average green FI of the mitochondria. The lower (inclusive) threshold was set at two thirds of this average. An upper threshold was not necessary since saturated images were carefully avoided during collection.

The FI values of PAGFPmt at each time point were normalized to the FI value immediately after photoconversion and then fitted to a hyperbolic function:

$$F(t) = 1 - F_{plateau}t/(t + T_{50})$$

F and $F_{plateau}$ denote FI at time t and in the plateau phase. T_{50} denote the time interval to a 50% decrease in normalized GFP FI ($[1 - F_{plateau}]/2$). All fitting procedures and statistical tests were conducted using Kaleida–Graph software (Synergy Software, Reading, PA). Paired student's T-tests were performed to calculate statistical significance.

4.3.2. Monitoring GFP spread

Using colocalization of PAGFPmt and TMRE as a metric for quantification is problematic for a number of reasons. The decrease in GFP intensity with each fusion event is so prominent that it affects the perceived colocalization and confounds the results. We have found that at later time points, the GFP intensity can become so weak that its colocalization with red pixels becomes unreliable. With photoconversion of 10–20% of the cell area, we typically find that the GFP intensity at equilibrium is on average 60% lower compared to the beginning of the trial. In addition, in order to perform the colocalization analysis, it is necessary to scan an interlaced z-series through the cell. This is because fusion events can occur in any orientation. Higher rates of image acquisition should be avoided in order to prevent artifacts caused by photobleaching. GFP intensity dilution can report fusion events occurring outside of the focal plane.

5. POTENTIAL ARTIFACTS AND IMPORTANT CONTROLS

There are a number of sources for potential artifacts that can lead to errors in the calculation of mitochondrial fusion measurements. This section will address these concerns and discuss ways to avoid these problems. It should be noted that any values for settings provided are for reference only and have only been tested on our system. The optimal settings may differ between systems, even from the same manufacturer.

5.1. Phototoxicity generated by 2-photon laser excitation

Photoconversion of PAGFPmt into its fluorescent form requires careful calibration of the 2-photon laser intensity. This potential problem has been addressed in detail in Section 2.1. We have observed that high 2-photon laser intensity can damage mitochondria and cause instability of $\Delta\psi$ as well as permanent depolarization. This could confound measurements of mitochondrial fusion rates because depolarized mitochondria are unable to undergo fusion (Griffin *et al.*, 2006; Meeusen *et al.*, 2004; Twig *et al.*, 2008). By using TMRE to co-stain mitochondria in the PAGFPmt fusion assay, it is possible to monitor if the photoconversion event itself caused depolarization of mitochondria. In order to determine the correct laser parameters to use for PAGFPmt photoconversion, increasing doses of laser intensity must be tested to determine if the TMRE fluorescence intensity is affected. It is important to consider that in order to use such low photoconversion stimuli, it is necessary to have sufficient expression of mitochondrial PAGFPmt. With our lentiviral delivery system, we find that increases in dosage of virus for transduction correlates with greater expression efficiency.

5.2. Photobleaching and saturation

During image acquisition, it is essential to carefully monitor the images for the effects of photobleaching or saturation. Photobleaching occurs when the 488 nm excitation laser is too strong. This can confound the measurements of PAGFPmt dilution and overestimate the level of mitochondrial fusion. To determine the laser intensity that does not cause bleaching, PAGFPmt intensity should be monitored over time in a system where mitochondrial fusion is blocked. It has been shown that MEF cells lacking MFN1 have mitochondria that are fragmented and unable to undergo fusion. These cells do not exhibit dilution of the mitochondrial PAGFPmt signal over time (Karbowski *et al.*, 2004). We have found that INS1 cells treated with high levels of fatty acid and glucose also exhibit mitochondrial fragmentation and generate a non-fusing mitochondrial sub-population. Using this system, we have been able to show that our image acquisition protocol does not cause photobleaching as reported by a photoconverted PAGFPmt signal that remains stable for the duration of the recording, up to 2 h. Alternatively, if a non-fusing condition can not be achieved, the whole cell PAGFPmt FI should be monitored over time. When appropriate 488 laser intensity is used, spreading of PAGFPmt signal should not result in the reduction of whole cell PAGFPmt FI. This can be measured by dividing the GFP fluorescence by the entire pixel area of the cell. On our Zeiss LSM 510 system, we find that using a 25 mW 488 nm argon laser set at 0.2–0.5% does not cause photobleaching even when six image z-stacks are obtained every 5 min for a recording time of 1 h.

PAGFPmt fluorescence saturation is also problematic because it can significantly limit the dynamic range of the fluorescence intensity curve. This would cause some fusion events, especially early in the recording time frame to go unrecognized. In addition to exceedingly strong 488 nm excitation, high gain settings for the image collection CCD camera are a likely culprit for saturation issues. Using the image acquisition software, it is important to ensure that the PAGFPmt image is not saturated after photo-conversion to its fluorescent form.

5.3. Thresholding

For image analysis, it is necessary to set a lower inclusive threshold in order to define which pixels are to be included in the quantification of intensity over time. The parameters we have chosen for the determination of this threshold have been described earlier in this chapter. Careful consideration must be applied when choosing this threshold value because picking one that is too low will introduce noise from non-mitochondrial fluorescence and one that is too high will limit the bottom end of the PAGFPmt intensity dynamic range. To prevent this issue, it is necessary to ensure that the chosen threshold value is suitable not only at time zero, right after photo-conversion, but also at the end time point. It is important to make sure that pixels are not lost toward the end of the recording time, when equilibrium has been reached.

6. COMPARISON WITH AN ALTERNATIVE METHOD

Polyethylene glycol (PEG) fusion has been used in a number of laboratories in order to study fusion activity within a population of mito-chondria in eukaryotic cells. Basically, cells expressing mitochondrial matrix GFP and mitochondrial matrix RFP are placed in coculture for 24 h.

Prior to fusion, *de novo* synthesis of proteins is halted with cyclohexamide. PEG is added in order to fuse cells and observations are performed at different time points in order to observe the mixing of the mitochondria GFP and RFP signals. A complete fusion event creates a yellow mitochondrion indicative of the mixing of green and red fluorophores. There are a number of limitations to this methodology. Firstly, dramatic alterations in the plasma membrane are necessary in order to fuse two cells together. This may lead to alterations in fusion ability simply due to the stress induced by the addition of PEG. The PEG fusion assay also requires that a cell with red mitochondria and another with green mitochondria end up in close enough proximity for fusion to occur between these cells. This requires precision at the time of plating to ensure that the plating densities of both cell types are optimal. With PEG fusion, the user cannot specify which mitochondrion or group of mitochondria is to be analyzed for fusion ability. Comparing fusion rates measured using the cell fusion and PAGFPmt-based assays reveals significant disparity. PEG fusion analyses in different cell types reveal that some fusion events occur within 90–120 min with complete equilibration taking place between 7 and 24 h (Chen *et al.*, 2003; Legros *et al.*, 2002; Mattenberger *et al.*, 2003). In comparison, the PAGFPmt-based assay, reports that full equilibration can be observed within 1 h with t ½ (average time required to reach a 50% reduction in fluorescence intensity) ranging from 30 min to 1 h (Karbowski *et al.*, 2004; Twig *et al.*, 2006, 2008). The advent of photoactivateable fluorescent proteins provides an alternative approach for monitoring dynamics that addresses some of the limitations of PEG fusion.

ACKNOWLEDGMENTS

We thank Dr Jennifer Lippincott-Schwartz for sharing PAGFP construct with us. This work was supported by National Institutes of Health Grants 5R01HL071629–03 and 5R01DK074778. We thank Drs Daniel Dagan, Gilad Twig, and Sarah Haigh for their comments and helpful advice in writing the manuscript. We thank Drs David Nicholls and Gyorgy Hajnoczky for helpful discussion and for their insights.

REFERENCES

Bereiter-Hahn, J., and Voth, M. (1994). Dynamics of mitochondria in living cells: Shape changes, dislocations, fusion, and fission of mitochondria. *Microsc. Res. Tech.* **27,** 198–219.

Chan, D. C. (2006). Mitochondria: Dynamic organelles in disease, aging, and development. *Cell* **125,** 1241–1252.

Chen, H., Detmer, S. A., Ewald, A. J., Griffin, E. E., Fraser, S. E., and Chan, D. C. (2003). Mitofusins Mfn1 and Mfn2 coordinately regulate mitochondrial fusion and are essential for embryonic development. *J. Cell Biol.* **160,** 189–200.

Detmer, S. A., and Chan, D. C. (2007). Functions and dysfunctions of mitochondrial dynamics. *Nat. Rev. Mol. Cell Biol.* **8,** 870–879.

Griffin, E. E., Detmer, S. A., and Chan, D. C. (2006). Molecular mechanism of mitochondrial membrane fusion. *Biochim. Biophys. Acta* **1763,** 482–489.

Herzig, S., and Martinou, J. C. (2008). Mitochondrial dynamics: To be in good shape to survive. *Curr. Mol. Med.* **8,** 131–137.

Hoffmann, H. P., and Avers, C. J. (1973). Mitochondrion of yeast: Ultrastructural evidence for one giant, branched organelle per cell. *Science* **181,** 749–751.

Jeong, S. Y., and Seol, D. W. (2008). The role of mitochondria in apoptosis. *BMB Rep.* **41,** 11–22.

Karbowski, M., Arnoult, D., Chen, H., Chan, D. C., Smith, C. L., and Youle, R. J. (2004). Quantitation of mitochondrial dynamics by photolabeling of individual organelles shows that mitochondrial fusion is blocked during the Bax activation phase of apoptosis. *J. Cell Biol.* **164,** 493–499.

Legros, F., Lombes, A., Frachon, P., and Rojo, M. (2002). Mitochondrial fusion in human cells is efficient, requires the inner membrane potential, and is mediated by mitofusins. *Mol. Biol. Cell* **13,** 4343–4354.

Malka, F., Guillery, O., Cifuentes-Diaz, C., Guillou, E., Belenguer, P., Lombes, A., and Rojo, M. (2005). Separate fusion of outer and inner mitochondrial membranes. *EMBO Rep.* **6,** 853–859.

Mattenberger, Y., James, D. I., and Martinou, J. C. (2003). Fusion of mitochondria in mammalian cells is dependent on the mitochondrial inner membrane potential and independent of microtubules or actin. *FEBS Lett.* **538,** 53–59.

Meeusen, S., McCaffery, J. M., and Nunnari, J. (2004). Mitochondrial fusion intermediates revealed *in vitro. Science* **305,** 1747–1752.

Mozdy, A. D., and Shaw, J. M. (2003). A fuzzy mitochondrial fusion apparatus comes into focus. *Nat. Rev. Mol. Cell Biol.* **4,** 468–478.

Nunnari, J., Marshall, W. F., Straight, A., Murray, A., Sedat, J. W., and Walter, P. (1997). Mitochondrial transmission during mating in *Saccharomyces cerevisiae* is determined by mitochondrial fusion and fission and the intramitochondrial segregation of mitochondrial DNA. *Mol. Biol. Cell* **8,** 1233–1242.

Partikian, A., Olveczky, B., Swaminathan, R., Li, Y., and Verkman, A. S. (1998). Rapid diffusion of green fluorescent protein in the mitochondrial matrix. *J. Cell Biol.* **140,** 821–829.

Patterson, G. H., and Lippincott-Schwartz, J. (2002). A photoactivatable GFP for selective photolabeling of proteins and cells. *Science* **297,** 1873–1877.

Scott, S. V., Cassidy-Stone, A., Meeusen, S. L., and Nunnari, J. (2003). Staying in aerobic shape: How the structural integrity of mitochondria and mitochondrial DNA is maintained. *Curr. Opin. Cell Biol.* **15,** 482–488.

Szabadkai, G., Simoni, A. M., Bianchi, K., De, S. D., Leo, S., Wieckowski, M. R., and Rizzuto, R. (2006). Mitochondrial dynamics and Ca^{2+} signaling. *Biochim. Biophys. Acta* **1763,** 442–449.

Twig, G., Elorza, A., Molina, A. J., Mohamed, H., Wikstrom, J. D., Walzer, G., Stiles, L., Haigh, S. E., Katz, S., Las, G., Alroy, J., Wu, M., *et al.* (2008). Fission and selective fusion govern mitochondrial segregation and elimination by autophagy. *EMBO J.* **27,** 433–446.

Twig, G., Graf, S. A., Wikstrom, J. D., Mohamed, H., Haigh, S. E., Elorza, A., Deutsch, M., Zurgil, N., Reynolds, N., and Shirihai, O. S. (2006). Tagging and tracking individual networks within a complex mitochondrial web with photoactivatable GFP. *Am. J. Physiol. Cell Physiol.* **291,** C176–C184.

Zuchner, S., Mersiyanova, I. V., Muglia, M., Bissar-Tadmouri, N., Rochelle, J., Dadali, E. L., Zappia, M., Nelis, E., Patitucci, A., Senderek, J., Parman, Y., Evgrafov, O., *et al.* (2004). Mutations in the mitochondrial GTPase mitofusin 2 cause Charcot-Marie-Tooth neuropathy type 2A. *Nat. Genet.* **36,** 449–451.

CHAPTER SEVENTEEN

DETERMINATION OF YEAST MITOCHONDRIAL KHE ACTIVITY, OSMOTIC SWELLING AND MITOPHAGY

Karin Nowikovsky,* Rodney J. Devenish,[†] Elisabeth Froschauer,* and Rudolf J. Schweyen*

Contents

1. Introduction	306
2. Disturbance of Mitochondrial K^+ Homeostasis by dox-Regulated Mdm38 Expression	307
3. Morphological Changes in Mitochondria in Mdm38-Depleted Cells: Confocal Microscopy and Electron Microscopy	309
4. Measurement of KHE Activity and Other Bioenergetic Parameters	310
4.1. KOAc-induced swelling	310
4.2. KHE measurements with fluorescent dyes entrapped in submitochondrial particles	311
4.3. Membrane potential	312
5. Following Mitophagy Using a Novel Biosensor and Fluorescence Microscopy	313
5.1. The novel biosensor	313
5.2. Targeting to mitochondria	314
6. Growth and Preparation of Cells for Fluorescence Microscopy Observations of Mitophagy	314
6.1. Transformation of yeast cells with biosensor constructs	314
6.2. Growth of $MDM38^{dox}$ yeast cells expressing biosensor constructs	315
7. Fluorescence Microscopy	315
Acknowledgments	316
References	317

* Max F. Perutz Laboratories, Department of Genetics, University of Vienna, Campus Vienna Biocenter, Wien, Austria
[†] Department of Biochemistry and Molecular Biology, and ARC Centre of Excellence in Structural and Functional Microbial Genomics, Monash University, Clayton campus, Melbourne, Victoria, Australia

Methods in Enzymology, Volume 457
ISSN 0076-6879, DOI: 10.1016/S0076-6879(09)05017-4

Abstract

The mitochondrial K^+/H^+ exchanger (KHE) is a key regulator of mitochondrial K^+, the most abundant cellular cation, and thus for volume control of the organelle. Downregulation of the mitochondrial KHE results in osmotic swelling and autophagic degradation of the organelle. This chapter describes methods to shut-off expression of Mdm38p, an essential factor of the mitochondrial KHE, and to observe the cellular consequences thereof, in particular changes in KHE activity and morphogenetic changes of mitochondria by applying new techniques developed in our laboratories.

ABBREVIATIONS

KHE	K^+/H^+ exchanger
$\Delta\psi$	membrane potential
KOAc	potassium acetate
SMPs	submitochondrial particles
TES	N-tris(hydroxymethyl)-methyl-2-aminoethane-sulfonic acid
A23187	4-bromo-calcium ionophore
JC-1	5,5′,6,6′-tetrachloro-1,1′,3,3′-tetraethylbenzimidazolylcarbocyanine iodide
FCCP	carbonylcyanide-p-trifluoromethoxyphenylhydrazone
RFP	red fluorescent protein
FM 4-64	N-(3-triethylammoniumpropyl)-4-(p-diethylaminophenyl-hexatrienyl) pyridinium dibromide
CMAC-Arg	7-amino-4-chloromethylcoumarinL-arginine amide

1. INTRODUCTION

Mitochondrial volume homeostasis is essential for maintaining the structural integrity of the organelle. Any dysbalance in the fluxes of the main intracellular ion, potassium (K^+), between the cytoplasm and mitochondria will thus affect the osmotic balance and promote water movement between these two compartments. The mitochondrial inner membrane has "low permeability to protons and to anions and cations generally" and "substrate-specific exchange–diffusion carrier systems that permit the effective reversible transmembrane exchange of anions against OH^- ions and of cations against H^+ ions" (Mitchell and Moyle, 1967). Besides electrophoretic diffusion, K^+ is known to enter the mitochondrial matrix through channels, driven by the inside negative mitochondrial membrane potential—and extruded by Na^+, K^+/H^+ exchange systems. The inner membrane protein of yeast mitochondria, Mdm38p, also referred to as Mkh1p (Nowikovsky et al., 2009) and its functional homolog LETM1 in mammals have been shown to be essential for function of an active KHE. Genetic manipulations that increase or decrease Mdm38p or LETM1p expression selectively up- or downregulate mitochondrial KHE (Nowikovsky et al., 2007). KHE also may be manipulated by drugs (Garlid, 1980; Manon and Guerin, 1992). However, in contrast to up- or downregulation of Mdm38 or LETM1 protein expression, drug treatment may not selectively affect mitochondrial KHE.

2. DISTURBANCE OF MITOCHONDRIAL K^+ HOMEOSTASIS BY DOX-REGULATED MDM38 EXPRESSION

To dissect the pleiotropic *MDM38* deletion phenotypes over-time, we use the W303 *mdm38Δ* null mutant strain transformed with the plasmid pCM189-*MDM38*[dox]. This plasmid allows expression of Mdm38p from a promoter that can be shut off by addition of doxycycline. A control strain, W303 wild-type, is transformed with the same plasmid not containing an expression cassette (pCM189 empty). Both strains are first grown in the synthetic growth medium (Sherman, 1991) containing 0.67% yeast nitrogen base w/o amino acids, 5 g/l NH_4SO_4, 2% glucose with all required amino acids then shifted to same medium containing 2% galactose (SGal, low glucose repression of mitochondria) in absence of doxycycline and grown to the same OD_{600}. Thereafter both cultures are diluted to OD_{600} of 1.6 in SGal with added doxycycline (dox, 5 $\mu g/ml$) and kept growing logarithmically by diluting them to an OD_{600} of 0.1 in prewarmed SGal + dox when they reach OD_{600} of 1.6. Cells were harvested about 3 h after dilution to minimize short-term effects of diluting the cultures. To follow Mdm38p depletion overtime after dox-mediated shut-off of *MDM38* transcription, cells (corresponding to 1 ml = OD_{600} of 2) are harvested at intervals of 4 h, trichloroacetic acid (TCA) precipitated and proteins separated on 12% SDS–PAGE. Western blotting and immunodetection with antibodies against Mdm38p and against the control protein (e.g., Hxk1p) allow determination of Mdm38p decay (cf. Fig. 17.1C).

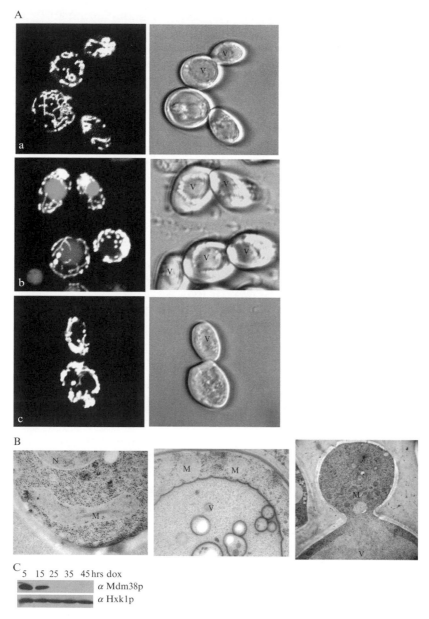

Figure 17.1 (A). Confocal microscopy of Mdm38p dependent mitochondrial turn-over. Wild-type cells and *MDM38*[dox] cells expressing Rosella were monitored for changes in green and red fluorescence using confocal microscopy. Cells were grown to the early logarithmic stage for 50 hrs in presence of doxycycline (a) wild-type, (b) *MDM38*[dox], or in presence of doxycycline and nigericin (c) *MDM38*[dox]. (a) Green and red fluorescence documents the normal branched reticular network of

3. MORPHOLOGICAL CHANGES IN MITOCHONDRIA IN MDM38-DEPLETED CELLS: CONFOCAL MICROSCOPY AND ELECTRON MICROSCOPY

Morphological changes of the organelle upon Mdm38p depletion can be followed in both wild-type and *MDM38*[dox] downregulated cells transformed with pYX232-mtGFP, expressing a mitochondrially targeted GFP (Westermann and Neupert, 2000). Cells are grown in SGal + dox medium to $OD_{600} = 1$ and observed using confocal microscopy over extended times from 0 to 50 h at 4 h intervals of dox-mediated Mdm38p depletion. Morphological changes of organelles can be correlated with the level of Mdm38p observed by immunoblotting. The percentage of cells showing changed morphology can be determined by examining statistically relevant numbers of cells (1000/experiment, with identity hidden and each experiment repeated three times). Since the mitochondrial GFP displays strong fluorescence, it is easy to differentiate and classify the mitochondrial shapes into tubular, fragmented and globular categories. The extent of organelle

---

mitochondria in wild-type cells. Red fluorescence co-localises with green fluorescence as expected in wild-type cells; the colocalization shown as overlay of green and red fluorescence is seen here as white colour. There is little evidence for mitochondrial uptake into the vacuole (v) since no red fluorescence is accumulated in the vacuoles. Exclusive red fluorescence would show in the black and white image as grey colour. Thus, growth of wild-type cells in presence of doxycycline for 50 hrs had no effect on mitochondrial morphology or turnover. (b) By contrast, in *MDM38*[dox] cells expressing Rosella in the presence of doxycycline, the alteration of mitochondrial morphology, from elongated to rounded forms, became evident. Similar to wild-type cells, red fluorescence first fully co-localised with green. However, at a later time point of growth in the presence of doxycycline correlating to the loss of mitochondrial K^+/H^+ exchange activity and ongoing swelling (see (Nowikovsky *et al.*, 2007) for details), mitochondrial fragmentation became apparent and some fragmented organelles were found lacking green fluorescence but retaining red fluorescence (grey). At later time points (e.g., 50 hrs) significant accumulation of red fluorescence in vacuoles was detected, reflecting turnover of mitochondria in vacuole. The appearance of red fluorescence in vacuoles is seen here in grey. (c) In a parallel experiment, *MDM38*[dox] cells expressing Rosella were grown for 50 hrs in the presence of doxycycline with nigericin added for the last cell doubling. Remarkably, the mitochondria fragmented to a much lesser extent and more strikingly, these cells displayed no accumulation of red fluorescence in the vacuoles suggesting that nigericin reversed the defect in mitochondrial morphology and consequently markedly reduced turnover. Overlay of green and red fluorescence is shown in white. (B). EM of wild-type (left panel) and *MDM38*[dox] (middle panel) grown in presence of doxycycline for 50 hrs to OD_{600} 0.5, or *MDM38*[dox] (right panel) grown in presence of doxycycline and nigericin (doxycycline for 50 hrs with nigericin added for the last cell doubling): mitochondria, V: vacuole. (C). Western Blot detection of Mdm38p in lysates of *MDM38*[dox] cells grown in the presence of doxycycline (time indicated in hrs); Hxk1p serves as a loading control.

swelling *in situ* is observed by electron microscopy (EM). Cells are in grown SGal + dox to $OD_{600} = 0.6$ for different times. Upon Mdm38p depletion EM photographs show enlarged organelles with few or no visible cristae (cf. Fig. 17.1B, middle panel) and much reduced electron density of the matrix, consistent with osmotic swelling due to K^+ accumulation in the matrix followed by massive uptake of water. This mitochondrial swelling can be largely reverted by the addition of 2 μM nigericin, a K^+/H^+ ionophore, to cells during the last 4 h of growth before cells are harvested for EM inspection (cf Fig. 17.1C, right panel).

4. Measurement of KHE Activity and Other Bioenergetic Parameters

Hallmarks of Mdm38p depletion are the reduction in KHE activity of mitochondria and the reduction in $\Delta\psi$. KHE activity may be observed indirectly by KOAc-induced swelling of isolated mitochondria, or more directly by use of submitochondrial particles (SMPs) with entrapped K^+ and H^+ sensitive dyes.

4.1. KOAc-induced swelling

When incubated in KOAc, isolated mitochondria rapidly take up the protonated form of acetic acid. Its ionization in the mitochondrial matrix leads to acidification of the matrix and activation of KHE resulting in the accumulation of K^+ for H^+ In nonrespiring mitochondria, this results in the net accumulation of potassium acetate, uptake of water, and swelling of the organelle, which can be measured as a decrease in light scattering or absorbance. The swelling experiment is carried out at room temperature on freshly isolated mitochondria. Isolated mitochondria are prepared as described in (Zinser and Daum, 1995) with the following changes (use of zymolyase at 2 mg/g cells to generate spheroplasts; carrying out dounce homogenization of cell lysates in the absence of protease inhibitors; and collection of mitochondria by centrifugation of the lysate at 35,000 rpm for 3 min followed by centrifugation of the resultant supernatant at 10,000 rpm for 10 min). Isolated mitochondria are re-suspended in breaking buffer (0.6 *M* Sorbitol, pH 7.4 Tris/Cl) at a concentration of 10 mg/ml protein. Aliquots of 100 μl are incubated for 5 min at room temperature in the presence of 5 μM Antimycin A to block proton pumping by the respiratory chain. Mitochondria are recovered by centrifugation at 10,000 rpm for 2 min and then re-suspended in 100 μl breaking buffer supplemented with 0.5 μM A23187 (divalent cation ionophore) and 10 m*M* EDTA (chelator of divalent cations with preference for Mg^{2+}) to deplete the mitochondria of endogenous Mg^{2+} (as Mg^{2+} inhibits the KHE). Thereafter, treated mitochondria are transferred to acryl-cuvettes (Sarstedt, Germany). The swelling reaction is started directly by rapidly adding 900 μl of swelling buffer (55 mM KOAc, 5 mM TES, 0.1 mM EGTA, 0.1 mM EDTA) followed by immediate measurement of OD_{540} in a standard photometer (HitachiU-2000 program Timescan, WL 540 nm). (Alternatively, swelling can be observed by use of a light scattering photometer equipped with a stopped flow device.) While wild-type mitochondria react with a rapid decrease of absorbance, reflecting swelling, mitochondria depleted of Mdm38p appear already swollen prior to addition of KOAc and then show no or very little additional swelling. As a control confirming that swelling of Mdm38p expressing mitochondria is dependent on the mitochondrial KHE, quinine, an inhibitor of KHE is added (final concentration of 0.2 mM) to the mitochondrial pellet before re-suspension in the breaking buffer containing A-23187 and EDTA, followed by measurement of OD_{540} as described above.

4.2. KHE measurements with fluorescent dyes entrapped in submitochondrial particles

The use of SMPs, which essentially are inner membrane vesicles having inside-out orientation, allows the investigator to establish defined environments (e.g. salt concentration, pH, $\Delta\psi$) inside and outside of these vesicles. Accordingly, transit of ions through this membrane is not disturbed by the complex composition of the matrix of intact mitochondria. Diffusion of protons or cations across SMP membranes prepared from yeast mitochondria is low and does not disturb H^+ or K^+ assays with their very fast kinetics (Froschauer et al., 2005). The first step in preparing SMPs consists of producing mitoplasts. For this purpose 10 mg of mitochondria (in 0.6 M sorbitol, 10 mM Tris/Cl pH 7.4) is diluted in six volumes of 10 mM Tris/Cl pH 7.4 for 20 min on ice. After centrifugation at 40,000g (Beckman TLA 100.4), the pellet is re-suspended in 1 ml 250 mM sucrose-Tris/Cl pH 7.4 buffer. To generate SMPs, 1 ml of mitoplast suspension on ice is pulse-sonified three times for 60 s with 90% intensity in a Bandelin sonicator UW70/GM70 for a total of 3 min with intervals of 30 s, followed by a clarifying centrifugation at 10,000 rpm for 10 min. The supernatant containing the SMPs is collected and centrifuged for 1 h at 100,000g. The pellet consisting of SMPs is re-suspended in 250 mM sucrose-Tris/Cl pH 7.4 buffer, resulting in equal milieus inside and outside the particles. The pH of buffers inside and outside of SMPs may be varied in the range from 6 to 8 without harming SMP quality or fluorescence of the entrapped indicators. SMPs can be stored frozen at this step at −80 °C. To entrap fluorescent dyes the SMPs are sonicated as described above in the presence of fluorescent K^+- or H^+-sensitive indicators (PBFI and BCECF, respectively, Invitrogen). Excess of nonentrapped dye is washed away in three centrifugation steps of 10 min each at 100,000g, and the resulting pellet is re-suspended in 250 mM sucrose-Tris/Cl pH 7.4 buffer and the SMP suspension (0.5 mg protein/ml) kept on ice. Usually there is no significant loss of the dyes from SMP up to at least 3 h after loading, but membrane integrity can be assessed by collecting a sample of SMPs by centrifugation and determining the fluorescence of the supernatant. K^+/H^+ exchange is essentially dependent on the concentration gradients of K^+ and/ or H^+ across the SMP membrane. Assays are started by placing cuvettes with 2 ml of the SMP suspension with stirring at 25 °C into a Perkin Elmer LS55 spectrofluorometer, (FL Winlab program: Fast Filter Accessory—FFA. MTH). Recording of PBFI or BCECF fluorescence is performed with excitation wavelength set at 340 and 380 nm, emission at 500 nm, 7 nm slit width for PBFI and excitation wavelength set at 440 and 490 nm, emission at 535 nm, 15 nm slit width for BCECF. Measurements are made before and after addition of KCl (or other potassium salts), or after acidification or alkalinization of the external buffer.

4.3. Membrane potential

To measure the membrane potential ($\Delta\psi$) of isolated mitochondria, we use JC-1, a lipophilic cationic vital dye (Reers *et al.*, 1995). The dye accumulates in mitochondria and binds to the membrane as a monomer at low concentrations (emission: 530 nm) and/or forms J-aggregates at higher concentrations (emission: 590 nm), as shown in Fig. 17.2. Isolated mitochondria (10 mg/ml) are re-suspended in $\Delta\psi$ measurement buffer (0.6 M sorbitol Tris/Cl pH7.4 containing 0.5 mM ATP, 0.2% succinate and 0.01% pyruvate freshly prepared). To 15 μl mitochondrial suspension 5 μM JC-1 is added and the suspension is shaken or vortexed until the dye is well dissolved. After addition of 1 ml $\Delta\psi$ measurement buffer, the suspension is incubated in the dark at room temperature for 7 min, washed three times, and then re-suspended in 2 ml $\Delta\psi$ measurement buffer. Fluorescence is measured using a Perkin Elmer LS55 spectrophotometer (FL Winlab program—Scan.MTH, excitation starting at 500 and ending at 600 nm, slit width 10.0 nm). The ratio of the reading at 590 nm to the reading at 530 nm (590:530 ratio) is considered as the relative $\Delta\psi$ value. $\Delta\psi_{max}$ is obtained in parallel control preparations that are hyperpolarized by addition of nigericin (1 μl of 10 mM stock/ml mitochondria) to the $\Delta\psi$ measurement buffer 10 min prior to incubation with JC-1 and measurement of fluorescence as described above (Fig. 17.2, dotted line). For $\Delta\psi_{min}$, mitochondria are re-suspended in the presence of 0.3 M sucrose 0.15 M KCl and 0.15 M $\Delta\psi$ measurement buffer plus valinomycin 30 min before JC-1 loading (4μl of 10 mM stock/ml mitochondria) (Fig. 17.2, broken line).

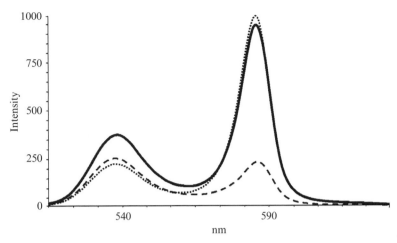

Figure 17.2 Typical measurment curve used for the quantification of the relative $\Delta\psi$ value with JC-1. JC-1 forms aggregates, or stays in the monomeric form depending on the polarization of the membrane. The monomeric form (representing depolarization) is detected at 540 nm whereas aggregates (representing hyperpolarisation) are detected at 590 nm. The ratio of fluorescence (590:540 nm) is seen as the relative $\Delta\psi$ value. For quantification of the relative $\Delta\psi$ value in percentage terms, the $\Delta\psi$ is set artificially to 0 by using FCCP (broken line) and to 100% by using nigericin (dotted line).

 ## 5. FOLLOWING MITOPHAGY USING A NOVEL BIOSENSOR AND FLUORESCENCE MICROSCOPY

5.1. The novel biosensor

By fusion of two fluorescent proteins we have developed a single reporter molecule that allows the autophagic uptake of specific cellular compartments into the vacuole (or lysosome) for degradation to be monitored. The high structural stability of fluorescent proteins renders them relatively resistant to proteolysis and the low of pH of the vacuole, and as such they can maintain a signal sufficiently persistent overtime that is suitable for following autophagy. The reporter is a dual color emission biosensor comprising a relatively pH-stable RFP fused at its C-terminus to a pH-sensitive GFP variant. The GFP component is the super ecliptic pHluorin (SEP; λ_{max} emission 508 nm) which has a pK of $\sim$7.0 such that the fluorescence emission is almost negligible at the vacuolar pH of 6.2. The RFP component is a fast maturing RFP variant, DsRed.T3 (λ_{max} emission 587 nm; Bevis and Glick, 2002), whose fluorescence emission is quite stable over a wide range of pH ($\sim$4.9–9) typical of these proteins (Heikal et al., 2000). Expression of the two fluorescent proteins, fused by a short nine amino acid linker, ensures their balanced relative expression which is essential for determining a fluorescence emission ratio that can be correlated with pH. The biosensor allows not only qualitative, but also quantitative analysis of cells in a population undergoing autophagy. The dual color nature of the biosensor, in contrast to a single color fluorescent reporter, allows unambiguous distinction between material delivered to the vacuole (pH-based loss of green fluorescence) and that localized in distinctly different cellular compartments such as the cytosol, mitochondria, and nucleus (green and red fluorescence retained). This distinction relies on the pH difference between the vacuole [or lysosome; nominally pH ∼6.2 (Preston *et al.*, 1989) or 4.8 (Ohkuma and Poole, 1978; Yoshimori *et al.*, 1991), respectively] and other compartments of the cell (pH >7.0). We have named the biosensor Rosella, after the brightly colored parrots native to the south-eastern region of Australia (Rosado *et al.*, 2008).

5.2. Targeting to mitochondria

In our studies on Mdm38p, Rosella was targeted to the mitochondrial matrix by use of the N-terminal leader of the *CIT1* gene encoding mitochondrial citrate synthase (CS). Rosella is expressed from a multicopy expression vector under control of the *PGK1* promoter (Nowikovsky *et al.*, 2007; Rosado *et al.*, 2008). On import into mitochondria the CS leader sequence is cleaved releasing Rosella into the matrix.

6. GROWTH AND PREPARATION OF CELLS FOR FLUORESCENCE MICROSCOPY OBSERVATIONS OF MITOPHAGY

6.1. Transformation of yeast cells with biosensor constructs

Yeast transformations using the "LiAc method" were performed by standard procedures (Gietz and Woods, 2002) using only minor modifications. The following are points worthy of note:

1. Cells of the strains to be transformed should be grown on plates containing a nonfermentable growth substrate (e.g., galactose, ethanol, lactate) to remove any accumulated petite cells (having deletions or complete loss of mitochondrial DNA).
2. The transformation procedure requires the cell suspension prepared to be subjected to heat shock at 42 °C for 30 min. The optimum time can vary for different yeast strains and should be determined empirically to achieve high efficiency transformation. As some strains accumulate petite cells in response to heat shock at 42 °C, it may be necessary to use an alternative transformation procedure such as that of Klebe *et al.* (1983).

3. Before undertaking experiments to determine mitophagy, cells from several transformant colonies should be viewed using a fluorescence microscope to select for further analysis those transformants that retain the ability to grow on nonfermentable substrate and also display strong fluorescence.

6.2. Growth of *MDM38*^{dox} yeast cells expressing biosensor constructs

1. *MDM38*^{dox} cells are grown aerobically at 30 °C in SGal medium as described above.
2. To deplete cells of Mdm38p, *MDM38*^{dox} cells are grown in the presence of doxycline (5 μg/ml) for 4, 10, 20, 25, 29, and 50 h. Control non-Mdm38p depleted cells are grown under the same conditions in the absence of doxycycline.
3. Cells are harvested at the desired intervals.

7. FLUORESCENCE MICROSCOPY

1. In preparation for microscopy, cells were washed in PBS to remove medium and mounted on slides coated with Concanavalin A (Sigma). Slides are incubated just before use with 5 μl Concanavalin A (6 mg/ml) at room temperature for 5 min, washed three times with distilled water and dried. Thereafter 4 μl yeast cell culture is spotted onto the coated slide. After 5 min nonadhered cell material is aspirated away.
2. Cells were then immediately imaged using confocal laser scanning microscopy.
3. When required, cell vacuoles were stained for imaging with *N*-(3-triethylammoniumpropyl)-4-(*p*-diethylaminophenyl-hexatrienyl) pyridinium dibromide (FM 4-64; Invitrogen, final concentration of 10 μM) according to a modified method as described previously by Vida and Emr (1995). An alternative stain for vacuoles is CMAC-Arg (7-amino-4-chloromethylcoumarin, L-arginine amide; Invitrogen). Incubate for 30 min at 28 °C at a final concentration of 100 mM, and then wash three times with sterile water before mounting.
4. Cells are imaged as soon as possible after mounting. For imaging purposes any wide-field or confocal fluorescence microscope equipped with a suitable objective can be used, e.g., a Zeiss Axiovert LSM 510 microscope; images are acquired in multitrack Z stack and projected with LSM5 Image Browser (Nowikovsky *et al.*, 2007). Alternatively, a Fluoview FV500

mounted on an inverted IX-81 microscope (Olympus, Australia). ImageJ software (version 1.36b) may be used for image analysis (http://rsb.info.nih. gov/ij/) (Rosado *et al.*, 2008). Green fluorescence emission (530 ± 30 nm; SEP) and red fluorescence emission (>590 nm; DsRed or FM 4-64) are acquired sequentially upon excitation with 488 or 543 nm laser light, respectively. If CMAC-Arg is used to stain vacuoles then blue fluorescence emission (469 nm) is acquired upon excitation with 405 nm laser light, but this should be prior to obtaining green and red fluorescence. Sequential (as opposed to simultaneous) scanning should be used in order to eliminate any bleed through from one channel to another. Transmitted light images can be acquired in order to provide additional information on the vacuole and its relative position in the cell.

Notes:

i. Yeast are among small eukaryotes ($\sim$5–10 μm in diameter); therefore, it is necessary to use high level of magnification in fluorescence microscopy. For visual observation, a 60× water-immersion (or 100× oil-immersion) objective lens with a numerical aperture of 1.2 and a 10× eyepiece lens is suitable.

ii. The confocal pinhole value determines how the mitochondria are visualized. In Mdm38p depletion experiments a pinhole value of 115 μm was used to depict the typical filamentous mitochondrial network distributed throughout the cell (and the changes it undergoes) (Nowikovsky *et al.*, 2007).

iii. It is important that the parameters selected for fluorescence microscopy (e.g., laser intensity, scan rate, pinhole value, or exposure for CCD camera) are maintained throughout an experiment. This will ensure that comparisons can be made between control (growing cells) and starved samples.

iv. Mitophagy as reported by delivery of Rosella to the vacuole can be quantified by scoring for the accumulation of red fluorescence in the vacuole concomitant with checking for the absence of green fluorescence (seen as gray dots in Fig. 17.1A). In each experiment at least 100–200 cells are scored.

ACKNOWLEDGMENTS

KN and RJS thank Siegfried Reipert who has performed the EM studies on mitochondrial swelling and mitophagy. RJD thanks his Monash colleagues, particularly Mark Prescott and Dalibor Mijaljica who have contributed to studies on mitophagy, for helpful discussions. This work was supported by the Austrian Research Fund and SYSMO/Translucent to RJS and by the Australian Research Council funding to RJD. This chapter is dedicated to the memory of our colleague, Rudolf J. Schweyen, who passed away in February 2009.

REFERENCES

Bevis, B. J., and Glick, B. S. (2002). Rapidly maturing variants of the Discosoma red fluorescent protein (DsRed). *Nat. Biotechnol.* **20,** 83–87.

Froschauer, E., Nowikovsky, K., and Schweyen, R. J. (2005). Electroneutral K^+/H^+ exchange in mitochondrial membrane vesicles involves Yol027/Letm1 proteins. *Biochim. Biophys. Acta* **1711,** 41–48.

Garlid, K. D. (1980). On the mechanism of regulation of the mitochondrial K^+/H^+ exchanger. *J. Biol. Chem.* **255,** 11273–11279.

Gietz, R. D., and Woods, R. A. (2002). Transformation of yeast by lithium acetate/singlestranded carrier DNA/polyethylene glycol method. *Methods Enzymol.* **350,** 87–96.

Heikal, A. A., Hess, S. T., Baird, G. S., Tsien, R. Y., and Webb, W. W. (2000). Molecular spectroscopy and dynamics of intrinsically fluorescent proteins: Coral red (dsRed) and yellow (Citrine). *Proc. Natl. Acad. Sci. USA* **97,** 11996–12001.

Klebe, R. J., Harriss, J. V., Sharp, Z. D., and Douglas, M. G. (1983). A general method for polyethylene-glycol-induced genetic transformation of bacteria and yeast. *Gene* **25,** 333–341.

Manon, S., and Guerin, M. (1992). K^+/H^+ exchange in yeast mitochondria: Sensitivity to inhibitors, solubilization and reconstitution of the activity in proteoliposomes. *Biochim. Biophys. Acta* **1108,** 169–176.

Mitchell, P., and Moyle, J. (1967). Chemiosmotic hypothesis of oxidative phosphorylation. *Nature* **213,** 137–139.

Nowikovsky, K., Reipert, S., Devenish, R. J., and Schweyen, R. J. (2007). Mdm38 protein depletion causes loss of mitochondrial K^+/H^+ exchange activity, osmotic swelling and mitophagy. *Cell Death Differ.* **14,** 1647–1656.

Nowikovsky, K., Schweyen, R. J., and Bernardi, P. (2009). Pathophysiology of mitochondrial volume homeostasis: Potassium transport and permeability transition. *Biochim. Biophys. Acta.* (In press).

Ohkuma, S., and Poole, B. (1978). Fluorescence probe measurement of the intralysosomal pH in living cells and the perturbation of pH by various agents. *Proc. Natl. Acad. Sci. USA* **75,** 3327–3331.

Preston, R. A., Murphy, R. F., and Jones, E. W. (1989). Assay of vacuolar pH in yeast and identification of acidification-defective mutants. *Proc. Natl. Acad. Sci. USA* **86,** 7027–7031.

Reers, M., Smiley, S. T., Mottola-Hartshorn, C., Chen, A., Lin, M., and Chen, L. B. (1995). Mitochondrial membrane potential monitored by JC-1 dye. *Methods Enzymol.* **260,** 406–417.

Rosado, C. J., Mijaljica, D., Hatzinisiriou, I., Prescott, M., and Devenish, R. J. (2008). Rosella: A fluorescent pH-biosensor for reporting vacuolar turnover of cytosol and organelles in yeast. *Autophagy* **4,** 205–213.

Sherman, F. (1991). Getting started with yeast. *Methods Enzymol.* **194,** 3–21.

Vida, T. A., and Emr, S. D. (1995). A new vital stain for visualizing vacuolar membrane dynamics and endocytosis in yeast. *J. Cell Biol.* **128,** 779–792.

Westermann, B., and Neupert, W. (2000). Mitochondria-targeted green fluorescent proteins: Convenient tools for the study of organelle biogenesis in *Saccharomyces cerevisiae*. *Yeast* **16,** 1421–1427.

Yoshimori, T., Yamamoto, A., Moriyama, Y., Futai, M., and Tashiro, Y. (1991). Bafilomycin A1, a specific inhibitor of vacuolar-type H(+)-ATPase, inhibits acidification and protein degradation in lysosomes of cultured cells. *J. Biol. Chem.* **266,** 17707–17712.

Zinser, E., and Daum, G. (1995). Isolation and biochemical characterization of organelles from the yeast, *Saccharomyces cerevisiae*. *Yeast* **11,** 493–536.

CHAPTER EIGHTEEN

IMAGING AXONAL TRANSPORT OF MITOCHONDRIA

Xinnan Wang *and* Thomas L. Schwarz

Contents

1. Introduction	320
1.1. Imaging mitochondrial movement in cultured rat hippocampal neurons	322
1.2. Imaging mitochondrial movement in Drosophila larval axons	322
2. Protocols to Image Axonal Transport of Mitochondria in Rat Hippocampal Neurons	324
2.1. Embryonic neuronal dissection	324
2.2. Primary neuronal culture	324
2.3. Constructs	325
2.4. Transient transfection and dye application	325
2.5. Image acquisition	325
2.6. Image analysis	325
2.7. Expected results	326
3. Protocols to Image Axonal Transport of Mitochondria in Drosophila Larval Neurons	328
3.1. Fly stocks and culture	328
3.2. Grape juice agar plates	329
3.3. Staging and dissection of live larvae	329
3.4. Image acquisition	329
3.5. Image analysis	329
3.6. Expected results	330
4. Conclusion	331
Acknowledgments	332
References	332

Abstract

Neuronal mitochondria need to be transported and distributed in axons and dendrites in order to ensure an adequate energy supply and provide sufficient

F. M. Kirby Neurobiology Center, Children's Hospital Boston, and Department of Neurobiology, Harvard Medical School, Boston, Massachusetts, USA

Methods in Enzymology, Volume 457
ISSN 0076-6879, DOI: 10.1016/S0076-6879(09)05018-6

Ca^{2+} buffering in each portion of these highly extended cells. Errors in mito-chondrial transport are implicated in neurodegenerative diseases. Here we present useful tools to analyze axonal transport of mitochondria both *in vitro* in cultured rat neurons and *in vivo* in Drosophila larval neurons. These methods enable investigators to take advantage of both systems to study the properties of mitochondrial motility under normal or pathological conditions.

1. INTRODUCTION

Neurons are exquisitely organized and compartmentalized. Their cytoskeleton and motor proteins are necessary to distribute essential com-ponents to axons, dendrites, and synapses in sufficient supply and to return organelles and signals from the periphery to the cell soma. Mitochondria are one such important organelle that neurons need to allocate properly. Absent adequate mitochondrial movement, the distal portions of neurons, which in a human can be a meter distant from the cell body, may be deprived of sufficient ATP production and may be incapable of adequately buffering cytosolic Ca^{2+}. Moreover, the involvement of mitochondria in generating reactive oxygen species and apoptotic signaling heightens the importance of keeping healthy mitochondria in the soma, axons, and dendrites.

Mitochondrial movement is therefore needed to deliver the organelles to the periphery. In addition, it is likely the movement from the periphery back to the soma is needed to provide for protein turnover, since most mitochondrial proteins are nuclear-encoded and peripheral proteins synthe-sis is likely to be limited. Damaged mitochondria can be highly destructive to the cell via the production of reactive oxygen species (Hoye *et al.*, 2008). Therefore, turnover of the organelle and refreshment of peripheral mito-chondria by fusion with newly minted mitochondria that arrive from the soma may be an important means of maximizing mitochondrial function and minimizing their deleterious potential. This may be one of the most significant reasons that 30–40% of the mitochondria in an axon or dendrite are in motion at any given time (Chen *et al.*, 2007; Overly *et al.*, 1996; Waters and Smith, 2003). Moreover, the optimal density of mitochondria differs from one neuron to another, even from one synapse from another within the same neuron, and is constantly changing (Hollenbeck and Saxton, 2005) as the activity of the neuron shifts the energetic needs of a subcellular domain. Activation of receptors and channels can also place an acute burden of Ca^{2+} influx on a subcellular region, and this too may require a shift in mitochondrial distribution. The necessity of tailoring mitochondrial density to changing needs may be an additional reason that mitochondria are so frequently in motion.

Mitochondria can congruently move a long distance without stopping, or travel a short distance and stop and start again. The majority of their movements are microtubule-based, either anterograde via kinesins, toward the (+)-ends of the microtubules such as in axonal terminal or retrograde via dynein, toward the (−)-ends in the cell bodies (Chang *et al.*, 2006; Hollenbeck and Saxton, 2005; Overly *et al.*, 1996). It is not surprising, therefore, that the dynamic nature of mitochondria is significant for neurons to maintain neuronal function and plasticity, and the abnormalities in mitochondrial distribution have been found in many neurodegenerative disease models such as Charcot–Marie–Tooth (Baloh *et al.*, 2007), Amyotrophic Lateral Sclerosis (De Vos *et al.*, 2007), Alzheimer's (Pigino *et al.*, 2003; Rui *et al.*, 2006; Thies and Mandelkow, 2007), Huntington's (Trushina *et al.*, 2004), and Hereditary Spastic Paraplegia (Ferreirinha *et al.*, 2004; McDermott *et al.*, 2003). To understand the means by which mitochondrial distribution is achieved and the dynamic regulation of that distribution, static imagery is insufficient. Only through live video microscopy can the mechanisms governing mitochondrial distribution be understood.

Although great progresses have been made in visualizing the transport of neuronal mitochondria, there remain significant challenges that include finding appropriate mitochondrial markers, an ideal perfusion/imaging system to keep neurons alive, and an amenable *in vivo* animal model. Offsetting these difficulties, neurons also offer advantages over other cell types for studying basic mechanisms of mitochondrial movement. The long parallel arrays of microtubules in neurites provide a simpler geometry in which long-distance movements of mitochondria can be followed. In addition, axons have largely uniform polarities of microtubules, with all (+)-ends oriented toward axon terminals. Therefore, the study of mitochondrial transport in axons provides the opportunity to distinguish kinesin-based movements from dynein based. In dendrites, microtubule polarity is typically more mixed, but the analysis of dendritic motility offers the potential to examine how that motility is affected by the activation of postsynaptic receptors on the dendrite surface and the association of mitochondria with postsynaptic specializations.

Here we describe two systems that have been used successfully in our laboratory and elsewhere to study axonal transport of mitochondria: cultured rat hippocampal neurons and Drosophila neurons. Both systems permit detailed measurements of the parameters of mitochondrial movement and the manipulation of the mechanism either via transgene expression or classical genetics. Separately, they provide the complementary advantages, detailed below, of either *in vitro* tissue culture or imaging in a semi-intact *in vivo* environment.

1.1. Imaging mitochondrial movement in cultured rat hippocampal neurons

Embryonic hippocampal neuronal cultures have been widely used to study protein distribution, neuronal function, and neuronal morphogenesis. Their utility derives from their elegant extension of neurites, their expression of key features of *in vivo* hippocampal neurons, and their suitability for transfection with transgenes (Kaech and Banker, 2006). Neurons dissociated from rat embryonic hippocampi contain few glial cells and can be cultured *in vitro* for up to a month. After 3–4 days *in vitro*, axons have differentiated from dendrites and exhibit unique morphological characteristics, such as long, thin, and uniform diameters, a lack of branching, and obvious growth cones. The identification of axons can be enhanced by transfecting the cells with a fluorescent axonal marker such as synaptophysin-YFP (Chang *et al.*, 2006), or marking dendrites with a marker such as PSD95-YFP (Chang *et al.*, 2006).

The axons of cultured hippocampal neurons can reach a few mm in length, enabling the observation of long-range movements. In these neuronal cultures, it is also easy to apply extracellular agonists, antagonists, toxins, or ionophores to regulate intracellular signaling, change intracellular ion balance, trigger neuronal cell death, or mimic a pathological cellular effect. The consequences of these manipulations on mitochondrial movements can be monitored acutely. These advantages have been exploited extensively for the study of mitochondrial motility (Chada and Hollenbeck, 2003, 2004; Chen *et al.*, 2007; Overly *et al.*, 1996; Waters and Smith, 2003).

In Section 2, we will describe the protocols to analyze mitochondrial movement in cultured rat hippocampal neurons. For optimal tracking of individual mitochondria, we have found it desirable to use cultures that were sparsely transfected with a construct encoding a fluorescent protein fused to a mitochondrial targeting sequence (RFP-mito). Under these circumstances, the processes of a single transfected neuron can be examined and thus the orientation of movement toward or away from the soma can be determined unambiguously. In densely transfected cultures or when membrane permeant dyes are used to image mitochondria, the bundling of neurites makes it impossible to determine the cell body or origin and the direction of movement. We and others have also found that some mitochondrial dyes may decrease movement of the organelles.

1.2. Imaging mitochondrial movement in Drosophila larval axons

As a model system for the genetic analysis of neurons, Drosophila remains unsurpassed. Existing mutant lines are available for dissecting many cell biological processes and ongoing mutant screens continue to identify additional proteins of interest. Despite progress with RNAi in mammalian systems, a null allele remains the best guaranty of complete loss of function. Large collections of RNAi lines are also available in the fly at low cost with which to screen candidate genes (http://stockcenter.vdrc.at/control/main). Additionally, the ease of combining different mutants and transgenes in an individual organism enables studies of double mutants and phenotypic rescue with modified transgenes in an intact organism. Moreover, many mutations whose early lethality would prevent their analysis in a knockout mouse can be studied in the fly through genetic tricks to make mutations homozygous or express RNAi selectively in small subsets of neurons. In consequence, Drosophila has been used to establish many human neurodegenerative disease models, an endeavor justified by the high degree of similarity of their genome to our own and similarity of the resultant cellular phenotypes to human neurodegeneration (Bolino *et al.*, 2004; Clark *et al.*, 2006; Gunawardena *et al.*, 2003; Wang *et al.*, 2007). Overexpression of a normal or a disease-related allele can be accomplished by crossing fly strains with cell type or tissue-specific GAL4 drivers to the fly strains harboring the gene-of-interest downstream of a GAL4 binding site (UAS) (Brand and Perrimon, 1993).

For live-imaging studies, the opaque cuticle of the adult fly poses an optical challenge and makes dissections more difficult. Therefore, most imaging studies have chosen the third-instar larval stage for analysis. These larvae are easy to dissect and many mutants and transgenic lines of interest survive to this stage. Mutations that severely impair the transport of mitochondria may die before this stage, however; *milton* null alleles, for example, die in the first instar (Stowers *et al.*, 2002). This need not preclude live imaging of mitochondria, however, though it does make the dissection more challenging and imaging in intact specimens through the cuticle may be preferred by those who have not mastered the dissection.

In Drosophila larvae, the cell bodies of central nervous system neurons form the cortex of the brain lobes and ventral nerve cord (VNC) and are bilaterally symmetrical. In the core of the VNC two parallel zones of neuropil lie to either side of the midline. It is in this neuropil where the axonal and dendritic projections of the cell bodies comingle and communicate. The complexity of the neuropil has, to date, made it unattractive for imaging studies of axonal transport. Instead, laboratories have focused on the nerves that issue from the VNC and extend to the body wall in each segment. These nerves include the axons of motorneurons that emanate from the VNC and terminate in well-characterized neuromuscular junctions on larval body walls. These segmental nerves also contain sensory axons whose cell bodies reside in the body wall and whose axons extend into the VNC, with the opposite orientation of the motorneurons. Therefore, to know the polarity of mitochondrial movement in these nerves, it is necessary to selectively label a subset of neurons of either the sensory or motorneuron class. To date, this has principally been done by the selective expression in motorneurons of a fusion of GFP with a mitochondrial import signal (Miller *et al.*, 2005; Pilling *et al.*, 2006). In Section 3, we will introduce the protocols to analyze the axonal transport of mitochondria in Drosophila neuronal axons at the late larval stage (third instar) and will describe a variation on the method of Pilling *et al.* (2006) in which we selectively label mitochondria in one peptidergic axon per segmental nerve.

2. PROTOCOLS TO IMAGE AXONAL TRANSPORT OF MITOCHONDRIA IN RAT HIPPOCAMPAL NEURONS

2.1. Embryonic neuronal dissection

Day 18 rat embryos were taken out from a euthanatized pregnant rat and decapitated. Hippocampal tissues were dissected and collected and washed with chilled Ca^{2+}, Mg^{2+}-free HBSS buffer (Hanks' Balanced Salt Solution, Invitrogen Catalog No. 14170–112). The hippocampal tissues were incubated with 15 ml Digestion Solution (stock: 30 ml HBSS, 75 μl 50 mM EDTA pH8.0, 1ml 15 mg/ml Papain) at 37 °C for 30 min and harvested in a tabletop centrifuge at 1300 rpm at room temperature for 5 min. Digestion solution was then replaced with 5 ml Digestion Inhibition Solution (stock: 27 ml HBSS, 3 ml Solution A, 30 μl 1% DNase (Sigma, MO)). Solution A stock: 100 ml HBAA, 1 g BSA, 1 g Trypsin Inhibitor (Sigma). The tissue was triturated 10 times with pipettes and added to 10 ml Digestion Inhibition Solution and 5 ml Solution A, before centrifugation at 1300 rpm for 10 min. The pellet of dissociated hippocampal neurons was re-suspended with Neurobasal medium (Invitrogen, Carlsbad, CA).

2.2. Primary neuronal culture

Coverslips (12 mm diameter, #1.5) were ethanol sterilized and autoclaved, and coated with filter-sterilized polyornithine (1 mg/50 ml, Sigma) and laminin (170 μg/50 ml, Invitrogen) in 24-well plates the night before plating. The next day, coated coverslips were washed twice with sterile water and once with Neurobasal medium (Invitrogen). Dissociated neurons were then plated onto the coated coverslips at 50–100,000/well in 24-well plates in Neurobasal supplemented with B27 (Invitrogen) and 100 units/ml penicillin and 100 μg/ml streptomycin for 1 h at 37 °C 5% CO_2 which allows most healthy neurons settle down and attach to the coverslips, and the medium was replaced with fresh medium to remove debris and dead cells. Neuronal cultures were maintained at 37 °C in 5% CO_2 for 8–11 days prior to transfection.

2.3. Constructs

The following constructs were used: pLPS-RFP-mito (Colwill *et al.*, 2006) (No. 11702, Addgene, Cambridge, MA); pEYFP-Synaptophysin, pEYFP-PSD95 (Chang *et al.*, 2006).

2.4. Transient transfection and dye application

For transient transfection, each well of cultured hippocampal neurons was washed twice with pre-warmed 500 μl plain Neurobasal (without B27 and antibiotics). 1 μg of pLPS-RFP-mito DNA and 200 ng of synaptophysin-YFP (or PSD95-YFP) DNA were then added together with 2.5 μl Lipofectamine 2000 (Invitrogen) in 100 μl plain Neurobasal. The DNA-lipofectamine 2000 mixture was removed 2 h later, and neurons were washed three times with plain Neurobasal before being returned to the conditioned medium (the medium that had been used to culture the neurons before transfection). Transfected neurons were imaged 2–4 days later. If a fluorescent dye was used rather than the RFP-mito, Rhodamine 123 (Invitrogen) was applied to the cultures at 10 μg/ml for 10 min at 37 °C 5% CO_2 and the cultures were washed with fresh Neurobasal three times and imaged.

2.5. Image acquisition

A coverslip with live neurons was placed into a 35 mm round dish and maintained in CO_2-independent medium (Hibernate E, Brainbits, IL, recipe online: http://www.brainbitsllc.com) on a 37 °C heated stage for less than an hour. Images were collected on a Zeiss LSM 510 confocal microscope (LSM 510 META/NLO; Carl Zeiss MicroImaging, Inc., Thornwood, NY) with LSM software 3.2 (Carl Zeiss MicroImaging, Inc.) using a 63×/N.A.0.90 water IR-Achroplan objective with excitation at 488 and 543 nm. Lasers were used at 10% to minimize damage, and pinholes were opened maximally to allow the entire thickness of the axon to be imaged. Images were captured every 2 s. Images were processed with Photoshop 10.0 (Adobe, San Jose, CA) using only linear adjustments of contrast and color, and Illustrator 13.0 (Adobe). The sensitivity of detecting small movements is potentially limited by two factors: the pixel size and drift of the image. Drift can be determined by measuring mitochondrial movement in fixed tissue and, assuming that this is minimal, the threshold for movement detection will be 1 pixel displacement per time interval between sweeps (typically 2 s).

2.6. Image analysis

To analyze the characteristics of movement of individual mitochondria, Metamorph software (MDC, CA) was used to generate kymographs from the time-lapse live-imaging movies. All the LSM pictures from one time-lapse file can be imported into Metamorph as an image stack. A region of the axon of a desired length (typically 70–150 μm) can be selected from the image stack and analyzed. In each kymograph, the x-axis represents the position along the axon and the y-axis represents time. Vertical lines indicate stationary mitochondria with no displacement during the time elapsed and diagonal lines represent moving mitochondria and their direction. Their velocity is reflected in the slope of the lines. From kymographs, one can also determine additional dynamic properties of movement, such as whether a mitochondrion is in constant motion, stops and starts again, or changes its directions. For a comprehensive description of the dynamics of mitochondria, it is advisable to calculate the number of mitochondria that are stationary, moving anterograde, or moving retrograde, the velocity of each mitochondrion that is in motion, for each direction of movement, and the percentage of time the mitochondria spend moving in each direction in the sample. The duration of continuous movements and the frequency of pauses or reversals of direction can give information about the processivity of mitochondrial movement. The length of each mitochondrion can also be determined. An example of such analysis can be found in Wang and Schwarz (2009).

2.7. Expected results

Neuronal mitochondria can be labeled by both mitochondrial dyes (such as MitoTrackers, Rhodamine 123, etc.) and fluorescent protein-tagged mitochondrial protein (such as RFP-mito used here). We found that Mito-Tracker dyes diminish mitochondrial motility and alter their morphology, probably by affecting mitochondrial membrane potential (Buckman et al., 2001), and therefore avoid MitoTracker for live imaging of neuronal mitochondria. Rhodamine 123 is another widely used mitochondrial dye (Morris and Hollenbeck, 1995). In cultured hippocampal neurons, Rhodamine 123 labeled mitochondria have normal morphology (Fig. 18.1) and are highly mobile (data not shown). Therefore, Rhodamine 123 might be a suitable choice for cultured neurons with uniform properties, such as neurons from knock-out animals. However, as mentioned above, by labeling all the mitochondria in the dish, the dye can make it difficult to discern which mitochondria are in a single identified neurite. The neurons therefore need to be plated at a very low density to distinguish individual axons and hence to know the orientation of movement. Tagging mitochondria by transfection with a construct encoding a mitochondrially targeted fluorescent protein (RFP-mito) can be a better choice than mitochondrial dyes for this reason, but also is preferable because it can be co-transfected with multiple proteins of interest into the neurons (also see Wang and Schwarz, 2009). Because neurons that have taken up the RFP-mito plasmid are very likely to have taken up the additional plasmids, one is selectively imaging from those neurons that are expressing the transgene of interests. Figure 18.2 shows an

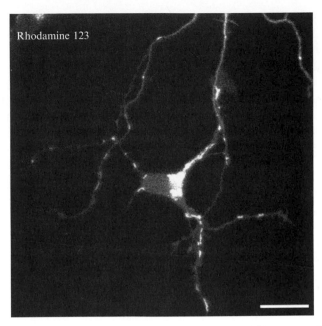

Figure 18.1 Rhodamine 123 staining of mitochondria in rat hippocampal neurons. A neuronal cell body with neurites, stained with Rhodamine 123 as in text. Neurons were dissected and dissociated from embryonic rat hippocampus, and cultured *in vitro* for 4 days. Scale bar, 10 μm. Data of X. Wang.

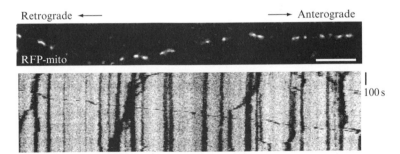

Figure 18.2 RFP-mito movement in cultured rat hippocampal neurons. The axon of a cultured hippocampal neuron transfected with RFP-mito and synaptophysin-YFP, with the axon terminal to the right and the cell body to the left. Neurons were dissociated and cultured as in text. The upper panel is the first frame of the time-lapse movie of RFP-mito and the lower panel is the kymograph generated from the movie and displaying the movements of RFP-mito during the time lapse. Time progresses from top to bottom of the kymograph. Thus vertical black lines in the kymograph represent mitochondria that did not change their position during the period of observation. Those mitochondria that moved appear as diagonal lines. Anterograde movements appear as lines slanting down to the right and retrograde movements as lines slanting down to the left. The slope reflects velocity. Scale bar, 10 μm. Data of X. Wang.

axon transfected with RFP-mito and synaptophysin-YFP. Mitochondrial movement labeled by RFP-mito was recorded and transformed to a kymograph. The majority of mitochondria are stationary. About 30% of mitochondria are moving during the interval shown (also see Wang and Schwarz, 2009). This method to analyze axonal transport of mitochondria in cultured neuronal system may be used to study the impact of overexpressed proteins, neurotrophic factors, toxins, stress conditions, or similar manipulations on the axonal transport and morphology of mitochondria. These approaches will not only illuminate the basic mechanisms of mitochondrial motion and the regulation of those dynamics, but will also provide a better understanding of the significance of mitochondrial motility in normal and pathological cellular processes.

3. PROTOCOLS TO IMAGE AXONAL TRANSPORT OF MITOCHONDRIA IN DROSOPHILA LARVAL NEURONS

3.1. Fly stocks and culture

The following fly stocks were used: *CCAP-GAL4* (Park *et al.*, 2003), *D42-GAL4* (Pilling *et al.*, 2006), and *UAS-mito-GFP* (Pilling *et al.*, 2006). CCAP-GAL4 drives protein expression in a particular class of neurosecretary neurons that make the neuropeptide CCAP (Crustacean Cardio-Active Peptide). This is a very sparse population of cells, thought to be just one per hemisegment of the VNC, whose cell bodies are laterally located in the VNC and which send an axon out in the segmental nerve to end in peptidergic boutons located on a single body wall muscle (Muscle 12) in each hemisegment (Park *et al.*, 2003; Vomel and Wegener, 2007). Thus, if this line is used to drive expression of a fluorescent marker, only the axon of this CCAP neuron will be visible in the segmental nerve. D42-GAL4 drives protein expression chiefly in motor neurons (Yeh *et al.*, 1995). Thus, if this line is used to drive expression of a fluorescent marker, all the efferent axons will be visible in the segmental nerve. Mito-GFP is a GFP-tagged mitochondrial targeting sequence (Pilling *et al.*, 2006). Placed, behind a UAS promoter, it will be expressed in whatever population of neurons is expressing GAL4.

Flies were cultured in a 12 h light–dark cycle at 25 °C on standard soft medium seeded with live yeast. *UAS-mito-GFP* was crossed to either *UAS-CCAP-GAL4* or *UAS-D42-GAL4* and the flies were placed in small fly cages with grape juice agar plates (see below) seeded with yeast as the only food source on which eggs could be laid. Grape juice plates were replaced every day after egg laying (AEL). The progeny of *UAS-mito-GFP* and *CCAP-GAL4* will have mito-GFP expression in CCAP neurons, and the progeny of *UAS-mito-GFP* and *D42-GAL4* will have mito–GFP expression in the broader pool of motor neurons.

3.2. Grape juice agar plates

A 473 ml Welch's Grape Juice, 455 ml sterile water, and 29.3 g Granulated Agar were mixed in a 2 l beaker, and heated in microwave for at least 8 min till all the agar was dissolved. It was stirred occasionally as it cooled down till 65 °C, and 9.9 ml 95% Ethanol and 9.5 ml Glacial Acetic Acid were added. Small tissue culture dishes (60 mm diameter) were filled with the mixed grape juice agar, and covered until the agar had hardened and prior to use.

3.3. Staging and dissection of live larvae

Drosophila larvae reach the third instar stage approximately 4–5 days AEL and live about 48 h before pupariation. Wandering third instar larvae were picked from grape juice agar plates and dissected promptly in fresh Schneider's medium (Invitrogen) with 5 mM EGTA. Larvae were dissected as described (Hurd and Saxton, 1996) on a glass slide (25 × 75 mm, 1.0 mm thick, VWR International, PA) that had been coated with a thin layer of Sylgard (Dow Corning Corporation, MI) and sealed with a silicon imaging chamber (Grace Bio-labs, OR) to hold medium. Dissection pins (FST, CA) that had been cut to 2–3 mm to avoid touching the lens were used to immobilize the larvae. The larvae are cut open along the dorsal midline, the body wall muscles are pinned back, and the major gut structures and salivary glands are carefully removed. The dissection needs to be completed within 5 min and should expose an intact VNC, segmental nerves and muscles.

3.4. Image acquisition

Live dissected larvae in the imaging chamber on the glass slide were maintained in fresh Schneider's medium with 5 mM EGTA at 22 °C for less than an hour. Images were collected on a Nikon E800 epifluorescence upright microscope with SPOT software (Diagnostic Instruments, MI) using a Nikon NIR Apo 60×/N.A.1.0 water objective with excitation at 488 nm. Images were captured every 2 s for 200 frames. Images were processed with Photoshop 10.0 (Adobe, San Jose, CA) using only linear adjustments of contrast and color, and Illustrator 13.0 (Adobe).

3.5. Image analysis

The original time-lapse pictures are saved as a tiff stack by the SPOT software. This tiff stack can be imported into Metamorph as described previously. Since the stage scale of the Nikon E800 microscope needs to be calibrated manually by the users, the actual displacements of these original tiff pictures will not be retained automatically when imported into Metamorph, but the pixel values of these pictures will still be displayed. Kymographs can be generated as in Section 2.6.

3.6. Expected results

The *in vivo* imaging of axonal transport of mitochondria by mito–GFP in *Drosophila* larvae was first established by Pilling *et al.* (2006) in motor neuron axons using D42–GAL4. Since mito–GFP is expressed in all motor neurons, segmental nerves that contain numerous axons are intensely fluorescent with mito–GFP (Fig. 18.3A and Pilling *et al.*, 2006), to the extent that individual mitochondria can be difficult to resolve and difficult to track for any distance. To circumvent this problem, regional photobleaching was required (Pilling *et al.*, 2006). After photobleaching a small area in an axonal bundle, only those mitochondria that move into the bleached zone from

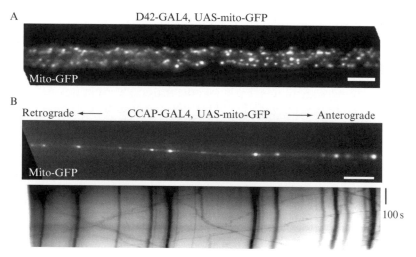

Figure 18.3   Mito–GFP expression and movement in Drosophila larval neurons. (A) A segmental nerve of a third instar Drosophila larva with the VNC to the left. Mito–GFP was expressed in all motor neurons by means of the D42-GAL4 driver. (B) A segmental nerve of a third instar Drosophila larva with the VNC to the left. Mito–GFP was expressed in the CCAP expressing peptidergic neurons by means of the CCAP-GAL4 driver and a single axon is therefore visible within the nerve. The upper panel shows the first frame of a time–lapse movie from this axon. The lower panel is the kymograph generated from the movie, showing the movements of individual mitochondria during the period of analysis. As in Fig. 18.2, vertical black lines correspond to stationary mitochondria and diagonal black lines represent those that were in motion. Data of X. Wang. Scale bars, 10 μm.

either end will be visible in that zone. Although this is an excellent way to analyze moving mitochondria and allows multiple mitochondria moving in either the anterograde or retrograde directions to be analyzed, information about the pool of stationary mitochondria in the original photobleaching area is necessarily lost. Thus it is not possible to determine the percent of mitochondria that are motile versus stationary. In addition, a strong laser is required for photobleaching; in Piling et al. (2006), a 488-nm light at full intensity from the MRC600 confocal laser ($60\times$ objective, zoom factor 4) was used to bleach the segment. Therefore we have modified that technique by expressing mito-GFP in CCAP neurons which selectively innervate a single muscle (Muscle 12) per hemisegment (Park et al., 2003; Vomel and Wegener, 2007). The content of mito-GFP in axonal bundles is therefore greatly reduced and analysis without photobleaching is possible (compare Fig. 18.3A and B). Mitochondria exhibit similar bidirectional movements as in cultured mammalian neurons (Fig. 18.3B, Pilling et al., 2006; Wang and Schwarz, 2009). This method will permit this semi-intact in vivo system to be applied to study axonal transport of mitochondria in conjunction with Drosophila genetics. The axonal transport of mitochondria can therefore be imaged and analyzed in different mutant backgrounds, and the effect of pathological proteins, expressed as transgenes with UAS promoters, can be examined because the transgene will be expressed in the same subset of neurons that are marked with *UAS-mito-GFP*. An additional modification of this method has been made (Miller et al., 2005) in which imaging can be done in intact, living larvae, without the need for a dissection. This method was used for examining the transport of synaptic vesicles, but should be applicable to the examination of mitochondria as well. The method has the advantage of a true in vivo record of organelle motility, but it can be more difficult to keep a significant length of axon in the plane of focus of the microscope and requires the use of chloroform to immobilize the larvae.

4. Conclusion

The advancements of modern microscopic techniques and neuronal culturing systems have greatly facilitated our ability to image the axonal transport of mitochondria in live neurons. Here we introduced two neuronal systems, one is in vitro cultured rat hippocampal neurons, and the other is in vivo Drosophila third instar larval neurons, to record live mitochondria in axons. Both systems have advantages and disadvantages, and both methods can be used together to establish mechanisms and regulatory systems underlying mitochondrial motility, and pathological influences that will figure in neurodegeneration.

ACKNOWLEDGMENTS

We thank Drs A. M. Craig and T. Pawson for reagents; Drs F. Sun, Z. Wills, M. Greenberg for assistance with hippocampal cultures; Dr L. Bu and the Developmental Disorders Research Center Imaging Core and E. Pogoda for technical assistance. This work was supported by NIH RO1GM069808.

REFERENCES

Baloh, R. H., Schmidt, R. E., Pestronk, A., and Milbrandt, J. (2007). Altered axonal mitochondrial transport in the pathogenesis of Charcot–Marie–Tooth disease from mitofusin 2 mutations. *J Neurosci*. **27,** 422–430.

Bolino, A., Bolis, A., Previtali, S. C., Dina, G., Bussini, S., Dati, G., Amadio, S., Del Carro, U., Mruk, D. D., Feltri, M. L., Cheng, C. Y., Quattrini, A., et al. (2004). Disruption of Mtmr2 produces CMT4B1-like neuropathy with myelin outfolding and impaired spermatogenesis. *J. Cell Biol*. **167,** 711–721.

Brand, A. H., and Perrimon, N. (1993). Targeted gene expression as a means of altering cell fates and generating dominant phenotypes. *Development* **118,** 401–415.

Buckman, J. F., Hernandez, H., Kress, G. J., Votyakova, T. V., Pal, S., and Reynolds, I. J. (2001). MitoTracker labeling in primary neuronal and astrocytic cultures: Influence of mitochondrial membrane potential and oxidants. *J. Neurosci. Methods* **104,** 165–176.

Chada, S. R., and Hollenbeck, P. J. (2003). Mitochondrial movement and positioning in axons: The role of growth factor signaling. *J. Exp. Biol*. **206,** 1985–1992.

Chada, S. R., and Hollenbeck, P. J. (2004). Nerve growth factor signaling regulates motility and docking of axonal mitochondria. *Curr. Biol*. **14,** 1272–1276.

Chang, D. T., Honick, A. S., and Reynolds, I. J. (2006). Mitochondrial trafficking to synapses in cultured primary cortical neurons. *J. Neurosci*. **26,** 7035–7045.

Chen, S., Owens, G. C., Crossin, K. L., and Edelman, D. B. (2007). Serotonin stimulates mitochondrial transport in hippocampal neurons. *Mol. Cell Neurosci*. **36,** 472–483.

Clark, I. E., Dodson, M. W., Jiang, C., Cao, J. H., Huh, J. R., Seol, J. H., Yoo, S. J., Hay, B. A., and Guo, M. (2006). Drosophila pink1 is required for mitochondrial function and interacts genetically with parkin. *Nature* **441,** 1162–1166.

Colwill, K., Wells, C. D., Elder, K., Goudreault, M., Hersi, K., Kulkarni, S., Hardy, W. R., Pawson, T., and Morin, G. B. (2006). Modification of the Creator recombination system for proteomics applications—improved expression by addition of splice sites. *BMC Biotechnol*. **6,** 13.

De Vos, K. J., Chapman, A. L., Tennant, M. E., Manser, C., Tudor, E. L., Lau, K. F., Brownlees, J., Ackerley, S., Shaw, P. J., McLoughlin, D. M., Shaw, C. E., Leigh, P. N., et al. (2007). Familial amyotrophic lateral sclerosis-linked SOD1 mutants perturb fast axonal transport to reduce axonal mitochondria content. *Hum. Mol. Genet*. **16,** 2720–2728.

Ferreirinha, F., Quattrini, A., Pirozzi, M., Valsecchi, V., Dina, G., Broccoli, V., Auricchio, A., Piemonte, F., Tozzi, G., Gaeta, L., Casari, G., Ballabio, A., et al. (2004). Axonal degeneration in paraplegin-deficient mice is associated with abnormal mitochondria and impairment of axonal transport. *J. Clin. Invest*. **113,** 231–242.

Gunawardena, S., Her, L. S., Brusch, R. G., Laymon, R. A., Niesman, I. R., Gordesky-Gold, B., Sintasath, L., Bonini, N. M., and Goldstein, L. S. (2003). Disruption of axonal transport by loss of huntingtin or expression of pathogenic polyQ proteins in Drosophila. *Neuron* **40,** 25–40.

Hollenbeck, P. J., and Saxton, W. M. (2005). The axonal transport of mitochondria. *J. Cell Sci.* **118,** 5411–5419.

Hoye, A. T., Davoren, J. E., Wipf, P., Fink, M. P., and Kagan, V. E. (2008). Targeting mitochondria. *Acc. Chem. Res.* **41,** 87–97.

Hurd, D. D., and Saxton, W. M. (1996). Kinesin mutations cause motor neuron disease phenotypes by disrupting fast axonal transport in Drosophila. *Genetics* **144,** 1075–1085.

Kaech, S., and Banker, G. (2006). Culturing hippocampal neurons. *Nat. Protoc.* **1,** 2406–2415.

McDermott, C. J., Grierson, A. J., Wood, J. D., Bingley, M., Wharton, S. B., Bushby, K. M., and Shaw, P. J. (2003). Hereditary spastic paraparesis: Disrupted intracellular transport associated with spastin mutation. *Ann. Neurol.* **54,** 748–759.

Miller, K. E., DeProto, J., Kaufmann, N., Patel, B. N., Duckworth, A., and Van Vactor, D. (2005). Direct observation demonstrates that liprin-α is required for trafficking of synaptic vesicles. *Curr. Biol.* **15,** 684–689.

Morris, R. L., and Hollenbeck, P. J. (1995). Axonal transport of mitochondria along microtubules and F-actin in living vertebrate neurons. *J. Cell Biol.* **131,** 1315–1326.

Overly, C. C., Rieff, H. I., and Hollenbeck, P. J. (1996). Organelle motility and metabolism in axons vs. dendrites of cultured hippocampal neurons. *J. Cell Sci.* **109**(Pt 5), 971–980.

Park, J. H., Schroeder, A. J., Helfrich-Forster, C., Jackson, F. R., and Ewer, J. (2003). Targeted ablation of CCAP neuropeptide-containing neurons of Drosophila causes specific defects in execution and circadian timing of ecdysis behavior. *Development* **130,** 2645–2656.

Pigino, G., Morfini, G., Pelsman, A., Mattson, M. P., Brady, S. T., and Busciglio, J. (2003). Alzheimer's presenilin 1 mutations impair kinesin-based axonal transport. *J. Neurosci.* **23,** 4499–4508.

Pilling, A. D., Horiuchi, D., Lively, C. M., and Saxton, W. M. (2006). Kinesin-1 and Dynein are the primary motors for fast transport of mitochondria in Drosophila motor axons. *Mol. Biol Cell.* **17,** 2057–2068.

Rui, Y., Tiwari, P., Xie, Z., and Zheng, J. Q. (2006). Acute impairment of mitochondrial trafficking by beta-amyloid peptides in hippocampal neurons. *J. Neurosci.* **26,** 10480–10487.

Stowers, R. S., Megeath, L. J., Gorska-Andrzejak, J., Meinertzhagen, I. A., and Schwarz, T. L. (2002). Axonal transport of mitochondria to synapses depends on milton, a novel Drosophila protein. *Neuron* **36,** 1063–1077.

Thies, E., and Mandelkow, E. M. (2007). Missorting of tau in neurons causes degeneration of synapses that can be rescued by the kinase MARK2/Par-1. *J. Neurosci.* **27,** 2896–2907.

Trushina, E., Dyer, R. B., Badger, J. D. 2nd, Ure, D., Eide, L., Tran, D. D., Vrieze, B. T., Legendre-Guillemin, V., McPherson, P. S., Mandavilli, B. S., Van Houten, B., Zeitlin, S., et al. (2004). Mutant huntingtin impairs axonal trafficking in mammalian neurons in vivo and in vitro. *Mol. Cell Biol.* **24,** 8195–8209.

Vomel, M., and Wegener, C. (2007). Neurotransmitter-induced changes in the intracellular calcium concentration suggest a differential central modulation of CCAP neuron subsets in Drosophila. *Dev. Neurobiol.* **67,** 792–808.

Wang, X., Shaw, W. R., Tsang, H. T., Reid, E., and O'Kane, C. J. (2007). Drosophila spichthyin inhibits BMP signaling and regulates synaptic growth and axonal microtubules. *Nat. Neurosci.* **10,** 177–185.

Wang, X., and Schwarz, T. L. (2009). The mechanism of Ca^{2+}-dependent regulation of kinesin-mediated mitochondrial motility. *Cell* **136,** 163–174.

Waters, J., and Smith, S. J. (2003). Mitochondria and release at hippocampal synapses. *Pflugers Arch.* **447,** 363–370.

Yeh, E., Gustafson, K., and Boulianne, G. L. (1995). Green fluorescent protein as a vital marker and reporter of gene expression in Drosophila. *Proc. Natl. Acad. Sci. USA* **92,** 7036–7040.

CHAPTER NINETEEN

GENERATION OF MTDNA-EXCHANGED CYBRIDS FOR DETERMINATION OF THE EFFECTS OF MTDNA MUTATIONS ON TUMOR PHENOTYPES

Kaori Ishikawa*,†,‡ *and* Jun-Ichi Hayashi*

Contents

1. Introduction	336
2. Establishment of ρ^0 Cells of Mice	337
2.1. Treatment of mouse cell lines with DC	338
2.2. Picking up the colonies and confirmation of their lack of mtDNA	338
3. Preparation and Enucleation of mtDNA-Donor Cells	340
3.1. Preparation of mtDNA-donor cells	340
3.2. Enucleation of mtDNA-donor cells	341
4. Cell Fusion between ρ^0 Cells and Cytoplasts	341
4.1. Preparation of cell suspension	342
4.2. Cell fusion	342
4.3. Selection	342
5. Concluding Remarks	343
References	344

Abstract

It has been proposed that mutations of mitochondrial DNA (mtDNA) and resultant mitochondrial dysfunction induce various phenotypes, such as mitochondrial diseases, aging, and tumorigenesis. However, it is difficult to conclude whether mtDNA mutations are truly responsible for these phenotypes due to the regulation of the mitochondrial functions by both mtDNA and nuclear DNA. The mtDNA-exchange techniques are very effective to exclude the influence of

* Graduate School of Life and Environmental Sciences, University of Tsukuba, Tennodai, Tsukuba, Ibaraki, Japan
† Japan Society for the Promotion of Science (JSPS), Chiyoda-ku, Tokyo, Japan
‡ Pharmaceutical Research Division, Takeda Pharmaceutical Company Limited, Tsukuba, Ibaraki, Japan

Methods in Enzymology, Volume 457   © 2009 Elsevier Inc.
ISSN 0076-6879, DOI: 10.1016/S0076-6879(09)05019-8   All rights reserved.

nuclear DNA mutations on expression of these phenotypes. Using these techniques, we recently showed that specific mtDNA mutations can regulate tumor cell metastasis. In this chapter, we describe the methods to establish the mtDNA-exchanged cell lines (cybrids). Applying this technique will reveal how mtDNA mutations are related to various biological phenomena.

1. INTRODUCTION

Accumulation of mutations in mitochondrial DNA (mtDNA) has been considered to be responsible for oncogenic transformation for many years (Petros et al., 2005; Shay and Werbin, 1987; Shidara et al., 2005). In fact, the mutation rate of mtDNA is higher than nuclear DNA, and a higher frequency of somatic mutations in mtDNA was reported in human tumors (Fliss et al., 2000; Polyak et al., 1998). However, this circumstantial evidence cannot exclude the possibility that these mtDNA mutations are not responsible for oncogenic transformation or tumor-related phenotypes. Actually, most mutations found in tumor mtDNA have not been proven to be pathogenic mutations that induce respiration defects. Moreover, there is no statistical evidence that shows an association between respiration defects and oncogenic transformation to develop tumors in patients with mitochondrial diseases expressing respiration defects. If some mtDNA mutations, irrespective of whether they are pathogenic or polymorphic, induce oncogenic transformation, all family members who share the same mother carrying such mutations should develop tumors due to the maternal inheritance of mtDNA (Kaneda et al., 1995; Shitara et al., 1998). However, no bias towards maternal inheritance of tumor development has been reported.

Because mitochondrial respiratory functions are controlled by both mtDNA and nuclear DNA, it is difficult to identify which genome, mitochondrial or nuclear, is responsible for the mitochondrial defects (Augenlicht and Heerdt, 2001). Therefore, establishment of mtDNA-exchanged cells (cybrids) is very effective to exclude the influence of nuclear DNA mutations in tumorigenicity and tumor growth, and to clarify the influence of mtDNA mutations in these processes. Intercellular transfer of mtDNA had revealed that mutations in nuclear DNA but not mtDNA are responsible for tumorigenicity in human cells (Hayashi et al., 1986, 1992) and mouse cells (Akimoto et al., 2005). Recently, we demonstrated that some pathogenic mutations in mtDNA can control the metastatic potential in certain tumor cell lines using the same technology (Ishikawa et al., 2008). Here, we introduce our methods for establishment of the trans-mitochondrial cybrids taking our recent study as an example. The outline of the process to establish the cybrids is shown in Fig. 19.1.

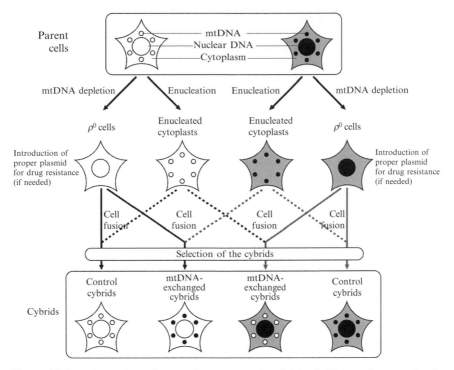

Figure 19.1 Scheme for isolation of trans-mitochondrial cybrids carrying completely exchanged mtDNA between two cell lines. The procedures to isolate mtDNA-exchanged cybrids are as follows; establishment of ρ^0 cells, introduction of drug-resistant plasmid into ρ^0 cells (this step can be omitted if ρ^0 cells are resistant to certain drug which mtDNA-donor cell are sensitive to), enucleation of mtDNA-donor cells, cell fusion of ρ^0 cells and mtDNA-donor cytoplasts, and selection using UP$^-$ (without uridine and pyruvate) medium containing proper drug. To obtain the reliable results, 3–5 cybrid clones containing the same mitochondrial and nuclear DNA should be isolated. The control cybrids, which contain the same mitochondrial and nuclear DNA with parent cells, are also should be isolated to exclude the influences of drug treatment for establishment of ρ^0 cells, chemical and physical irritations during cell fusion, and so on.

2. ESTABLISHMENT OF ρ^0 CELLS OF MICE

The previous mtDNA-transfer system relied on chloramphenicol (CAP) resistant phenotypes caused by mtDNA mutations in rRNA genes for selective isolation of the cybrids (Iwakura *et al.*, 1985). In this system, only the cells that are resistant to CAP can be used as mtDNA donors. This condition limits the variation of mtDNA-donor cells. Moreover, CAP treatment cannot completely remove endogenous wild-type mtDNA in the recipient cells (Hayashi *et al.*, 1982; Oliver and Wallace, 1982). Therefore, the influence of the remaining mtDNA derived from the mtDNA-recipient cells cannot be excluded completely.

Pretreatment of the mtDNA-recipient cells with rhodamine-6G (R6G) seemed to be effective to exclude endogenous mtDNA in the recipient cells (Trounce and Wallace, 1996). However, it was difficult to confirm complete elimination of endogenous mtDNA. Consequently, even when R6G-pretreated cells are used as mtDNA recipients for isolation of the cybrids, the influence of the remaining mtDNA derived from mtDNA-recipient cells again cannot be excluded completely.

These problems could be resolved by using mtDNA-less (ρ^0) cells as the mtDNA recipients. In case of human cell lines, ρ^0 cells can be isolated by treating cells with ethidium bromide (EtBr) (Hayashi *et al.*, 1991; King and Attardi, 1989). Although it has been reported that mouse ρ^0 cells also can be established by exposing cells to high concentrations of EtBr for a long time (Acín-Pérez *et al.*, 2003), our previous study showed that treatment of mouse cell lines with EtBr frequently induced EtBr-resistant cell lines containing mtDNA, instead of isolating ρ^0 cells (Hayashi *et al.*, 1990). Then, we carried out screening of the drugs expected to decrease mtDNA content in mouse cell lines, and found that ditercalinium (DC, an anticancer bis–intercalating agent) was effective to eliminate mtDNA in mouse cells (Inoue *et al.*, 1997).

2.1. Treatment of mouse cell lines with DC

The cells are seeded at 1×10^4 cells/dish, then incubated in 5% CO_2 at 37 °C in Dulbecco's modified Eagle's medium (DMEM) supplemented with 10% fetal calf serum (FCS), 50 μg/ml uridine, and 0.1 mg/ml pyruvate. Twenty-four hours after plating, the cells are treated with 1.5 μg/ml of DC for at least 1 month. The medium containing DC should be changed every 3–4 days. If all of the cells cannot survive in the medium, the lower concentration of DC should be used. On the contrary, if too many colonies were grown for clonal isolation of ρ^0 cells, higher concentration of DC should be used to obtain ρ^0 cells.

2.2. Picking up the colonies and confirmation of their lack of mtDNA

The colonies grown in the DC medium are clonally isolated by the cylinder method. Then, cloned cells are cultivated in the normal medium (with FCS, uridine and pyruvate, and without DC).

To confirm their complete mtDNA depletion, mtDNA-specific PCR should be carried out. In the case of mouse cells, the nucleotide sequences from n.p. 15,497 to 15,523 and 15,713 to 15,690 (sequence ref. AY172335)

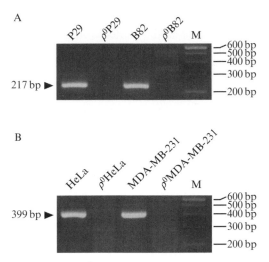

Figure 19.2 Confirmation of ρ^0 cells lacking their mtDNA completely by PCR. Amplifications of mtDNA by PCR were carried out using primer sets specific to mouse mtDNA (**A**) and human mtDNA (**B**), respectively. While each parent cell line showed the signal of their mtDNA, ρ^0 cells did not. P29, a Lewis lung carcinoma cell line derived from C57BL/6 mouse; B82, a fibrosarcoma cell line derived from C3H/An mouse; HeLa and MDA-MB-231, human tumor cell lines derived from cervical and breast cancer, respectively. M represents DNA mass ladder.

are used as oligonucleotide primers. If the cloned cells possess their mtDNA, 217-bp fragments would be amplified (Fig. 19.2A). In the case of human cells, we use the nucleotide sequences from n.p. 3153 to 3174 and 3551 to 3528 (sequence ref. AC_000021) as oligonucleotide primers. This primer set amplifies 399-bp fragments if the cells contain mtDNA (Fig. 19.2B). Since nuclear DNA in both mouse and human cells contain sequences that are homologous or completely identical to those in part of the mtDNA, it should be noted that the primer sets specific to only mtDNA sequences are required for confirmation of ρ^0 cells. When the presumptive ρ^0 cells possessed a very small copy number of mtDNA that was not detectable by the PCR amplification, the reduced copy number would be completely recovered during long-term cultivation. Therefore, confirmation of ρ^0 cells by PCR should be repeated after 3–5 months of their cloning. When mtDNA signals are not detected, the cells are considered to be ρ^0 cells. For cultivation and proliferation of the ρ^0 cells, the medium supplemented with both uridine and pyruvate (UP$^+$ medium) is indispensable (King and Attardi, 1989).

To exclude mtDNA-donor cells and isolate the cybrids (mtDNA-exchanged cells) selectively, ρ^0 cells should be resistant to some drugs. For example, mouse fibrosarcoma B82 cells derived from L929 cells, are deficient in thymidine kinase, resulting in simultaneous expression of resistant and sensitive phenotypes to BrdU (5-bromo-2'-deoxyuridine) and HAT (hypoxanthine, aminopterin, and thymidine), respectively. On the contrary, P29 cells derived mouse Lewis lung carcinoma cells possess normal thymidine kinase activity, resulting in expression of HAT-resistant and BrdU-sensitive phenotypes. Thus, the selection medium with BrdU can be used for selective isolation of the cybrids possessing nuclear DNA from B82 cells and mtDNA from P29 cells (about selection, see also the Section 4.3). However, if the phenotypes of drug resistance of both mtDNA-donor and mtDNA-recipi-ent cells are identical, resistance to another drug (e.g., neomycin) should be introduced artificially into ρ^0 cells before cell fusion. For this, some plasmid vectors (e.g., pcDNA3.1 (Invitrogen), pSV2-neo (ATCC), etc.) are avail-able. Our recent study (Ishikawa *et al.*, 2008) used low-metastatic P29 and high-metastatic A11 cell lines derived from the same Lewis lung carcinoma cells. Since their drug resistance characteristics are identical, we transfected the pSV2-neo plasmid into ρ^0 P29 cells and ρ^0 A11 cells, and fused them with enucleated A11 and P29 cells, respectively, for isolation of the cybrids with mtDNA exchanged between P29 and A11 cells.

3. Preparation and Enucleation of mtDNA-Donor Cells

Enucleation is carried out by high-speed centrifugation of the mtDNA-donor cells in the presence of cytochalasin B, which blocks the formation of microfilaments and shortens actin filaments. Although several protocols to enucleate the mtDNA-donor cells and isolate the cytoplasts have been established (King and Attardi, 1989; Trounce *et al.*, 1994), we would like to show our own methods in this section.

3.1. Preparation of mtDNA-donor cells

Appropriate numbers of mtDNA-donor cells are seeded onto glass discs in culture dishes (eight discs can be placed in a normal culture dish). We use the discs with 25 mm diameter and 1 mm thickness fitting to the inside diameter of our centrifugation tubes (NALGENE, Cat. No. 3117–9500). The number of the cells for seeding should be determined depending on the growth rates of mtDNA-donor cells. For effective enucleation, the culture conditions in which cells reach 80–90% confluence in 2–3 days after seeding are preferable. In case of the P29 cells, we seed 1.2×10^5 cells/dish, 3 days before the enucleation, whereas we seed 5.0×10^5 cells/dish for the B82 cells due to their slower growth rate.

3.2. Enucleation of mtDNA-donor cells

All the equipment used for centrifugation (centrifuge tubes with caps and plastic rings) are autoclaved before the enucleation. The plastic rings (made with polycarbonate, with an inside diameter, outside diameter, and height are 18, 21.5, and 12 mm, respectively) are needed to set of glass discs into the centrifuge tubes.

The enucleation medium containing 10 μg/ml of cytochalasin B (SIGMA) is prepared by dissolving the stock solution of cytochalasin B (1 mg/ml in ethanol) into normal medium. The medium of mtDNA-donor cells is replaced by enucleation medium, and cells are incubated for 10 min at 37 °C. After this process, cells appear shrunken due to their relaxed cytoskeletons. Then, glass discs are placed (the cell-attached side should be downward) onto the plastic rings in the centrifuge tubes, containing 10 ml of enucleation medium. The centrifugation conditions vary depending on the cell types, since they express different adhesiveness to the discs. For example, centrifugation at 7500g for 10 min at 37 °C corresponds to effective enucleation condition of P29 and A11 cells. On the other hand, suitable enucleation condition of B82 cells is 22,500g centrifugation for 30 min at 37 °C.

After the centrifugation, the discs are placed into a culture dish containing 10 ml of PBS. When the cells are enucleated effectively, the enucleated cells with separating nuclei like "toy balloons with strings" (spherical nuclei tied to cytoplasts with thin membranes) are frequently observed. If most cells are detached from the glass discs, then the centrifugation speed is too fast or its time is too long. On the contrary, when cells with nuclei like balloons are rarely observed, faster and/or longer centrifugation should be carried out.

Then, the glass discs with enucleated cells are put into normal medium (without cytochalasin B) in a culture dish, and incubated for more than 30 min. During this "recovery" step, the enucleated cells attach themselves tightly to the discs again showing shapes similar to those observed before the cytochalasin B treatment. These resultant enucleated cells (cytoplasts) should not be left longer than 3 h.

4. CELL FUSION BETWEEN ρ^0 CELLS AND CYTOPLASTS

Originally, cell fusion has been carried out using Sendai virus (SeV or HVJ; Hemagglutinating virus of Japan). At present, polyethylene glycol (PEG) or an electric pulse are used for cell fusion. In this section, we introduce the representative protocol using PEG for fusion between ρ^0 cells and mtDNA-donor cytoplasts.

4.1. Preparation of cell suspension

The mtDNA-donor cytoplasts and mtDNA-recipient ρ^0 cells are trypsinized, and then collected into tubes. To minimize the loss of cytoplasts and harvest them effectively, glass discs are washed at least twice with the medium. If the culture media of ρ^0 cells and mtDNA donors are different, the medium of ρ^0 cells should be used to harvest both ρ^0 cells and the cytoplasts.

After calculation of the cell numbers, ρ^0 cells with one-fifteenth the number of the cytoplasts are added into the cytoplast suspension. Then, a part of the mixed suspension (about one-tenth of total volume) is seeded into a culture dish before fusion. These simple mixtures (PEG$^-$) are used as unfused negative controls. Please see the Section 4.3 for more information about selection.

4.2. Cell fusion

The remaining cell mixtures are centrifuged at 350g for 5 min, and the supernatant is removed. Then, the pellet is spun-down again to obtain dry pellets by removing the supernatant completely. This step is important because the remaining supernatant dilutes PEG, and weakens the effect of PEG on the cell fusion. Next, 100 μl of PEG (MW1500, Boehringer Mannheim) is added onto the pellet, then the tube is allowed to stand for 15 s and then tapped strongly for 15 s. Then, 5 ml of the medium is added immediately, and the pellet is suspended gently. The cell suspension is centrifuged and the resultant pellet is centrifuged again to remove the medium containing PEG completely. Finally, the pellet is resuspended into 5 ml of medium, and the cells are cultivated for selection. Since the efficiencies of cell fusion differ widely in cell types, cells should be seeded into 3–4 dishes at various concentrations to pick up the cybrid colonies clonally. Normally, we split the 5 ml of cell suspension into four dishes (2.0, 1.5, 1.0, and 0.5 ml suspension/dish). The medium is added into all the dishes up to 10 ml, and cells are incubated at 37 °C, 5% CO_2.

Although the condition of cell fusion described above (15 s for standing, next 15 s for tapping) is the most effective standard for various cell types, slight modification of the protocol may be required in some specific cell lines.

4.3. Selection

Selection is started 2–3 days after fusion. Culture medium in each dish (including the dish of PEG$^-$) is aspirated and 3–5 ml of fresh medium without uridine and pyruvate (UP$^-$ medium) is added to wash the remaining uridine and pyruvate, then, the medium is aspirated again carefully.

Table 19.1 A list of representative drugs for selection of cybrids

Drug	Selected (resistant) cells	Killed (sensitive) cells	Suggested concentration
HAT	TK$^+$ cells	TK$^-$ cells	Hypoxanthine, 100 μM
	HPRT$^+$ cells	HPRT$^-$ cells	Aminopterin, 0.4 μM
			Thymidine, 16 μM
BrdU	TK$^-$ cells	TK$^+$ cells	30 $\mu g/ml$
Tg	HPRT$^-$ cells	HPRT$^+$ cells	20 μM

Abbreviations: HAT, hypoxanthine/aminopterin/thymidine; BrdU, 5-bromo-2′-deoxyuridine; Tg, 6-thioguanine; TK, thymidine kinase; HPRT, hypoxanthine guanine phosphoribosyl transferase.

After washing, 10 ml of UP$^-$ medium and the drug which ρ^0 cells are resistant to and mtDNA-donor cells are sensitive to are added. Unfused ρ^0 cells cannot survive in UP$^-$ medium and intact mtDNA-donor cells (without enucleation) cannot survive in the presence of the drug. Therefore, this treatment (UP$^-$ + drug) allows only the fused cybrids to survive and grow. The representative drugs for selection are listed in Table 19.1.

The expected cybrid colonies usually grow to form sufficient sizes for cloning 15–20 days after starting the selection. It should be confirmed that no colonies grow in the PEG$^-$ dish. If some colonies are still alive in the PEG$^-$ dish, selection should be extended for several days before cloning. The resultant cybrid colonies are isolated by the cylinder method and transfered into a 24-well dish, and cultivated for further growth. To avoid the contamination of remaining ρ^0 cells or intact mtDNA-donor cells, treatment of UP$^-$ and drug should be continued for additional 5–7 days.

The genotyping of the mtDNA in the presumptive cybrids should be carried out to examine whether they possess mtDNA derived from mtDNA-donor cells using proper methods such as RFLP (restriction fragment length polymorphism) analysis of PCR products or Southern blot analysis. A representative example for mtDNA genotyping of cybrid by RFLP is shown in Fig. 19.3. Then, the cybrids containing introduced mtDNA are ready for further investigations or experiments.

5. CONCLUDING REMARKS

Because mitochondria are multifunctional organelle, genetic and reverse-genetic approaches are needed to understand the role of mitochondria in each biological phenomenon. However, at the present time, no technique or method is available to manipulate mtDNA (such as knocking-out of the gene or gene transfer) directly. Therefore, the mtDNA transfer

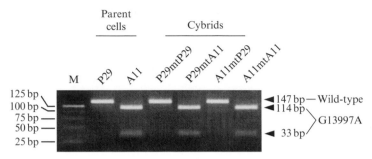

Figure 19.3 Identification of the mtDNA genotypes by means of restriction enzyme digestion of the PCR products. Because only the mtDNA from high-metastatic A11 cells possesses G13997A substitution, the PCR products of the mtDNA from P29 and A11 cells can be distinguished by RFLP analysis. The PCR products of the mtDNA with the G13997A mutation from A11 cells produce 114- and 33-bp fragments because of the gain of an *Afl* II site by a G13997A substitution, whereas those of the mtDNA from P29 cells produce a 147-bp fragment due to the absence of the *Afl* II site (Ishikawa *et al.*, 2008). P29mtP29, a control cybrid possessing both nuclear and mitochondrial genome from P29 cells; P29mtA11, a mtDNA-exchanged cybrid possessing nuclear DNA from P29 cells but mtDNA from A11 cells; A11mtP29, a mtDNA-exchanged cybrid possessing nuclear DNA from A11 cells but mtDNA from P29 cells; A11mtA11, a control cybrid possessing both nuclear and mitochondrial genome from A11 cells. M represents DNA mass ladder.

technique is a very valuable tool to investigate whether and/or how mtDNA and its mutations interact with various phenomena. In fact, this technique has been used in basic research on aging (Isobe *et al.*, 1998; Ito *et al.*, 1999), disease phenotypes caused by mtDNA mutations (Brown *et al.*, 2000; Inoue *et al.*, 2000; Malfatti *et al.*, 2007), and cancer biology (Akimoto *et al.*, 2005; Ishikawa *et al.*, 2008). The "cybrid work" can be applied to other life science fields where mtDNA may play a role. Without this technique, we cannot even distinguish which genomes, nuclear or mitochondrial, is responsible for certain phenomena and phenotypes.

REFERENCES

Acín-Pérez, R., Bayona-Bafaluy, M. P., Bueno, M., Machicado, C., Fernández-Silva, P., Pérez-Martos, A., Montoya, J., López-Pérez, M. J., Sancho, J., and Enríquez, J. A. (2003). An intragenic suppressor in the cytochrome *c* oxidase I gene of mouse mitochondrial DNA. *Hum. Mol. Genet.* **12,** 329–339.

Akimoto, M., Niikura, M., Ichikawa, M., Yonekawa, H., Nakada, K., Honma, Y., and Hayashi, J.-I. (2005). Nuclear DNA but not mtDNA controls tumor phenotypes in mouse cells. *Biochem. Biophys. Res. Commun.* **327,** 1028–1035.

Augenlicht, L. H., and Heerdt, B. G. (2001). Mitochondria: Integrators in tumorigenesis? *Nat. Genet.* **28,** 104–105.

Brown, M. D., Trounce, I. A., Jun, A. S., Allen, J. C., and Wallace, D. C. (2000). Functional analysis of lymphoblast and cybrid mitochondria containing the 3460, 11778, or 14484 Leber's hereditary optic neuropathy mitochondrial DNA mutation. *J. Biol. Chem.* **275,** 39831–39836.

Fliss, M. S., Usadel, H., Caballero, O. L., Wu, L., Buta, M. R., Eleff, S. M., Jen, J., and Sidransky, D. (2000). Facile detection of mitochondrial DNA mutations in tumors and bodily fluids. *Science* **287,** 2017–2019.

Hayashi, J.-I., Gotoh, O., Tagashira, Y., Tosu, M., Sekiguchi, T., and Yoshida, M. C. (1982). Mitochondrial DNA analysis of mouse-rat hybrid cells: Effect of chloramphenicol selection on the relative amounts of parental mitochondrial DNAs. *Somatic Cell Genet.* **8,** 67–81.

Hayashi, J.-I., Ohta, S., Kikuchi, A., Takemitsu, M., Goto, Y., and Nonaka, I. (1991). Introduction of disease-related mitochondrial DNA deletions into HeLa cells lacking mitochondrial DNA results in mitochondrial dysfunction. *Proc. Natl. Acad. Sci. USA* **88,** 10614–10618.

Hayashi, J.-I., Takemitsu, M., and Nonaka, I. (1992). Recovery of the missing tumorigenicity in mitochondrial DNA-less HeLa cells by introduction of mitochondrial DNA from normal human cells. *Somat. Cell Mol. Genet.* **18,** 123–129.

Hayashi, J.-I., Tanaka, M., Sato, W., Ozawa, T., Yonekawa, H., Kagawa, Y., and Ohta, S. (1990). Effects of ethidium bromide treatment of mouse cells on expression and assembly of nuclear-coded subunits of complexes involved in the oxidative phosphorylation. *Biochem. Biophys. Res. Commun.* **167,** 216–221.

Hayashi, J.-I., Werbin, H., and Shay, J. W. (1986). Effects of normal human fibroblast mitochondrial DNA on segregation of HeLaTG Mitochondrial DNA and on tumorigenicity of HeLaTG cells. *Cancer Res.* **46,** 4001–4006.

Inoue, K., Ito, S., Takai, D., Soejima, A., Shisa, H., LePecq, J. B., Segal-Bendirdjian, E., Kagawa, Y., and Hayashi, J.-I. (1997). Isolation of mitochondrial DNA-less mouse cell lines and their application for trapping mouse synaptosomal mitochondrial DNA with deletion mutations. *J. Biol. Chem.* **272,** 15510–15515.

Inoue, K., Nakada, K., Ogura, A., Isobe, K., Goto, Y., Nonaka, I., and Hayashi, J.-I. (2000). Generation of mice with mitochondrial dysfunction by introducing mouse mtDNA carrying a deletion into zygotes. *Nat. Genet.* **26,** 176–181.

Ishikawa, K., Takenaga, K., Akimoto, M., Koshikawa, N., Yamaguchi, A., Imanishi, H., Nakada, K., Honma, Y., and Hayashi, J.-I. (2008). ROS-generating mitochondrial DNA mutations can regulate tumor cell metastasis. *Science* **320,** 661–664.

Isobe, K., Ito, S., Hosaka, H., Iwamura, Y., Kondo, H., Kagawa, Y., and Hayashi, J.-I. (1998). Nuclear-recessive mutations of factors involved in mitochondrial translation are responsible for age-related respiration deficiency of human skin fibroblasts. *J. Biol. Chem.* **273,** 4601–4606.

Ito, S., Ohta, S., Nishimaki, K., Kagawa, Y., Soma, R., Kuno, S. Y., Komatsuzaki, Y., Mizusawa, H., and Hayashi, J.-I. (1999). Functional integrity of mitochondrial genomes in human platelets and autopsied brain tissues from elderly patients with Alzheimer's disease. *Proc. Natl. Acad. Sci. USA* **96,** 2099–2103.

Iwakura, Y., Nozaki, M., Asano, M., Yoshida, M. C., Tsukada, Y., Hibi, N., Ochiai, A., Tahara, E., Tosu, M., and Sekiguchi, T. (1985). Pleiotropic phenotypic expression in cybrids derived from mouse teratocarcinoma cells fused with rat myoblast cytoplasts. *Cell* **43,** 777–791.

Kaneda, H., Hayashi, J., Takahama, S., Taya, C., Lindahl, K. F., and Yonekawa, H. (1995). Elimination of paternal mitochondrial DNA in intraspecific crosses during early mouse embryogenesis. *Proc. Natl. Acad. Sci. USA* **92,** 4542–4546.

King, M. P., and Attardi, G. (1989). Human cells lacking mtDNA: Repopulation with exogenous mitochondria by complementation. *Science* **246,** 500–503.

Malfatti, E., Bugiani, M., Invernizzi, F., de Souza, C. F., Farina, L., Carrara, F., Lamantea, E., Antozzi, C., Confalonieri, P., Sanseverino, M. T., Giugliani, R., Uziel, G., *et al.* (2007). Novel mutations of ND genes in complex I deficiency associated with mitochondrial encephalopathy. *Brain* **130,** 1894–1904.

Oliver, N. A., and Wallace, D. C. (1982). Assignment of two mitochondrially synthesized polypeptides to human mitochondrial DNA and their use in the study of intracellular mitochondrial interaction. *Mol. Cell Biol.* **2,** 30–41.

Petros, J. A., Baumann, A. K., Ruiz-Pesini, E., Amin, M. B., Sun, C. Q., Hall, J., Lim, S., Issa, M. M., Flanders, W. D., Hosseini, S. H., Marshall, F. F., and Wallace, D. C. (2005). mtDNA mutations increase tumorigenicity in prostate cancer. *Proc. Natl. Acad. Sci. USA* **18,** 719–724.

Polyak, K., Li, Y., Zhu, H., Lengauer, C., Willson, J. K., Markowitz, S. D., Trush, M. A., Kinzler, K. W., and Vogelstein, B. (1998). Somatic mutations of the mitochondrial genome in human colorectal tumours. *Nat. Genet.* **20,** 291–293.

Shay, J. W., and Werbin, H. (1987). Are mitochondrial DNA mutations involved in the carcinogenic process? *Mutat. Res.* **186,** 149–160.

Shitara, H., Hayashi, J.-I., Takahama, S., Kaneda, H., and Yonekawa, H. (1998). Maternal inheritance of mouse mtDNA in interspecific hybrids: Segregation of the leaked paternal mtDNA followed by the prevention of subsequent paternal leakage. *Genetics* **148,** 851–857.

Shidara, Y., Yamagata, K., Kanamori, T., Nakano, K., Kwong, J. Q., Manfredi, G., Oda, H., and Ohta, S. (2005). Positive contribution of pathogenic mutations in the mitochondrial genome to the promotion of cancer by prevention from apoptosis. *Cancer Res.* **65,** 1655–1663.

Trounce, I., Neill, S., and Wallace, D. C. (1994). Cytoplasmic transfer of the mtDNA nt 8993 T→G (ATP6) point mutation associated with Leigh syndrome into mtDNA-less cells demonstrates cosegregation with a decrease in state III respiration and ADP/O ratio. *Proc. Natl. Acad. Sci. USA* **91,** 8344–8348.

Trounce, I., and Wallace, D. C. (1996). Production of transmitochondrial mouse cell lines by cybrid rescue of rhodamine-6G pretreated L-cells. *Somat. Cell Mol. Genet.* **22,** 81–85.

MITOCHONDRIAL FUNCTION IN PANCREATIC BETA CELLS AND INSULIN-RESPONSIVE TISSUES

Functional Assessment of Isolated Mitochondria In Vitro

Ian R. Lanza *and* K. Sreekumaran Nair

Contents

1. Introduction	350
2. Mitochondrial Isolation Procedures	354
2.1. Tissue homogenization	355
2.2. Separation of mitochondria by differential centrifugation	357
2.3. Mitochondrial yield and integrity	358
3. Mitochondrial ATP Production	360
3.1. Principles of the bioluminescent approach	360
3.2. Preparation of substrates and ADP	361
3.3. Running the experiment and quantitation of ATP	362
4. Mitochondrial Respiration	365
4.1. Overview	365
4.2. An oxygraph protocol for mitochondrial assessment	365
5. Summary	369
References	369

Abstract

Mitochondria play a pivotal role in cellular function, not only as a major site of ATP production, but also by regulating energy expenditure, apoptosis signaling, and production of reactive oxygen species. Altered mitochondrial function is reported to be a key underlying mechanism of many pathological states and in the aging process. Functional measurements of intact mitochondria isolated from fresh tissue provides distinct information regarding the function of these organelles that complements conventional mitochondrial assays using previously frozen tissue as well as *in vivo* assessment using techniques such as magnetic resonance and near-infrared spectroscopy. This chapter describes the process by which mitochondria are isolated from small amounts of human skeletal muscle obtained by needle biopsy and two approaches used to assess mitochondrial oxidative capacity and other key components of mitochondrial physiology.

Division of Endocrinology, Endocrinology Research Unit, Mayo Clinic College of Medicine, Rochester, Minnesota, USA

Methods in Enzymology, Volume 457
ISSN 0076-6879, DOI: 10.1016/S0076-6879(09)05020-4

We first describe a bioluminescent approach for measuring the rates of mitochondrial ATP production. Firefly luciferase catalyzes a light-emitting reaction whereby the substrate luciferin is oxidized in an ATP-dependent manner. A luminometer is used to quantify the light signal, which is proportional to ATP concentration. We also review a method involving polarographic measurement of oxygen consumption. Measurements of oxygen consumption, which previously required large amounts of tissue, are now feasible with very small amounts of sample obtained by needle biopsy due to recent advances in the field of high-resolution respirometry. We illustrate how careful attention to substrate combinations and inhibitors allows an abundance of unique functional information to be obtained from isolated mitochondria, including function at various energetic states, oxidative capacity with electron flow through distinct complexes, coupling of oxygen consumption to ATP production, and membrane integrity. These measurements, together with studies of mitochondrial DNA abundance, mRNA levels, protein expression, and synthesis rates of mitochondrial proteins provide insightful mechanistic information about mitochondria in a variety of tissue types.

1. INTRODUCTION

Mitochondria are believed to have originated in eukaryotic cells by endosymbiosis∼2 billion years ago (Gray *et al.*, 1999). Although most mitochondrial proteins are encoded by nuclear DNA and imported into the organelle, mitochondria contain ribosomes and between 2 and 10 copies of circular DNA containing 13 protein-encoding regions and 22 tRNA-encoding genes. Mitochondria are an essential part of normal cellular function, particularly in their role in oxidizing carbon substrates to satisfy cellular energy requirements. Hydrolysis of ATP to ADP or AMP releases free energy to drive energy-requiring processes such as cross–bridge cycling, maintenance of membrane potentials, sarcoplasmic reticulum calcium uptake, signal transduction, membrane transport, and protein synthesis. Nearly 90% of cellular ATP demand is satisfied by mitochondrial ATP synthesis, which, as illustrated in (Fig. 20.1), is driven by a proton gradient across the inner mitochondrial membrane as a result of oxidation of carbon substrates in the tricarboxylic acid (TCA) cycle and electron flow through various enzymes and proteins at the inner mitochondrial membrane. These organelles not only produce chemical energy in the form of ATP, but also liberate energy through thermogenic uncoupling, regulate apoptosis, and are a major source of reactive oxygen species (ROS). While ROS production plays an important role in cell signaling, increased ROS levels can result in oxidative damage to mitochondrial DNA, particularly when ROS production exceeds the capacity of antioxidant defense systems (glutathione peroxidase, catalase, and superoxide dismutase) and DNA repair. Indeed, accumulated oxidative damage to various molecules has been proposed as a mechanism by which the aging

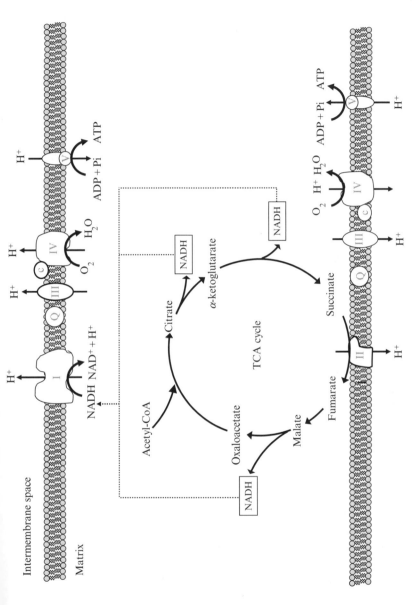

Figure 20.1 Overview of mitochondrial oxidative phosphorylation. Phosphorylation of ADP to ATP at complex V (ATP synthase) is driven by a proton gradient across the inner mitochondrial membrane. Oxidation of carbon substrates in the TCA cycle generates reducing equivalents that subsequently provide electron flow to the electron transport system. Electrons are transferred from NADH through complex I (NADH dehydrogenase) and oxidation of succinate by complex II (succinate dehydrogenase). Electron flow from other sources such as electron transferring flavoprotein is not shown.

process is accelerated (Harman, 1956). The importance of mitochondria in cellular function and energy balance has spawned much interest in their role in the aging process and metabolic diseases such as type 2 diabetes. Accurate, precise assessment of tissue mitochondrial function is necessary to understand the aging process and the underlying mechanisms of many diseases. Various tools, each with distinct advantages and caveats, have been used to probe mitochondrial properties *in vitro* and *in vivo*. For example, magnetic resonance spectroscopy permits noninvasive *in vivo* assessment of muscle oxidative capacity (Argov et al., 1987; Kent-Braun and Ng, 2000; Lanza et al., 2007), steady-state mitochondrial ATP synthesis rate (Befroy et al., 2008; Lebon et al., 2001; Petersen et al., 2003), and TCA cycle flux (Befroy et al., 2008; Lebon et al., 2001; Petersen et al., 2004). The ability to assess mitochondrial function under conditions where circulatory and regulatory systems are intact is a strength of these *in vivo* approaches. Experiments performed *ex vivo* permit a reductionist approach to probe specific molecular levels within the complex pathway of mitochondrial ATP synthesis; information which is crucial to understanding mechanisms by which mitochondrial function is altered with aging and disease and pathways by which exercise and pharmacological interventions exert beneficial effects. Moreover, the ability to quantify mito-chondrial DNA copy numbers (Asmann et al., 2006; Barazzoni et al., 2000; Chow et al., 2007; Short et al., 2005), expression (mRNA and protein) of various mitochondrial proteins (Lanza et al., 2008; Short et al., 2005), mito-chondrial enzyme activities (Rooyackers et al., 1996; Short et al., 2003), and stable isotope-based synthesis of mitochondrial proteins (Rooyackers et al., 1996) allows an enormous amount of complementary data to be obtained from the same biopsy sample that is used for measurements of mitochondrial function. In addition, a recent approach to measure synthesis rates of individ-ual skeletal muscle mitochondrial proteins has been reported (Jaleel et al., 2008), which offers an opportunity to determine the translational efficiency of gene transcripts, especially when combined with large scale gene transcript measurements (Asmann et al., 2006). Together, these methods permit com-prehensive investigation of mitochondrial function and its regulation at various molecular and cellular levels to understand the underlying mechan-isms of mitochondrial changes that occur in conditions related to aging, disease, and physical activity.

Historically, the maximal activities of key mitochondrial enzymes have been widely used as indices of mitochondrial oxidative capacity (Houmard et al., 1998; Rooyackers et al., 1996; Wicks and Hood, 1991). Citrate synthase is a common matrix enzyme marker, while succinate dehydrogenase and cytochrome *c* oxidase are frequently measured as representative enzymes from the inner mitochondrial membrane. Since only small amounts of previ-ously frozen tissue are required, spectrophotometric-based enzyme activity assays are well-suited for human studies where tissue quantities are limited. Although it is not unreasonable to relate maximal activities of these marker enzymes to mitochondrial oxidative capacity, it is unlikely that a single enzyme can accurately reflect the collective function of an organelle as complex as the mitochondrion. Thus, there is a critical need to implement more direct functional measurements of mitochondrial function and oxidative capacity. Measurement of oxygen consumption in isolated mitochondria, pioneered by Britton Chance over 50 years ago (Chance and Williams, 1956), has long been used to assess function of freshly isolated mitochondria. Accurate respiration measurements using conventional respirometers necessitated large quantities of tissue, limiting its practicality for routine human studies. However, recent technological advances in the field of high-resolution respirometry allow these types of measurements using very small amounts of sample (Gnaiger, 2001; Haller et al., 1994), permitting investigators to obtain a wealth of information regarding mitochondrial function from small amounts of human biopsy tissue. In addition to respiration-based measurements, it is also possible to assess mitochondrial function by measuring synthesis of ATP, the ultimate end-product of oxidative phosphorylation. The ability to measure ATP production in freshly isolated intact mitochondria offers an alternative approach for assessing mitochondrial function that provides distinct, complementary information to traditional measurements of oxygen consumption. By exploiting a photon-emitting reaction involving ATP, luciferin, and firefly luciferase, it is possible to quantify the rates of ATP production under various conditions.

For many years, luciferase-based measurements of skeletal muscle mitochondrial ATP production rates (MAPR) (Wibom and Hultman, 1990) have become an integral part of studies in our laboratory. We have applied this method in various ways, including to demonstrate an age-related decline in mitochondrial oxidative capacity (Short et al., 2005) and the absence of this trend in individuals who engage in chronic endurance exercise (Lanza et al., 2008). Using this technique, we have observed paradoxically elevated MAPR in Asian Indian individuals in spite of stark insulin resistance compared to people of northern European descent (Nair et al., 2008). To help elucidate the controversial relationship between insulin resistance and mitochondrial function, we have demonstrated that insulin stimulates MAPR in healthy controls but not in insulin resistant individuals (Stump et al., 2003). These studies point to insulin as a key regulator of mitochondrial function and impaired insulin signaling as a potential mechanism by which mitochondrial dysfunction manifests in insulin resistant people (Asmann et al., 2006). Combining measurements of maximal ATP production with measures of mitochondrial respiration using various substrates offer an opportunity to asses defects at specific levels within the process of mitochondrial energy metabolism.

This chapter describes methodology for assessing the function of mitochondria isolated from skeletal muscle, with particular attention to the procedures for isolating mitochondria from skeletal muscle samples, a luciferase-based bioluminescent measurement of mitochondrial ATP production, and mitochondrial oxygen consumption measurements using high-resolution respirometry. Although we describe the detailed procedures for functional measurements in isolated mitochondria from skeletal muscle, similar approaches with relatively minor modifications could be applied to investigate other tissues such as liver, adipose, kidney, heart, brain, permeabilized fibers, and cell culture.

2. MITOCHONDRIAL ISOLATION PROCEDURES

Skeletal muscle contains two distinct populations of mitochondria (Cogswell et al., 1993). Subsarcolemmal mitochondria ($\sim$20% of total mitochondrial content) are defined as those within 2 μm from the sarcolemma, often densely clustered in the proximity of nuclei and easily liberated by gentle, mechanical homogenization (Elander et al., 1985). Intermyofibrillar mitochondria ($\sim$80% of total mitochondria) are imbedded amongst the contractile machinery and are best liberated by softening the tissue with proteolytic enzymes prior to homogenization (Elander et al., 1985). Improper use of proteolytic enzymes may potentially damage mitochondria; however, careful restriction of proteolytic action prevents damage to the organelles while effectively liberating the organelles from contractile machinery. The two distinct populations of mitochondria can be readily visualized and quantified using electron microscopy. The two populations appear to supply ATP exclusively to their respective subcellular regions with subsarcolemmal mitochondria supplying ATP to the membrane ion and substrate transporters and intermyofibrillar mitochondria supplying ATP primarily to myosin ATPases (Hood, 2001). Furthermore, subsarcolemmal mitochondria have been shown to be more sensitive to changes induced by exercise and disuse (Hood, 2001) and more susceptible to the detrimental effects of aging, type 2 diabetes, and obesity (Menshikova et al., 2005; Ritov et al., 2005). The possibility that distinct mitochondrial populations may be affected in different ways by aging and physical activity has prompted some groups to develop methods for isolating distinct mitochondrial fractions (Krieger et al., 1980; Menshikova et al., 2006; Ritov et al., 2005). Muscle tissue is crudely homogenized to liberate subsarcolemmal mitochondria, which are separated by centrifugation. The remaining intermyofibrillar components are liberated by a more aggressive procedure involving protease or a potassium chloride buffer, which partially dissolves myosin. This approach is complicated by uncertain purity of each isolated fraction and the potential for compromised integrity of the mitochondria. Since our group is interested in characterizing the entire population of skeletal muscle mitochondria regardless of cellular location, we employ an isolation procedure to maximize the yield of both subsarcolemmal and intermyofibrillar mitochondria. A high-fractional yield typically requires harsh homogenization methods that often damage the organelles. In contrast, gentle homogenizing procedures preserves the integrity of the outer membrane at the expense of lower yield. To optimize both yield and integrity, our isolation methods are largely based on the careful work of Rasmussen and colleagues (Rasmussen and Rasmussen, 1997; Rasmussen et al., 1997) who have extensively published on this topic.

2.1. Tissue homogenization

Human muscle tissue is obtained from the vastus lateralis muscle using a modified Bergstrom needle under suction (Edwards et al., 1980; Nair et al., 1988). About 50–100 mg of muscle tissue is immediately transferred to gauze soaked in ice-cold saline for the short transport (<3 min) from the patient room to the laboratory. An illustration of the homogenization procedure is shown in (Fig. 20.2) to accompany the following description. The tissue is trimmed of any obvious fat and connective tissue, blotted with gauze, and weighed before transfer to an ice-cold glass Petri dish with 1–2 ml of cold homogenization buffer. While some groups use nonionic homogenization buffers containing sucrose, mannitol, or sorbitol (Drew and Leeuwenburgh, 2003; Trounce et al., 1996), an ionic medium containing KCl is often used to minimize agglutination of mitochondria in tissues that otherwise become gelatinous when homogenized (Nedergaard and Cannon, 1979). A drawback of using a homogenization buffer with high-ionic strength is the potential for loss of cytochrome c should the outer mitochondrial membrane be damaged during isolation. For skeletal muscle, we use a homogenization buffer (buffer A) containing 100 mM KCl, 50 mM Tris, 5 mM MgCl$_2$, 1.8 mM ATP, and 1 mM EDTA, which is mixed in large batches, adjusted to pH 7.2, and frozen in 10 ml aliquots. EDTA is added as a heavy metal chelator with high affinity for Ca^{2+} to minimize uncoupling and phospholipase activity (Nedergaard and Cannon, 1979). Magnesium chloride provides essential Mg^{2+} ions to maintain the catalytic activity of ATP synthase. Muscle tissue is cut into small pieces ($\sim$1 mm^3) using a scalpel and sharp forceps. A glass Pasteur pipette is used to carefully draw off buffer, and the minced tissue is incubated for 2 min in 1 ml of protease medium (1 ml buffer A + 5.66 mg protease from Bacillus licheniformis, 10.6 U/mg) to effectively disrupt the muscle cells with negligible damage to mitochondria (Rasmussen et al., 1997). Following the protease treatment, the tissue is immediately washed twice with 3 ml of buffer A to dilute and rinse away the protease to avoid excessive tissue breakdown and damage to mitochondria. Any excess buffer is drawn off before transferring tissue to an ice-cold 5 ml homogenizing vessel with 4 ml of buffer A. Typical ratios of tissue weight to homogenizing buffer volume is 1:5 to 1:10 (w/v) Nedergaard and Cannon, 1979).

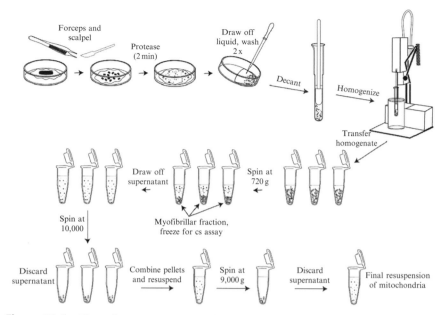

Figure 20.2 Tissue homogenization and isolation of mitochondria. About 40–100 mg of muscle tissue is placed on an ice-cold Petri dish and finely minced with forceps and scalpel. A 2-min protease incubation is used to soften the tissue to help liberate mitochondria that are tightly bound to the contractile filaments. The protease is removed by diluting and washing with buffer. An all glass Potter-Elvehjem tissue grinder (0.3 mm clearance) is used to gently homogenize the tissue in an ice bath. The mitochondria are isolated by differential centrifugation. A low-speed spin first removed the myofibrillar portion. A series of two high-speed spins generates a mitochondrial pellet, which is resuspended in buffer for functional measurements.

Custom-made glass Potter-Elvehjem tissue grinders (Kimble/Kontes, Vineland, New Jersey) with 0.3 mm clearance between the outer diameter of the pestle and inner diameter of the vessel are ideal for gentle yet effective liberation of both subsarcolemmal and intermyofibrillar mitochondria in skeletal muscle (Rasmussen *et al.*, 1997). Softer tissues, such as liver, are easily homogenized using a Dounce or glass-Teflon Potter-Elvehjem homogenizer without the use of proteolytic enzymes. Although various custom-built motor-driven homogenizer units have been described in the literature, we favor a commercially available homogenizer (Potter S, Sartorius AG, Goettingen, Germany) that is well-suited for mitochondrial isolation. Precise alignment between the plunger and homogenizer vessel minimizes grinding between the two surfaces. An integrated cooling vessel allows chilled water to circulate without losing the ability to visually monitor the tissue. Variable rotation speed and careful manual control of the vertical action of the plunger allows the homogenization procedure to be carefully controlled. As extensively discussed (Rasmussen *et al.*, 1997), these are all key factors that are essential to successful isolation of mitochondria with high yield and integrity. The tissue is homogenized at 150 rpm for 10 min with slow vertical plunger movements ($\sim$5 cm in 30 s) to minimize the annular flow shear forces that are known to damage the mitochondrial membranes (Rasmussen et al., 1997). The tissue is carefully monitored through the clear walls of the cooling vessel during the initial 30–60 s of homogenizing where short pestle strokes are used to fragment the bulk of the tissue. Thereafter, the pestle is only momentarily paused at the bottom of the vessel with each vertical plunge to break up any remaining pieces of tissue that settle to the bottom.

2.2. Separation of mitochondria by differential centrifugation

The small size of mitochondria compared to other components of the muscle homogenate allows separation by differential centrifugation. Myofibrillar components form a pellet at low-speed centrifugation, whereas sedimentation of mitochondria requires greater centrifugal force. The exact centrifugation speeds to maximize yield and minimize contamination will vary across tissues and species due to variation in sedimentation coefficients (Nedergaard and Cannon, 1979) and must be determined empirically for each application. The muscle homogenate is transferred from the homogenizer vessel into 1.5 ml microcentrifuge tubes that have been chilled in an ice bath. Conical, tapered tubes are preferable as they are well-suited to forming pellets that are easily separated from the supernatant. All centrifugation steps are performed at 3–4 °C using a microcentrifuge (Eppendorf 5417, Westbury, New York). In the absence of a temperature-controlled environment for the centrifuge, Rasmussen and colleagues describe a clever method for centrifugation of "tip-tubes" in individual ice baths (Rasmussen et al., 1997). The tubes are centrifuged at low speed ($720g$) for 5 min to pellet nuclei, myofibrillar components, and other fragments, which are frozen for later analysis to determine the homogenization efficiency (described below). Resuspension of the pellet followed by a second low-speed centrifugation can slightly improve the yield since inevitably some mitochondria will sediment in the first pellet. The supernatant containing the mitochondrial fraction is transferred to clean, chilled microcentrifuge tubes and centrifuged at $10,000g$ for 5 min to pellet mitochondria. The supernatant is carefully drawn away using a glass Pasteur pipette to avoid disturbing the mitochondrial pellet. The pellets from each tube are combined and resuspended in 1 ml of buffer A by gentle stirring and pipetting. Resuspension must be done slowly using a 1 ml pipette tip to avoid excessive turbulence and shear stress that would otherwise damage the mitochondria. A second high-speed centrifugation is performed at $9000g$ for 5 min, which removes some of the lighter contaminants. The supernatant is discarded and the pellet is gently resuspended in a nonionic buffer containing 225 mM sucrose, 44 mM KH$_2$PO$_4$, 12.5 mM Mg acetate, and 6 mM EDTA (buffer B) at a volume of 4 μl of buffer per milligram of muscle tissue, which can be stored on ice for up to 1 h without any noticeable loss of function. For longer-term storage, mitochondria are suspended in a buffer containing 0.5 mM EGTA, 3 mM MgCl$_2$·6H$_2$O, 60 mM potassium lactobionate, 20 mM taurine, 10 mM KH$_2$PO$_4$, 20 mM HEPES, 110 mM sucrose, 1 g/l fatty acid free BSA, 20 mM histidine, 20 μM vitamin E succinate, 3 mM glutathione, 1 μM leupeptine, 2 mM glutamate, 2 mM malate, and 2 mM Mg-ATP (MiPO2, Oroboros, Innsbruck, Austria). Experiments in our lab have found that the function of isolated mitochondria remains unchanged up to 18 h following homogenization when stored in this buffer at 4 °C.

2.3. Mitochondrial yield and integrity

The fractional yield (i.e., efficiency) of the homogenization procedure is determined by measuring the mitochondrial marker enzyme citrate synthase remaining in the myofibrillar fraction following the homogenization procedure (Rasmussen et al., 1997). The previously frozen myofibrillar pellet from the initial low-speed centrifugation is thawed and vigorously homogenized using a tight-fitting tissue grinder to liberate any mitochondria that were not released during the initial homogenization. Homogenization is performed in a buffer containing 20 mM HEPES, 1 mM EDTA, 250 mM sucrose, and 0.1% Triton X-100 to liberate enzymes. Citrate synthase activity is measured spectrophotometrically based on the absorption at 412 nm by the product thionitrobenzoic acid (TNB), which, in the presence of saturating concentrations of substrates acetyl-CoA, oxaloacetate, and dithionitrobenzoic acid (DNTB), is a function of the activity of citrate synthase according to the following two-step reaction (Srere, 1969):

$$Acetyl\text{-}CoA + oxaloacetate + H_2O \rightarrow citrate + CoA$$

$$CoA + DTNB \rightarrow TNB + CoA\text{-}TNB$$

The homogenization yield is determined by expressing citrate synthase measured in the mitochondrial fraction as a percentage of the total homogenate (myofibrillar fraction + mitochondrial fraction) (Rasmussen et al., 1997). This homogenization method results in yields of 40–50% (Rasmussen et al., 1997; Rasmussen and Rasmussen, 1997).

Mitochondrial integrity is an important consideration when assessing the function of isolated mitochondria. The outer membrane is easily damaged during homogenization procedures, in which case cytochrome c would exit into the buffer and become rate limiting to oxygen consumption and ATP synthesis. Furthermore, improper use of protease or high levels of fatty acids and their esters may induce uncoupling. We have confirmed the integrity and function of our mitochondrial preparations in three ways. First, a polarographic oxygen electrode (Oxygraph 2K, Oroboros Instruments, Innsbruck, Austria) is often used to measure respiration with serial additions of 10 mM glutamate and 2 mM malate (state 2), 2.5 mM ADP (state 3), and 10 μM cytochrome c, as shown in (Fig. 20.3). Exogenous cytochrome c does not enhance mitochondrial respiration in these preparations, indicating minimal damage to the outer mitochondrial membrane during isolation procedures. A typical example of a preparation with compromised mitochondrial integrity, reflected by enhanced respiration with exogenous cytochrome c is shown in Panel A of (Fig. 20.3). Panel B illustrates the absence of cytochrome c-stimulated respiration in a mitochondrial preparation with high integrity. The inner mitochondrial membrane is impermeable to NADH, allowing inner membrane integrity to be assessed from the increment in the rate of oxygen consumption in the presense of 2.8 mM NADH (Puchowicz et al., 2004). A crude index of mitochondrial integrity is the respiratory control ratios (state 3:state 4). State-3 respiration is measured in the presence of substrates and ADP while state 4 is induced by addition of oligomycin to inhibit ATP synthase. We routinely observe respiratory control ratio values greater than 8, characteristic of high-mitochondrial integrity and energy-conserving capacity. Alternatively, mitochondrial integrity may be assessed by measuring citrate synthase activity in isolated mitochondria before and

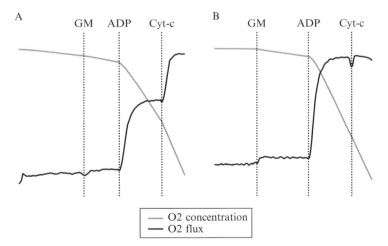

Figure 20.3 Testing the integrity of the outer mitochondrial membrane. Oxygen concentration (grey line) and respiration rate (black line) measured using a polarographic oxygen electrode. Respiration is stimulated by the addition of substrates glutamate and malate (GM) and ADP to stimulate state-3 respiration. 10 μM cytochrome c is added to assess outer membrane integrity. An increment in respiration with exogenous cytochrome c is apparent if the outer mitochondrial membrane is damaged (Panel A). Panel B illustrates a preparation where membrane integrity is high.

after membrane disruption by freeze–thaw cycles and extraction of enzyme by Triton X-100. Citrate synthase activity is measured using standard spectrophotometric techniques, as described above. Consistent with the polarographic-based cytochrome c test, we typically find that mitochondrial preparations are 90–95% intact (Asmann *et al.*, 2006).

3. Mitochondrial ATP Production

3.1. Principles of the bioluminescent approach

Luciferase from the firefly *Photinus pyralis* catalyzes a two-step reaction that oxidizes luciferin in an ATP-dependent reaction that generates a light signal in proportion to ATP concentration (DeLuca and McElroy, 1974). Luciferyl adenylate and inorganic pyrophosphate are formed when ATP is added to luciferin in the presence of luciferase. Oxyluciferin and AMP are subsequently formed when luciferyl adenylate reacts with oxygen, and light is generated when the products are released from the surface of the enzyme:

$$\text{luciferin} + \text{ATP} \rightarrow \text{luciferyl adenylate} + \text{PPi}$$

$$\text{luciferyl adenylate} + O_2 \rightarrow \text{oxyluciferin} + \text{AMP} + \text{light}$$

In nature, these insects use bioluminescence to attract mates. In the laboratory, recombinant luciferase is a powerful analytical tool that can be used to quantify ATP using a luminometer, provided that ATP is the limiting substrate in the reaction (Lundin and Thore, 1975). Since isolated mitochondria respire and generate ATP when provided with appropriate substrates *in vitro*, it is possible to use a luciferase-based bioluminescent approach to measure the rates of ATP synthesis. Our technique is built on the pioneering methods of Wibom and colleagues who first demonstrated that mitochondrial ATP production could be measured in small amounts of human muscle biopsy tissue.

A microplate luminometer (Veritas, Turner Biosystems) with a highly sensitive photon-counting photomultiplier tube is used to measure bioluminescence from samples in a 96-well plate format (Costar 3912, Corning Life Sciences). A 50-mg muscle sample will yield 200 μl of mitochondrial suspension with protein concentrations ranging 2–5 μg/μl, depending on species and muscle phenotype. A working dilution of 40 μl mitochondrial suspension to 1000 μl buffer B is used for measuring ATP production rates. Since reliable measurements can be achieved with very small amounts of mitochondrial protein (< 4 μg), it is possible to perform measurements under a variety of conditions in triplicate to take full advantage of the 96-well plate format. Different combinations of substrates and inhibitors permit assessment of the capacity of distinct respiratory chain complexes and different pathways that generate electron flow into the mitochondrial electron transport system, as illustrated in (Fig. 20.1). The substrate combination of glutamate and malate (GM) provides carbon sources for dehydrogenase reactions in the TCA cycle that generate NADH which is subsequently oxidized by complex I (NADH dehydrogenase). Furthermore, since malate is in equilibrium with fumarate, high-malate concentrations will result in product-inhibition of the succinate dehydrogenase reaction, which would otherwise provide direct electron flow into the quinone pool through $FADH_2$. Thus addition of high concentrations of glutamate and malate provides an opportunity to assess the capacity for ATP production with electron flow exclusively through complex I. Similarly, it is possible to assess complex II (succinate dehydrogenase) function in the presence of the substrate succinate and rotenone (SR). By selectively inhibiting complex I, rotenone induces a redox shift that effectively inhibits all of the NADH-linked dehydrogenases in the TCA cycle. Thus, in the presence of rotenone, succinate selectively stimulates electron flow through complex II. ATP production as a result of fatty acid β-oxidation can be assessed using the substrates palmitoyl carnitine and malate (PCM), which provide electron flow through electron-transferring flavoprotein. Finally, a combination of pyruvate, palmitoyl-L-carnitine, α-ketoglutarate, and malate (PPKM) provides an index of the additive effects of convergent electron flow from multiple sites along the electron transport system. ADP is a potent regulator of mitochondrial ATP synthesis, both as a substrate and through allosteric enzyme regulation. When ADP is added in saturating concentrations in combination with other substrates, it becomes possible to assess oxidative capacity (state 3) through distinct complexes that shuttle electrons into the Q junction.

3.2. Preparation of substrates and ADP

Stock solutions of substrate/inhibitor combinations of GM, SR, PCM, and PPKM are prepared in advance and stored at −20 °C in small aliquots. With the exception of rotenone, which is dissolved in 50% ethanol, all substrate solutions are made using ultra pure water (Milli-Q, Millipore). The final volume of each well containing substrates (25 μl), ADP (25 μl), sample (25 μl), and luciferin–luciferase reagent (175 μl) is 250 μl. Final concentrations of substrates in each well are as follows:

GM : 10 mM glutamate + 5 mM malate
SR : 20 mM succinate + 0.1 mM rotenone
PCM : 0.05 mM palmitoyl-L-carnitine + 2 mM malate
PPKM : 1 mM pyruvate + 0.05 mM palmitoyl-L-carnitine
            +10 mM α-ketoglutarate + 1 mM malate

A 10-mM stock solution of ADP gives a final concentration of 1 mM in each well. Since the ADP concentration required to elicit half-maximal respiration (K_m) is 10–30 μM in isolated mitochondria (Chance and Williams, 1955; Slater and Holton, 1953), 1 mM ADP is sufficiently high to induce state 3 in the presence of substrates and oxygen. Commercially available ADP is ~95% pure with up to 3% contamination with ATP. Even this small ATP contamination substantially increases the background bioluminescence signal and confounds the measurement of mitochondrial ATP levels that are orders of magnitude lower. Therefore, ADP stock solutions must be depleted of ATP, which can be done using high-performance liquid chromatography or enzymatically. The hexokinase reaction can be exploited to enzymatically deplete ATP from ADP stock (Mahaut-Smith *et al.*, 2000) according to the following reaction:

$$glucose + ATP \rightarrow glucose\text{-}6\text{-}phosphate + ADP$$

A 10-mM solution of ADP in Tris acetate buffer (20 mM Tris acetate, 0.2 mM Mg acetate) is incubated for 1 h at 37 °C with 22 mM glucose and 3 U/ml hexokinase. By phosphorylating glucose at the expense of ATP, this reaction effectively depletes ATP from the solution. The solution is then heated to 99 °C for 3 min to deactivate hexokinase. Lyophilized firefly luciferin–luciferase reagent is commercially available (BioThema 11–501-M) and reconstituted according to the manufacturer's instructions.

3.3. Running the experiment and quantitation of ATP

Substrates and sample are added to the wells in triplicate, and the plate is loaded into the luminometer. Luciferase reagent and ADP are added to the wells using programmable injectors that are integrated into the luminometer system. A basic macro controls the volume, timing, and which wells receive the injections. Immediately following the injections, a light reading is taken from each well with a 2-s integration time. Measurements are repeated seven times on each well with 6 min between subsequent readings. Light units are converted to ATP concentration using equations derived from a series of ATP standard curves that are run on the same plate (Fig. 20.4). Eight ATP standards ranging from 0.1 to 50 μM are measured in duplicate for each combination of substrates. Standard curves deviate from linearity at high-ATP concentrations, necessitating their fitting with second-order polynomial equations with correlation coefficients greater than 0.95. It is necessary to generate an independent standard curve for each substrate condition as some substrate combinations may interfere with the activity of luciferase or absorb a portion of the bioluminescence signal according to

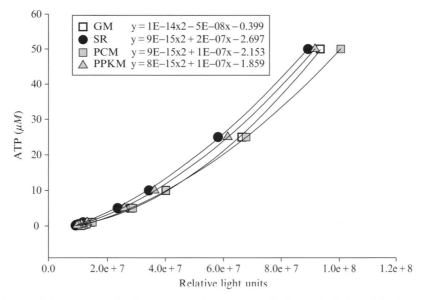

Figure 20.4 ATP standard curves. A series of known ATP standards from 0.1 to 50 μM are used to quantify ATP synthesis rates in isolated mitochondria. Individual standard curves for each substrate combination (GM: mlutamate + malate, SR: succinate + rotenone, PCM: palmitoyl carnitine + malate, PPKM: pyruvate + palmitoyl carnitine + α-ketoglutarate + malate) are generated by measuring light production (relative light units) following the addition of the luciferin–luciferase reagent.

the Beer–Lambert law. Since the standard curves are being measured under conditions identical to wells containing functional mitochondria, there is no need to account for background signal.

ATP synthesis rates are calculated for each substrate combination using values obtained over the seven measurement cycles. ATP concentration is plotted as a function of time and the overall rate of ATP synthesis is determined as the linear slope through all seven data points (Fig. 20.5). The ATP synthesis rates are commonly normalized using a denominator that allows direct comparison across groups or conditions. For example, expressing rates per tissue wet weight is ideal if an investigator wishes to assess oxidative capacity per unit muscle mass, regardless of potential differences in mitochondrial content. Normalizing to citrate synthase activity, mitochondrial protein content, or mitochondrial DNA copy number accounts for differences in mitochondrial content and provides an index of the inherent mitochondrial properties, independent of differences in tissue mitochondrial content. We have found excellent reproducibility in repeated experiments using isolated mitochondria from rat liver and skeletal muscle (CV <6%).

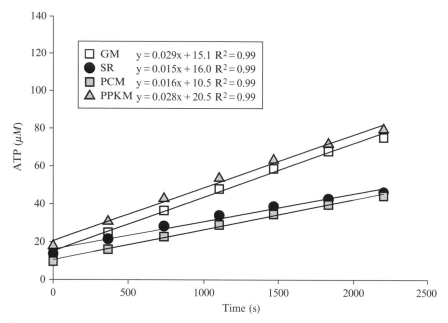

Figure 20.5 ATP synthesis rates. Light units are measured at seven time points for each substrate combination. Standard curves are used to convert light units into ATP concentration. ATP concentration is plotted as a function of time, of which the linear slope represents the rate of ATP production.

Mitochondrial DNA abundance is measured using real-time quantitative PCR using primer-probe sequences for two marker genes in mitochondrial DNA. A QIAamp DNA mini kit (QIAGEN, Chatworth, California) is used to extract DNA from frozen muscle samples. The abundance of two marker genes NADH dehydrogenase subunits 1 and 4 are measured using a PCR system (PE Biosystems, Foster City, California). Samples are run in duplicate and normalized to 28S ribosomal DNA (Lanza *et al.*, 2008; Short *et al.*, 2005). Mitochondrial protein concentration is measured using a colorometric assay (Bio-Rad *DC* Protein Assay) using a microplate spectrophotometer. There are numerous other measurements that can be performed using frozen tissue that will complement the information obtained from freshly isolated mitochondria. These methods include, but are certainly not limited to, histological analysis, maximal enzyme activities, mRNA expression of nuclear-encoded mitochondrial proteins, protein expression of both nuclear and mitochondrial-encoded proteins, and electron microscopy-based mitochondrial determinations. When combined with functional assessment of freshly isolated mitochondria described below, these methods provide a wealth of complementary mechanistic information.

4. MITOCHONDRIAL RESPIRATION

4.1. Overview

Oxygen in solution can be measured polarographically with a Clark-type oxygen electrode. Clark electrodes have gold or platinum cathodes and silver or silver/silver chloride anodes, which are connected by a salt bridge and covered by an oxygen-permeable membrane. As oxygen diffuses across the membrane, it is reduced by a fixed voltage between the cathode and anode which generates current in proportion to the concentration of oxygen in solution. By calibrating the voltage with known oxygen concentrations, it is possible to measure the rate of oxygen consumption in a medium containing actively respiring mitochondria. Since reduction of oxygen is a critical step in the process of mitochondrial electron transport and ATP synthesis, measurement of mitochondrial oxygen consumption provides a convenient way to assess mitochondrial function. Although numerous Clark electrode systems are commercially available, our group favors the system manufactured by Oroboros Instruments (Innsbruck, Austria). The combination of twin-2 ml chambers with large (2 mm diameter) cathodes results in high signal-to-noise and very low signal drift. Furthermore, careful choice of inert materials for the chambers, stoppers, and stir bars results in residual oxygen diffusion of less than 2 pmol/s/ml at a PO_2 of zero. The sample chambers are housed in an insulated copper block with precise temperature control by Peltier thermopiles. This system enables high-resolution measurements of oxygen consumption with very small amounts of tissue (1 mg permeabilized muscle fibers, 0.01 mg mitochondrial protein) with a limit of detection of 0.5 pmol/s/ml at steady-state over 5 min (Gnaiger, 2001).

4.2. An oxygraph protocol for mitochondrial assessment

A 2 ml oxygraph (Oxygraph-2k; Oroboros) chamber is washed with 70% ethanol, rinsed three times with distilled water, then filled with respiration medium containing 0.5 mM EGTA, 3 mM MgCl$_2$·6H$_2$O, 60 mM potassium lactobionate, 20 mM taurine, 10 mM KH$_2$PO$_4$, 20 mM HEPES, 110 mM sucrose, and 1 g/l fatty acid-free BSA (MiR05; Oroboros, Innsbruck, Austria). The chamber is allowed to equilibrate with an ambient gas phase at 37 °C with a stirrer speed of 750 rpm for >10 min to allow air saturation of the respiration medium. Mitochondria are isolated as described above, a 90-μl aliquot of the mitochondrial suspension containing ~300 μg mitochondrial protein is added to the chamber. The polyvinylidene fluoride stopper is inserted to generate a closed system with a final volume of 2 ml. Oxygen concentration is recorded at 0.5 Hz and converted from voltage to oxygen concentration using a two-point calibration. The calibration points include ambient oxygen saturation from room air and a zero point where oxygen is completely depleted from the chamber by respiring sample or chemically using a sodium hydrosulfite solution. Respiration rates (O_2 flux) are calculated as the negative time derivative of oxygen concentration (Datlab Version 4.2.1.50, Oroboros Instruments). The O_2 flux values are corrected for the small amount of back-diffusion of oxygen from materials within the chamber, any leak of oxygen from outside of the vessel, and oxygen consumed by the polarographic electrode. Detailed description of standardized instrumental and chemical calibrations is beyond the scope of this chapter and is discussed in detail elsewhere (Gnaiger, 2001). A protocol involving serial additions of various substrates, inhibitors, and uncouplers allows comprehensive assessment of mitochondrial function as shown in (Fig. 20.6) and the corresponding step-wise description below:

1. *Equilibration with ambient oxygen.* Oxygen concentration is measured while in equilibration with a gas phase for air saturation
2. *Baseline respiration.* An aliquot of mitochondrial suspension is added to the chamber, the stopper is closed, and respiration is measured in the absence of exogenous substrates.
3. *State-2 respiration.* Glutamate (10 mM) and malate (2 mM) are injected through a small capillary tube in the chamber stopper using Hamilton syringes. In the absence of adenylates, oxygen consumption with GM reflects state-2 respiration specific to complex I.
4. *ADP pulse titration.* Small pulses of ADP below saturation levels (15 μM) are used to transiently stimulate respiration above state 2. Integration of oxygen flux over the duration of the peak allows quantitation of oxygen consumed per unit of ADP phosphorylated (i.e., ATP/O flux ratio). The ADP/O ratio is defined as the total amount of ADP added divided by the amount of oxygen consumed until respiration returns to a constant low rate. This approach for measuring mitochondrial phosphorylation efficiency, as well as the use of ADP-injection respirometry is described in detail by Gnaiger and colleagues (Gnaiger *et al.*, 2000).
5. *State-3 respiration (complex I).* ADP is added at a saturating level (2.5 mM) to maximally stimulate respiration in the presence of glutamate and malate.
6. *Outer mitochondrial membrane integrity.* Mitochondrial integrity may become compromised during isolation procedures, resulting in damage to the outer membrane and release of cytochrome *c*, which then becomes a limiting factor in mitochondrial respiration. Since cytochrome *c* cannot penetrate an intact outer mitochondrial membrane, addition of exogenous cytochrome *c* is a convenient qualitative confirmation of mitochondrial integrity as a quality control measure (Gnaiger and Kuznetsov, 2002;

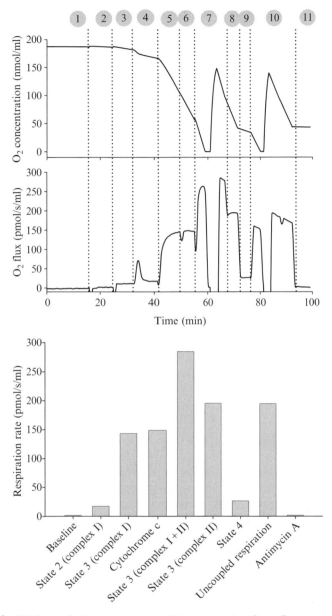

Figure 20.6 High-resolution respirometry. Representative plots of oxygen concentration (top panel) and oxygen flux rate (bottom panel) during a step-wise protocol designed for functional assessment of isolated mitochondria. (1) gas-phase equilibration, (2) baseline mitochondrial respiration, (3) substrates glutamate and malate (state 2, complex I), (4) submaximal ADP pulse for assessment of P:O, (5) saturating ADP (state 3, complex I), (6) cytochrome c (membrane integrity), (7) succinate (state 3, complex I + II), (8) rotenone (state 3, complex II), (9) oligomycin (state 4), (10) FCCP (uncoupled respiration), (11) antimycin A (background). Average rates of oxygen consumption for each condition are shown in the bottom panel.

Puchowicz *et al.*, 2004). The absence of a stimulatory effect of cytochrome c on O_2 flux rates is indicative of a high-quality preparation.

7. *State-3 respiration (complex I + II).* The addition of 10 mM succinate stimulates respiration above GM-stimulated state 3 since succinate provides additional electron flow through complex II.

8. *State-3 respiration (complex II).* Subsequent addition of 0.5 μM rotenone inhibits complex I. Under these conditions, electron input to complex I is eliminated, providing an index of state-3 respiration through complex II exclusively.

9. *State-4 respiration.* The addition of oligomycin (2 $\mu g/\mu l$) inhibits the Fo unit of ATP synthase and induces state-4 respiration by blocking the proton channel and effectively eliminating ATP synthesis. The residual oxygen consumption in the absence of ADP phosphorylation is attributable to proton leak across the inner mitochondrial membrane. Thus, oligomycin inhibited respiration serves as an indicator of the degree of uncoupled respiration or proton leak under these conditions.

10. *Uncoupled respiration.* Step-wise titration of the protonophore carbonyl-cyanide-4-(trifluoromethoxy)-phenylhydrazone (FCCP; 0.05 mM titrations) induces an uncoupled state by dissipating the proton gradient across the inner mitochondrial membrane. Under these conditions, mitochondrial oxidative capacity through complex II can be determined in the absence of the potential control exerted by ATP synthase, adenine nucleotide translocase, or phosphate transporters. Increments in O_2 flux above state-3 respiration with rotenone are indicative of a limitation in respiratory capacity by ATP synthase, adenine nucleotide translocases, or phosphate transporters.

11. *Nonmitochondrial respiration.* Antimycin A is added at a concentration of 2.5 μM to inhibit cytochrome bc_1 complex (complex III). By inhibiting the reduction of cytochrome c, antimycin A allows nonmitochondrial respiration to be measured.

Respiration rates are measured for each condition by an average value of the oxygen flux within a region of stable, consistent oxygen consumption (Fig. 20.5, bottom panel). Respiration rates are then normalized to the amount of mitochondrial protein to allow comparison across groups or conditions without the confounding influence of differences in the amount of mitochondria loaded into the chamber. A standardized, step-wise protocol such as this allows an enormous amount of information to be obtained from a relatively small sample. As with our bioluminescent measures of mitochondrial ATP production, we have found excellent reproducibility (CV ~6%). Although this chapter is focused on measurements using isolated mitochondria, there are numerous excellent references that describe the application of high-resolution respirometry to cultured cells and permeabilized muscle fibers (Boushel *et al.*, 2007; Kay *et al.*, 2000; Stadlmann *et al.*, 2006).

5. SUMMARY

The function and capacity of intact mitochondria isolated from small amounts of tissue can be objectively determined using polarographic-based measurements of oxygen consumption and luciferase-based bioluminescent measurements of ATP synthesis. Careful selection of substrate combinations and inhibitors enable quantitative measurement of mitochondrial function in various energetic states, oxidative capacity with electron flow through distinct complexes, coupling of oxygen consumption to ATP production, and membrane integrity. These methods provide direct assessment of the function of the organelle in its entirety, as opposed to inferences made from maximal activities of single enzymes. These techniques have wide applicability in both human and animal research and cell culture experiments.

A limitation of the described bioluminescent and polarographic-based measures of mitochondrial function is that the isolated mitochondria are typically exposed to nonphysiological conditions. First, isolated mitochondria *in vitro* are exposed to oxygen concentrations that are much higher than what is present in the cellular microenvironment *in vivo*. Furthermore, respiration and ATP synthesis rates are measured in response to substrate and ADP concentrations that are saturating and dramatically exceed levels that are endogenous to the cell. Thus, a strength of measuring mitochondrial function in intact muscle fibers or from organisms *in vivo* is that all circulatory and regulatory systems are intact. Notwithstanding, *in vitro* measurements from isolated mitochondria permit detailed assessment of the capacity and function of mitochondria, independent of the influence of external variables such as blood flow, oxygen or substrate delivery, or mitochondrial volume density. Furthermore, measurements conducted in isolated mitochondria have the distinct advantage of providing information relevant to the function of distinct components of the electron transport system. Ideally, both *in vitro* and *in vivo* methodologies would be used in concert to assess mitochondrial function at the level of the intact organism as well as the individual constituents of the mitochondrial machinery responsible for generating ATP.

REFERENCES

Argov, Z., Bank, W. J., Maris, J., Peterson, P., and Chance, B. (1987). Bioenergetic heterogeneity of human mitochondrial myopathies: Phosphorus magnetic resonance spectroscopy study. *Neurology* **37**, 257–262.

Asmann, Y. W., Stump, C. S., Short, K. R., Coenen-Schimke, J. M., Guo, Z., Bigelow, M. L., and Nair, K. S. (2006). Skeletal muscle mitochondrial functions, mitochondrial DNA copy numbers, and gene transcript profiles in type 2 diabetic and nondiabetic subjects at equal levels of low or high insulin and euglycemia. *Diabetes* **55**, 3309–3319.

Barazzoni, R., Short, K. R., and Nair, K. S. (2000). Effects of aging on mitochondrial DNA copy number and cytochrome *c* oxidase gene expression in rat skeletal muscle, liver, and heart. *J. Biol. Chem.* **275,** 3343–3347.

Befroy, D. E., Falk, P. K., Dufour, S., Mason, G. F., Rothman, D. L., and Shulman, G. I. (2008). Increased substrate oxidation and mitochondrial uncoupling in skeletal muscle of endurance-trained individuals. *Proc. Natl. Acad. Sci. USA* **105,** 16701–16706.

Boushel, R., Gnaiger, E., Schjerling, P., Skovbro, M., Kraunsoe, R., and Dela, F. (2007). Patients with type 2 diabetes have normal mitochondrial function in skeletal muscle. *Diabetologia* **50,** 790–796.

Chance, B., and Williams, G. (1955). Respiratory enzymes in oxidative phosphorylation. I. Kinetics of oxygen utilization. *J. Biol. Chem.* **217,** 383–393.

Chance, B., and Williams, G. (1956). Respiratory enzymes in oxidative phosphorylation. VI. The effects of adenosine diphosphate on azide-treated mitochondria. *J. Biol. Chem.* **221,** 477–489.

Chow, L. S., Greenlund, L. J., Asmann, Y. W., Short, K. R., McCrady, S. K., Levine, J. A., and Nair, K. S. (2007). Impact of endurance training on murine spontaneous activity, muscle mitochondrial DNA abundance, gene transcripts, and function. *J. Appl. Physiol.* **102,** 1078–1089.

Cogswell, A. M., Stevens, R. J., and Hood, D. A. (1993). Properties of skeletal muscle mitochondria isolated from subsarcolemmal and intermyofibrillar regions. *Am. J. Physiol.* **264,** C383–C389.

DeLuca, M., and McElroy, W. D. (1974). Kinetics of the firefly luciferase catalyzed reactions. *Biochemistry* **13,** 921–925.

Drew, B., and Leeuwenburgh, C. (2003). Method for measuring ATP production in isolated mitochondria: ATP production in brain and liver mitochondria of Fischer-344 rats with age and caloric restriction. *Am. J. Physiol. Regul. Integr. Comp. Physiol.* **285,** R1259–R1267.

Edwards, R., Young, A., and Wiles, M. (1980). Needle biopsy of skeletal muscle in the diagnosis of myopathy and the clinical study of muscle function and repair. *N. Engl. J. Med.* **302,** 261–271.

Elander, A., Sjostrom, M., Lundgren, F., Schersten, T., and Bylund-Fellenius, A. C. (1985). Biochemical and morphometric properties of mitochondrial populations in human muscle fibres. *Clin. Sci. (Lond.)* **69,** 153–164.

Gnaiger, E. (2001). Bioenergetics at low oxygen: Dependence of respiration and phosphorylation on oxygen and adenosine diphosphate supply. *Respir. Physiol.* **128,** 277–297.

Gnaiger, E., and Kuznetsov, A. V. (2002). Mitochondrial respiration at low levels of oxygen and cytochrome *c. Biochem. Soc. Trans.* **30,** 252–258.

Gnaiger, E., Mendez, G., and Hand, S. C. (2000). High phosphorylation efficiency and depression of uncoupled respiration in mitochondria under hypoxia. *Proc. Natl. Acad. Sci. USA* **97,** 11080–11085.

Gray, M. W., Burger, G., and Lang, B. F. (1999). Mitochondrial evolution. *Science* **283,** 1476–1481.

Haller, T., Ortner, M., and Gnaiger, E. (1994). A respirometer for investigating oxidative cell metabolism: Toward optimization of respiratory studies. *Anal. Biochem.* **218,** 338–342.

Harman, D. (1956). Aging: A theory based on free radical and radiation chemistry. *J. Gerontol.* **11,** 298–300.

Hood, D. A. (2001). Invited review: Contractile activity-induced mitochondrial biogenesis in skeletal muscle. *J. Appl. Physiol.* **90,** 1137–1157.

Houmard, J. A., Weidner, M. L., Gavigan, K. E., Tyndall, G. L., Hickey, M. S., and Alshami, A. (1998). Fiber type and citrate synthase activity in the human gastrocnemius and vastus lateralis with aging. *J. Appl. Physiol.* **85,** 1337–1341.

Jaleel, A., Short, K. R., Asmann, Y. W., Klaus, K., Morse, D., Ford, G. C., and Nair, K. S. (2008). *In vivo* measurement of synthesis rate of individual skeletal muscle mitochondrial proteins. *Am. J. Physiol. Endocrinol. Metab.* **295**, E1255–E1268.

Kay, L., Nicolay, K., Wieringa, B., Saks, V., and Wallimann, T. (2000). Direct evidence for the control of mitochondrial respiration by mitochondrial creatine kinase in oxidative muscle cells *in situ. J. Biol. Chem.* **275**, 6937–6944.

Kent-Braun, J. A., and Ng, A. V. (2000). Skeletal muscle oxidative capacity in young and older women and men. *J. Appl. Physiol.* **89**, 1072–1078.

Krieger, D. A., Tate, C. A., McMillin-Wood, J., and Booth, F. W. (1980). Populations of rat skeletal muscle mitochondria after exercise and immobilization. *J. Appl. Physiol.* **48**, 23–28.

Lanza, I. R., Larsen, R. G., and Kent-Braun, J. A. (2007). Effects of old age on human skeletal muscle energetics during fatiguing contractions with and without blood flow. *J. Physiol.* **583**, 1093–1105.

Lanza, I. R., Short, D. K., Short, K. R., Raghavakaimal, S., Basu, R., Joyner, M. J., McConnell, J. P., and Nair, K. S. (2008). Endurance exercise as a countermeasure for aging. *Diabetes* **57**, 2933–2942.

Lebon, V., Dufour, S., Petersen, K. F., Ren, J., Jucker, B. M., Slezak, L. A., Cline, G. W., Rothman, D. L., and Shulman, G. I. (2001). Effect of triiodothyronine on mitochondrial energy coupling in human skeletal muscle. *J. Clin. Invest.* **108**, 733–737.

Lundin, A., and Thore, A. (1975). Analytical information obtainable by evaluation of the time course of firefly bioluminescence in the assay of ATP. *Anal. Biochem.* **66**, 47–63

Mahaut-Smith, M. P., Ennion, S. J., Rolf, M. G., and Evans, R. J. (2000). ADP is not an agonist at P2X(1) receptors: Evidence for separate receptors stimulated by ATP and ADP on human platelets. *Br. J. Pharmacol.* **131**, 108–114.

Menshikova, E. V., Ritov, V. B., Fairfull, L., Ferrell, R. E., Kelley, D. E., and Goodpaster, B. H. (2006). Effects of exercise on mitochondrial content and function in aging human skeletal muscle. *J. Gerontol. A. Biol. Sci. Med. Sci.* **61**, 534–540.

Menshikova, E. V., Ritov, V. B., Toledo, F. G., Ferrell, R. E., Goodpaster, B. H., and Kelley, D. E. (2005). Effects of weight loss and physical activity on skeletal muscle mitochondrial function in obesity. *Am. J. Physiol. Endocrinol. Metab.* **288**, E818–E825.

Nair, K. S., Bigelow, M. L., Asmann, Y. W., Chow, L. S., Coenen-Schimke, J. M., Klaus, K. A., Guo, Z. K., Sreekumar, R., and Irving, B. A. (2008). Asian Indians have enhanced skeletal muscle mitochondrial capacity to produce ATP in association with severe insulin resistance. *Diabetes* **57**, 1166–1175.

Nair, K. S., Halliday, D., and Griggs, R. C. (1988). Leucine incorporation into mixed skeletal muscle protein in humans. *Am. J. Physiol.* **254**, E208–E213.

Nedergaard, J., and Cannon, B. (1979). Overview—Preparation and properties of mitochondria from different sources. *Methods Enzymol.* **55**, 3–28.

Petersen, K. F., Befroy, D., Dufour, S., Dziura, J., Ariyan, C., Rothman, D. L., DiPietro, L., Cline, G. W., and Shulman, G. I. (2003). Mitochondrial dysfunction in the elderly: Possible role in insulin resistance. *Science* **300**, 1140–1142.

Petersen, K. F., Dufour, S., Befroy, D., Garcia, R., and Shulman, G. I. (2004). Impaired mitochondrial activity in the insulin-resistant offspring of patients with type 2 diabetes. *N. Engl. J. Med.* **350**, 664–671.

Puchowicz, M. A., Varnes, M. E., Cohen, B. H., Friedman, N. R., Kerr, D. S., and Hoppel, C. L. (2004). Oxidative phosphorylation analysis: Assessing the integrated functional activity of human skeletal muscle mitochondria—Case studies. *Mitochondrion* **4**, 377–385.

Rasmussen, H. N., Andersen, A. J., and Rasmussen, U. F. (1997). Optimization of preparation of mitochondria from 25 to 100 mg skeletal muscle. *Anal. Biochem.* **252**, 153–159.

Rasmussen, H. N., and Rasmussen, U. F. (1997). Small scale preparation of skeletal muscle mitochondria, criteria of integrity, and assays with reference to tissue function. *Mol. Cell. Biochem.* **174**, 55–60.

Ritov, V. B., Menshikova, E. V., He, J., Ferrell, R. E., Goodpaster, B. H., and Kelley, D. E. (2005). Deficiency of subsarcolemmal mitochondria in obesity and type 2 diabetes. *Diabetes* **54,** 8–14.

Rooyackers, O. E., Adey, D. B., Ades, P. A., and Nair, K. S. (1996). Effect of age on *in vivo* rates of mitochondrial protein synthesis in human skeletal muscle. *Proc. Natl. Acad. Sci. USA* **93,** 15364–15369.

Srere, P. A. (1969). Citrate synthase. *Methods Enzymol.* **13,** 3–5.

Short, K. R., Bigelow, M. L., Kahl, J., Singh, R., Coenen-Schimke, J., Raghavakaimal, S., and Nair, K. S. (2005). Decline in skeletal muscle mitochondrial function with aging in humans. *Proc. Natl. Acad. Sci. USA* **102,** 5618–5623.

Short, K. R., Vittone, J. L., Bigelow, M. L., Proctor, D. N., Rizza, R. A., Coenen-Schimke, J. M., and Nair, K. S. (2003). Impact of aerobic exercise training on age-related changes in insulin sensitivity and muscle oxidative capacity. *Diabetes* **52,** 1888–1896.

Slater, E., and Holton, F. (1953). The adeninenucleotide specificity of oxidative phosphorylation. *Biochem. J.* **53,** ii.

Stadlmann, S., Renner, K., Pollheimer, J., Moser, P. L., Zeimet, A. G., Offner, F. A., and Gnaiger, E. (2006). Preserved coupling of oxidative phosphorylation but decreased mitochondrial respiratory capacity in IL-1beta-treated human peritoneal mesothelial cells. *Cell Biochem. Biophys.* **44,** 179–186.

Stump, C. S., Short, K. R., Bigelow, M. L., Schimke, J. M., and Nair, K. S. (2003). Effect of insulin on human skeletal muscle mitochondrial ATP production, protein synthesis, and mRNA transcripts. *Proc. Natl. Acad. Sci. USA* **100,** 7996–8001.

Trounce, I. A., Kim, Y. L., Jun, A. S., and Wallace, D. C. (1996). Assessment of mitochondrial oxidative phosphorylation in patient muscle biopsies, lymphoblasts, and transmitochondrial cell lines. *Methods Enzymol.* **264,** 484–509.

Wibom, R., and Hultman, E. (1990). ATP production rate in mitochondria isolated from microsamples of human muscle. *Am. J. Physiol.* **259,** E204–E209.

Wicks, K. L., and Hood, D. A. (1991). Mitochondrial adaptations in denervated muscle: Relationship to muscle performance. *Am. J. Physiol.* **260,** C841–C850.

CHAPTER TWENTY-ONE

Assessment of *In Vivo* Mitochondrial Metabolism by Magnetic Resonance Spectroscopy

Douglas E. Befroy,* Kitt Falk Petersen,† Douglas L. Rothman,‡ *and* Gerald I. Shulman§

Contents

1. Introduction	374
1.1. Mitochondrial fatty acid oxidation and insulin resistance	374
2. *In Vivo* Magnetic Resonance Spectroscopy	375
2.1. Examining exchange reactions by magnetization-transfer MRS	377
2.2. Measuring muscle ATP synthesis *in vivo* by ^{31}P saturation-transfer MRS	379
2.3. Monitoring metabolic fluxes using ^{13}C MRS	380
2.4. Metabolic modeling of the TCA cycle flux	384
3. Insulin Resistance and Resting Mitochondrial Metabolism	386
3.1. Resting mitochondrial metabolism versus maximal capacity	388
4. Conclusions	389
References	389

Abstract

Magnetic resonance spectroscopy (MRS), a companion technique to the more familiar MRI scan, has emerged as a powerful technique for studying metabolism noninvasively in a variety of tissues. In this article, we review two techniques that we have developed which take advantage of the unique characteristics of ^{31}P and ^{13}C MRS to investigate two distinct parameters of muscle mitochondrial metabolism; ATP production can be estimated by using the ^{31}P saturation-transfer technique, and oxidation via the TCA cycle can be modeled from ^{13}C MRS data obtained during the metabolism of a ^{13}C-labeled substrate.

* Departments of Diagnostic Radiology and Internal Medicine, Yale University School of Medicine, New Haven, Connecticut, USA
† Department of Internal Medicine, Yale University School of Medicine, New Haven, Connecticut, USA
‡ Departments of Diagnostic Radiology and Biomedical Engineering, Yale University School of Medicine, New Haven, Connecticut, USA
§ Departments of Internal Medicine and Cellular and Molecular Physiology, Howard Hughes Medical Institute, Yale University School of Medicine, New Haven, Connecticut, USA

Methods in Enzymology, Volume 457
ISSN 0076-6879, DOI: 10.1016/S0076-6879(09)05021-6

We will also examine applications of the techniques to investigate how these parameters of muscle mitochondrial metabolism are modulated in insulin resistant and endurance trained individuals.

1. INTRODUCTION

The prevalence of type 2 diabetes has dramatically increased over the past 25 years, and has been predicted to maintain this expansion over the next two decades (Wild *et al.*, 2004). Although causal mechanisms remain to be elucidated, the development of insulin resistance assumes a key, early role in the pathogenesis of most cases of diabetes (Haffner *et al.*, 1990; Lillioja *et al.*, 1987; Warram *et al.*, 1990). Elevated, fasting plasma fatty acid concentrations are a hallmark of insulin resistance in obese and diabetic individuals (Boden and Shulman, 2002), and an inverse relationship between insulin sensitivity and circulating fatty acid levels has also been observed in normal-weight offspring of type 2 diabetic patients (Perseghin *et al.*, 1997). Skeletal muscle triglyceride (TG) content, estimated from biopsy samples, was found to be negatively correlated with insulin sensitivity (Pan *et al.*, 1997) and more recently, several studies have demonstrated an even stronger association with intramyocellular lipid (IMCL) content, estimated using magnetic resonance (MR) techniques (Krssak *et al.*, 1999; Perseghin *et al.*, 1999; Szczepaniak *et al.*, 1999), in healthy, as well as insulin resistant individuals. This relationship is conserved in other insulin-responsive tissues, for example, liver (Petersen *et al.*, 2002, 2005; Seppala-Lindroos *et al.*, 2002), suggesting that dysregulation of lipid metabolism may contribute to insulin resistance in these individuals. Substantiated by numerous animal studies, mechanisms for lipid-induced insulin resistance have been proposed (Morino *et al.*, 2006; Savage *et al.*, 2007; Shulman, 2000), that attribute impaired insulin action in muscle and liver to the presence of lipid intermediates, in particular diacylglycerol (DAG), rather than intracellular TG *per se*, which interfere with the insulin-signaling cascade.

1.1. Mitochondrial fatty acid oxidation and insulin resistance

The net accumulation of intracellular lipid metabolites in muscle could either be caused by increased delivery and uptake of fatty acids, by decreased utilization, or a combination of these factors (Shulman, 2000). Intriguingly, a number of muscle biopsy studies demonstrated that the occurrence of obesity and/or diabetes was accompanied by a decrease in the content and activity of mitochondrial enzymes (Colberg *et al.*, 1995; He *et al.*, 2001; Kelley *et al.*, 1999; Ritov *et al.*, 2005; Simoneau and Kelley, 1997;

Simoneau *et al.*, 1999) suggesting that impaired fatty acid oxidation may contribute to the pathogenesis of these conditions. Mitochondrial density was also found to be reduced in the muscle of obese and type 2 diabetics accompanied by smaller and morphologically damaged mitochondria (Kelley *et al.*, 2002) and analysis of gene-chip datasets indicated that down-regulation of genes associated with oxidative metabolism may occur in a coordinate manner in obese, type 2 diabetic individuals (Mootha *et al.*, 2003; Patti *et al.*, 2003) and overweight, nondiabetic subjects with a family history of diabetes (Mootha *et al.*, 2003; Patti *et al.*, 2003). However, one limitation of these muscle biopsy studies is that they provide an index of mitochondrial capacity but may not represent function. We therefore sought to implement techniques by which muscle mitochondrial function could be determined *in vivo*.

2. *IN VIVO* MAGNETIC RESONANCE SPECTROSCOPY

Since its initial demonstration as a viable methodology for *in vivo* research (Hoult *et al.*, 1974), Magnetic resonance spectroscopy (MRS) has emerged as a powerful technique for studying metabolic function in a variety of tissues in both animals and humans (Dobbins and Malloy, 2003; Prompers *et al.*, 2006; Shulman *et al.*, 1996). MRS is a companion technique to the more familiar magnetic resonance imaging (MRI) scan, which is in widespread use providing clinical anatomical images. Unlike MRI, which measures the spatial distribution of water in the region of interest, MRS determines the content of MR visible nuclei. Isotopes of many biologically relevant elements are MR visible, including hydrogen, carbon, and phosphorus. Each has a characteristic resonance frequency in the presence of a strong static magnetic field, i.e., within the MR scanner. What makes MRS of particular utility is that the chemical environment of a nucleus also influences its precise frequency, a phenomenon which gives rise to the MRS spectrum. Therefore, within a tissue of interest, the content of specific compounds and their constituent chemical groups can be resolved and determined using *in vivo* MRS. Figure 21.1 illustrates examples of *in vivo* spectra obtained from muscle showing two nuclei of metabolic relevance, phosphorus (^{31}P) and carbon (^{13}C).

The advantages of MRS as a tool for investigating muscle metabolism are considerable. It is safe, noninvasive, involves no ionizing radiation and permits repeated measures of metabolites during a single session, or over an extensive study. It is, therefore, ideal for monitoring dynamic metabolic processes. Multinuclear capabilities enable a plethora of information to be collected and provide the flexibility to study independent parameters of metabolism during the same experiment. However, MRS is a relatively

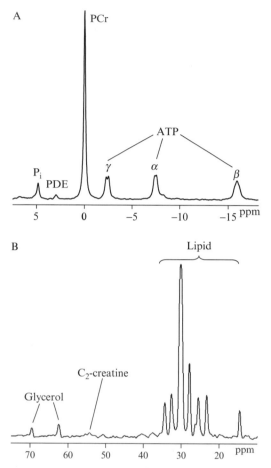

Figure 21.1 Typical ³¹P (A) and ¹³C (B) MRS spectra acquired from the soleus-gastrocnemius muscle complex *in vivo*.

insensitive technique compared to other imaging/spectrometry modalities, for example, PET (Positron Emission Tomography) or mass spectrometry, and metabolite concentrations in the millimolar range are required to be observed *in vivo*. Specialized equipment is necessary, although spectroscopy packages are now available for the new generation of clinical MR scanners.

As can be seen in panel A, a ³¹P spectrum obtained from muscle is characterized by peaks due to phosphocreatine (PCr), the three (α, β, and γ) phosphate moieties of adenosine tri-phosphate (ATP), unbound inorganic phosphate, and mono- and di-phosphoesters (PME, PDE). With the exception of adenosine di-phosphate (ADP), which is present in too low a concentration to be observed in resting muscle, the principal metabolites involved in energy transduction can be observed which makes *in vivo* ³¹P MRS particularly useful for studying muscle energetics.

Since all organic compounds contain carbon atoms, ^{13}C MRS can provide information on a huge number of compounds involved in muscle metabolism. High resolution ^{13}C spectra of muscle extracts contain a myriad of overlapping peaks corresponding to different metabolites and their constituent carbons. *In vivo*, this situation is somewhat simplified, due to the low natural abundance of [13]C (1.1% of total carbon) and the high concentration of lipids compared to other metabolites. An *in vivo* muscle ^{13}C spectrum (Fig. 21.1B) is dominated by peaks due to the fatty acid and glycerol components of lipid, accompanied by peaks of other metabolites present in relatively high concentrations. However, this potential drawback can be turned to an advantage by infusing a ^{13}C-labeled compound and using ^{13}C MRS to monitor the metabolism of this tracer.

This article focuses on two techniques we have developed which take advantage of the unique characteristics of ^{31}P and ^{13}C MRS to investigate different facets of human muscle mitochondrial metabolism. The terminal stage of mitochondrial energy production can be estimated by using a ^{31}P magnetization-transfer technique to determine the unidirectional flux of $P_i \rightarrow$ ATP ("ATP synthesis"), and oxidation via the TCA cycle can be assessed by applying a metabolic model to ^{13}C MRS data obtained during the oxidation of a ^{13}C-labeled substrate. After providing a thorough description of the implementation of the techniques, we will describe how these parameters of muscle metabolism are modulated in insulin resistant and endurance trained individuals.

2.1. Examining exchange reactions by magnetization-transfer MRS

Although ATP can be synthesized from several sources, the principal synthetic reaction is from the combination of ADP and P_i which occurs primarily in mitochondria. ATP synthesis is in equilibrium with ATP hydrolysis as shown in Eq. (21.1), forming a phosphate exchange reaction with forward and reverse rate constants, k_f and k_r. The rate of ATP synthesis (V_{ATP}) is therefore equivalent to the forward rate constant multiplied by the concentrations of ADP and P_i, Eq. (21.2). Under steady–state conditions, where the concentration of ADP can be assumed to remain constant, this relationship can be further simplified such that V_{ATP} is the product of the pseudo-first-order rate constant (for P_i), k'_f, and the P_i concentration, Eq. (21.3).

$$\text{ADP} + \text{P}_i \underset{k_r}{\overset{k_f}{\rightleftarrows}} \text{ATP} \qquad (21.1)$$

$$V_{ATP} = k_f[\text{ADP}][\text{P}_i] \qquad (21.2)$$

$$V_{ATP} = k'_f[\text{P}_i] \qquad (21.3)$$

The unidirectional fluxes involved in an exchange reaction can be studied using MRS since the magnetic equilibrium of the participant metabolites can be perturbed and monitored, while the chemical equilibrium remains intact. This effect can be accomplished using magnetization-transfer techniques (Forsen and Hoffman, 1963; Hoffman and Forsén, 1966). As mentioned above, a ^{31}P MRS spectrum detects signals from the "high-energy phosphates" involved in cellular energetics. Therefore, the ATP synthesis flux ($P_i \rightarrow$ ATP) can be estimated, since both the ATP and P_i metabolite pools are visible.

During an MRS experiment, the signal of a particular metabolite can be expressed using the Bloch equations (Bloch, 1946), which describe the variation of the MRS signal over time. These can be modified to incorporate chemical exchange (McConnell, 1958). The response of the P_i signal involved in $P_i \leftrightarrow$ ATP exchange is shown in Eq. (21.4):

$$dM_{Pi}/dt = \left(M_0^{Pi} - M_{Pi}\right)/T_1^{Pi} - k_f'[M_{Pi}] + k_r[M_{ATP}] \qquad (21.4)$$

where M_{Pi} is the P_i magnetization, M_0^{Pi} is the equilibrium P_i magnetization, M_{ATP} is the γ-ATP magnetization, and T_1^{Pi} is the longitudinal relaxation time of P_i.

During a magnetization-transfer experiment, the γ-phosphate peak of ATP (i.e., that which participates in exchange) can be selectively irradiated using a long-duration, frequency-selective, saturation pulse to equalize its spin-states and eliminate its signal, i.e., $M_{ATP} = 0$, which simplifies the Bloch equation to Eq. (21.5).

$$dM_{Pi}/dt = \left(M_0^{Pi} - M_{Pi}\right)/T_1^{Pi} - k_f'[M_{Pi}] \qquad (21.5)$$

Effectively this states that under conditions of γ-ATP saturation, the signal of P_i will decrease due to $P_i \rightarrow$ ATP flux. A solution to this equation is, Eq. (21.6):

$$M_{Pi}' = M_0^{Pi}/\left(1 + k_f'T_1^{Pi}\right) - M_0^{Pi}\left\{1 - 1/\left(1 + k_f'T_1^{Pi}\right)\right\}\exp\left\{-\left(k_f' + 1/T_1^{Pi}\right)t\right\}$$

$$(21.6)$$

Steady-state saturation is attained following a long duration saturation pulse. Under these conditions $t \rightarrow \infty$, the second term in Eq. (21.6) vanishes, and it simplifies to:

$$M_{Pi}' = M_0^{Pi}/\left(1 + k_f'T_1^{Pi}\right) \qquad (21.7)$$

The T_1 relaxation times of P_i under γ-ATP saturated (T_1') and non-saturated $\left(T_1^{Pi}\right)$ conditions are related by the expression: $1/T_1' = 1/T_1^{Pi} + k_f'$. Substitution into Eq. (21.7) and rearranging gives the classic equation of saturation transfer, Eq. (21.8).

$$k_f' = \left(1 - M_{Pi}'/M_0^{Pi}\right)/T_1' \qquad (21.8)$$

Therefore, the pseudo-first-order rate constant for ATP synthesis is equivalent to the change in the P_i signal from equilibrium conditions (without saturation, M_0^{Pi}) to γ-ATP saturated conditions $\left(M_{Pi}'\right)$ expressed in units of the T_1 of P_i under conditions of γ-ATP saturation $\left(T_1'\right)$.

2.2. Measuring muscle ATP synthesis *in vivo* by ^{31}P saturation-transfer MRS

The application of the ^{31}P saturation-transfer technique to a biological system was initially used to measure the kinetics of the bacterial ATPase (Brown *et al.*, 1977), and has since been implemented in a variety of *in vitro* and *in vivo* systems (Alger *et al.*, 1982; Brindle *et al.*, 1989; Lei *et al.*, 2003; Matthews *et al.*, 1981). We have successfully used the technique to investigate the synthesis of ATP in resting human muscle using 2.1 T and 4 T whole-body MR systems. Experimentally, our methodology incorporates three separate elements: (1) calculation of the saturation-transfer effect (i.e., M_{Pi}'/M_0^{Pi}) caused by γ-ATP saturation on P_i; (2) calibration of the T_1 of P_i under conditions of γ-ATP saturation (i.e., T_1'), together these parameters can be inserted into Eq. (21.8) to derive the rate constant for ATP synthesis; and (3) determination of the *in vivo* P_i concentration (i.e., $[P_i]$), which enables the unidirectional $P_i \rightarrow$ ATP flux to be calculated by solving Eq. (21.3).

The saturation-transfer effect can be determined by using a simple pulse-acquire MRS sequence preceded by a long, low power, frequency-selective saturation pulse (Fig. 21.2A). *In vivo*, muscle ^{31}P spectra are typically acquired using surface coils, therefore, adiabatic (B_1 insensitive) pulses are preferred for optimal signal excitation (de Graaf, 2008). Pairs of spectra are acquired with the saturation pulse centered on the γ-ATP resonance or at a downfield frequency equidistant from P_i, to partially compensate for off-resonance effects of the saturation pulse, which can be significant, particularly at lower field strengths (Kingsley and Monahan, 2000a,b). Precise acquisition parameters are system dependent; at 4 T, we use a 9 cm surface coil for ^{31}P spectroscopy of the soleus/gastrocnemius muscle complex, an adiabatic half passage (90°) pulse for excitation and a 15 s "soft" pulse for saturation. A typical pair of saturation-transfer spectra acquired with these parameters is shown in Fig. 21.3. The T_1 of P_i under conditions of γ-ATP

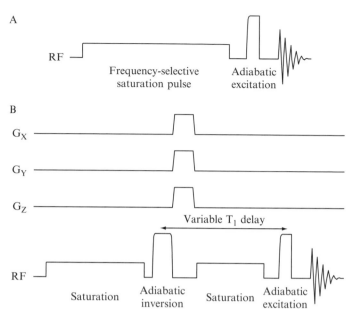

Figure 21.2 Pulse sequences for the ^{31}P saturation-transfer experiment. (A) Pulse-acquire sequence with RF saturation. A long, low power, frequency-selective saturation pulse precedes signal excitation/detection. Adiabatic excitation pulses are optimal for *in vivo* MRS using surface coils. (B) Inversion-recovery with RF saturation for T_1' calibration. Frequency-selective saturation of the γ-ATP peak is applied prior to the adiabatic inversion pulse, and during the variable inversion-delay to maintain steady-state spin saturation throughout the inversion-recovery measurement.

saturation is measured by an 8-point inversion-recovery calibration using a customized sequence (Fig. 21.2B), with γ-ATP saturation not only prior to the adiabatic full passage (180°) inversion pulse, but also during the variable inversion delay to maintain steady-state saturation of the γ-ATP spins. Inversion recovery data is fitted by a 3-parameter model (MacFall *et al.*, 1987), optimized using a nonlinear least squares algorithm. A fully relaxed ^{31}P spectrum acquired without saturation is used to determine intracellular $[P_i]$, assuming a constant muscle ATP concentration of 5.5 mmol/kg (Kemp *et al.*, 2007).

2.3. Monitoring metabolic fluxes using ^{13}C MRS

As mentioned previously, ^{13}C MRS offers a method to examine the metabolism of carbon-based compounds *in vivo*. The metabolic fate of a ^{13}C-labeled substrate can be monitored in real time, providing that it is metabolized sufficiently rapidly to induce a significant, dynamic enrichment in one of its metabolic products, that this product can be resolved from

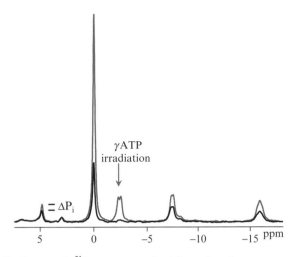

Figure 21.3 *In vivo* muscle ^{31}P spectra acquired from the soleus-gastrocnemius muscle complex during a saturation-transfer experiment. Spectra are acquired with (black spectrum) and without (gray spectrum) frequency-selective irradiation of the γ-ATP peak. The rate constant of exchange $\left(k'_f\right)$ is equivalent to the decrease in the P_i peak upon γ-ATP saturation $\left(\Delta P_i \Rightarrow 1 - M_{P_i}/M_0^{P_i}\right)$ divided by the T_1 relaxation time $\left(T'_1\right)$. $V_{ATP} = k'_f[P_i]$.

other metabolites with ^{13}C MRS and that it is of high enough concentration to be detected *in vivo*. For mitochondrial metabolism, the TCA cycle, which participates in the oxidation of lipids and carbohydrate, is ideally suited for examination by this technique.

Substrates derived from either source enter the cycle via the common intermediate, acetyl CoA, and are metabolized through a series of condensation, isomerization and oxidation (decarboxylation) reactions in the mitochondrial matrix ultimately producing carbon dioxide, NADH and FADH$_2$, the latter of which are oxidized by the electron transport chain. The ^{13}C label can be introduced into the cycle as shown in Fig. 21.4 by infusing [2-^{13}C]-labeled acetate as an MR-visible tracer, which is rapidly taken up by muscle and other tissues and converted into [2-^{13}C]-acetyl-CoA that enters the TCA cycle by condensing with oxaloacetate to form [4-^{13}C]-citrate. The position of the ^{13}C label is conserved through the initial steps of the TCA cycle, labeling α-ketoglutarate at the C$_4$ position. The principal TCA-cycle intermediates (TCAI) are present in concentrations that are too low to be observed *in vivo* using ^{13}C MRS. However, glutamate, which is present in millimolar concentrations in muscle, is in relatively rapid exchange with α-ketoglutarate and will also become ^{13}C enriched at C$_4$. As the TCA cycle progresses, the ^{13}C label becomes scrambled between the C$_2$ and C$_3$ positions due to the symmetry of the

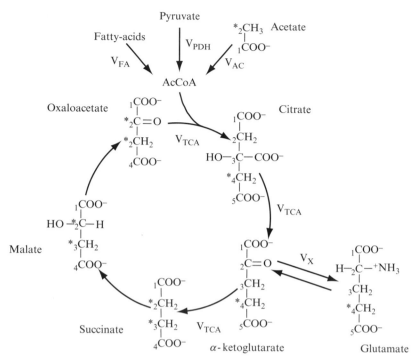

Figure 21.4 Oxidation via the TCA cycle: substrate derived from carbohydrate and lipid enters the cycle via the common intermediate acetyl CoA. Infused plasma [2-^{13}C]-acetate is converted to acetyl CoA, and the ^{13}C label (denoted by *) becomes incorporated into the muscle TCA-cycle intermediate and glutamate pools forming [4-^{13}C] glutamate on the first turn of the TCA cycle. The ^{13}C label becomes scrambled between the C_2 and C_3 positions due to the symmetry of the succinate molecule; a second turn of the cycle forms [2-^{13}C] and [3-^{13}C]-glutamate. Reaction rates incorporated into the metabolic model to calculate the TCA cycle flux (V_{TCA}) are shown and described in the text. Adapted and reprinted from Befroy *et al.* (2008a); © 1993–2008 by The National Academy of Sciences of the United States of America, all rights reserved.

succinate molecule; a second turn of the TCA cycle yields [2-^{13}C] and [3–^{13}C]-glutamate.

Unlike brain, where natural abundance [4-^{13}C]-glutamate can be resolved in an *in vivo* ^{13}C MRS spectra, the C_4 glutamate peak at 34.3 ppm in muscle is obscured by overlapping lipid resonances as shown in Fig. 21.1. The contribution from sub-cutaneous fat can be eliminated by using a localized MRS sequence to selectively acquire signal from muscle only. The IMCL contribution is also significant, but can be minimized by an inversion–null strategy which takes advantage of the more rapid T_1 relaxation times of lipid moieties compared to other metabolites as shown in Fig. 21.5. At 4 T, we use an adiabatic ^{1}H–^{13}C polarization-transfer (PT) sequence (Burum and Ernst, 1980; Shen *et al.*, 1999) for ^{13}C acquisition

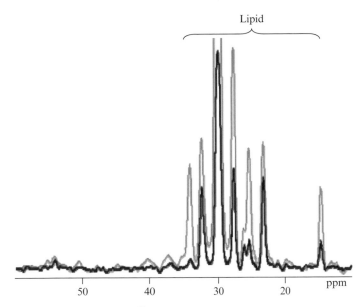

Figure 21.5 Natural abundance ^{13}C spectra acquired from the soleus-gastrocnemius muscle complex demonstrating the use of lipid suppression techniques to minimize the contribution of subcutaneous fat and IMCL resonances to the spectrum.

with WALTZ 16 decoupling. PT echo-times are optimized for C_4 glutamate detection. Spectra are acquired using a 9 cm diameter ^{13}C surface coil positioned underneath the soleus-gastrocnemius muscle complex of the calf, with twin, orthogonal 13 cm diameter ^{1}H surface coils arrayed in quadrature for decoupling and excitation/inversion on the ^{1}H channel. Voxel localization is achieved using 2-dimensional outer volume suppression (OVS) to select a ~90 cm^3 volume within the calf muscles. OVS enables the maximal ^{13}C metabolite signal to be acquired, although an ISIS-based localization scheme could also be implemented. IMCL is suppressed using adiabatic T_1-selective nulling, as described above, optimized for the lipid resonance at 34.7 ppm. Previously, we have used direct ^{13}C detection with decoupling at 2.1 T (Lebon *et al.*, 2001); however, the acquired ^{13}C signal is enhanced approximately 2-fold *in vivo* by using a PT sequence, and localization and lipid suppression can be performed on the ^{1}H channel where there is less error due to chemical shift dispersion.

To determine the TCA cycle flux, the time course of ^{13}C incorporation into C_4-glutamate is monitored during a 120-min infusion of 99% enriched [2-^{13}C]-acetate (350 mmol/l sodium salt) at a rate of 3.0 mg/kg/min. ^{13}C spectra are acquired with ~5 min time resolution for 20 min prior to the infusion to establish a baseline, and continuously until the end of the experiment. Plasma samples are obtained at 10 min intervals throughout the experiment for the measurement of plasma acetate concentration and fractional enrichment by gas chromatography/mass spectrometry (Befroy *et al.*, 2007). Steady-state enrichment of plasma acetate is attained within 5 min of originating the infusion and reaches ~85%; with the concentration of plasma acetate approximately 0.80 mM, consistent with functioning as a tracer-level substrate. To eliminate any residual contribution of lipid, the increment in the C_4-glutamate peak during the infusion is determined from difference spectra created by subtracting averaged baseline spectra from the ^{13}C spectrum for each time-point. The maximal enrichment in the [4-^{13}C]-glutamate peak is calculated relative to that of the C_2-glutamate peak assuming partial (5%) dilution of the C_2 pool due to anaplerosis. The absolute enrichment of the C_2-glutamate peak at the end of infusion is determined relative to the natural abundance enrichment (1.1%) of the baseline spectra.

2.4. Metabolic modeling of the TCA cycle flux

The TCA cycle flux (V_{TCA}) is determined by using CWave software to model the incorporation of ^{13}C label from plasma [2-^{13}C]-acetate into the muscle [4-^{13}C]-glutamate pool. The CWave model consists of isotopic mass balance equations shown below that describe the metabolic fate of the plasma [2-^{13}C]-acetate. The corresponding reaction rates are noted in Fig. 21.4. The ^{13}C label enters the TCA cycle following the conversion of [2-^{13}C]-acetate into [2-^{13}C]-acetylCoA (V_{AC}). Entry of unlabeled substrates via acetyl CoA is incorporated into the model as a separate reaction ($V_{PDH + FA}$). CWave determines V_{TCA} as the rate of total carbon flow from acetyl CoA to α-ketoglutarate using a non-linear least-squares algorithm to fit the curve of C_4-glutamate enrichment. The rate of α-ketoglutarate/glutamate exchange has been determined in previous studies and is fixed at 150 nmol/g(muscle)/min which is significantly faster than the TCA cycle flux, and therefore non-limiting. In this model, V_{TCA} is equal to the sum of ($V_{AC} + V_{PDH + FA}$) but is independent of the absolute fractional enrichment at C_4-glutamate ($V_{AC}/V_{PDH + FA}$). The intracellular concentration of glutamate has been determined by enzymatic analysis of muscle biopsy samples from healthy individuals, and is assumed to be 2.41 mmol/l in all subject groups (Lebon *et al.*, 2001). In our *in vivo* studies to date, the ratio of natural abundance C_2-glutamate/C_2-creatine is unaffected by the study population indicating that the intracellular concentration of glutamate is stable (Befroy *et al.*, 2007, 2008a). The concentrations of the other TCAI are very low and therefore not critical to the accuracy of the model and have been taken from the literature (Hawkins and Mans, 1983). The curve of muscle [4-^{13}C]-glutamate enrichment for a typical healthy individual is shown in Fig. 21.6 with the fit of the CWave metabolic model of these data superimposed, along with the calculated V_{TCA}.

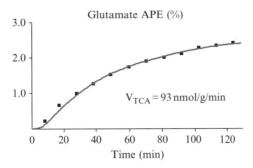

Figure 21.6 Time course of muscle [4-^{13}C]-glutamate enrichment measured by ^{13}C MRS in a typical healthy individual during a [2-^{13}C]-acetate infusion. The fit of the data using the C Wave metabolic model of the TCA cycle is superimposed along with the calculated TCA cycle flux (V_{TCA}).

2.4.1. Isotopic mass balance equations
2.4.1.1. Mass balance

$$d(\text{Citrate})/dt = V_{TCA} - V_{TCA}$$
$$d(\alpha KG)/dt = V_{TCA} + V_X - (V_{TCA} + V_X)$$
$$d(\text{Glutamate})/dt = V_X - V_X$$
$$d(\text{AcetylCoA})/dt = V_{AC} + V_{PDH+FA} - V_{TCA}$$

2.4.1.2. Isotope balance

$$d(C_4\text{-Citrate})/dt = V_{TCA}(C_2\text{-AcetylCoA}/\text{AcetylCoA})$$
$$- V_{TCA}(C_4\text{-Citrate}/\text{Citrate})$$
$$d(C_4\text{-}\alpha KG)/dt = V_{TCA}(C_4\text{-Citrate}/\text{Citrate})$$
$$+ V_X(C_4\text{-Glutamate}/\text{Glutamate})$$
$$- (V_{TCA} + V_X)(C_4\text{-}\alpha KG/\alpha KG)$$
$$d(C_4\text{-Glutamate})/dt = V_X(C_4\text{-}\alpha KG/\alpha KG)$$
$$- V_X(C_4\text{-Glutamate}/\text{Glutamate})$$
$$d(C_2\text{-AcetylCoA})/dt = V_{AC}(C_2\text{-Acetate}/\text{Acetate})$$
$$+ V_{PDH+FA}(C_0\text{-FFA}/\text{FFA})$$
$$- V_{TCA}(C_2\text{-AcetylCoA}/\text{AcetylCoA})$$

2.4.1.3. Muscle metabolite concentrations

$$\text{AcetylCoA} = 0.05 \text{ umol/g}$$
$$\text{Citrate} = 0.2 \text{ umol/g}$$
$$\text{Glutamate} = 2.41 \text{ umol/g}$$
$$\alpha\text{KG} = 0.05 \text{ umol/g}$$

This relatively simple metabolic model, which incorporates only a single turn of the TCA cycle, and requires only plasma C_2-acetate and muscle C_4-glutamate enrichment, performs well in resting muscle. The advantage of the single-turn model is that the ^{13}C MRS methods can be optimized to detect [4-^{13}C]-glutamate with the maximum possible sensitivity and time-resolution and therefore obtain the most accurate time course of enrichment. This scheme generates the same rates of V_{TCA} as a more robust 2-turn model that also requires data for muscle C_2-glutamate which is enriched during a second turn of the TCA cycle. We have previously used this more complex model to estimate V_X which confirmed our assumption that $V_X > V_{TCA}$ (Befroy *et al.*, 2007; Lebon *et al.*, 2001). We also investigated the influence of anaplerosis on the calculated rate of TCA cycle flux by using a model that incorporated an anaplerotic flux via pyruvate carboxylase (V_{ana}). The calculated V_{TCA} is unaffected by incremental increases in V_{ana} up to 20% of V_{TCA} (Befroy *et al.*, 2007), which is significantly greater than is likely in resting muscle. We have also found that the absolute fractional enrichment of C_2-glutamate at the end of the [2-^{13}C]-acetate infusion is not different between subject groups in our studies, indicating that rates of anaplerosis in quiescent muscle are similar between study populations (Befroy *et al.*, 2007, 2008a). However, this condition may not be preserved for all subject cohorts or in other tissues.

3. INSULIN RESISTANCE AND RESTING MITOCHONDRIAL METABOLISM

Using the techniques outlined above, we have examined how these parameters of resting mitochondrial metabolism are affected in two different cohorts of insulin-resistant individuals. Both V_{TCA} and V_{ATP} were decreased by ~40% in the muscle of insulin-resistant elderly subjects (mean age ~70 years) compared with activity-matched young adults (mean age ~27 years) (Petersen *et al.*, 2003), demonstrating that these older individuals had decreased resting mitochondrial function. Similarly, both parameters were decreased ~30% in insulin-resistant offspring of type 2 diabetic patients compared with insulin-sensitive controls (Befroy *et al.*, 2007; Petersen *et al.*, 2004), consistent with the ~40% reduced mitochondrial density observed in a similar cohort of subjects (Morino *et al.*, 2005). Average curves of muscle [4-^{13}C]-glutamate enrichment for each group are shown in Fig. 21.7, with the fits of the CWave model superimposed, demonstrating the difference in the rates of enrichment and the resultant TCA cycle flux between these cohorts. It is worth noting that these studies were performed in lean, otherwise healthy individuals and the subject groups were matched for body composition, to prevent the confounding factor of obesity, and to try to isolate effects occurring early in the pathogenesis of insulin resistance. Activity levels were also matched, since many of the age-related effects on muscle metabolism can be attributed to the decrease in physical activity that typically occurs in older individuals (Chilibeck *et al.*, 1998; Kent-Braun and Ng, 2000). These spectroscopy techniques are also independent of the mass of tissue from which the MRS signal is detected and are therefore not directly affected by the muscle atrophy that is commonly associated with aging (Frontera *et al.*, 1991).

We have also investigated how resting mitochondrial metabolism is modulated in endurance trained muscle, which is characterized by increased mitochondrial number and content of oxidative enzymes. Although V_{TCA} was increased in the muscle of trained individuals, V_{ATP} was unaltered (Befroy *et al.*, 2008a) suggesting a decrease in the efficiency of energy production at rest. This elevated "idling" rate of the mitochondria may enable a more rapid response to the increased energy demand at the onset of exercise. We did not observe any significant effect on whole-body energy expenditure demonstrating that whole-body indirect calorimetry techniques are too insensitive to detect significant changes in muscle oxidation rates.

Comparison of the absolute rates of $V_{ATP,}$ calculated by the ^{31}P-saturation-transfer technique, and of oxygen consumption, estimated by a number of

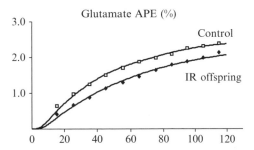

Figure 21.7 Muscle [4-^{13}C]-glutamate enrichment (averaged data) measured by ^{13}C MRS in a group of healthy, insulin-sensitive control subjects and insulin-resistant (IR) offspring of type 2 diabetic patients. CWave fits of the data are superimposed demonstrating the difference in the rates of enrichment and the resultant TCA cycle flux between the groups. Adapted and reprinted from Befroy *et al.* (2007); © American Diabetes Association (ADA).

techniques (including V_{TCA} by ^{13}C MRS), give rise to P:O ratios in resting muscle which are higher than the theoretical maximum (Brindle *et al.*, 1989; Kemp, 2008). This is in contrast to other tissues which have higher respiration rates, for example, kidney (Freeman *et al.*, 1983) and brain (Du *et al.*, 2008; Lei *et al.*, 2003), where good agreement is observed between these parameters. The total $P_i \rightarrow$ ATP flux from all cellular contributions (oxidative + glycolytic) is evaluated using the saturation-transfer technique; although the glycolytic contribution in resting muscle is low (Jucker *et al.*, 2000); and unpublished data) it has been postulated that in quiescent muscle reversible $P_i \leftrightarrow$ ATP exchange via the mitochondrial F_1F_0 ATPase may contribute to the elevated absolute V_{ATP} (Brindle and Radda, 1987; Campbell-Burk *et al.*, 1987). It is worth noting that even if exchange is the dominating contribution, it is likely to be proportional to the F_1 ATPase activity. We have successfully miniaturized the saturation-transfer technique to study mouse muscle and have found that the $P_i \rightarrow$ ATP flux responds as expected to perturbations in the expression of mitochondrial genes in transgenic mice including UCP3 knock-out (Cline *et al.*, 2001) and UCP3 overexpression (Befroy unpublished data) and PGC1α overexpression (Choi *et al.*, 2008) confirming that the rate of ATP synthesis determined by the saturation-transfer technique is an effective biomarker of mitochondrial function.

3.1. Resting mitochondrial metabolism versus maximal capacity

In counterpoint to our *in vivo* MRS studies investigating resting muscle mitochondrial metabolism, other groups have observed that mitochondrial oxidative capacity, estimated *in vivo* from PCr kinetics during exercise, was unaffected by aging (Chilibeck *et al.*, 1998; Lanza *et al.*, 2005) when activity patterns were accounted for, and was unchanged throughout the development of insulin-resistance and diabetes in the ZDF rat (De Feyter *et al.*, 2008). Muscle biopsy studies also suggest that mitochondrial capacity is increased in Asian-Indians despite severe insulin resistance (Nair *et al.*, 2008). The disparity between these studies highlights the fact that capacity and resting function are different parameters of mitochondrial metabolism, likely to be regulated or limited by different factors, and that the term mitochondrial "dysfunction," which has become commonly used in the field and the source of some controversy, is misleading since it has no precise definition and could refer to reduced capacity, function or a combination of both. Furthermore, the findings of these studies will also have been influenced by the precise characteristics of the populations studied and the muscle chosen for analysis. Different muscle groups exhibit distinct metabolic and functional phenotypes (Essen *et al.*, 1975; Pette and Spamer, 1986) and respond differently to aging (Houmard *et al.*, 1998) and potentially disease.

Therefore, it is highly likely that there are muscle-group specific differences in mitochondrial function and capacity in healthy subjects and in the modulation of energy production by aging/disease. With this issue in mind, we have recently implemented the saturation-transfer methodology in a localized manner to obtain muscle-group specific rates of $P_i \rightarrow ATP$ flux (Befroy *et al.*, 2008b).

4. CONCLUSIONS

Because of its distinct characteristics, MRS is a unique tool for investigating metabolism across a wide range of systems, encompassing human, animal and cellular models, a wide array of tissues, and *in vivo* as well as *in vitro* applications. Multinuclear capabilities further extend these possibilities and since the technique is non-invasive, dynamic processes can be studied using repeated measurements in addition to static estimates of metabolite concentrations. By taking advantage of these unique properties, we have developed techniques that allow us to monitor muscle mitochondrial metabolism *in vivo* and investigate how it is modulated in health and disease. The implementation of spectroscopy packages on clinical MR scanners will allow these methodologies to reach a wider audience of investigators. With the progression to higher-field MR systems, which increase spectral resolution and sensitivity, and the continued development of enhanced detection techniques, there are a multitude of possibilities for future research.

REFERENCES

Alger, J. R., den Hollander, J. A., and Shulman, R. G. (1982). *In vivo* phosphorus-31 nuclear magnetic resonance saturation transfer studies of adenosinetriphosphatase kinetics in *Saccharomyces cerevisiae*. *Biochemistry* **21**, 2957–2963.

Befroy, D. E., Petersen, K. F., Dufour, S., Mason, G. F., de Graaf, R. A., Rothman, D. L., and Shulman, G. I. (2007). Impaired mitochondrial substrate oxidation in muscle of insulin-resistant offspring of type 2 diabetic patients. *Diabetes* **56**, 1376–1381.

Befroy, D. E., Petersen, K. F., Dufour, S., Mason, G. F., Rothman, D. L., and Shulman, G. I. (2008a). Increased substrate oxidation and mitochondrial uncoupling in skeletal muscle of endurance-trained individuals. *Proc. Natl. Acad. Sci. USA* **105**, 16701–16706.

Befroy, D. E., Petersen, K. F., Shulman, G. I., and Rothman, D. L. (2008b). Localized 31P saturation transfer reveals differences in gastrocnemius and soleus rates of ATP synthesis *in-vivo*. *Proc. Int. Soc. Mag. Reson. Med.* 2565.

Bloch, F. (1946). Nuclear induction. *Phys. Rev.* **70**, 460.

Boden, G., and Shulman, G. I. (2002). Free fatty acids in obesity and type 2 diabetes: Defining their role in the development of insulin resistance and beta-cell dysfunction. *Eur. J. Clin. Invest.* **32**(Suppl. 3), 14–23.

Brindle, K. M., Blackledge, M. J., Challiss, R. A., and Radda, G. K. (1989). ^{31}P NMR magnetization-transfer measurements of ATP turnover during steady-state isometric muscle contraction in the rat hind limb *in vivo*. *Biochemistry* **28**, 4887–4893.

Brindle, K. M., and Radda, G. K. (1987). ^{31}P-NMR saturation transfer measurements of exchange between Pi and ATP in the reactions catalysed by glyceraldehyde-3-phosphate dehydrogenase and phosphoglycerate kinase *in vitro*. *Biochim. Biophys. Acta* **928**, 45–55.

Brown, T. R., Ugurbil, K., and Shulman, R. G. (1977). ^{31}P nuclear magnetic resonance measurements of ATPase kinetics in aerobic *Escherichia coli* cells. *Proc. Natl. Acad. Sci. USA* **74**, 5551–5553.

Burum, D. P., and Ernst, R. R. (1980). Net polarization transfer via a J-ordered state for signal enhancement of low-sensitivity nuclei. *J. Magn. Reson. (1969)* **39**, 163–168.

Campbell-Burk, S. L., Jones, K. A., and Shulman, R. G. (1987). ^{31}P NMR saturation-transfer measurements in *Saccharomyces cerevisiae*: Characterization of phosphate exchange reactions by iodoacetate and antimycin A inhibition. *Biochemistry* **26**, 7483–7492.

Chilibeck, P. D., McCreary, C. R., Marsh, G. D., Paterson, D. H., Noble, E. G., Taylor, A. W., and Thompson, R. T. (1998). Evaluation of muscle oxidative potential by ^{31}P-MRS during incremental exercise in old and young humans. *Eur. J. Appl. Physiol. Occup. Physiol.* **78**, 460–465.

Choi, C. S., Befroy, D. E., Codella, R., Kim, S., Reznick, R. M., Hwang, Y., Liu, Z., Lee, H., Distefano, A., Samuel, V. T., Zhang, D., Cline, G. W., *et al.* (2008). Paradoxical effects of increased expression of PGC-1α on muscle mitochondrial function and insulin-stimulated muscle glucose metabolism. *Proc. Natl. Acad. Sci. USA* **105**, 19926–19931.

Cline, G. W., Vidal-Puig, A. J., Dufour, S., Cadman, K. S., Lowell, B. B., and Shulman, G. I. (2001). *In vivo* effects of uncoupling protein-3 gene disruption on mitochondrial energy metabolism. *J. Biol. Chem.* **276**, 20240–20244.

Colberg, S. R., Simoneau, J. A., Thaete, F. L., and Kelley, D. E. (1995). Skeletal muscle utilization of free fatty acids in women with visceral obesity. *J. Clin. Invest.* **95**, 1846–1853.

De Feyter, H. M., Lenaers, E., Houten, S. M., Schrauwen, P., Hesselink, M. K., Wanders, R. J., Nicolay, K., and Prompers, J. J. (2008). Increased intramyocellular lipid content but normal skeletal muscle mitochondrial oxidative capacity throughout the pathogenesis of type 2 diabetes. *FASEB J.* **22**, 3947–3955.

de Graaf, R. A. (2008). *In Vivo* NMR Spectroscopy: Principles and Techniques. Wiley, Chichester.

Dobbins, R. L., and Malloy, C. R. (2003). Measuring *in-vivo* metabolism using nuclear magnetic resonance. *Curr. Opin. Clin. Nutr. Metab. Care* **6**, 501–509.

Du, F., Zhu, X. H., Zhang, Y., Friedman, M., Zhang, N., Ugurbil, K., and Chen, W. (2008). Tightly coupled brain activity and cerebral ATP metabolic rate. *Proc. Natl. Acad. Sci. USA* **105**, 6409–6414.

Essen, B., Jansson, E., Henriksson, J., Taylor, A. W., and Saltin, B. (1975). Metabolic characteristics of fibre types in human skeletal muscle. *Acta Physiol. Scand.* **95**, 153–165.

Forsen, S., and Hoffman, R. A. (1963). Study of moderately rapid chemical exchange reactions by means of nuclear magnetic double resonance. *J. Chem. Phys.* **39**, 2892–2901.

Freeman, D., Bartlett, S., Radda, G., and Ross, B. (1983). Energetics of sodium transport in the kidney. Saturation transfer ^{31}P-NMR. *Biochim. Biophys. Acta* **762**, 325–336.

Frontera, W. R., Hughes, V. A., Lutz, K. J., and Evans, W. J. (1991). A cross-sectional study of muscle strength and mass in 45- to 78-yr-old men and women. *J. Appl. Physiol.* **71**, 644–650.

Haffner, S. M., Stern, M. P., Mitchell, B. D., Hazuda, H. P., and Patterson, J. K. (1990). Incidence of type II diabetes in Mexican Americans predicted by fasting insulin and glucose levels, obesity, and body-fat distribution. *Diabetes* **39**, 283–288.

Hawkins, R. A., and Mans, A. M. (1983). Intermediary metabolism of carbohydrates and other fuels. *In* "Handbook of Neurochemistry" (A. Lajtha, ed.), **3**, pp. 259–294. Plenum, New York.

He, J., Watkins, S., and Kelley, D. E. (2001). Skeletal muscle lipid content and oxidative enzyme activity in relation to muscle fiber type in type 2 diabetes and obesity. *Diabetes* **50**, 817–823.

Hoffman, R. A., and Forsén, S. (1966). High resolution nuclear magnetic double and multiple resonance. *Prog. NMR Spectrosc.* **1**, 15–204.

Hoult, D. I., Busby, S. J., Gadian, D. G., Radda, G. K., Richards, R. E., and Seeley, P. J. (1974). Observation of tissue metabolites using ^{31}P nuclear magnetic resonance. *Nature* **252**, 285–287.

Houmard, J. A., Weidner, M. L., Gavigan, K. E., Tyndall, G. L., Hickey, M. S., and Alshami, A. (1998). Fiber type and citrate synthase activity in the human gastrocnemius and vastus lateralis with aging. *J. Appl. Physiol.* **85**, 1337–1341.

Jucker, B. M., Ren, J., Dufour, S., Cao, X., Previs, S. F., Cadman, K. S., and Shulman, G. I. (2000). ^{13}C/^{31}P NMR assessment of mitochondrial energy coupling in skeletal muscle of awake fed and fasted rats. Relationship with uncoupling protein 3 expression. *J. Biol. Chem.* **275**, 39279–39286.

Kelley, D. E., Goodpaster, B., Wing, R. R., and Simoneau, J. A. (1999). Skeletal muscle fatty acid metabolism in association with insulin resistance, obesity, and weight loss. *Am. J. Physiol. Endocrinol. Metab.* **277**, E1130–E1141.

Kelley, D. E., He, J., Menshikova, E. V., and Ritov, V. B. (2002). Dysfunction of mitochondria in human skeletal muscle in type 2 diabetes. *Diabetes* **51**, 2944–2950.

Kemp, G. J. (2008). The interpretation of abnormal ^{31}P magnetic resonance saturation transfer measurements of Pi/ATP exchange in insulin-resistant skeletal muscle. *Am. J. Physiol. Endocrinol. Metab.* **294**, E640–E642; author reply E643–E644.

Kemp, G. J., Meyerspeer, M., and Moser, E. (2007). Absolute quantification of phosphorus metabolite concentrations in human muscle *in vivo* by 31P MRS: A quantitative review. *NMR Biomed.* **20**, 555–565.

Kent-Braun, J. A., and Ng, A. V. (2000). Skeletal muscle oxidative capacity in young and older women and men. *J. Appl. Physiol.* **89**, 1072–1078.

Kingsley, P. B., and Monahan, W. G. (2000a). Corrections for off-resonance effects and incomplete saturation in conventional (two-site) saturation-transfer kinetic measurements. *Magn. Reson. Med.* **43**, 810–819.

Kingsley, P. B., and Monahan, W. G. (2000b). Effects of off-resonance irradiation, cross-relaxation, and chemical exchange on steady-state magnetization and effective spin-lattice relaxation times. *J. Magn. Reson.* **143**, 360–375.

Krssak, M., Falk Petersen, K., Dresner, A., DiPietro, L., Vogel, S. M., Rothman, D. L., Roden, M., and Shulman, G. I. (1999). Intramyocellular lipid concentrations are correlated with insulin sensitivity in humans: A ^{1}H NMR spectroscopy study. *Diabetologia* **42**, 113–116.

Lanza, I. R., Befroy, D. E., and Kent-Braun, J. A. (2005). Age-related changes in ATP-producing pathways in human skeletal muscle *in vivo*. *J. Appl. Physiol.* **99**, 1736–1744.

Lebon, V., Dufour, S., Petersen, K. F., Ren, J., Jucker, B. M., Slezak, L. A., Cline, G. W., Rothman, D. L., and Shulman, G. I. (2001). Effect of triiodothyronine on mitochondrial energy coupling in human skeletal muscle. *J. Clin. Invest.* **108**, 733–737.

Lei, H., Ugurbil, K., and Chen, W. (2003). Measurement of unidirectional Pi to ATP flux in human visual cortex at 7 T by using *in vivo* ^{31}P magnetic resonance spectroscopy. *Proc. Natl. Acad. Sci. USA* **100**, 14409–14414.

Lillioja, S., Young, A. A., Culter, C. L., Ivy, J. L., Abbott, W. G., Zawadzki, J. K., Yki-Jarvinen, H., Christin, L., Secomb, T. W., and Bogardus, C. (1987). Skeletal muscle capillary density and fiber type are possible determinants of *in vivo* insulin resistance in man. *J. Clin. Invest.* **80**, 415–424.

MacFall, J. R., Wehrli, F. W., Breger, R. K., and Johnson, G. A. (1987). Methodology for the measurement and analysis of relaxation times in proton imaging. *Magn. Reson. Imaging* **5**, 209–220.

Matthews, P. M., Bland, J. L., Gadian, D. G., and Radda, G. K. (1981). The steady-state rate of ATP synthesis in the perfused rat heart measured by ^{31}P NMR saturation transfer. *Biochem. Biophys. Res. Commun.* **103,** 1052–1059.

McConnell, H. M. (1958). Reaction rates by nuclear magnetic resonance. *J. Chem. Phys.* **28,** 430–431.

Mootha, V. K., Lindgren, C. M., Eriksson, K. F., Subramanian, A., Sihag, S., Lehar, J., Puigserver, P., Carlsson, E., Ridderstrale, M., Laurila, E., Houstis, N., Daly, M. J., *et al.* (2003). PGC-1alpha-responsive genes involved in oxidative phosphorylation are coordinately downregulated in human diabetes. *Nat. Genet.* **34,** 267–273.

Morino, K., Petersen, K. F., Dufour, S., Befroy, D., Frattini, J., Shatzkes, N., Neschen, S., White, M. F., Bilz, S., Sono, S., Pypaert, M., and Shulman, G. I. (2005). Reduced mitochondrial density and increased IRS-1 serine phosphorylation in muscle of insulin-resistant offspring of type 2 diabetic parents. *J. Clin. Invest.* **115,** 3587–3593.

Morino, K., Petersen, K. F., and Shulman, G. I. (2006). Molecular mechanisms of insulin resistance in humans and their potential links with mitochondrial dysfunction. *Diabetes* **55** (Suppl. 2), S9–S15.

Nair, K. S., Bigelow, M. L., Asmann, Y. W., Chow, L. S., Coenen-Schimke, J. M., Klaus, K. A., Guo, Z. K., Sreekumar, R., and Irving, B. A. (2008). Asian Indians have enhanced skeletal muscle mitochondrial capacity to produce ATP in association with severe insulin resistance. *Diabetes* **57,** 1166–1175.

Pan, D. A., Lillioja, S., Kriketos, A. D., Milner, M. R., Baur, L. A., Bogardus, C., Jenkins, A. B., and Storlien, L. H. (1997). Skeletal muscle triglyceride levels are inversely related to insulin action. *Diabetes* **46,** 983–988.

Patti, M. E., Butte, A. J., Crunkhorn, S., Cusi, K., Berria, R., Kashyap, S., Miyazaki, Y., Kohane, I., Costello, M., Saccone, R., Landaker, E. J., Goldfine, A. B., *et al.* (2003). Coordinated reduction of genes of oxidative metabolism in humans with insulin resistance and diabetes: Potential role of PGC1 and NRF1. *Proc. Natl. Acad. Sci. USA* **100,** 8466–8471.

Perseghin, G., Ghosh, S., Gerow, K., and Shulman, G. I. (1997). Metabolic defects in lean nondiabetic offspring of NIDDM parents: A cross-sectional study. *Diabetes* **46,** 1001–1009.

Perseghin, G., Scifo, P., De Cobelli, F., Pagliato, E., Battezzati, A., Arcelloni, C., Vanzulli, A., Testolin, G., Pozza, G., Del Maschio, A., and Luzi, L. (1999). Intramyocellular triglyceride content is a determinant of *in vivo* insulin resistance in humans: A ^{1}H-^{13}C nuclear magnetic resonance spectroscopy assessment in offspring of type 2 diabetic parents. *Diabetes* **48,** 1600–1606.

Petersen, K. F., Befroy, D., Dufour, S., Dziura, J., Ariyan, C., Rothman, D. L., DiPietro, L., Cline, G. W., and Shulman, G. I. (2003). Mitochondrial dysfunction in the elderly: Possible role in insulin resistance. *Science* **300,** 1140–1142.

Petersen, K. F., Dufour, S., Befroy, D., Garcia, R., and Shulman, G. I. (2004). Impaired mitochondrial activity in the insulin-resistant offspring of patients with type 2 diabetes. *N. Engl. J. Med.* **350,** 664–671.

Petersen, K. F., Dufour, S., Befroy, D., Lehrke, M., Hendler, R. E., and Shulman, G. I. (2005). Reversal of nonalcoholic hepatic steatosis, hepatic insulin resistance, and hyperglycemia by moderate weight reduction in patients with type 2 diabetes. *Diabetes* **54,** 603–608.

Petersen, K. F., Oral, E. A., Dufour, S., Befroy, D., Ariyan, C., Yu, C., Cline, G. W., DePaoli, A. M., Taylor, S. I., Gorden, P., and Shulman, G. I. (2002). Leptin reverses insulin resistance and hepatic steatosis in patients with severe lipodystrophy. *J. Clin. Invest.* **109,** 1345–1350.

Pette, D., and Spamer, C. (1986). Metabolic properties of muscle fibers. *Fed. Proc.* **45,** 2910–2914.

Prompers, J. J., Jeneson, J. A., Drost, M. R., Oomens, C. C., Strijkers, G. J., and Nicolay, K. (2006). Dynamic MRS and MRI of skeletal muscle function and biomechanics. *NMR Biomed.* **19,** 927–953.

Ritov, V. B., Menshikova, E. V., He, J., Ferrell, R. E., Goodpaster, B. H., and Kelley, D. E. (2005). Deficiency of subsarcolemmal mitochondria in obesity and type 2 diabetes. *Diabetes* **54,** 8–14.

Savage, D. B., Petersen, K. F., and Shulman, G. I. (2007). Disordered lipid metabolism and the pathogenesis of insulin resistance. *Physiol. Rev.* **87,** 507–520.

Seppala-Lindroos, A., Vehkavaara, S., Hakkinen, A. M., Goto, T., Westerbacka, J., Sovijarvi, A., Halavaara, J., and Yki-Jarvinen, H. (2002). Fat accumulation in the liver is associated with defects in insulin suppression of glucose production and serum free fatty acids independent of obesity in normal men. *J. Clin. Endocrinol. Metab.* **87,** 3023–3028.

Shen, J., Petersen, K. F., Behar, K. L., Brown, P., Nixon, T. W., Mason, G. F., Petroff, O. A., Shulman, G. I., Shulman, R. G., and Rothman, D. L. (1999). Determination of the rate of the glutamate/glutamine cycle in the human brain by *in vivo* ^{13}C NMR. *Proc. Natl. Acad. Sci. USA* **96,** 8235–8240.

Shulman, G. I. (2000). Cellular mechanisms of insulin resistance. *J. Clin. Invest.* **106,** 171–176.

Shulman, R. G., Rothman, D. L., and Price, T. B. (1996). Nuclear magnetic resonance studies of muscle and applications to exercise and diabetes. *Diabetes* **45**(Suppl. 1), S93–S98.

Simoneau, J. A., and Kelley, D. E. (1997). Altered glycolytic and oxidative capacities of skeletal muscle contribute to insulin resistance in NIDDM. *J. Appl. Physiol.* **83,** 166–171.

Simoneau, J. A., Veerkamp, J. H., Turcotte, L. P., and Kelley, D. E. (1999). Markers of capacity to utilize fatty acids in human skeletal muscle: Relation to insulin resistance and obesity and effects of weight loss. *FASEB J.* **13,** 2051–2060.

Szczepaniak, L. S., Babcock, E. E., Schick, F., Dobbins, R. L., Garg, A., Burns, D. K., McGarry, J. D., and Stein, D. T. (1999). Measurement of intracellular triglyceride stores by H spectroscopy: Validation *in vivo*. *Am. J. Physiol. Endocrinol. Metab.* **276,** E977–E989.

Warram, J. H., Martin, B. C., Krolewski, A. S., Soeldner, J. S., and Kahn, C. R. (1990). Slow glucose removal rate and hyperinsulinemia precede the development of type II diabetes in the offspring of diabetic parents. *Ann. Intern. Med.* **113,** 909–915.

Wild, S., Roglic, G., Green, A., Sicree, R., and King, H. (2004). Global prevalence of diabetes: Estimates for the year 2000 and projections for 2030. *Diabetes Care* **27,** 1047–1053.

CHAPTER TWENTY-TWO

METHODS FOR ASSESSING AND MODULATING UCP2 EXPRESSION AND FUNCTION

Lício A. Velloso, Giovanna R. Degasperi,
Aníbal E. Vercesi, *and* Mário A. Saad

Contents

1. Introduction	396
2. Analysis of UCP2 Expression by Immunoblot	397
2.1. Isolating pancreatic islets	397
2.2. Preparing mitochondria from hypothalamus and pancreatic islets	398
2.3. Immunoblotting of UCP2 in samples from hypothalamus and pancreatic islet mitochondria	398
3. Analysis of UCP2 Expression by Real-Time PCR	399
3.1. Preparation of cDNA	399
3.2. Real-time PCR	399
4. Mitochondria Respiration Assay	400
5. Modulating UCP2 Expression by Cold Exposure	400
6. Inhibiting UCP2 Expression by Antisense Oligonucleotide	401
7. Inhibiting PGC-1α as an Indirect Means of Controlling UCP2 Expression	402
References	403

Abstract

Uncoupling protein 2 (UCP2) is a member of the uncoupling protein family. It is expressed in the inner mitochondrial membrane and plays a role in the control of free radical production, oxidative damage, insulin secretion, and fatty-acid peroxide exportation. Although UCP2 expression occurs in several tissues, some of its most remarkable functions are exerted in organs of difficult experimental access, such as the central nervous system, particularly the hypothalamus and the pancreatic islets. In addition, due to its low levels of expression in the

Departments of Internal Medicine and Clinical Pathology, University of Campinas, Campinas, SP, Brazil

Methods in Enzymology, Volume 457
ISSN 0076-6879, DOI: 10.1016/S0076-6879(09)05022-8

mitochondrial membrane, studying UCP2 expression and function depends on specific- and well-established methods. This chapter describes methods for directly assessing UCP2 expression and function in different tissues. Purified mitochondria preparations are used for enhancing the capacity of detection of UCP2 protein or for evaluating the role of UCP2 in mitochondria respiration. Exposure of experimental animals to cold environment leads to increased UCP2 expression, while reduction of its expression can be achieved directly by targeting its mRNA with antisense oligonucleotides, or indirectly by targeting PGC-1α expression with antisense oligonucleotides.

1. INTRODUCTION

The inner mitochondrial membrane protein, uncoupling protein-2 (UCP2), belongs to the UCP family and is expressed in several tissues of the body, including the central nervous system and the pancreatic islets. In contrast to UCP1, which is the prototype protein of the UCP family, and has been well characterized as a modulator of brown adipose tissue thermogenesis, the functions of UCP2 are still a matter of investigation (Cannon *et al.*, 2006; Simoneau *et al.*, 1998). Although UCP2 shares high similarity with UCP1, most studies suggest that its roles in thermogenesis are considerably smaller; if they exist at all (Esteves and Brand, 2005). Three facts support this concept; first, UCP2 is expressed at a very low concentration in the mitochondrial membrane (less than 1% of UCP1 levels); secondly, it depends on specific activators in order to act as a transporter of protons; and thirdly, UCP2 knockout mice do not present any major defect of thermogenesis (Brand and Esteves, 2005). Nevertheless, under specific stimuli, UCP2 can transport protons across the mitochondrial membrane. At least two distinct outcomes may result from this transport. Mitochondria are the main source of reactive oxygen species (ROS); increased ROS production has damaging effects in cells and a number of diseases, such as Alzheimer's, atherosclerosis, obesity, and diabetes may result, at least in part, from ROS accumulation in distinct tissues (Brand *et al.*, 2004). The production of ROS in mitochondria is very sensitive to minimal changes in proton transport and several groups have shown that activating UCP2 protonmotive activity results in a substantial reduction of ROS production, which may act as a protective factor against the progression of these diseases (Brand *et al.*, 2004).

In pancreatic β cells, the secretion of insulin depends on the ratio of ATP/ADP, which is mostly controlled by the extracellular levels of glucose. When blood glucose is high, it enters the β cell through the low-K_m GLUT2, resulting in increased ATP production. A high ATP/ADP ratio closes ATP-sensitive K-channels, leading to depolarization of the cell's membrane and opening of voltage-sensitive Ca^{2+} channels, influx of Ca^{2+}

and consequently, insulin secretion. Increased expression of UCP2 in the β cells hamper insulin secretion (De Souza *et al.*, 2003), while reduction of UCP2 expression by distinct methods improve insulin secretion and can lead to a better metabolic control of animal models of diabetes (De Souza *et al.*, 2003, 2007; Zhang *et al.*, 2001).

An important issue regarding studies of UCP2 is the fact that its protein levels in tissues are extremely low. Most studies evaluate UCP2 expression by real-time PCR; however, it is as yet unknown whether protein and mRNA levels of this protein are coordinately regulated. Here, we describe methods that allow quantitative and functional analysis of UCP2. In addition, we present methods that allow *in vivo* modulation of UCP2 expression. Most protocols were tested in pancreatic islets and hypothalamus of rats and mice.

2. ANALYSIS OF UCP2 EXPRESSION BY IMMUNOBLOT

Regular immunoblot techniques fail to detect UCP2 in several tissues, even if protein loading is very high (Degasperi *et al.*, 2008; De Souza *et al.*, 2007). Two of the tissues on which UCP2 exerts some of its most remarkable effects, pancreatic islets and hypothalamus, exist in small amounts in the body, which makes it even more difficult to detect the expression of UCP2. Since UCP2 is expressed only in the inner mitochondrial membrane, performing immunoblot using purified mitochondria preparation provides an easily reproductive method to enhance the capacity to detect this protein.

2.1. Isolating pancreatic islets

Rats (or optionally mice) are anesthetized with halothane, the abdomen is opened and the common bile duct is cannulated under nonsterile conditions. The pancreas is distended by infusion of 10 ml (3.0 ml for mice) sterile Krebs–Ringer–Hepes solution containing 10% bovine serum albumin (BSA fraction V, Sigma, St. Louis, Missouri). Subsequently, the pancreas is excised and brought into a laminar flow cabinet. Excised pancreata are chopped and digested using a two-stage incubation of 20 min at 37 °C with successive 1.0 and 0.7 mg/ml collagenase P (Boehringer Mannheim, Mannheim, Germany). Islets are separated from exocrine tissue by centrifugation over a discontinuous dextran (Sigma) gradient (van Suylichem *et al.*, 1990) and further purification by handpicking under a stereo microscope. Islets are plated in nontreated Petri dishes in portions of 100 islets/ml in CMRL 1066 (Invitrogen, Carlsbad, California), containing 10% fetal calf serum (Invitrogen), 8.3 mM glucose, 10 mM Hepes, and 1% penicillin/streptomycin (Invitrogen), at 37 °C in humidified air containing 5% CO_2. From this stage, islets can be used in procedures for mitochondria isolation, evaluation of protein expression by immunoblot, or evaluation of RNA expression by real-time PCR.

2.2. Preparing mitochondria from hypothalamus and pancreatic islets

Hypothalami obtained from four rats (or optionally eight mice) or 8000 isolated pancreatic islets (from either rats or mice) are placed into ice-cold isolation buffer containing 75 mM sucrose, 1.0 mM K^+-EGTA, 0.1% bovine serum albumin (BSA; free fatty acid), 10 mM K^+-HEPES, pH 7.2 and 5.0 mg protease inhibitor (P-8038, Sigma). The tissue is finely minced with scissors and washed three times with this buffer. Thereafter, the tissue is homogenized by hand in an all-glass Dounce homogenizer using six up-and-down strokes with the pestle. The homogenate is centrifuged at 4000 rpm for 5 min in a Beckman JA 20 rotor (Beckman, Palo Alto, California). The supernatant is carefully decanted and the pellet resuspended in half of the original volume. The homogenate is once more centrifuged as above, the supernatant retained, and the pellet discarded. The pooled supernatant is centrifuged at 16,000 rpm for 10 min. The decanted supernatant is discarded and the pellet resuspended in 15% Percoll (Sigma). The discontinuous gradient is prepared in 15 × 100 mm centrifuge tubes by layering 3.0 ml fractions of the resuspended pellet onto two preformed layers consisting of 3.5 ml of 23% Percoll over 3.5 ml of 40% Percoll. The tubes are centrifuged for 5 min at 19,000 rpm (acceleration time, 50 s; deceleration with brake, 80 s). Three major bands of material are obtained, and the material banding near the interface of the lower two Percoll layers is diluted 1:4 by gently mixing with isolation buffer and then centrifuged at 14,000 rpm for 10 min. The supernatant is removed with a Pasteur pipette and the remaining material is gently resuspended. Fatty acid-free bovine serum albumin (10 mg/ml) is added and the mixture is diluted in isolation buffer. After centrifugation at 10,000 rpm for 10 min, the supernatant is rapidly decanted and the pellet gently resuspended in isolation buffer. The protein concentration is determined by the Bradford method (Bradford, 1976). Isolated mitochondria are stored on ice for further use in immunoblot or respiration assays (Sims, 1990).

2.3. Immunoblotting of UCP2 in samples from hypothalamus and pancreatic islet mitochondria

Twenty-microliter samples containing 30 μg mitochondria from hypothalamus or pancreatic islets are diluted 1:1 (vol:vol) in Laemmli buffer (Laemmli, 1970), boiled (100 °C) for 1 min and separated in 10% nongradient SDS-PAGE. This method was originally optimized for the Mini-protean apparatus from Bio-Rad (Hercules, California). Running begins with 70 mV and once samples are completely aligned into the gel it is increased to 100 mV.

Once separation is concluded, proteins are transferred to nitrocellulose membranes and blotted with anti–UCP2 antibody (sc-6526, goat polyclonal) using 10 μl antibody in 10 ml blotting buffer. Specific bands are detected by chemiluminescence and visualization is performed by exposure of the membranes to RX-films. Quantification of the bands is performed using the Scion software (ScionCorp, Frederick, Maryland).

3. ANALYSIS OF UCP2 EXPRESSION BY REAL-TIME PCR

3.1. Preparation of cDNA

One hypothalamus from rat (or optionally, two hypothalami from mice), or 2000 pancreatic islets (from either rats or mice), is/are homogenized in 1.0 ml Trizol reagent (Invitrogen) using a power homogenizer (Polytron, Glenn Mills, Clifton, New Jersey). The homogenate is then centrifuged at 12,000g for 10 min at 4 °C and the supernatant transferred to a new tube and incubated at room temperature for 5 min. Chloroform (0.2 ml) is then added and the mixture is hand shaken vigorously for 15 s. The suspension is incubated for another 5 min at room temperature, centrifuged at 12,000g for 15 min at 4 °C and the upper aqueous phase is transferred to a new tube. A half volume of isopropyl alcohol is added, and the solution is incubated at room temperature for 10 min and centrifuged at 12,000g for 10 min at 4 °C. The supernatant is removed and 1.0 ml 75% ethanol added before vortexing and centrifuging at 7500g for 5 min at 4 °C. The pellet is air dried and RNA is dissolved in 50 μl RNAse free water. The volume containing 1.0 μg total RNA is then transferred to a new tube and 50 ng random hexamers added together with 1.0 μl (10 mM) dNTP mix and the volume completed to 10 μl with DEPC-treated water. The mixture is incubated at 65 °C for 5 min and then incubated on ice for 1 min. Nine microliters of reaction mixture is added (composed of 4.0 μl 25 mM MgCl$_2$, 0.1 μl DTT, 1.0 μl recombinant ribonuclease inhibitor, and 2.0 μl reverse transcription 10× buffer), and incubated at 25 °C for 2 min. One microliter (50 units) of SuperScript (Invitrogen) is added, mixed and incubated at 25 °C for 10 min before transferring to 42 °C and incubating for 50 min. The reaction is terminated by incubating the sample at 70 °C for 15 min, before chilling on ice, adding 1.0 unit RNAse H and incubating at 37 °C for 20 min. cDNA is quantified and stored at −20 °C until used in real-time PCR amplification.

3.2. Real-time PCR

Real-time quantitative PCR analysis for UCP2 was tested on an ABI Prism 7700 sequence detection system (Applied Biosystems, Foster City, California), using the SYBR Green I dye detection. The reactions are assembled as follows: 20 μl of reaction mixture contain 1× reaction buffer (Applied Biosystems), 1.0 μM forward and reverse primers, and 3.0 ng of the cDNA template. The PCR conditions are 10 min at 95 °C, followed by 45 cycles of 2 s at 95 °C, 2 s at annealing temperature, and 12 s at 72 °C. The specificity of PCR can be evaluated by measuring the melting curve of the PCR product at the end of the reaction. Fluorescent data are specified for collection during primer extension. The relative cDNA ratio is calculated using GAPDH as a reference calibrator to normalize equal loading of template (Invitrogen ID numbers, Rn01775763_m1 and Mm03302249_g1, for rat and mouse, respectively). The PCR primer sequences for rat UCP2 are designed under the ID number Rn01754856_m1; and for mouse UCP2 are designed under the ID number Mm00627599_m1, both from Invitrogen.

4. Mitochondria Respiration Assay

Oxygen uptake by isolated mitochondria (2.0 mg/ml) is determined polarographically at 28 °C using a Clark-type oxygen electrod (Hansatech Instruments, Pentney King's Lynn, UK) (Sims and Blass, 1986). Samples are added to an air-saturated calcium free standard respiration medium containing 125 mM sucrose, 65 mM KCl, 2.0 mM inorganic phosphate, 1.0 mM MgCl$_2$, 10 mM HEPES buffer (pH 7.2), and a cocktail of NAD$^+$-linked substrates at 5.0 mM (malate, pyruvate, and α-ketoglutarate) or 5.0 mM succinate (FAD-linked substrate) to a final volume of 1.0 ml in a glass chamber equipped with magnetic stirring. Incubation conditions for determining state-3 respiration (ADP and substrate present) and state-4 respiration (after ADP was depleted) are as defined by Chance and Williams (1956). Uncoupled respiration is determined from the maximal oxygen uptake rate after addition of CCCP 1.0 μM. Respiration rates are given in nmol O$_2$/mg protein/min. The experimental approach is calibrated using the oxygen content of air saturated medium of 425 ng atoms/ml at 28 °C (Reynafarje *et al.*, 1985).

5. Modulating UCP2 Expression by Cold Exposure

Cold exposure induces a significant increase of UCP2 expression in the hypothalamus and pancreatic islets (Degasperi *et al.*, 2008; De Souza *et al.*, 2003). Rats (or optionally mice) are accommodated in individual cages with water and chow *ad libitum*, and cages are placed in a cold room (4 °C) maintaining artificial light/dark cycles of 12 h. Increased UCP2

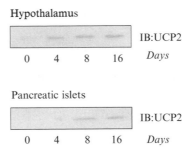

Figure 22.1 Male Wistar rats were placed in individual cages and exposed to a cold environment ($+4\,^\circ$C) for 0–16 days. At the end of the experimental period, the rats were deeply anesthetized and hypothalami and pancreatic islets were obtained for mitochondria preparation. Samples containing 30 μg mitochondria were separated by SDS-PAGE, transferred to nitrocellulose membranes and blotted (IB) with anti-UCP2 specific antibody. Figures are representative of typical experiments.

expression in hypothalamus is optimally detected after 4 days cold exposure (increase of UCP2 protein level of ~40%), while in pancreatic islets, an optimal increase occurs after 8 days (increased UCP2 protein level by ~50%) (Fig. 22.1). At the end of the experimental periods, animals are rapidly removed from the cold environment and used for preparation of protein or RNA from hypothalami and pancreatic islets, as described above.

6. Inhibiting UCP2 Expression by Antisense Oligonucleotide

Phosphorthioate-modified sense and antisense oligonucleotides (produced by Invitrogen) are diluted to a final concentration of 10 nmol/μl (for storage) in dilution buffer containing 10 mmol/l Tris–HCl and 1.0 mmol/l EDTA. All except the last $3'$-terminal base is subjected to phosphorthioate modification. For inhibiting UCP2 expression, the rats (or optionally mice) are injected with one daily intraperitoneal dose of 200 μl of dilution buffer containing UCP2 sense (control) or antisense oligonucleotides. Typically, 2.0 or 4.0 nmol/day oligonucleotide provides reproducible inhibition of UCP2. The oligonucleotides were designed according to the *Rattus norvegicus* (or *Mus musculus*) UCP2 sequence deposited at the NIH-NCBI (www.ncbi.nlm.nih.gov) under the designations NM 019354, for rats; and NM 011671, for mice; and were composed of $5'$-TGC ATT GCA GAT CTC A-$3'$ (sense), and $5'$-TGA GAT CTG CAA TGC A-$3'$ (antisense). Total daily doses of 4.0 nmol produce up to 60% reduction of UCP2 expression (Fig. 22.2).

Hypothalamus

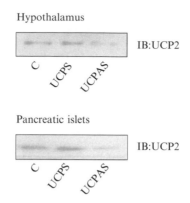

IB:UCP2

C   UCPS   UCPAS

Pancreatic islets

IB:UCP2

C   UCPS   UCPAS

Figure 22.2   Male Wistar rats were placed in individual cages and exposed to cold environment (+4 °C) for 8 days. Beginning on the fourth day of cold exposure, the rats were
treated with one daily intraperitoneal dose (200 μl) of saline, sense (UCPS, 4.0 nmol) or
antisense (UCPAS, 4.0 nmol) phosphorthioate modified oligonucleotides for 4 days. At
the end of the experimental period, the rats were deeply anesthetized and hypothalami
and pancreatic islets were obtained for mitochondria preparation. Samples containing
30 μg mitochondria were separated by SDS-PAGE, transferred to nitrocellulose membranes and blotted (IB) with anti-UCP2 specific antibody. Figures are representative of
typical experiments.

7. Inhibiting PGC-1α as an Indirect Means of Controlling UCP2 Expression

Proliferator-activated receptor gamma coactivator-1α (PGC-1α) is a
transcriptional coactivator known to play an important role in the control of
thermogenesis, and in the metabolism of glucose and lipids (Handschin and
Spiegelman, 2006). In pancreatic islets of cold-exposed rodents, PGC-1α
controls the expression of UCP2 (De Souza et al., 2003). Therefore,
modulating the expression of PGC-1α leads to indirect control of UCP2
expression. Phosphorthioate-modified sense and antisense oligonucleotides
(Invitrogen, Carlsbad, California) are diluted to a final concentration of
5.0 μM in dilution buffer containing 10 mmol/l Tris–HCl and 1.0 mmol/l
EDTA. The mice are injected with one daily intraperitoneal dose of 200 μl
dilution buffer containing, or not, 4.0 nmol sense or antisense PGC-1α
oligonucleotides. The oligonucleotides were designed according to the
M. musculus and *R. norvegicus* Ppargc1a sequences deposited at the NIH-
NCBI (www.ncbi.nlm.nih.gov) under the designations BC 066868 and
NM 031347, and are composed of 5′-TCA GGA GCT GGA TGG C-3′,
sense; and 5′-GCC ATC CAG CTC CTG A-3′, antisense. The antisense
oligonucleotides reduce the expression of PGC-1α by ∼60% in pancreatic
islets (De Souza et al., 2003, 2005) (Fig. 22.3) and around 70% in skeletal
muscle (Oliveira et al., 2004).

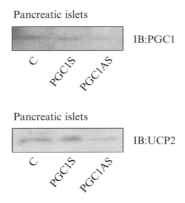

Figure 22.3 Male Wistar rats were placed in individual cages and exposed to cold environment ($+4\,^{\circ}C$) for 8 days. Beginning on the fourth day of cold exposure the rats were treated with one daily intraperitoneal dose (200 μl) of saline, sense (PGC1S, 4.0 nmol) or antisense (PGC1AS, 4.0 nmol) phosphorthioate modified oligonucleotides for 4 days. At the end of the experimental period, the rats were deeply anesthetized and the pancreatic islets were obtained for total protein extract (A) or mitochondria preparation (B). Samples containing 200 μg total protein (A), or 30 μg (B) mitochondria were separated by SDS-PAGE, transferred to nitrocellulose membranes and blotted (IB) with anti-PGC-1α (A) or UCP2 (B) specific antibody. Figures are representative of typical experiments.

REFERENCES

Bradford, M. M. (1976). A rapid and sensitive method for the quantitation of microgram quantities of protein utilizing the principle of protein-dye binding. *Anal. Biochem.* **72,** 248–254.

Brand, M. D., Affourtit, C., Esteves, T. C., Green, K., Lambert, A. J., Miwa, S., Pakay, J. L., and Parker, N. (2004). Mitochondrial superoxide: Production, biological effects, and activation of uncoupling proteins. *Free Radic. Biol. Med.* **37,** 755–767.

Brand, M. D., and Esteves, T. C. (2005). Physiological functions of the mitochondrial uncoupling proteins UCP2 and UCP3. *Cell Metab.* **2,** 85–93.

Cannon, B., Shabalina, I. G., Kramarova, T. V., Petrovic, N., and Nedergaard, J. (2006). Uncoupling proteins: A role in protection against reactive oxygen species—or not? *Biochim. Biophys. Acta* **1757,** 449–458.

Chance, B., and Williams, G. R. (1956). Respiratory enzymes in oxidative phosphorylation. VI. The effects of adenosine diphosphate on azide-treated mitochondria. *J. Biol. Chem.* **221,** 477–489.

Degasperi, G. R., Romanatto, T., Denis, R. G., Araujo, E. P., Moraes, J. C., Inada, N. M., Vercesi, A. E., and Velloso, L. A. (2008). UCP2 protects hypothalamic cells from TNF-alpha-induced damage. *FEBS Lett.* **582,** 3103–3110.

De Souza, C. T., Araujo, E. P., Prada, P. O., Saad, M. J., Boschero, A. C., and Velloso, L. A. (2005). Short-term inhibition of peroxisome proliferator-activated receptor-gamma coactivator-1alpha expression reverses diet-induced diabetes mellitus and hepatic steatosis in mice. *Diabetologia* **48,** 1860–1871.

De Souza, C. T., Araujo, E. P., Stoppiglia, L. F., Pauli, J. R., Ropelle, E., Rocco, S. A., Marin, R. M., Franchini, K. G., Carvalheira, J. B., Saad, M. J., Boschero, A. C., Carneiro, E. M., *et al.* (2007). Inhibition of UCP2 expression reverses diet-induced diabetes mellitus by effects on both insulin secretion and action. *FASEB J.* **21**, 1153–1163.

De Souza, C. T., Gasparetti, A. L., Pereira-da-Silva, M., Araujo, E. P., Carvalheira, J. B., Saad, M. J., Boschero, A. C., Carneiro, E. M., and Velloso, L. A. (2003). Peroxisome proliferator-activated receptor gamma coactivator-1-dependent uncoupling protein-2 expression in pancreatic islets of rats: A novel pathway for neural control of insulin secretion. *Diabetologia* **46**, 1522–1531.

Esteves, T. C., and Brand, M. D. (2005). The reactions catalysed by the mitochondrial uncoupling proteins UCP2 and UCP3. *Biochim. Biophys. Acta* **1709**, 35–44.

Handschin, C., and Spiegelman, B. M. (2006). Peroxisome proliferator-activated receptor gamma coactivator 1 coactivators, energy homeostasis, and metabolism. *Endocr. Rev.* **27**, 728–735.

Laemmli, U. K. (1970). Cleavage of structural proteins during the assembly of the head of bacteriophage T4. *Nature* **227**, 680–685.

Oliveira, R. L., Ueno, M., de Souza, C. T., Pereira-da-Silva, M., Gasparetti, A. L., Bezzera, R. M., Alberici, L. C., Vercesi, A. E., Saad, M. J., and Velloso, L. A. (2004). Cold-induced PGC-1alpha expression modulates muscle glucose uptake through an insulin receptor/Akt-independent, AMPK-dependent pathway. *Am. J. Physiol. Endocrinol. Metab.* **287**, E686–E695.

Reynafarje, B., Costa, L. E., and Lehninger, A. L. (1985). O2 solubility in aqueous media determined by a kinetic method. *Anal. Biochem.* **145**, 406–418.

Simoneau, J. A., Kelley, D. E., Neverova, M., and Warden, C. H. (1998). Overexpression of muscle uncoupling protein 2 content in human obesity associates with reduced skeletal muscle lipid utilization. *FASEB J.* **12**, 1739–1745.

Sims, N. R. (1990). Rapid isolation of metabolically active mitochondria from rat brain and subregions using Percoll density gradient centrifugation. *J. Neurochem.* **55**, 698–707.

Sims, N. R., and Blass, J. P. (1986). Expression of classical mitochondrial respiratory responses in homogenates of rat forebrain. *J. Neurochem.* **47**, 496–505.

van Suylichem, P. T., Wolters, G. H., and van Schilfgaarde, R. (1990). The efficacy of density gradients for islet purification: A comparison of seven density gradients. *Transpl. Int.* **3**, 156–161.

Zhang, C. Y., Baffy, G., Perret, P., Krauss, S., Peroni, O., Grujic, D., Hagen, T., Vidal-Puig, A. J., Boss, O., Kim, Y. B., Zheng, X. X., Wheeler, M. B., *et al.* (2001). Uncoupling protein-2 negatively regulates insulin secretion and is a major link between obesity, beta cell dysfunction, and type 2 diabetes. *Cell* **105**, 745–755.

CHAPTER TWENTY-THREE

MEASURING MITOCHONDRIAL BIOENERGETICS IN INS-1E INSULINOMA CELLS

Charles Affourtit* *and* Martin D. Brand*,[†]

Contents

1. Introduction	406
2. Quantification of Cellular Bioenergetics—Theoretical Aspects	407
3. INS-1E Cells—A Valuable Pancreatic Beta Cell Model	411
3.1. General tissue culture procedures	411
3.2. Detection of UCP2 protein in isolated INS-1E mitochondria and whole-cell lysates	413
3.3. UCP2 knockdown by RNAi	415
4. Measurement of Coupling Efficiency in Trypsinized INS-1E Cells	416
5. Noninvasive Measurement of Cellular Bioenergetics	418
5.1. Measurement of coupling efficiency in attached INS-1E cells— Probing the effect of UCP2 knockdown	419
5.2. Modular-kinetic measurement of INS-1E bioenergetics— Practical considerations for future experiments	420
Acknowledgment	422
References	423

Abstract

Pancreatic beta cells secrete insulin in response to raised blood glucose levels. This glucose-stimulated insulin secretion (GSIS) depends on mitochondrial function and is regulated by the efficiency with which oxidative metabolism is coupled to ATP synthesis. Uncoupling protein-2 (UCP2) affects this coupling efficiency and is therefore a plausible pathological and physiological regulator of GSIS. In this respect, it is important to be able to measure coupling efficiencies accurately. Here, we describe experimental protocols to determine the coupling efficiency of trypsinized INS-1E cells, a popular beta cell model, and we present practical details of our RNA interference studies to probe the effect of UCP2 knockdown on this efficiency. We also introduce a method to determine

* MRC Dunn Human Nutrition Unit, Cambridge, United Kingdom
† Buck Institute for Age Research, Novato, California, USA

Methods in Enzymology, Volume 457
ISSN 0076-6879, DOI: 10.1016/S0076-6879(09)05023-X

coupling efficiencies noninvasively in attached cells and discuss theoretical and practical aspects of a modular-kinetic approach to describe and understand cellular bioenergetics.

1. INTRODUCTION

Pancreatic beta cells help to maintain euglycemia by secreting insulin when blood glucose levels are high. This glucose-stimulated insulin secretion (GSIS) depends to a great extent on mitochondrial function: when the plasma glucose concentration is high, beta cells boost their oxidative metabolism and thereby increase their mitochondrial protonmotive force (pmf) and rate of ATP synthesis. The consequent rise in the cytoplasmic ATP/ADP ratio then causes closure of ATP-sensitive potassium channels, depolarization of the plasma membrane potential, opening of voltage-gated calcium channels, calcium influx, and the eventual exocytosis of insulin-containing granules (Rutter, 2001). In this canonical view of events, the ATP/ADP poise and GSIS are controlled by a variable substrate supply. This unusual bioenergetic design is often attributed to the beta cell's particular hexokinase isozyme, glucokinase, which is widely believed to govern ATP/ADP and GSIS exclusively because of its exceptionally high control over glycolytic flux (Sweet et al., 1996). However, we have previously identified additional control over ATP/ADP, which comes from the efficiency with which mitochondrial substrate oxidation is coupled to ATP synthesis (Affourtit and Brand, 2006). Moreover, it seems likely that this coupling efficiency of oxidative phosphorylation is related to substrate supply as it appears to be controlled relatively strongly by the mitochondrial respiratory capacity (Affourtit and Brand, 2006). It may thus be clear that the mitochondrial coupling efficiency of beta cells is a good potential regulatory step over ATP/ADP and hence GSIS.

Beta cell mitochondria contain an uncoupling protein, UCP2, the activity of which is anticipated to uncouple mitochondrial respiratory activity from ATP synthesis. As such, UCP2 is a plausible (patho)physiological GSIS regulator. The experimental evidence to support the notion that UCP2 indeed modulates GSIS is extensive (see Affourtit and Brand (2008a), for review), but it has not been established conclusively that the protein's attenuating effect on insulin secretion is secondary to uncoupling of oxidative phosphorylation. Importantly in this respect, we have shown recently that UCP2 contributes significantly to the relatively high mitochondrial proton leak activity of INS-1E insulinoma cells (Affourtit and Brand, 2008b), a widely used model of pancreatic beta cells (Merglen et al., 2004). This contribution to proton leak was demonstrated to lower the coupling efficiency of INS-1E cells and impair GSIS (Affourtit and Brand, 2008b).

In this chapter, we describe the experimental protocols that enabled us to measure the coupling efficiency of trypsinized INS-1E cells. We also present details of our RNA interference (RNAi) and immuno-blotting procedures to achieve and confirm, respectively, a significant knock-down of UCP2 protein in INS-1E cells. In addition, we introduce a method to determine coupling efficiency noninvasively in attached cells. First, however, we discuss coupling efficiency as well as respiratory capacity from a modular-kinetic perspective of oxidative phosphorylation, thereby providing a useful theoretical framework within which to consider the relation between these two important bioenergetic parameters.

2. QUANTIFICATION OF CELLULAR BIOENERGETICS— THEORETICAL ASPECTS

Cellular respiration is the sum of mitochondrial oxidative phosphory-lation and nonmitochondrial oxygen uptake, as depicted from a top-down perspective in Fig. 23.1. In mammalian hepatocytes, nonmitochondrial respiratory activity is generally responsible for 10–20% of total cellular oxygen consumption (e.g., Nobes et al., 1990; Rolfe et al., 1999) and is most likely accounted for by peroxisomal oxidases (van den Bosch et al., 1992) and by enzymes located in the endoplasmic reticulum, for example, xanthine oxidase (Harrison, 1997). The remaining oxygen uptake in cells is due to mitochondrial respiration, and can be coupled to ATP synthesis. This oxidative phosphorylation is a process that can be divided conceptually into events that either generate or dissipate the mitochondrial membrane poten-tial (Brand, 1998). Such a modular view of mitochondrial energy transduc-tion (Fig. 23.1) emphasizes that steady-state respiratory activities and membrane potential ($\Delta\psi$) levels result from the kinetic interplay between substrate oxidation, terminating in the mitochondrial respiratory chain ($\Delta\psi$-producer), and the combined activities of ATP turnover (including the ADP phosphorylation machinery) and the proton leak across the mito-chondrial inner membrane ($\Delta\psi$-consumers). The kinetic dependency of $\Delta\psi$-establishing and $\Delta\psi$-dissipating modules on $\Delta\psi$ is modeled in Fig. 23.2A in which the two $\Delta\psi$-dissipating modules are assumed to respond exponentially to $\Delta\psi$ and the $\Delta\psi$-establishing module is assumed to respond hyperbolically. In resting cells, the respiratory steady state is attained at the point where the rates of these processes are equal (R). Also presented in Fig. 23.2A are steady states attained upon either inhibition of ADP phosphorylation with oligomycin (O) or dissipation of the mitochon-drial pmf with excess chemical uncoupler, FCCP (F). Nonmitochondrial respiratory activity is independent of $\Delta\psi$ and is shown in this system-kinetic plot as a constant oxygen-uptake activity (Fig. 23.2A, shaded).

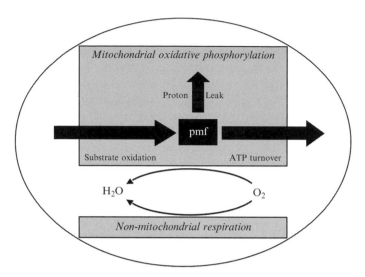

Figure 23.1 A bird's-eye view of cellular oxygen consumption. Both mitochondrial and nonmitochondrial respiration contribute to oxygen uptake in intact cells. Oxidases that reside in the endoplasmic reticulum and in peroxisomes are most likely responsible for nonmitochondrial respiratory activity, whilst oxidative phosphorylation accounts for mitochondrial reduction of molecular oxygen to water (thin arced arrows). Oxidative phosphorylation can be considered as an interaction between processes that establish a mitochondrial protonmotive force (pmf) and those that dissipate it. As indicated by the thick straight arrows, a pmf is "produced" by the oxidative catabolism of carbon fuels, which culminates in mitochondrial respiratory chain activity (substrate oxidation), whereas it is "consumed" by mitochondrial proton leak and by cellular ATP turnover, which involves pmf-driven ATP synthesis.

The coupling efficiency of oxidative phosphorylation can be defined as the proportion of mitochondrial respiratory rate that is used to drive ATP synthesis in any particular condition. The steady-state coupling efficiency can thus be derived from the relative rates of the phosphorylation apparatus and proton leak, as illustrated in Fig. 23.2B for modeled resting respiratory conditions under which 80% of respiration is used to synthesize ATP. Modular-kinetic visualization of oxidative phosphorylation also allows ready calculation of maximum and spare respiratory capacities, which are the absolute respiratory activity and the increase in respiratory activity, respectively, that are observed upon addition of FCCP, assuming that FCCP does not indirectly affect the kinetics of substrate oxidation (Fig. 23.2C).

When only rates of oxygen consumption are measured, and the modular kinetics cannot be plotted, coupling efficiency is generally approximated as the proportion of respiration that is sensitive to oligomycin (Fig. 23.2D). Because inhibition of phosphorylation with oligomycin causes a small

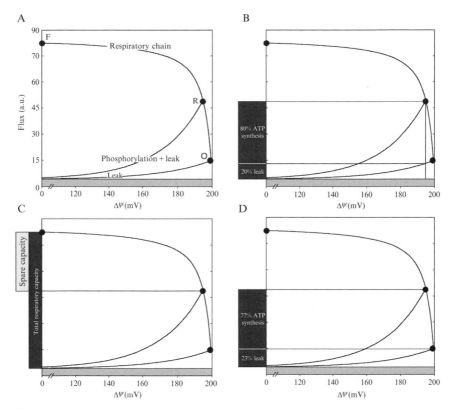

Figure 23.2 Modeled modular-kinetic behavior of the oxidative phosphorylation system. Panel (A) The kinetic dependency of both substrate oxidation and the sum of proton leak and ATP turnover activities (fluxes expressed in arbitrary units) on the mitochondrial pmf was modeled using exponential (proton leak, ATP turnover) and hyperbolic (substrate oxidation) equations. The kinetic model assumes zero ΔpH so that the pmf equals the mitochondrial membrane potential ($\Delta\psi$). A resting steady state is achieved when $\Delta\psi$-establishing flux equals total $\Delta\psi$-dissipating flux, which is reflected by the intersection of the kinetic curves describing the behavior of the respective activities (R). Also depicted are steady states modeling the effect of oligomycin (O) and FCCP (F). Nonmitochondrial respiratory activity is reflected by the shaded box below the kinetic curves. The curves and steady states in panels B, C, and D are identical to those shown in panel A.

increase in $\Delta\psi$, stimulating the proton leak rate, this approach will inevitably slightly underestimate the true coupling efficiency (Fig. 23.2D). However, the underestimation is usually small. To quantify it, we used published modular-kinetic data to calculate coupling efficiencies accurately and to approximate them from oligomycin-inhibited respiratory activity (Table 23.1). A direct comparison of the respective values reveals that approximation causes an underestimation of less than 10% on average (Table 23.1).

Table 23.1 Calculation and approximation of coupling efficiency

System	Coupling efficiency (%)[a]		Underestimate (%)	Reference(s)
	$\Delta\psi$-Corrected	Approximated		
Rat hepatocytes[b]	75	71	5.3	Brand (1990)
	77	55	29	Brown et al. (1990)
	62	51	18	Harper and Brand (1993)
	59	52	12	Harper and Brand (1993)
	64	62	3.1	Harper and Brand (1995)
	77	75	2.6	Porter and Brand (1995)
	74	69	6.8	Rolfe et al. (1999)
Rat thymocytes[c]	89	84	5.6	Buttgereit et al. (1994)
Other hepatocytes:				Porter and Brand (1995)
Mouse	73	68	6.8	
Ferret	69	58	16	
Sheep	65	63	3.1	
Pig	75	73	2.7	
Horse	82	74	9.8	
Average			9.2	
SEM			2.1	
Median			6.8	

[a] Coupling efficiencies of oxidative phosphorylation (% respiratory activity used to synthesize ATP) were calculated and approximated from published modular-kinetic data as described in the text.
[b] Oxidative phosphorylation in hepatocytes was measured under resting conditions except for the study by Rolfe and co-workers in which respiration was stimulated with lactate, glutamate, and ornithine.
[c] Thymocytes were stimulated by concanavalin-A.

Pancreatic beta cells increase their oxidative metabolism in response to glucose, which implies that mitochondrial respiratory activity is glucose-dependent. Previously, we have predicted a regulatory link between respiratory capacity and coupling efficiency based on the relatively strong control exerted by the mitochondrial electron transfer chain over the ratio of phosphorylation and proton leak activities in isolated INS-1E mitochondria (Affourtit and Brand, 2006). A connection between these bioenergetic parameters is also apparent from the modeled modular-kinetic data shown in Fig. 23.3A, which demonstrate that coupling efficiency is likely to decrease slightly when respiratory activity is relatively low. From Fig. 23.3B, it is furthermore clear, based on the modeled kinetics shown in Figs. 23.2 and 23.3A, that coupling efficiency is correlated positively with the mitochondrial membrane potential when this is altered by large changes in the activity of substrate oxidation. This suggests that respiratory capacity and efficiency of oxidative phosphorylation are related causally, such that the efficiency increases when substrate oxidation rises, as a simple consequence of the underlying kinetics. Possible concomitant glucose-dependent stimulation of phosphorylation or proton leak activities (the latter for example through activation of UCP2) would amplify or dampen, respectively, this substrate-driven increase in coupling efficiency.

3. INS-1E Cells—A Valuable Pancreatic Beta Cell Model

Generally, it is deemed better to study primary than cultured clonal cells to unravel the physiology and biochemistry of pancreatic beta cells. However, the use of primary beta cells in our bioenergetic studies would require isolation of vast numbers of pancreatic islets, which is a highly laborious process and demands many animals. Moreover, the additional presence of alpha, delta, and F cells in pancreatic islets potentially hampers a conclusive interpretation of experimental islet data. Therefore, our research involves predominantly INS-1E cells, a widely-used pancreatic beta cell model that was kindly provided by Drs Pierre Maechler and Claus Wollheim from the Department of Internal Medicine at the University Medical Centre in Geneva, Switzerland. INS-1E is a clonal cell line that was derived from parental INS-1 cells (Janjic et al., 1999), which were originally isolated from a radiation-induced rat insulinoma (Asfari et al., 1992). INS-1E cells have been characterized in detail and have been shown to exhibit all the important GSIS characteristics of primary beta cells (Merglen et al., 2004). Importantly from a bioenergetic point of view, the stimulatory effect of UCP2-ablation on GSIS seen in isolated mouse islets (Zhang et al., 2001), is also observed in INS-1E cells (Affourtit and Brand, 2008b). INS-1E cells are therefore a convenient and appropriate beta cell model.

3.1. General tissue culture procedures

INS-1E cells are cultured under humidified atmospheric conditions at 5% CO_2 in RPMI-1640 medium (Sigma R0883) that contains 11 mM glucose and is supplemented with 5% heat-inactivated fetal calf serum

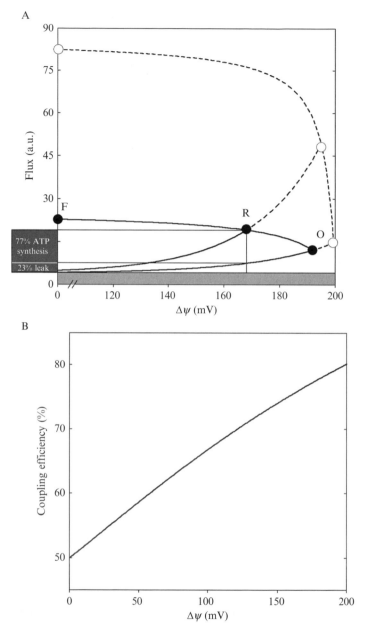

Figure 23.3 Respiratory capacity and coupling efficiency are related parameters. Kinetics of the oxidative phosphorylation system were modeled as described in the legend to (Fig. 23.2). Panel (A) Steady states are modeled under resting conditions (R) or in the presence of oligomycin (O) or FCCP (F) in a system with relatively low respiratory capacity. The dashed lines are shown to facilitate a direct comparison with the kinetic system modeled in (Fig. 23.2A). Panel (B) Dependence of coupling efficiency on mitochondrial membrane potential ($\Delta\psi$) based on modeled proton leak and ATP turnover kinetics shown in (Figs. 23.2 and 23.3A).

(Sigma F9665), 10 mM HEPES, 1 mM sodium pyruvate, 50 μM β-mercaptoethanol, 2 mM glutamine, 100 U/mL penicillin, and 100 μg/mL streptomycin. Note that RPMI-1640 contains 2 g/L sodium bicarbonate, which buffers the medium at pH 7.2 in a 5% CO_2 atmosphere, but will cause the medium to become progressively more alkaline until a new equilibrium is reached under air. Cells are grown on various scales using either 500 cm^2 Nuclon^{TM} trays, 175 cm^2 BD Falcon^{TM} flasks, 24-well BD Falcon^{TM}, or Seahorse Bioscience XF24 cell culture microplates. The XF24 plates have 24 conically-shaped wells with growth surface areas usually found in typical 96-well plates. Cultures are maintained in 175 cm^2 flasks and passaged by trypsinization: cells grown to roughly 90% confluence are washed with 10 mL phosphate-buffered saline (PBS) and then incubated for 2 min with 3 mL trypsin-EDTA (Sigma T3924); following gentle tapping to aid cell detachment, 7 mL RPMI-1640 is added and cells are resuspended by pipetting up-and-down; subsequently, cells are harvested by 3 min centrifugation at 220g and resuspended in $\sim$6 mL RPMI-1640 to remove trypsin, and then divided over 3 $\times$ 175 cm^2 flasks.

3.2. Detection of UCP2 protein in isolated INS-1E mitochondria and whole-cell lysates

3.2.1. Preparation of mitochondrial samples

Cells are seeded either in 175 cm^2 flasks (1.5 $\times$ 10^7 cells in $\sim$25 mL) or in 500 cm^2 trays (3.75 $\times$ 10^7 cells in $\sim$150 mL) and grown to reach $\sim$90% confluence, which typically takes 4–5 days. Cells are harvested from flasks by trypsinization, with final cell pellets resuspended in ice-cold buffer containing 0.25 M sucrose, 20 mM Hepes (pH 7.4), 2 mM EGTA, 10 mM KCl, 1.5 mM MgCl$_2$, and 0.1% (w/v) defatted BSA (referred to as SHE). Cells grown on trays are washed twice with SHE, scraped off using a razor blade, and collected on ice in SHE. The protocols for flasks and trays are identical from this point: cells are washed once in SHE by centrifugation at 350g (10 min, 4 °C) and the resulting pellets are resuspended in 2–5 mL (volume adjusted to pellet size) SHE supplemented with a protease inhibitor cocktail (Calbiochem 539131) at one-third of the dose recommended by the manufacturer. Cells are disrupted with a 7-mL glass homogeniser (Wheaton, USA) by moving a tight-fitting plunger down-and-up at least 20 times to maximize mitochondrial yield. (Note that homogenizing movements should be restricted to 10 if mitochondrial membranes are to remain intact.) Following 15$\times$ dilution in supplemented SHE and removal of cell debris (3 min centrifugation at 1100g followed by filtration through muslin), mitochondria are pelleted by centrifugation at 12,000g (11 min, 4 °C) and resuspended in the small amount of SHE that remains upon discarding the supernatant. Mitochondrial protein concentration is determined with the "Bio-Rad D_C Protein Assay" following the manufacturer's instructions.

Mitochondrial aliquots are centrifuged for 10 min in a microfuge at maximum speed, supernatants are aspirated, and pellets are resuspended in gel-loading buffer (10% (w/v) SDS, 250 mM Tris–HCl (pH 6.8), 5 mM EDTA, 50% (v/v) glycerol, 5% (v/v) β-mercaptoethanol, 0.05% (w/v) bromophenol blue) at a concentration of 2 μg/μL. Complete and homogenous protein solubilization is ensured by repeated up-and-down pipetting followed by a 30-s vortex step and a 5-min incubation at 100 °C. After a 2-min cool-down period, samples can be used for Western blotting immediately or may be stored at −80 °C for at least several weeks.

3.2.2. Preparation of whole-cell lysates

Cells are seeded on 24-well Falcon plates (3×10^5 cells in 1 mL/well) or Seahorse XF24 plates (4×10^4 cells in 200 μL/well) and grown to ∼80% confluence, which typically takes 2–3 days. For Falcon plates, medium is aspirated, cells are washed by adding and removing 1 mL PBS, and 100 μL trypsin–EDTA is added to each well. After 2-min incubation, 400 μL PBS is added, cells are resuspended by pipetting, and the yields of three wells are pooled. For Seahorse plates, cells are washed with PBS by adding 500 μL and then removing 650 μL, leaving a notional volume of 50 μL. Another 500 μL PBS is added to the first well and cells are resuspended by pipetting up-and-down vigorously. The resuspension is transferred to collect cells from the second well and this process is repeated to pool all 24 wells. The protocols for both plate types are identical from this point: cells are pelleted by 5-min centrifugation in a microfuge at maximum speed, resuspended in 1 mL PBS, and counted using an "Improved Neubauer" hemocytometer (Weber Scientific International Ltd.). After another 2-min microfuge spin, cells are resuspended in gel-loading buffer at 5.6×10^3 cells/μL. To ensure complete and homogenous protein solubilization, this resuspension step involves vigorous pipetting, 30-s vortexing, and a 5-min incubation at 100 °C. After a 2-min cool-down period, the sample is divided into 25 μL aliquots, equivalent to 1.4×10^5 cells each, which can be used for immuno-blotting procedures immediately, but may be stored at −80 °C for at least several weeks.

3.2.3. Western analysis

To detect UCP2, either 20 μg mitochondrial protein (i.e., 10 μL sample) or lysate equivalent to 1.12×10^5 cells (i.e., 20 μL sample) is separated on a 12% SDS polyacrylamide gel by running it for 90 min at 150 V. Proteins are transferred to a nitrocellulose membrane using a Trans-Blot® SD (BioRad) semi-dry transfer cell applying 20 V for 30 min. The membrane is blocked for 2 h at room temperature in Tris-buffered saline containing 0.1% (v/v) Tween 20 and 5% (w/v) dried skimmed milk (Marvel), and then probed overnight at 4 °C with goat anti-UCP2 1° antibodies (Santa Cruz Biotechnology, Santa Cruz, California) diluted to 0.2 μg/mL in this blocking buffer. We emphasize that the quality of these Santa Cruz UCP2 antibodies

(Sc-6525) is strongly lot-dependent and needs to be checked, ideally using purified recombinant UCP2 protein or, alternatively, by comparing cross-reaction in paired samples from UCP2-containing and -ablated systems (cf. Azzu *et al.*, 2008). After incubation with 1° antibodies, the membrane is incubated for 1 h at room temperature with peroxidase-conjugated 2° antibodies (Pierce Biotechnology, Rockford, Illinois) diluted to 0.1 μg/mL in blocking buffer. Cross-reacted proteins are visualized by ECL® (enhanced chemiluminescence, Amersham Biosciences) and the membrane is stained with GelCode® Blue reagent (Pierce Biotechnology) to confirm equal protein loading. Protein can be quantified by densitometry using public *ImageJ* software (http://rsb.info.nih.gov/ij/).

3.3. UCP2 knockdown by RNAi

The small interfering RNA (siRNA) oligonucleotides that we use routinely to knock down UCP2 in INS-1E cells were predesigned by Ambion (Huntingdon, UK) and are targeted at rat *Ucp2* exons 3, 5, and 8 (Table 23.2). In typical RNAi experiments, cells are seeded in Seahorse microplates (4×10^4 cells in 200 μL/well), 24-well Falcon plates (3×10^5 cells in 1 mL/well), 175 cm^2 flasks (1.5×10^7 cells in $\sim$25 mL), or 500 cm^2 trays (3.75×10^7 cells in $\sim$150 mL), and are cultured in RPMI-1640 lacking antibiotics to roughly 50% confluence. This takes $\sim$24 h for all set-ups except the 500 cm^2 trays that require incubation for 48 h. At this point, the growth medium of the trays is replaced such that the total post-transfection volume is about 100 mL. Cells are transfected with siRNA fragments that have been allowed to form complexes with 1.7 μg/mL Lipofectamine™ 2000 (Invitrogen, Paisley, UK) for 20 min at room temperature in RPMI-1640 medium lacking fetal calf serum and antibiotics; 200 n*M* siRNA is used for the multiwell plates and 50 n*M* for the flasks and trays. To identify potential nonspecific effects, cells are transfected in parallel with scrambled siRNA (Ambion, Silencer® Negative Control 1). Following 2–3 days' further growth, cells are harvested and prepared for immuno-blotting analysis as described in the previous section.

Figure 23.4A shows a typical Western blot that demonstrates the identification of a single protein of $\sim$32 kDa in INS-1E mitochondria. This protein almost fully disappears after transfection with *Ucp2* siRNA targeted at exon 8. The same protein is detected in cell lysates (Fig. 23.4B) and again its level is lowered considerably following transfection. However, the UCP2 antibody appears to be less specific with cell lysates than isolated mitochondria, as additional proteins with a higher molecular weight are detected in whole cells; the level of these proteins is not affected by *Ucp2* siRNA.

Table 23.2   siRNA oligonucleotide sequences (5'–3' sense) targeted at rat *Ucp2*

Target	Sequence
Exon 3	GAUCUCAUCACUUUCCCUCtt
Exon 5	GCACUGUCGAAGCCUACAAtt
Exon 8	CGUAGUAAUGUUUGUCACCtt

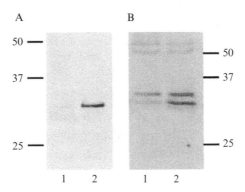

Figure 23.4   UCP2 knockdown in INS-1E cells grown on a small and large scale. Protein was detected by Western analysis using mitochondria isolated from cells grown in 175 cm² flasks (panel A) or whole-cell lysate prepared from cells grown in Seahorse XF24 microplates (panel B). Cells had been transfected with *Ucp2*-targeted or scrambled siRNA (lanes 1 and 2, respectively).

4. MEASUREMENT OF COUPLING EFFICIENCY IN TRYPSINIZED INS-1E CELLS

Nontransfected INS-1E cells are seeded in 175 cm² flasks (1.5×10^7 cells in ~25 mL) and grown for typically 3 days to ~80% confluence. In RNAi experiments, cells are cultured overnight in 175 cm² flasks to roughly 50% confluence, transfected as described in Section 3.3, and then grown for another 3 days. Cells are harvested by trypsinization (Section 3.1), resuspended in RPMI-1640, and stored on ice. Cell aliquots (600 µL) are put in glass vessels, flushed with carbogen (5% CO₂-95% air), sealed with plastic stoppers, incubated for 30 min in a shaking (100 cycles/min) water bath at 37 °C, and then transferred to an oxygraph. These incubations are also performed in the presence of 7 µM oligomycin to inhibit ATP synthase activity, and 9 µM myxothiazol to inhibit mitochondrial respiration. Both these effectors of oxidative phosphorylation are dissolved in DMSO (final concentration ≤ 0.2%) at levels that give maximum effects as checked by doubling concentrations. Note that, despite the shaking, INS-1E cells tend to settle to the bottom of the glass vessels during the 30-min incubation, which may cause transient hypoxia. Oxygen consumption is measured at 37 °C using a Clark-type electrode (Rank Brothers, Cambridge, UK) calibrated with air-saturated RPMI-1640 assumed to contain 406 nmol atomic oxygen mL^{-1} (Reynafarje et al., 1985). Respiratory rates are calculated from the initial steady parts of the oxygen-uptake traces and expressed per 10^6 viable cells (counted using a hemocytometer). The myxothiazol-insensitive oxygen-uptake rate is subtracted from all other rates to correct for non-mitochondrial respiratory activity. The coupling efficiency is approximated as the percentage of myxothiazol-sensitive (i.e., mitochondrial) respiration that is sensitive to oligomycin.

Following the protocol described above, we have recently discovered that INS-1E cells exhibit an exceptionally high mitochondrial proton leak activity: up to 75% of total respiration is used to drive seemingly futile leak (Affourtit and Brand, 2008b). This high leak activity is reflected by a coupling efficiency of 25–30%, which is very low compared to the values reported for other mammalian (and nonmammalian) cells including hepatocytes, thymocytes, and neurons (Table 23.3). It is unlikely that this relatively low coupling efficiency of INS-1E cells is merely due to their clonal origin since oxidative phosphorylation in cultured C2C12 myoblasts has high coupling efficiency (Table 23.3). Similarly, the carcinomal

Table 23.3 Coupling efficiencies in primary cells and tissues from various organisms and several cultured cell lines

System	Coupling efficiency	Reference(s)
Mammalian		
Hepatocytes (cf. Table 23.1)	60–80	Nobes et al. (1990), Porter and Brand (1995)
Rat muscle	50–65	Rolfe et al. (1999)
Rat thymocytes	80–90	Buttgereit et al. (1994)
Rat cerebellar granule neurons	80	Jekabsons and Nicholls (2004)
C2C12 myoblasts	90	Affourtit and Brand (2008b)
A549 human lung carcinoma cells	85	Amo et al. (2008), Brand et al. (2008)
INS-1E insulinoma cells	25–30	Affourtit and Brand (2008b)
Nonmammalian		
Avian hepatocytes	> 80	Else et al. (2004)
Crocodile hepatocytes	70–85	Hulbert et al. (2002)
Lizard hepatocytes	> 70	Brand et al. (1991)
Frog hepatocytes	> 75	Brand et al. (2000)
Lamprey hepatocytes	50–75	Savina and Gamper (1998)
Snail hepatopancreatic cells	75–85	Bishop and Brand (2000)

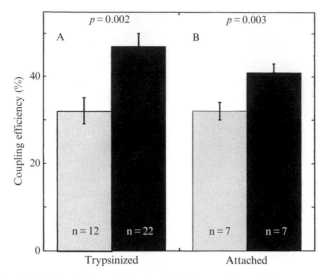

Figure 23.5   UCP2 knockdown increases INS-1E coupling efficiency. Coupling efficiencies were measured in trypsinized (panel A) and attached (panel B) cells that were transfected with scrambled (grey bars) or *Ucp2*-targeted (black bars) siRNA.

background of INS-1E cells does not provide an explanation for low coupling either, as A459 human lung carcinoma cells exhibit a coupling efficiency of 85% (Table 23.3).

Interestingly, we find that RNAi-mediated UCP2 knock-down results in a significant increase in INS-1E coupling efficiency (Fig. 23.5A), implying that UCP2 contributes considerably to the high proton leak activity of INS-1E cells (Affourtit and Brand, 2008b). Furthermore, we have demonstrated that UCP2 depletion improves GSIS in INS-1E cells (Affourtit and Brand, 2008b), which agrees qualitatively with the observations of Zhang and co-workers in islets from *Ucp2*-ablated mice (Zhang *et al.*, 2001). We, therefore, think that high proton leak is of physiological importance for pancreatic beta cells and suggest that it amplifies the effect of UCP2 effectors on the ATP/ADP ratio and hence insulin secretion.

 ## 5. Noninvasive Measurement of Cellular Bioenergetics

As described in Section 4, we have established that there is a relatively low coupling efficiency in INS-1E cells using an experimental procedure that involves detachment of cells from their growth surface by trypsinization. Such methodology raises the possibility that the observed high mitochondrial proton leak activity is part of a stress response of INS-1E cells to trypsin treatment. Although this eventuality is rather unlikely, because we demonstrated that proton leak activity in trypsinized C2C12 myoblasts is relatively low (Table 23.3, Affourtit and Brand, 2008b), it is nonetheless conceivable that for example the effect of UCP2 knockdown on coupling efficiency in INS-1E cells is facilitated by a trypsin–induced stress response. More generally, it seems important to measure cellular bioenergetics under conditions that mimic the growth situation of the cultured cells as closely as possible.

The energy metabolism of attached cells can be investigated using a Seahorse XF24 Extracellular Flux Analyzer (Seahorse Bioscience, USA). This instrument is described in full detail elsewhere (Wu et al., 2007) but, essentially, facilitates real-time measurement of uptake and secretion of important metabolic analytes by cells cultured on 24-well plates. Currently, these analytes include protons and molecular oxygen that are both detected by specially designed disposable fluorescent biosensors, all 24 pairs of which are coupled to a fiber-optic waveguide. This set-up thus enables continuous measurement of extracellular pH and oxygen tension and therefore allows the rate at which these parameters change to be calculated readily. The resultant extra-cellular acidification and oxygen depletion rates are indicative of glycolytic activity (lactate, the acidic end product of glycolysis, is secreted) and cellular respiratory activity, respectively. Clearly, coupling efficiency can be determined from the respiration data obtained in attached cells (see e.g., Amo et al., 2008; Brand et al., 2008).

5.1. Measurement of coupling efficiency in attached INS-1E cells—Probing the effect of UCP2 knockdown

Cells are seeded in Seahorse XF24 microplates (4×10^4 cells in 200 μL/well) and cultured overnight to allow full attachment. As described in Section 3.3, cells are transfected with 200 nM siRNA (scrambled or targeted at $Ucp2$ exon 8), which raises the total incubation volume to 240 μL/well. After two more days' growth, RPMI-1640 medium containing a low glucose concentration (2 mM) and lacking antibiotics, sodium pyruvate and sodium bicarbonate, is used to wash and assay the cells. With an electronic multichannel pipette, 850 μL medium is carefully added and 1000 μL is subsequently removed making sure that the cell monolayer is not disturbed. Notionally, this leaves 90 μL medium but the real volume will be closer to 50 μL given likely evaporation of growth medium during culture. Another 1000 μL wash/assay medium is pipetted in-and-out, following which a final 650 μL is added; note that the original growth medium is diluted more than 1500 times during this washing procedure. Cells are then incubated for 70 min at 37 °C under air before being transferred to the Seahorse instrument. Note that the pH of RPMI-1640 lacking bicarbonate remains constant under air by virtue of 10 mM HEPES.

This incubation under air appears to be an essential step in the protocol, most likely because the cells need to recover from any stress incurred upon washing; without incubation, most cells detach during the Seahorse assay. During the incubation, appropriate amounts of concentrated glucose, oligomycin, myxothiazol, and rotenone stocks are diluted into 75 μL wash/assay medium and loaded into the addition ports of a disposable Seahorse measuring cartridge that also harbors the fluorescent oxygen probes (cf. Wu et al., 2007). Following a 30-min calibration of these probes, the tissue culture plate containing the cells is transferred from the air incubator to the Seahorse analyzer and assay cycles of 1-min sample mixing, 2-min waiting, and 3-min measuring oxygen consumption are initiated. After three of these assay cycles, the glucose concentration is raised by automatic pneumatic addition of an extra 8 mM glucose (i.e., 10 mM total). Following assay cycle 8, oligomycin is added at a final level of 1 μg/mL to inhibit ADP phosphorylation, and cells are eventually (after cycle 13) subjected to a rotenone/myxothiazol mixture (final concentrations of 1 and 2 μM, respectively) to inhibit mitochondrial respiration fully. The resultant nonmitochondrial respiratory activity is measured during four assay cycles and is subtracted from all other rates. A typical time-resolved oxygen consumption rate trace is shown in Fig. 23.6. Coupling efficiency is calculated as the percentage of total respiration at 10 mM glucose that is sensitive to oligomycin. Note that coupling efficiency is a normalized parameter that corrects for any differences in cell density between wells.

From the data shown in Fig. 23.5, a direct comparison can be made between coupling efficiencies determined in trypsinized and attached cells. Whilst the values observed in cells transfected with scrambled siRNA are almost identical in these systems, coupling efficiency appears to be slightly greater in the trypsinized than attached state when cells are transfected with siRNA targeted at Ucp2. Importantly, however, UCP2 knockdown causes a statistically significant increase in coupling efficiency in attached cells, which demonstrates that this effect is not merely an artifact due to trypsin-induced stress.

5.2. Modular-kinetic measurement of INS-1E bioenergetics — Practical considerations for future experiments

Sections 4 and 5.1 provide protocols to determine coupling efficiencies in trypsinized and attached INS-1E cells, respectively. However, these methods approximate coupling efficiency by comparing respiratory activity in the presence and absence of oligomycin and, as argued in Section 2, this results in an underestimation of the true coupling efficiency. Approximation is inevitable if just respiratory data are available, since in that case the respiration flux cannot be dissected into its component parts, that is, the fractions that are used to drive ATP synthesis and mitochondrial proton leak. Such dissection is possible, however, if the common intermediate through

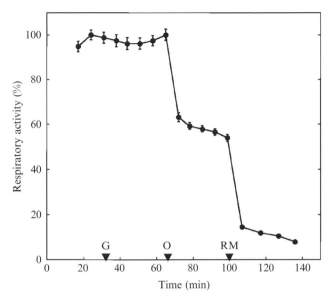

Figure 23.6 Typical Seahorse oxygen consumption rate trace. Oxygen uptake rates of attached INS-1E cells transfected with *Ucp2*-targeted siRNA were measured in RPMI-1640 medium using a Seahorse XF24 analyzer and are expressed as a percentage of measurement 8, that is, the last data point prior to addition of oligomycin. 8 mM glucose (G), 1 μg/mL oligomycin (O), and a mixture of 1 μM rotenone and 2 μM myxothiazol (RM) were added at times indicated by the arrow heads. Note that INS-1E respiration does not appear to increase in response to glucose when measured in RPMI-1640. Coupling efficiency is calculated as the percentage of mitochondrial respiratory activity at 10 mM glucose that is sensitive to oligomycin (see text). For this calculation, the eighth data point from the start of the experiment was used as a measure of total oxygen uptake rate, and the averaged 3 and 4 points after oligomycin and rotenone/myxothiazol addition, respectively, as measures of rate in the cumulative presence of these effectors. The coupling efficiency in this particular experiment was 45 $\pm$ 1%. Data are averages ($\pm$ SEM) of oxygen uptake rates measured in 5 separate wells.

which respiration, phosphorylation, and leak interact, that is, $\Delta\psi$, is measured concomitantly with the oxygen consumption rate. When both flux and intermediate concentration data are gathered, the kinetic dependency of the three separate fluxes on $\Delta\psi$ can be determined, and true steady-state coupling efficiencies can consequently be calculated (cf. Section 2). A complete modular-kinetic description of the oxidative phosphorylation system would furthermore yield information on total and "spare" respiratory capacities (cf. Section 2).

The kinetic behavior of a "$\Delta\psi$-producer" (i.e., the mitochondrial respiratory chain) can be established by specific modulation of a "$\Delta\psi$-consumer" (i.e., either phosphorylation or proton leak activity) and, reciprocally, the behavior of a consumer is revealed upon modulation of a producer (Brand, 1998). In other words, if resting respiration is titrated with subsaturating amounts of myxothiazol or rotenone, the kinetics with respect to $\Delta\psi$ of the total $\Delta\psi$-dissipating activity (i.e., leak + phosphorylation) are revealed. If this titration is performed in the presence of oligomycin, the kinetics of proton leak alone are obtained, which may be subtracted from the overall $\Delta\psi$-dissipating kinetics to reveal the behavior of ATP turnover. If respiration is titrated with FCCP, either in the presence or the absence of oligomycin, the kinetic dependency of substrate oxidation on $\Delta\psi$ is established. During these respiration titrations, successive steady states will be attained in which both the oxygen uptake rate (flux) and $\Delta\psi$ (intermediate concentration) will need to be measured. Plotting of oxygen consumption rates as a function of the concomitant $\Delta\psi$ values allows empirical system-kinetic plots to be constructed such as shown in Figs. 23.2 and 23.3A.

Oxygen consumption measurements are relatively straightforward as these can be made noninvasively using a Seahorse XF24 Analyzer. The biggest experimental challenge of a modular-kinetic approach is the accurate and quantitative determination of $\Delta\psi$. Generally, fluorescent membrane-permeable cations are used as "indicators" of $\Delta\psi$ since these molecules accumulate in the mitochondrial matrix in a $\Delta\psi$-dependent manner. However, accumulation is also affected by the magnitude of the plasma membrane potential and possible changes during titrations in this parameter have to be accounted for if true $\Delta\psi$ values are to be established. The best currently available approach that provides a quantitative means of compensating the fluorescent signal of $\Delta\psi$ indicators for changes in plasma membrane potential, involves the simultaneous fluorometric measurement of both cationic and anionic probes (Nicholls, 2006). This method has been described recently for cerebellar granule neurons and has the added advantage that accurate mitochondrial and plasma membrane potentials are obtained simultaneously and continuously (Nicholls, 2006).

It may be clear that modular-kinetic investigations of oxidative phosphorylation yield accurate information on the coupling efficiency of this process. The approach has further merits: (1) it will allow identification of the site(s) at which suspected effectors of mitochondrial energy metabolism act and (2) it will provide conclusive evidence as to the role played by poorly characterized proteins in oxidative phosphorylation. In this respect, it is important to stress that modular-kinetic insight is essential to establish unequivocally that UCP2 lowers the coupling efficiency of INS-1E cells by increasing proton leak activity as its name would indeed suggest. Without this insight, the effect of UCP2 on the efficiency of oxidative phosphorylation could formally be explained by indirect consequences of a UCP2-modulation of substrate oxidation (cf. Fig. 23.3).

ACKNOWLEDGMENT

The research described in this chapter is supported by the Medical Research Council, United Kingdom.

REFERENCES

Affourtit, C., and Brand, M. D. (2006). Stronger control of ATP/ADP by proton leak in pancreatic beta-cells than skeletal muscle mitochondria. *Biochem. J.* **393,** 151–159.

Affourtit, C., and Brand, M. (2008a). On the role of uncoupling protein-2 in pancreatic beta cells. *Biochim. Biophys. Acta* **1777,** 973–979.

Affourtit, C., and Brand, M. D. (2008b). Uncoupling protein-2 contributes significantly to high mitochondrial proton leak in INS-1E insulinoma cells and attenuates glucose-stimulated insulin secretion. *Biochem. J.* **409,** 199–204.

Amo, T., Yadava, N., Oh, R., Nicholls, D. G., and Brand, M. D. (2008). Experimental assessment of bioenergetic differences caused by the common European mitochondrial DNA haplogroups H and T. *Gene* **411,** 69–76.

Asfari, M., Janjic, D., Meda, P., Li, G., Halban, P. A., and Wollheim, C. B. (1992). Establishment of 2-mercaptoethanol-dependent differentiated insulin-secreting cell lines. *Endocrinology* **130,** 167–178.

Azzu, V., Affourtit, C., Breen, E., Parker, N., and Brand, M. D. (2008). Dynamic regulation of uncoupling protein 2 content in INS-1E insulinoma cells. *Biochim. Biophys. Acta* **1777,** 1378–1383.

Bishop, T., and Brand, M. D. (2000). Processes contributing to metabolic depression in hepatopancreas cells from the snail *Helix aspersa. J. Exp. Biol.* **203,** 3603–3612.

Brand, M. D. (1990). The proton leak across the mitochondrial inner membrane. *Biochim. Biophys. Acta* **1018,** 128–133.

Brand, M. D. (1998). Top-down elasticity analysis and its application to energy metabolism in isolated mitochondria and intact cells. *Mol. Cell. Biochem.* **184,** 13–20.

Brand, M. D., Affourtit, C., Amo, T., Choi, S. W., Cornwall, E. J., Gerencser, A. A., Nicholls, D. G., Oh, R., Parker, N., and Yadava, N. (2008). The mechanisms, regulation and functions of mitochondrial uncoupling. 4th CPB Meeting in Africa: Mara 2008. Molecules to migration: The pressures of life (S. Morris and A. Vosloo, eds.) pp. 331–338. Medimond publishing Co, via Maserati 6/2, 40124 Bologna, Italy.

Brand, M. D., Bishop, T., Boutilier, R. G., and St-Pierre, J. (2000). Mitochondrial proton conductance, standard metabolic rate and metabolic depression. *In* "Life in the Cold" (G. Heldmaier and M. Klingenspor, eds.), pp. 413–430. Springer, Berlin.

Brand, M. D., Couture, P., Else, P. L., Withers, K. W., and Hulbert, A. J. (1991). Evolution of energy metabolism. Proton permeability of the inner membrane of liver mitochondria is greater in a mammal than in a reptile. *Biochem. J.* **275,** 81–86.

Brown, G. C., Lakin-Thomas, P. L., and Brand, M. D. (1990). Control of respiration and oxidative phosphorylation in isolated rat liver cells. *Eur. J. Biochem.* **192,** 355–362.

Buttgereit, F., Grant, A., Müller, M., and Brand, M. D. (1994). The effects of methylpred-nisolone on oxidative phosphorylation in Concanavalin-A-stimulated thymocytes. Top-down elasticity analysis and control analysis. *Eur. J. Biochem.* **223,** 513–519.

Else, P. L., Brand, M. D., Turner, N., and Hulbert, A. J. (2004). Respiration rate of hepatocytes varies with body mass in birds. *J. Exp. Biol.* **207,** 2305–2311.

Harper, M. E., and Brand, M. D. (1993). The quantitative contributions of mitochondrial proton leak and ATP turnover reactions to the changed respiration rates of hepatocytes from rats of different thyroid status. *J. Biol. Chem.* **268,** 14850–14860.

Harper, M. E., and Brand, M. D. (1995). Use of top-down elasticity analysis to identify sites of thyroid hormone-induced thermogenesis. *Proc. Soc. Exp. Biol. Med.* **208,** 228–237.

Harrison, R. (1997). Human xanthine oxidoreductase: In search of a function. *Biochem. Soc. Trans.* **25,** 786–791.

Hulbert, A. J., Else, P. L., Manolis, S. C., and Brand, M. D. (2002). Proton leak in hepatocytes and liver mitochondria from archosaurs (crocodiles) and allometric relation-ships for ectotherms. *J. Comp. Physiol. B. Biochem. Syst. Environ. Physiol.* **172,** 387–397.

Janjic, D., Maechler, P., Sekine, N., Bartley, C., Annen, A. S., and Wolheim, C. B. (1999). Free radical modulation of insulin release in INS-1 cells exposed to alloxan. *Biochem. Pharmacol.* **57,** 639–648.

Jekabsons, M. B., and Nicholls, D. G. (2004). *In situ* respiration and bioenergetic status of mitochondria in primary cerebellar granule neuronal cultures exposed continuously to glutamate. *J. Biol. Chem.* **279,** 32989–33000.

Merglen, A., Theander, S., Rubi, B., Chaffard, G., Wollheim, C. B., and Maechler, P. (2004). Glucose sensitivity and metabolism-secretion coupling studied during two-year continuous culture in INS-1E insulinoma cells. *Endocrinology* **145,** 667–678.

Nicholls, D. G. (2006). Simultaneous monitoring of ionophore- and inhibitor-mediated plasma and mitochondrial membrane potential changes in cultured neurons. *J. Biol. Chem.* **281,** 14864–14874.

Nobes, C. D., Brown, G. C., Olive, P. N., and Brand, M. D. (1990). Non-ohmic proton conductance of the mitochondrial inner membrane in hepatocytes. *J. Biol. Chem.* **265,** 12903–12909.

Porter, R. K., and Brand, M. D. (1995). Causes of differences in respiration rate of hepatocytes from mammals of different body mass. *Am. J. Physiol.* **269,** R1213–R1224.

Reynafarje, B., Costa, L. E., and Lehninger, A. L. (1985). O_2 solubility in aqueous media determined by a kinetic method. *Anal. Biochem.* **145,** 406–418.

Rolfe, D. F. S., Newman, J. M., Buckingham, J. A., Clark, M. G., and Brand, M. D. (1999). Contribution of mitochondrial proton leak to respiration rate in working skeletal muscle and liver and to SMR. *Am. J. Physiol.* **276,** C692–C699.

Rutter, G. A. (2001). Nutrient-secretion coupling in the pancreatic islet beta-cell: Recent advances. *Mol. Aspects Med.* **22,** 247–284.

Savina, M. V., and Gamper, N. L. (1998). Respiration and adenine nucleotides of Baltic lamprey (*Lampetra fluviatilis* l.) hepatocytes during spawning migration. *Comp. Biochem. Phys. B. Biochem. Mol. Biol.* **120,** 375–383.

Sweet, I. R., Li, G., Najafi, H., Berner, D., and Matschinsky, F. M. (1996). Effect of a glucokinase inhibitor on energy production and insulin release in pancreatic islets. *Am. J. Physiol.* **271,** E606–E625.

van den Bosch, H., Schutgens, R. B., Wanders, R. J., and Tager, J. M. (1992). Biochemistry of peroxisomes. *Annu. Rev. Biochem.* **61,** 157–197.

Wu, M., Neilson, A., Swift, A. L., Moran, R., Tamagnine, J., Parslow, D., Armistead, S., Lemire, K., Orrell, J., Teich, J., Chomicz, S., and Ferrick, D. A. (2007). Multiparameter metabolic analysis reveals a close link between attenuated mitochondrial bioenergetic function and enhanced glycolysis dependency in human tumor cells. *Am. J. Physiol. Cell Physiol.* **292,** C125–C136.

Zhang, C. Y., Baffy, G., Perret, P., Krauss, S., Peroni, O., Grujic, D., Hagen, T., Vidal-Puig, A. J., Boss, O., Kim, Y. B., Zheng, X. X., Wheeler, M. B., *et al.* (2001). Uncoupling protein-2 negatively regulates insulin secretion and is a major link between obesity, beta cell dysfunction, and type 2 diabetes. *Cell* **105,** 745–755.

CHAPTER TWENTY-FOUR

INVESTIGATING THE ROLES OF MITOCHONDRIAL AND CYTOSOLIC MALIC ENZYME IN INSULIN SECRETION

Rebecca L. Pongratz, Richard G. Kibbey, *and* Gary W. Cline

Contents

1. Introduction	426
2. Malic Enzyme mRNA Expression in Rat Insulinoma INS-1 832/13 Cells, Rat Islets, and Mouse Islets	428
2.1. RNA isolation	428
2.2. Reverse transcription of RNA to cDNA	429
2.3. RTqPCR for ME1 and ME2 mRNA expression in cells and islets	430
3. siRNA Knock-Down of ME1 and ME2 in INS-1 832/13 β-Cells	432
3.1. ME1 and ME2 siRNA design	433
3.2. siRNA transfection optimization in INS-1 832/13 β-cells	434
3.3. Protocol for experimental siRNA KD of ME1 and ME2 in INS-1 832/13 β-cells	434
4. Enzymatic Assays to Determine Activity of Cytosolic and Mitochondrial Malic Enzymes	435
4.1. Cytosolic malic enzyme activity assay	435
4.2. Mitochondrial malic enzyme activity assay	437
5. Calculating Relative Rates of Anaplerotic Pathways from ^{13}C-Glutamate Isotopomer Distribution	442
5.1. ^{13}C-labeling of cells for isotopomer studies of anaplerotic pathways	444
5.2. ^{13}C NMR analysis of the ^{13}C isotopic isomers of glutamate	444
5.3. Calculation of metabolic fluxes with tcacalc	446
6. Discussion	448
References	449

Abstract

Glucose homeostasis depends upon the appropriate release of insulin from pancreatic islet β-cells. Postpandrial changes in circulating nutrient concentrations are coupled with graded release of stored insulin pools by the proportional

Department of Internal Medicine, Yale University School of Medicine, New Haven, Connecticut, USA

Methods in Enzymology, Volume 457
ISSN 0076-6879, DOI: 10.1016/S0076-6879(09)05024-1

changes in mitochondrial metabolism. The corresponding increased synthesis rates of both ATP and of anaplerotic metabolites have been shown to be mediators for nutrient-stimulated insulin secretion. Anaplerosis leads to the export of malate or citrate from the mitochondria, both of which can be recycled through metabolic pathways to reenter the Kreb's cycle. These metabolic cycles have the net effect of either transferring mitochondrial reducing equivalents to the cytosol, or of efficiently providing pyruvate to facilitate responsive changes in the Kreb's cycle flux in proportion to increased availability of glutamate and anaplerotic flux through glutamate dehydrogenase. Here, we describe siRNA knock-down and isotopic labeling strategies to evaluate the role of cytosolic and mitochondrial isoforms of malic enzyme in facilitating malate–pyruvate cycling in the context of fuel-stimulated insulin secretion.

1. INTRODUCTION

Secretion of insulin from pancreatic β-cells plays a central role in regulating whole-body metabolism after a meal. The ability of the β-cell to respond to the postprandial increases in plasma glucose and amino acids with graded rates of insulin release rests in the coupling of cellular metabolism with exocytosis of the insulin stores. Overarching the mechanistic explanations for nutrient-stimulated insulin secretion is the influx of Ca^{2+} that follows increases in glycolytic and mitochondrial ATP production and the depolarization of the cellular membrane after closure of ATP-sensitive potassium channels (K_{ATP}). Delving further into the coupling mechanisms though, many researchers have shown that stimulated insulin secretion is enhanced by a K_{ATP}-independent, yet metabolism-dependent, mechanism to generate other "second messengers." Despite intensive research by many groups, the identification of these second messengers is still uncertain and controversial. Several plausible pathways and metabolites have been proposed to fill the role of second messenger, and it is generally accepted that mitochondrial anaplerotic pathways contribute to their production (for a review: MacDonald *et al.*, 2005).

Anaplerosis increases the concentrations of several Kreb's cycle intermediates generating a large surplus of mitochondrial malate and citrate that is exported to the cytosol. The relatively high activity of both pyruvate carboxylase (PC) and cytosolic malic enzyme in the β-cell supports the concept that the mechanisms linking metabolism with insulin secretion may include a β-cell pyruvate–malate cycle (Fig. 24.1) (Liu *et al.*, 2002; Lu *et al.*, 2002; MacDonald, 1995; Pongratz *et al.*, 2007). Malate that is exported from the mitochondria to the cytosol is regenerated to pyruvate by cytosolic malic enzyme for cycling back to the mitochondria. Cytosolic malic enzyme, together with ATP citrate lyase and malate dehydrogenase, is also central to recycling of citrate back to pyruvate. Farfari *et al.* (2000) suggested

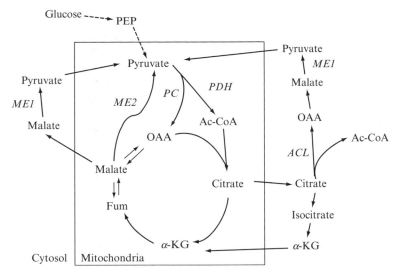

Figure 24.1 Anaplerotic substrate cycling pathways of insulin secreting cells: insulinoma INS-1 cells and pancreatic islet β-cells. Pyruvate cycling can occur via several redundant and complementary pathways. Cytosolic malic enzyme cycles malate derived from the export of mitochondrial malate, or derived indirectly from the exported citrate that is converted to oxaloacetate (OAA) by ATP:citrate lyase (ACL), and then malate by malate dehydrogenase. Pyruvate cycling can also occur entirely within the mitochondria by the conversion of mitochondrial malate into mitochondrial pyruvate by malic enzyme 2 (ME2). An alternative cycling pathway exists for return of citrate to α-ketoglutarate (α-KG) via the conversion of exported citrate to isocitrate and then α-KG. Each of these pathways can be considered a mechanism for the transfer of mitochondrial reducing equivalents to the cytosol in the form of NADPH.

that this citrate–malate–pyruvate cycle serves to regenerate NAD^+ and maintain glycolytic flux. Cytosolic citrate can also be cycled independent of malic enzyme via isocitrate for input to the Kreb's cycle at the level of α-ketoglutarate. These pyruvate cycles all lead to the exchange of reducing equivalents from mitochondrial NADH to cytosolic NADPH. The shift in redox state towards increased concentration of NADPH with increased pyruvate cycling may couple increased mitochondrial activity with downstream events in the cytosol leading to insulin secretion (Ashcroft and Christie, 1979; Ishihara et al., 1999; Ivarsson et al., 2005).

There also exists the potential for an alternative malate–pyruvate cycle within the mitochondrial matrix via mitochondrial-localized isoforms of malic enzyme. In addition to the cytosolic $NADP^+$-dependent isoform, malic enzyme 1 (ME1), two mammalian mitochondrial isoforms exist; malic enzyme 2 (ME2) with a preference for NAD^+, and an $NADP^+$-dependent malic enzyme 3 (ME3). Mandella and Sauer (1975) characterized the enzymatic properties of the mitochondrial isoforms, and suggested that the high K_m value of ME2 for malate and NAD^+ will minimize the conversion of malate to pyruvate under conditions of ample pyruvate supply, but will provide an alternative source of pyruvate from fumarate precursors such as glutamine when glycolytic flux is low. In other detailed studies of the kinetic regulation of malic enzyme by Teller *et al.* (1992), the results indicate that heteroenzyme interactions of ME2 and the PDH complex make ME2-generated pyruvate a better substrate for PDH than the pool of free pyruvate. Thus, mitochondrial malic enzyme may enhance the flux of glutamine as a respiratory fuel by shunting glutamate-derived malate towards the formation of pyruvate, and provide a mechanism to couple the metabolism of glutamine, in the presence of leucine, with enhanced Kreb's cycle flux.

This chapter describes strategies used in our laboratory to investigate the hypothesis that malic enzyme is integral to the coupling of metabolism with insulin secretion. Herein are described the methods used to confirm the expression and activity of the cytosolic and mitochondrial isoforms of malic enzyme in insulin secreting cells, to determine whether nutrient-stimulated insulin secretion is affected by selective changes in the activities of each isoform, and to identify metabolic pathways that may be regulated by the activity of malic enzyme isoforms.

2. Malic Enzyme mRNA Expression in Rat Insulinoma INS-1 832/13 Cells, Rat Islets, and Mouse Islets

Three malic enzyme isoforms are known to be expressed in mammalian tissue, differing in substrate specificity, $NADP^+$ or NAD^+, and cytosolic (ME1) or mitochondrial (ME2 and ME3) compartmentation. We used quantitative real time PCR (RTqPCR) to determine the relative mRNA expression levels of the malic enzyme isoforms in INS-1 cells, mouse islets, and rat islets.

2.1. RNA isolation

RNA can be obtained from cells and isolated islets using RNA extraction kits such as the RNeasy Mini Kit (Qiagen, Germantown, Maryland). Cells ($+/-$ siRNA transfection) are grown in 6-well plates until they are 80–90% confluent. The cells are treated with 350 μl of lysis buffer containing fresh β-mercaptoethanol (BME) and homogenized using a 1 ml syringe with a 20 G needle and passed 8–10 times as per the manufacturer's protocol. The remainder of the steps is followed as suggested by the manufacturer's protocol either on bench top or using the automated Qiacube system (Qiagen). Both manual and automated extractions yield the same quality of RNA. The DNase step which is suggested in the manual is highly recommended. For cells, RNA is eluted off the column with 50 μl of RNase free water and placed on ice for spectrophotometer analysis.

If extracting RNA from isolated rat or mouse islets, 100–150 islets with no exocrine tissue contamination are hand-picked into 1.7 ml tubes on ice using a glass pipette with a flamed tip to smooth the edges so that the islets do not shear. Islets are then centrifuged at a low speed of 100 rcf for 5 min at 4 °C and any extra medium is removed carefully by aspiration. The samples are then placed on bench top at room temperature and 350 μl of lysis buffer containing BME is added to each islet sample tube. Islets are left at room temperature for 10–15 min then vortexed briefly to ensure that all membranes are broken. The vortex step should not exceed 5 s. The islet lysates are then homogenized as described above for the cells and the remainder of the protocol steps are the same except islet RNA is eluted off the column using 30 μl of water in order to yield RNA at a sufficient concentration for the reverse transcription step.

2.2. Reverse transcription of RNA to cDNA

A fraction of the RNA sample is diluted 1:10 and read on the spectrophotometer to determine RNA concentration and purity. Five microliters of cell and islet RNA are brought to 50 μl using RNase/DNase free water and placed in a 50-μl quartz cuvette. The spectrophotometer (Biochrom Ultraspec 2100 *pro*, General Electric, Waukesha, Wisconsin) should be turned on for at least 15 min prior to the spectrophotometer reading to ensure that the tungsten (or deuterium for some spectrometers) bulb is warmed up. The background absorbance at 260 and 280 nm of RNase/DNase free water is measured as a blank. Absorbance of the samples is then measured and the concentration ($\mu g/\mu l$) of RNA is calculated from the absorbance at 260 nm where 1 Optical Density unit at 260 nm for RNA molecules is equal to 40 ng/μl of RNA. RNA purity is determined from the 260/280 nm ratio of absorbances. A ratio of 1.9 is optimal (range from 1.8 to 2.0) for successful RTqPCR. Islet RNA concentrations range from ∼0.04 to 0.08 $\mu g/\mu l$ for ∼125 mouse islets, ∼0.08 to 0.2 $\mu g/\mu l$ for ∼125 rat islets, and 0.2 to 0.6 $\mu g/\mu l$ for confluent cells in 1 well (35 mm diameter) of a 6-well plate. At these concentrations, there is enough total RNA to proceed with the reverse transcription procedure. The reverse transcription reaction works well with a total of 2 μg RNA in the final RT reaction volume. We use reverse transcriptase with a volume of 50 μl per sample reaction mixture (44 μl from Step 1 plus 6 μl from Step 2) as follows.

For Step 1, combine the following in a PCR tube, 2 μl of 100 μM 9-mer random primers (Agilent Technologies, Santa Clara, California) yielding a final reaction concentration of 4 μM, 5 μl of 40 mM DNTPs (10 mM per nucleotide) (Denville Scientific, Metuchen, New Jersey) yielding a final reaction concentration of 4 μM, and the purified RNA to yield a total of 2 μg RNA. However, the volume of purified RNA should not exceed 37 μl. For example, if the RNA concentration is 0.1 $\mu g/\mu l$, then 20 μl of the RNA solution would be needed (i.e., 2 μg total RNA $\div$ 0.1 μg RNA/μl = 20 μl). Adjust to a final volume of 44 μl with RNase free/DNase free water (in this example, 17 μl would be needed: 44 $\mu l - 2 \mu l - 5 \mu l - 20 \mu l$ = 17 μl). Mix and spin briefly. This initial mixture is heated at 70 °C for 5 min so that the primers can anneal to the RNA sequence.

For Step 2, the remainder of the mix, consisting of 5 μl of 10× RT enzyme buffer (NEB, Ipswich, Massachusetts) and 1 μl of 200 U/μl M-MuLV reverse transcriptase (NEB), is added to the sample PCR tube and the RNA is reverse transcribed to complimentary DNA (cDNA) using the following reverse transcriptase (RT-PCR) protocol: incubate for 1 h at 42 °C, inactivate the enzyme for 10 min at 90 °C, and then cool to 4 °C. The cDNA from the reverse transcriptase reaction can be stored at 4 °C or immediately analyzed using RTqPCR.

2.3. RTqPCR for ME1 and ME2 mRNA expression in cells and islets

The cDNA prepared in Section 2.2 from the cells or the islets is diluted 1:10 with RNase free/DNase free water for RTqPCR analysis. Standard curves for RTqPCR amplification efficiencies are obtained from four sequential 1:5 serial dilutions (1:5 to 1:625) of the cDNA from either the control cells or the islets. Efficiency of the primers is calculated from the slope of the standard curve where the efficiency is equal to $-1 + 10^{(-1/\text{slope})}$. Efficiencies between 1.9 and 2.1 are generally accepted for a doubling with each amplification cycle. If the efficiency is beyond this range, then new primer sequences should be designed.

The sensitivity of the RTqPCR instrument will determine the reaction size. For a 10-μl SYBR green reaction, components include; 0.5 μl of 10 μM forward primer for the target gene, 0.5 μl of 10 μM reverse primer for the target gene, 5 μl of 2× Fast SYBR green mix containing Taq polymerase (Applied Biosystems, Foster City, California) and 4 μl of 1:10 diluted cDNA. For 10 μl TaqMan internal florescent probe reaction, the components include; 1 μl custom TaqMan (Applied Biosystems) 10× primer/probe mix (10× primer/probe master mix stock is prepared as follows; 50 μl of 10 μM forward primer, 50 μl of 10 μM reverse primer and 25 μl of 10 μM probe and can be stored at -20 °C), 5 μl of 2× internal probe enzyme mix (Applied Biosystems (TaqMan Fast RT PCR Universal Master mix)) containing Taq polymerase and 4 μl of 1:10 diluted cDNA. Reactions are run in optically clear plates on a real-time PCR instrument (Applied Biosystems 7500 software version 4.1). The RTqPCR protocol will be specific to the primer mix. The protocol that we used is as follows.

Step 1 is 95 °C for 20 s with one repeat. Step 2 is 95 °C for 3 s, then increased to 60 °C for 30 s, with 40 repetitions. For SYBR Green, there is an additional step for the dissociation curve, as follows. Step 3 is 95 °C for 15 s, increased to 60 °C for 60 s, increased to 95 °C for 15 s, then cooled to 60 °C for 15 s, with one repeat.

Primers for either TaqMan or SYBR green RTqPCR analysis for ME1 and ME2 are as follows:

ME1 forward	5′-ATGGAGAAGGAAGGTTTATCAAAG-3′
ME1 reverse	5′-GGCTTCTAGGTTCTTCATTTCTTC-3′
ME2 forward	5′-GGCTTTAGCTGTTATTCTCTGTGA-3′
ME2 reverse	5′-TGAATATTAGCAAGTGATGGGTAAA-3′
ME3 forward	5′-AGATAAGTTCGGAATAAATTGCCT-3′
ME3 reverse	5′-CATCATTGAACATGCAGTATTTGT-3′

Internal probes for both ME1 and ME2 for TaqMan analysis are as follows:

ME1 probe	FAM-GGGCGTGCTTCTCTCACAGAAGA-TAMRA
ME2 probe	FAM-CCCGACACATCAGTGACACCGTTT-TAMRA
ME3 probe	FAM-GACTTTGCCAATGCCAATGCCTTC-TAMRA

The advantage of using internal probe technology is that the probe sequence, which anneals to the template downstream of the sequence primers, has a reporter dye (i.e., FAM) on the 5′ end and the quencher dye (i.e., TAMRA) on the 3′ end. As the primer extends, laying down complimentary base pairs on the template, it will reach the probe which the Taq DNA polymerase in the enzyme mix cleaves and the reporter dye is released and then quenched. Each amplification cycle can then be quantified upon the release of the reporter dye. When using SYBR green, since the dsDNA fluorophores are not specific to the gene target sequence, dissociation curves must be checked for each primer set to ensure that the amplification copies are quantified for the target sequence only, and not other nonspecific PCR products such as primer dimers. The dissociation curve should have a single peak at the appropriate melting temperature (T_m). Primer dimers result in two peaks in the dissociation curve. When determining small interfering ribonucleic acid (siRNA) knock-down, all mRNA transcript copies (CTs) are analyzed using $\Delta/\Delta CT$ analysis where message is normalized to an endogenous housekeeping gene such as beta actin and then also to the non-treated control. When determining overall relative expression of a gene product in a particular tissue type, mRNA expression is calculated using $\Delta/\Delta CT$ analysis where message is normalized

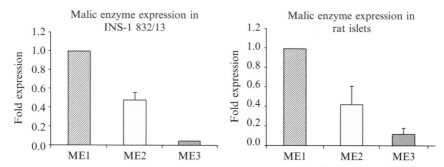

Figure 24.2   Malic enzyme isoform mRNA expression in INS-1 832/13 cells, and in isolated rat islets. INS-1 832/13 cells were cultured until they reached 80–90% conflu-ence. 100–150 islets were picked after a 24 hour recovery in culture media. RNA was extracted in RNA extraction lysis buffer and isolated from the cells and islets. Purified RNA was reverse transcribed to cDNA and expression of the malic enzyme was deter-mined using RTqPCR technology. mRNA expression was determined using $\Delta/\Delta CT$ analysis where expression of each malic enzyme isoform was normalized to β-actin then compared of the expression of cytosolic ME1 where expression levels of the mitochondrial isoforms ME2 and ME3 are shown relative to ME1. Data are mean $\pm$ S.E.

to an endogenous housekeeping gene and then relative fold expression is compared to a second gene of interest. Figure 24.2 shows the relative mRNA expression of ME1, ME2, and ME3 in INS-1 832/13 cells and rat islets.

3. siRNA Knock-Down of ME1 and ME2 in INS-1 832/13 β-Cells

In order to assess the relevance of ME1 and ME2 for nutrient-stimulated insulin secretion from pancreatic islet β-cells, we used the strategy of effec-tively blocking or "knocking-down" gene expression with double stranded siRNA sequences. Because of the low mRNA expression levels of ME3, we focused only on ME1 and ME2 in our studies; however, a similar strategy could be followed to assess the role of ME3. The key to RNA interference is the double stranded RNA configuration where nonendogenous dsRNA is recognized by the cell initiating a cascade of events which ultimately leads to silencing of endogenous RNA that is homologous to the dsRNA. siRNA 21-mer duplexes, consisting of a sense strand and its complementary antisense strand, are characterized by a 2-nucleotide dithymidine overhang at each 3′ terminus and 5′ phosphate groups. siRNAs are assembled into ribonuclease induced silencing complexes (RISC) which then unwind the siRNA to form single stranded RNA (ssRNA). The unwinding of the double stranded siRNA requires ATP and it is this configuration that activates the RISC complex which then seeks out homologous mRNA targets. Once this occurs, the RISC complex cleaves the mRNA, exonucleases degrade the sequence and the message is silenced (Bernstein *et al.*, 2001; Dalmay *et al.*, 2000; Hammond *et al.*, 2001). We chose two siRNA sequences for each malic enzyme isoform targeting two distinct locations on the gene to rule out any off pathway effects of RNA interference.

3.1. ME1 and ME2 siRNA design

The National Center for Biotechnology Information (NCBI) provides a database where specific sequences for a particular gene in an organism can be searched in order to serve as DNA target sequences for siRNA construction. The selected nucleotide sequence can then be analyzed using the Basic Local Alignment Search Tool (BLAST) program in order to verify that it is not homologous to any other gene in the genome except for the target gene under investigation. Commercially available custom 21-mer siRNA sequences can be purchased through various vendors and designed using the specific requirements for each vendor's design tools. We found that the siRNAs from Qiagen were very successful for transfection and decreased mRNA expression and function of ME1 and ME2. The siRNA duplex consists of the sense strand and its complementary antisense strand to provide a 2-nucleotide dithymidine overhang at each 3' terminus. The DNA target sequences for ME1 (a and b) and ME2 (a and b) are as follows:

ME1a: AACCAGGAGATCCAGGTCCTT and ME1b:
   AAGCCAAGAGGCCTCTTTATC,
ME2a: AACGGCTTGCTAGTTAAGGGC and ME2b:
   AAAGCCATGGCCGCTATCAAC.

The 19-base pair siRNA consisting of dsRNA for each target sequence are as follows:

ME1a: (CCAGGAGAUCCAGGUCCUU)d(TT)
   (AAGGACCUGGAUCUCCUGG)d(TT)
ME1b: (GCCAAGAGGCCUCUUUAUC)d(TT)
   (GAUAAAGAGGCCUCUUGGC)d(TT)
ME2a: (CGGCUUGCUAGUUAAGGGC)d(TT)
   (GCCCUUAACUAGCAAGCCG)d(TT)
ME2b: (AGCCAUGGCCGCUAUCAAC)d(TT)
   (GUUGAUAGCGGCCAUGGCU)d(TT)

The duplexes are resuspended and annealed as per manufacturer's recommendation and require a simple preparation of heating to 90 °C in a water bath for 1 min to disrupt any hairpins that may have formed during synthesis and then cooling to 37 °C for 1 h to allow the RNA to reanneal into an uninterrupted double stranded complimentary sequence configuration.

3.2. siRNA transfection optimization in INS-1 832/13 β-cells

Success in transfection of mammalian cells either in stable cell cultures or primary cell lines requires consideration of multiple interacting factors. Although stable cell cultures tend to transfect at a much higher efficiency than primary cells, optimization of the conditions for effective siRNA gene silencing requires an iterative approach. In addition to the cell type, other factors such as passage number, confluence, transfection reagent, and siRNA concentration can influence the effectiveness of siRNA transfection and function in the stable cell line. Positive and negative siRNA controls are useful for optimizing transfection efficiency for each particular cell model. Fluorescent labeled nonspecific siRNAs are commercially available and function as both a positive control and negative control (Qiagen). The fluorescent labeled siRNA functions as a positive control for transfection efficiency since siRNA incorporation into a cell can be visualized using fluorescence microscopy or quantified using fluorescence spectroscopy. The nonspecific siRNA sequence functions as a negative control to ensure the absence of nonspecific knock-down of the target gene, since the siRNA duplex has been created and tested to have no sequence homology to a functional mammalian gene. Once transfection efficiency is optimized for siRNA concentration and incorporation into the cell using control siRNAs, it is then necessary to transfect multiple siRNA sequences corresponding to various locations of the selected gene to test for mRNA down-regulation or "knock-down" and the effect, if any, on protein expression. It is also important to note that even with successful knock-down of mRNA expression, the effects upon the translation and function of the protein should be determined. It is also essential to monitor the duration of transfection with regard to protein turnover since rates of protein turnover vary for different proteins (McManus and Sharp, 2002). INS-1 832/13 cells, like many other mammalian cell lines, can incorporate siRNAs through lipid-mediated transfection (Qiagen RNAifect). This method requires specially designed lipophilic reagents which carry the siRNAs across the lipid bilayer of the plasma membrane and into the cell.

3.3. Protocol for experimental siRNA KD of ME1 and ME2 in INS-1 832/13 β-cells

INS-1 cells are grown in 6-well plates until they reach 50% for transfection. Plates are set up so that independent experiments evaluating ME1 and ME2 functional assays such as glucose and amino acid stimulated insulin secretion or enzyme activity assays, as well as, RNA extraction for mRNA expression can be performed simultaneously on the same passage of cells to ensure that siRNA knock down of message translates to functional down regulation of the ME1 and ME2 proteins. The transfection mixture is prepared based on optimization of the manufacturer's suggested protocol in 1 ml of serum-free OptiMEM with Glutamax (Gibco (Invitrogen), Carlsbad, California) transfection media and then placed on the cells. This will vary from cell type to cell type. In our hands, a 1:5 ratio of siRNA to RNAifect reagent (Qiagen) where 6 μl (1.8 μg) of 20 μM siRNA is brought up to 100 μl with EC transfection buffer provided in the kit and 9 μl of transfection reagent is applied to each well of a 6-well plate in 1 ml OptiMEM. The cells are incubated in the mixture at 37 °C for 6–15 h and then replenished with nutrient-rich RPMI media. The transfected cells are incubated for an additional 48 h and glucose/amino acid stimulation and RNA extraction for mRNA expression is performed.

4. ENZYMATIC ASSAYS TO DETERMINE ACTIVITY OF CYTOSOLIC AND MITOCHONDRIAL MALIC ENZYMES

In addition to assessing the effectiveness of siRNA to reduce mRNA expression levels, relative enzyme activity levels were used to confirm changes in protein levels. Cytosolic malic enzyme is $NADP^+$-dependent and can be determined in the whole cell lysate without interference from malate dehydrogenase which uses NAD^+ for oxidation of malate to oxaloacetate. In contrast, a similar approach to measure the activity of mitochondrial malic enzyme by following the rate of reduction of NAD^+ in the presence of malate is confounded by the presence and high activity of MDH in both the cytosol and mitochondria. Previously, this problem has been resolved by purification of mitochondrial malic enzyme on an affinity column. Our concern that differences in recovery may mask differences in activity, prompted us to develop an alternate isotopic–labeling approach that would allow us to measure mitochondrial malic enzyme activity in intact isolated mitochondria. The rationale for these experiments is that in the absence of PEPCK activity, the only mechanism for the appearance of uniformly ^{13}C-labeled pyruvate ($[^{13}C_3]$pyruvate) when incubating mitochondria with either uniformly ^{13}C-labeled fumarate ($[^{13}C_4]$fumarate or glutamate ($[^{13}C_5]$glutamate) is by the activity of the mitochondrial isoforms of malic enzyme. To evaluate any contribution of PEPCK activity, we include phenylalanine (1 mM final concentration) to inhibit pyruvate kinase (PK) activity and conversion of PEP to pyruvate.

4.1. Cytosolic malic enzyme activity assay

The assay to determine cytosolic enzyme activity consisted of extraction of whole cell lysate, incubation of the cell lysate in the presence of malate and $NADP^+$, monitoring the reduction of $NADP^+$ spectrophotometrically, and normalizing to total cellular protein. We describe the protocol for the assay in INS-1 cells; however, the methodology can be readily used for determining malic enzyme activity in the extracts of isolated islets. The method as described requires ~1–2 mg total cell protein in a volume of 6 ml.

4.1.1. Extraction of whole cell lysate

INS-1 cells are cultured in 6-well plates until they reach 80–90% confluence, typically, ~5.0 × 10^5 cells per well of a 6-well plate are extracted. Protein concentrations for each well range between 200 and 400 μg. On the day of extraction, cells are placed on ice, washed with ice-cold PBS then lysed using 1 ml of ice-cold 0.1% triton per well. The cells remain on the ice for roughly 15–30 min and wells are then scraped using a plastic cell lifter (Corning Incorporated, Corning, New York), mixed by pipetting and transferred into 1.7 ml plastic tubes on ice.

4.1.2. Malic Enzyme 1 assay reaction mixture stock components

All stocks are prepared prior to the experiment and stored as separate solutions at room temperature unless otherwise indicated. The assay components include; 250 mM Tris/HCl pH 7.4, 50 mM MnCl$_2$, 40 mM NH$_4$Cl, 1 M KCl, and 20 mM NADP$^+$ which must be prepared fresh and immediately placed on ice. Since NADP$^+$ powder is hydroscopic, the chemical was brought to room temperature prior to weighing in order to ensure an accurate concentration. One hundred millimolar of the enzyme reaction substrate, L-malate, is prepared, brought to pH 7.4 and stored at 4 °C. On the day of the experiment, the enzyme assay reagent mixture is prepared as described in Table 24.1.

A 150-μl aliquot of this reaction mixture is placed into each well of a 96-well clear flat bottom plate compatible for spectrophotometric analysis on a 96-well plate reader. (We used NUNC (Thermo Fisher Scientific), Rockford, Illinois, Immuno MicroWell 96-well plates and a Dynex, Chantilly, Virginia, MRX Revalation Plate Reader.) Test and control wells are set up in duplicate for each cell sample where 50 μl of H$_2$O is

Table 24.1   Enzyme assay reagent mixture for malic enzyme 1

Substance	Stock conc. (mM)	Volume of stock (ml)	Final conc. (mM) in mix
Tris/HCl	250	10	155
MnCl$_2$	50	0.5	1.55
NH$_4$Cl	40	0.625	1.55
KCl	1000	2.5	155
NADP$^+$	20	1.25	1.55

placed in test wells; 75 μl of H_2O is placed in control wells. Twenty-five microliters of the enzyme substrate L-malate is added to the test wells only. The final concentrations of the assay components in each well are as follows: $\sim$100 mM (93 mM) Tris/HCl, $\sim$1 mM (0.93) $MnCl_2$, $\sim$1 mM (0.93) NH_4Cl, $\sim$100 mM KCl, $\sim$1 mM (0.93) $NADP^+$, and 10 mM L-malate. The reaction plate is then brought to room temperature on bench-top. Twenty-five microliters of the cell lysate samples are then added to each of the test and control wells using a multidispensing pipette (Rainin, Oakland, California, EPD-3 PLUS) where the control wells are seeded first to avoid cross contamination of the substrate. The plate is mixed by shaking briefly and immediately read at 340 nm for 40 min with data collected every minute beginning immediately after an initial 5 s shaking step. The optical densities (OD) for each minute of data collection can then be plotted and slopes determined for cytosolic malic enzyme 1 activity in both test and control wells where control wells are subtracted from the test wells and data are then normalized to whole cell protein concentration. Activity $= \Delta OD/$min/mg protein. If desired, this can be expressed in terms of reduction rates of NAD^+ or $NADP^+$ with a suitable standard curve. For the purposes of these experiments, the relative changes in ME activity in control and siRNA-treated cells, as assessed by ΔOD, are sufficient to evaluate the relation between changes in mRNA expression and enzyme activity.

4.1.2.1. Protein assay   Protein concentrations in the whole cell lysates are assayed using the Micro BCA assay (Pierce, Thermo Fisher Scientific, Rockford, Illinois) following the manufacturer's protocol and measured at an absorbance of 595 nm. The standard curves are made from 1:2 serial dilutions of the 2 mg/ml BSA standard provided in the kit with final concentrations ranging from 20 to 0.625 μg/ml.

4.2. Mitochondrial malic enzyme activity assay

The assay to determine mitochondrial enzyme activity consisted of extraction and purification of intact viable mitochondria, incubation of the mitochondria in the presence of [U-^{13}C]-labeled substrates, fumarate or glutamate, followed by extraction of the mitochondrial metabolites and analysis by LC/MS/MS to determine the ^{13}C-enrichment of pyruvate and other Kreb's cycle intermediates. We describe the protocol for the assay in INS-1 cells; however, the methodology can be readily used for determining malic enzyme activity in the extracts of isolated islets. The method as described requires $\sim$1.5 mg mitochondrial protein to perform the kinetic analysis using LC/MS/MS to determine the time course of ^{13}C incorporation into pyruvate.

4.2.1. Isolation of mitochondria

About 48–72 h prior to isolation of the mitochondria, INS-1 cells are seeded at a density of $\sim 1.0 \times 10^7$ cells/plate into 150 cm^2 cell culture dishes. Each dish yields ~ 0.5 mg mitochondrial protein. Two dishes are needed for the LC/MS/MS kinetic studies. The cells should be cultured in RPMI-1640 complete media with 11.1 mM D-glucose supplemented with 10% (v/v) fetal bovine serum, 10 mM HEPES, 2 mM L-glutamine, 1 mM sodium pyruvate, 50 μM BME, 10,000 U/ml penicillin, and 10 mg/ml streptomycin. After 2-days culture in 5% CO$_2$ at 37 °C, the cell density should be $\sim 90\%$ confluent, and ready for isolation of the mitochondria.

Mitochondria are isolated under conditions to preserve function but to minimize enzymatic activity. All media, isolation buffers, centrifuge tubes, and centrifuges should be pre-cooled to between 0 and 4 °C. Fifteen milliliters of mitochondrial extraction buffer is needed for extraction of mitochondria from three 150 cm^2 plates (5 ml/plate). The plates are placed on ice and the growth medium is removed by aspiration. The cells in each plate are then quickly washed with 5 ml of ice-cold Phosphate Buffered Saline, PBS. The PBS is immediately removed by aspiration and replaced with 5 ml of ice-cold isotonic mitochondrial extraction buffer containing 65 mM sucrose, 215 mM D-mannitol, 5 mM HEPES, 3 mM MgCl$_2$, 5 mM KH$_2$PO$_4$, and 5 mM KHCO$_3$. The mitochondrial extraction buffer should be prepared fresh on the day of the extraction, as detailed in Table 24.2. The adherent cells are quickly scraped from the plates using a sturdy cell lifter (Corning), and the cell suspensions are transferred into one 15-ml conical centrifuge tubes for three plates. The cell suspensions are centrifuged at a low speed of ~ 100 rcf for 5 min at 4 °C. The supernatant is then removed by aspiration, and the cell pellet is gently resuspended in ice-cold mitochondrial buffer (1 ml/culture plate) until a homogeneous mixture is obtained.

Table 24.2 Mitochondrial Iso-osmotic buffer (2X)

Concentration	Substance	Molecular weight	200 mL of 2X stock
65 mM	Sucrose	342.3	8.90 g
215 mM	D-Mannitol	182.18	15.66 g
5 mM	HEPES	238.31	0.238 g
3 mM	MgCl$_2$	203.30	0.122 g
5 mM	KH$_2$PO$_4$	136.09	0.136 g

The 2X buffer can be stored up to one month at 4 °C.

For 100 ml of 1X Mitochondria buffer made fresh on the day of isolation:

Add 0.05 g KHCO$_3$ to 50 ml of 2X solution

1. Adjust the pH to 7.4 with 30% KOH (very little needed)
2. Add water to total final volume of 100 ml
3. Filter sterilize through 0.22 micron filter

Transfer 3 ml of the cell suspension to a pre-cooled (4 °C) 5-ml glass–teflon Potter Elvehjem homogenizer, and homogenize with the vertical passing of 50 strokes between teflon pestle and glass vial/mortar. We found that to obtain a cleaner mitochondrial preparation with minimal cytosolic contamination, it is best to resuspend the cell pellet in 1 ml/culture plate, and do multiple rounds of douncing rather than to concentrate the cell pellet with less volume per culture plate. The homogenate is then split into 1.7-ml microcentrifuge tubes (1 ml/tube), and the sheered cells are centrifuged for 3.5 min at 1800 rcf at 4 °C to pellet the nucleus, sheered plasma membrane, and other cellular fragments. The supernatant, containing the mitochondria, is transferred to new 1.7-ml tube and placed on ice while additional mitochondria trapped in the pellet are harvested two more times to yield the maximum amount of mitochondria as follows. The pellet is resuspended in 0.5 ml of mitochondrial extraction buffer and centrifuged for 3.5 min at 1800 rcf at 4 °C. The supernatant, with the mitochondria, is again transferred to a new tube and kept on ice. The process is repeated once more. The supernatants from the second and third spins are combined with the first mitochondrial supernatant in a 1.7-ml tube. (The pellet containing the membrane fragments can now be discarded.) The mitochondria are then pelleted by centrifugation of all the supernatant tubes for 5 min at a higher speed of 10,800 rcf at 4 °C. Carefully aspirate the supernatant from the yellowish mitochondrial pellet. At higher speeds, you may see a white fluffy band above the pellet. If this occurs, remove the white band with the rest of the supernatant by aspiration. The mitochondrial pellet from one tube is resuspended in 1.0 ml of extraction buffer and transferred to the next tube containing the second mitochondria pellet, and then resuspended to obtain one homogeneous suspension of mitochondria. The mitochondria are pelleted with a final high-speed centrifugation for 5 min at 4 °C. Aspirate the supernatant and resuspend the pellet in 300 μl of the mitochondrial extraction buffer for the kinetic labeling experiments for assessment of mitochondrial malic enzyme activity. A 30-μl aliquot of the mitochondrial suspension is taken for protein analysis by the Bradford method (Bio-Rad Laboratories, Hercules, California, reagent).

4.2.1.1. Protein Assay

Protein concentrations of the mitochondrial suspension are assayed at this point to determine the dilution needed to achieve a concentration of 1 mg protein/ml. The Bradford method (Bio-Rad protein reagent) must be used as sucrose interferes with the micro BCA assay. Protein assays are performed following manufacturers protocol measuring absorbance at 595 nm with a standard curve from 40 to 1.2 μg/ml.

4.2.1.2. Cytochrome c assay

Preliminary experiments should be performed to assess the integrity of the mitochondria isolated from the cell lysate as described above. The cytochrome c oxidase assay kit (Sigma Aldrich,

St. Louis, Missouri) based on cytochrome c oxidase enzyme (EC 1.9.3.1) assay allows for the detection of intact mitochondria by assessing outer membrane integrity via cytochrome c. Pure fractions of intact mitochondria are assayed following the manufacturer's protocol at 550 nm where changes in absorption of cytochrome c when reduced and then reoxidized determines cytochrome c activity. Pure mitochondrial fractions containing intact mitochondria will yield a positive change in absorption. A positive control is included in the assay kit. Negative controls include pure cytosolic fractions from the isolation and lysed mitochondria subjected to brief sonication.

4.2.1.3. Glycerol-3-P dehydrogenase assay   In order to be sure that the mitochondria fraction from the isolation is not contaminated with cytosolic components, both the cytosolic and mitochondrial fractions from the isolation were saved and assayed for α-glycerophosphate dehydrogenase (EC 1.1.1.8) activity, following the manufacturer's (Sigma Aldrich) recommendation monitoring the change in absorption at 340 nm to determine glycerol-3-P dehydrogenase activity. Pure mitochondrial fractions, free of cytosolic contamination, should have no activity as evidenced by no change in absorption. Glycerol-3-P dehydrogenase activity should only be evident in the cytosolic fraction from the supernatants saved from the high-speed spins.

4.2.2. Kinetic analysis of mitochondrial malic enzyme activity

To validate the activity of mitochondrial malic enzyme, we use electrospray tandem mass spectroscopy (LC/MS/MS) to measure the appearance of [$^{13}C_3$]pyruvate when freshly isolated mitochondria are incubated with NAD^+, ADP^+, and either [$^{13}C_4$]fumarate or [$^{13}C_5$]glutamate with leucine. A 96-well format with eight time points and four mitochondrial dilutions is used to obtain the time course of [$^{13}C_3$]pyruvate synthesis. In order to collect aliquots at the timed intervals, we make the following preparations.

First prepare eight identical 1.7-ml tubes prepared as listed in Table 24.3.

Next prepare two sequential 1:5 dilutions of the undiluted mitochondrial suspension ($\sim$1.5 mg mitochondrial protein in 300 μl prepared from three confluent 150 cm^2 dishes) that was prepared as described above in Section 4.2.1. For the 1:5 mitochondrial dilution, 150 μl of undiluted mitochondria plus 600 ml mitochondrial buffer is prepared. For the 1:25 mitochondrial dilution, 150 μl of 1:5 diluted mitochondria plus 600 ml mitochondrial buffer is prepared.

Place the eight previously prepared 1.7 ml tubes with the enzyme assay reagent mixtures (Table 24.3) in a 37 °C water bath for 3–5 min with the lids open (be sure that no water splashes in). Have a 96-well plate on ice ready and labeled so that each column corresponds to a time point and each row corresponds to the incubation tube. Add 150 μl of 1 mM acetic acid to

Table 24.3 Enzyme assay reagent mixture for malic enzyme 2

Solutions	Quantity (μl) Total volume: 500 μl	Final concentration
Mitochondrial Iso-osmotic buffer (2X): from Table 24.2	250	1X
ADP	2.5 of 100 mM	0.5 mM
GDP	2.5 of 100 mM	0.5 mM
[$^{13}C_4$]Fumarate*	12.5 of 100 mM	2.5 mM
MnCl2	50 of 10 mM	1 mM
H_2O	182.5	
[$^{13}C_5$]Glutamate plus leucine	12.5 of mixture of 100 mM ^{13}C-gln and 200 mM leu	2.5 mM 13C-gln, 5 mM leu

Prepare 8–1.7 ml reaction tubes with the above reagents for the kinetic analysis of formation of ^{13}C-pyruvate by mitochondrial malic enzyme 2; determined by LC/MS/MS.

* Alternatively to [$^{13}C_4$]Fumarate, use a mixture of [$^{13}C_5$]Glutamate and leucine, as below.

all 96 wells to immediately quench the reaction once an incubated aliquot is taken. Prior to adding the mitochondria, take 50 μl from each 1.7 ml tube and place in the corresponding 0 min column of the 96-well plate. Add 50 μl of the mitochondria to each tube, using the undiluted and each dilution sample for two replicate tubes (e.g., undiluted mitochondria in tubes 1 and 2, 1:5 diluted mitochondria in tubes 3 and 4, 1:25 diluted mitochondria in tubes 5 and 6). Immediately mix by pipetting and collect 50 μl from each tube in the same order as mitochondria samples were added. This will be time 0.5 min and spike into appropriate well in the 96-well plate. Follow this procedure for all of the sample tubes for each of the remaining time-points: 2.5, 5, 10, 15, 30, and 60 min.

4.2.2.1. Electrospray tandem mass spectrometry for determining ^{13}C-enrichment of Kreb's cycle intermediates and pyruvate

The samples are then thawed and transferred to 96-well filter plates, the supernatent is collected by centrifugation through the filter, and analyzed by electrospray tandem mass spectrometry (LC/MS/MS) for ^{13}C-enrichment of pyruvate, malate, citrate, and glutamate. In our laboratory, LC/MS/MS analyses of metabolites are performed on an Applied Biosystems API4000 QTrap interfaced to a Shimadzu, Kyoto, Japan, HPLC (LC-20AD, SIL-20AC, CTO-20A). Metabolites were eluted from a Dionex, Sunnyvale, California, Aclaim

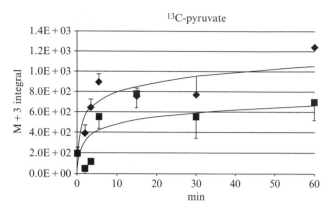

Figure 24.3 Time course of synthesis of [U-¹³C]pyruvate by mitochondria isolated from INS-1 cells overexpressing a dominate-negative HNF-1α mutation (filled squares), and control INS-1 cells (filled diamonds). [U-¹³C]pyruvate are determined by LC/MS/MS analysis; integrated intensity are normalized to mitochondrial protein. Mitochondria were incubated with 2.5 mM [¹³C₄]fumarate and 0.5 mM ADP and GDP in iso-osomotic buffer as described in the text and Table 24.3.

Polar Advantage (C16, 5 μm, 120 Å, 4.5 × 250 mm) column (40 °C) with acetonitrile:water buffered with 2 mM ammonium acetate using a linear gradient from 5% to 95% acetonitrile at a flow rate of 600 μl/min. Metabolite concentrations and enrichments are determined by electrospray ionization monitoring of the positive product ion transition pairs of pyruvate (87.0/32.0), citrate (191.0/87.0), succinate (117.0/73.0), glutamate (146.0/128.0), and malate (133.0/71.0). Isotopic labeling is monitored with incremental increases in the masses of the parent and daughter ions to account for the ensemble of potential isotopic isomers. To monitor the synthesis rate of [U-¹³C]pyruvate from [U-¹³C]fumarate, the transition pair for m + 3 pyruvate of 90.0/32.0 is shown in Fig. 24.3.

5. Calculating Relative Rates of Anaplerotic Pathways from ¹³C-Glutamate Isotopomer Distribution

The NMR ¹³C-isotopomer and modeling approach that we used has the advantage that a comprehensive picture of the entry into the Kreb's cycle of multiple substrates through multiple pathways can be determined in a single experiment. The metabolism of [U-¹³C]glucose results in labeling of Kreb's cycle intermediates that provides an experimentally convenient means of simultaneously quantifying relative flux rates through multiple pathways from a single isotopic tracer input. From the NMR analysis of the molecular distribution of ¹³C-label within glutamate, the contributions of glucose (through PDH) and fatty acids (by β-oxidation) into acetyl CoA are calculated. The analysis also provides the relative flux rates of substrates entering through several anaplerotic entry points including glucose through PC, and glutamate through glutamate dehydrogenase (GDH) when glutamine plus leucine are supplied as secretagogues (Fig. 24.4).

Metabolism of [U-^{13}C]glucose results in labeling of Kreb's cycle intermediates that are determined by relative rates of PC and PDH, as well as the entry of other metabolites into the cycle. PDH flux initially labels both

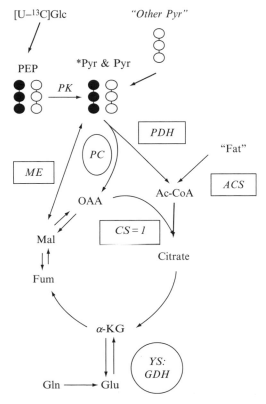

Figure 24.4 Model of Kreb's cycle used to determine relative flux rates through acetyl-CoA synthetase (ACS), pyruvate dehydrogenase complex (PDH), pyruvate carboxylase (PC), and glutamate dehydrogenase (GDH). Labeled substrate originates from exogenous [U-^{13}C$_6$]glucose. Unlabeled substrates originate from exogenous glutamine, leucine, and from endogenous sources feeding the substrate pools designated as "fat", and "other pyruvate". Malic enzyme (ME) provides one possible pathway for cycling of anaplerotic flux of pyruvate via PC. Anaplerotic entry of glutamate (from glutamine) is at the level of α-ketoglutarate (α-KG) by the action of GDH, calculated as YS by "tcacalc". Acetyl-CoA is derived from either fat via ACS, or pyruvate via the PDH. Pyruvate can be derived from [U-^{13}C]glucose (labeled pyr: filled circles), or from endogenous or cycling via Malic enzyme (unlabeled pyr: open circles). All fluxes are normalized by "tcacalc" to citrate synthase (CS) flux.

Glu 4 and 5 (Glu4 + 5). Further turns of the Kreb's cycle randomize the doubly labeled isotopomer to either Glu4 + 5 or Glu1 + 2. When PDH is the only entry point for Kreb's cycle flux, the mixture of glutamate isotopomers quickly reaches a steady-state distribution of glutamate labeled in all five carbons. In contrast, PC flux of [U-^{13}C]pyruvate (with unlabeled acetyl–CoA), initially labels Glu2 + 3 and Glu1 + 2 + 3. At steady-state, ~70% of the isotopomers consist of Glu1 + 2, Glu2 + 3, and Glu1 + 2 + 3. Because of these differences in steady-state isotopomer distributions, the proportion of PC to PDH flux can be readily determined from the glutamate isotopomer distribution. Furthermore, the isotopomer distribution also provides a measure of the pathways and relative flux rates of unlabeled substrates (fat, other three carbon precursors, or GDH) contributing to Kreb's cycle flux. The relative rates of labeled and unlabeled substrates are calculated as the product of the precursor enrichment and the relative rates for each step as determined by the program "tcacalc" (Malloy *et al.*, 1990).

5.1. ^{13}C-labeling of cells for isotopomer studies of anaplerotic pathways

INS-1 cells (+/− siRNA KD of malic enzyme), at 80% confluence, are pre–incubated for 2 h at low glucose (2 m*M*) in KRB to establish basal conditions. The cells are then washed to remove unlabeled glucose, and then incubated for 2 h in DMEM, or KRB, with [U-^{13}C]glucose (Cambridge Isotope Laboratories, Miamisburg, Ohio, 99% ^{13}C, 2.5 m*M* or 15 m*M*), alone or supplemented with 4 m*M* glutamine and 10 m*M* leucine (singly or in combination). At the end of the 2 h isotopic labeling period, the media is removed and frozen for later analysis. The monolayer of cells is quenched with ice-cold water, quickly scraped with a cell lifter, and transferred to a 50 ml conical tube with a 200 ml aliquot saved for protein and insulin assays. The remainder of the supernatant (and media) is frozen on dry ice, lyophilized, and redissolved in 400 μl D$_2$0 for NMR analysis.

5.2. ^{13}C NMR analysis of the ^{13}C isotopic isomers of glutamate

The ^{13}C positional isotopomer distribution of glutamate is determined by ^{13}C NMR spectroscopy. In our laboratory, we used an AVANCE 500–MHz spectrometer (Bruker Instruments, Inc. Billerica, Massachusetts). Spectra were acquired with TR = 0.5 s, NS = 10,000, 16 K data, and Waltz-16 broadband proton decoupling. Correction factors for differences in T$_1$ relaxation times should be determined from fully relaxed spectra of standard solutions (Cline *et al.*, 2004).

An example of the ^{13}C NMR spectrum of glutamate isotopomers that are generated by INS-1 cells after incubation with 15 m*M* [U-^{13}C$_6$]glucose

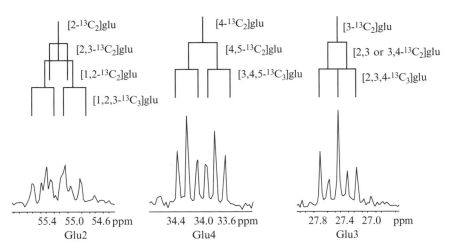

Figure 24.5 Proton-decoupled ^{13}C-NMR spectrum of glutamate carbons 2, 3, and 4. Extracts of ~30×10^6 cells were prepared after a 2 hour pre-incubation in KRB buffer containing 2 mM glucose, followed by 2 hour incubation at 37 °C in KRB with 15 mM [U-^{13}C$_6$]glucose. Isotopic isomers are identified from the ^{13}C-^{13}C $^1J_{CC}$ coupling constants as indicated above each carbon resonance, and the integrated areas for each isotopomer were determined by peak fitting for metabolic pathway modeling.

in KRB is shown in Fig. 24.5. The isotopic isomers are best identified from the ^{13}C–^{13}C J-coupling constants, and quantified by integration of line fitting (NUTS, Acorn, Inc) to the glutamate resonances at C2 (~55 ppm), C3 (~28 ppm), and C4 (~34 ppm).

5.2.1. Analysis of glutamate 3

The ^{13}C isotopic isomers determined from analysis of C3 are glutamate triply labeled at C2–C3–C4 (triplet), glutamate doubly labeled at C2–C3 or C3–C4 (doublet). The center resonance of C3 is the sum of the C2–C3–C4 isotopomer, and glutamate labeled at C3, but not C2 or C3. The triplet at C3 is due to the similarity of the coupling constants of 35 Hz for both C2–C3 and C3–C4.

5.2.2. Analysis of glutamate 4

In contrast to the triplet seen at glutamate C3 (for C2–C3–C4), when glutamate is triply labeled at C3–C4–C5, the differences in the coupling constants of 35 Hz for C3–C4, and 51 Hz for C4–C5, results in a quartet. From the coupling constant of 51 Hz, it can be determined that only the doubly labeled isotopomer, glutamate C4–C5, is present. The doubly labeled isotopomer, glutamate C3–C4 ($J = 35$ Hz) is undetectable in this spectrum.

5.2.3. Analysis of glutamate 2

Glutamate 2 is interpreted in the same way as glutamate 4. In contrast to the spectrum of glutamate 4, both of the doubly labeled isotopomers C2–C3 ($J = 35$ Hz), and C1–C2 ($J = 50$ Hz) can be observed.

5.3. Calculation of metabolic fluxes with tcacalc

Steady-state flux rates, relative to Kreb's cycle flux, are calculated from the isotopomers of glutamate observed at carbon positions 2, 3, and 4 in the ^{13}C NMR spectra, and modeled using the program tcacalc (Malloy *et al.*, 1990), and is available for download from the UTSW Medical Center Rogers NMR Center (http://www4.utsouthwestern.edu/rogersnmr/software.htm).

Using the nomenclature adopted for tcacalc (Fig. 24.4), the model provides the best fit to our NMR data by calculating the relative fluxes of acetyl-CoA synthetase (ACS), lactate dehydrogenase (LDH), pyruvate dehydrogenase (PDH), pyruvate carboxylase (PC), pyruvate kinase (PK), anaplerosis leading to succinyl-CoA (YS), and enrichments of "lactate" and "fat." Fat includes all metabolite inputs into acetyl-CoA other than glucose (e.g., leucine). All flux rates are referenced to a citrate synthase (CS) flux of 1, and is equivalent to Kreb's cycle flux. CS flux is modeled as the sum of PDH flux of labeled and unlabeled pyruvate, and of ACS flux, as presented in Fig. 24.4. YS represents unidirectional anaplerotic flux of substrates that lead to succinyl-CoA. As originally designed, YS represents flux of propionate into the Kreb's cycle. In our experiments with glutamine plus leucine as the secretagogues, YS represents the rate of synthesis of α-ketoglutarate from glutamate (Figs. 24.1 and 24.4: GDH).

An example input file with initial parameters and sequential addition of the pathways of pdh, pk, and ys, used to calculate fluxes is as follows:

fat12	0.5	0.1
LAC123	0.5	0.1
YPC	0.7	0.2
Add pdh	0.3	0.2
Add pk	0.3	0.2
Add ys	0.5	0.1

An example of a data file with fractional isotopomer distribution determined from the ^{13}C NMR spectrum of INS-1 cells incubated for 2 h in KRB with 15 m*M* glucose is

CO Created using tcasim and altered to have 3% error
CO FAT2 0.30 FAT12 0.40 YS 0.05 AS3 0.50 CYCLES 35.00

GLU2S	0.039
GLU2D23	0.333
GLU2D12	0.096
GLU2Q	0.531
GLU3S	0.037
GLU3D	0.232
GLU3T	0.730
GLU4S	0.000
GLU4D34	0.000
GLU4D45	0.319
GLU4Q	0.680

Define C3C4 GLU3F/GLU4F 0.804
Define C2C4 GLU2F/GLU4F 0.753

The above input file represents the fractional area for the multiplets at each of carbon resonances.

Using these inputs, tcacalc calculated the following fluxes:

ACS	0.857
Fat0	0.123
Fat12	0.877
LDH	0.593
Lac0	0.008
Lac123	0.992
PDH	0.143
YPC	1.230
YS	0.379
ASO	1.000
PK	0.780

From which we calculated the following Kreb's cycle parameters:

Relative flux rates: Acetyl-CoA entry: ACS + PDH = 1.00
Relative flux rates: Anaplerotic entry: YPC + YS = 1.61
Acetyl-CoA from [U-^{13}C]glucose: ACS × Fat12 + PDH × Lac123 = 0.89
Acetyl-CoA from other sources: ACS × Fat0 + PDH × Lac0 = 0.11

The relative rates of labeled and unlabeled substrates can be calculated as the product of the precursor enrichment and the relative rates.

Anaplerotic flux from [U-^{13}C]glc: YPC × Lac123 = 1.22
Anaplerotic flux from unlabeled pyr: YPC × Lac0 = 0.01
Anaplerotic flux from glutamate: YS × ASO = 0.38

The percent anaplerosis to total Kreb's cycle flux can be calculated as:

% Anaplerosis: $100 \times (PC + YS)/(PC + YS + PDH + ACS) = 62\%$

Although open to interpretation, Lu *et al.* (2002) proposed that an index of pyruvate cycling can be calculated as the average of the flux rates of PC and of PK.

Pyruvate cycling: $(PC + PK)/2 = 1.01$.

6. DISCUSSION

The pancreatic islet β-cell is uniquely specialized to tightly couple its metabolism and rates of insulin secretion with the levels of circulating nutrient fuels. Unlike other metabolically responsive cells, such as myocytes and hepatocytes, β-cells cannot accommodate increased glycolytic flux by storing the excess glucose as glycogen or eliminate it as lactate. Without an increase in energy demand that matches the increased glycolytic flux, feedback inhibition would limit the ability of the metabolic responsiveness of the cell. Substrate cycling, mediated by malic enzyme, provides a mechanism to allow glycolytic and mitochondrial fluxes to increase in direct proportion to circulating concentrations of glucose and other fuel secretagogues. We have used tcacalc to evaluate changes in the flux through glycolytic and Kreb's cycle pathways that are believed to play a role in the coupling of metabolism with insulin exocytosis. It should be noted though, that tcacalc represents a simplified model of metabolism within the β-cell, and has limited adaptability to incorporate the numerous intertwined pathways and substrate cycles that serve to modulate nutrient-stimulated insulin secretion. A pressing need exists to develop modeling programs that have the flexibility to include new pathways as they are revealed, and to be able to incorporate the wealth of data available from both NMR isotopomer and mass spectrometric isotopolog analysis. For example, substrate cycling of the anaplerotic metabolites malate and citrate is modeled by tcacalc based on the assumption that the average of unidirectional fluxes through PK and PC into the Kreb's cycle is a valid approach for calculating a cycling flux mediated by malic enzyme. Our use of siRNA to alter the activity of malic enzyme was undertaken to directly test the validity of this assumption, and the inference that the rate of insulin secretion was influenced by malate/citrate/pyruvate cycling, as opposed to PC flux *per se*. Our approach rests on the premise that malic enzyme exerts measurable control over the rate of these substrate cycles, and that this cycle is tied to the regulation of insulin secretion. Our results, as well as others, indicate that both the cytosolic and mitochondrial isoforms of malic enzyme exert metabolic control over the rate of pyruvate cycling in the rat insulinoma INS-1 832/13 cell line. However, recent results suggest that this may not be true in the context of the rat or mouse islets. These seemingly contradictory results highlight the need for a rigorous study of pyruvate cycling using metabolic control analysis. A low control coefficient for malic enzyme would result in little observable effect unless the activity of malic enzyme was virtually eliminated. In most cases, siRNA KD of malic enzyme is not absolute and a significant level of enzyme activity remains. Thus, one would observe little effect upon glucose-stimulated insulin secretion, despite the necessary involvement of malic enzyme and malate/pyruvate cycling in the coupling of metabolism with secretion. In conclusion, the redundancy of several complementary substrate cycles, the variance in their degree of metabolic control, and the species-dependency of expression and activity of malic enzyme in islet β-cells, makes it imperative to evaluate the role of malic enzyme in fuel-stimulated insulin secretion in human islet β-cells.

REFERENCES

Ashcroft, S. J. H., and Christie, M. R. (1979). Effects of glucose on the cytosolic ration of reduced/oxidized nicotinamide-adenine dinucleotide phosphate in rat islets of Langerhans. *Biochem. J.* **184,** 697–700.

Bernstein, E., Caudy, A. A., Hammond, S. M., and Hannon, G. J. (2001). Role for a bidentate ribonuclease in the initiation step of RNA interference. *Nature* **409,** 363–366.

Cline, G. W., LePine, R., Papas, K. K., and Shulman, G. I. (2004). ^{13}C NMR isotopomer analysis of anaplerotic pathways in INS-1 cells. *J. Biol. Chem.* **279,** 44370–44375.

Dalmay, T., Hamilton, A., Rudd, S., Angell, S., and Baulcombe, D. C. (2000). An RNA-dependent RNA polymerase gene in *Arabidopsis* is required for posttranscriptional gene silencing mediated by a transgene but not a virus. *Cell* **101,** 543–553.

Farfari, S., Schulz, V., Corkey, B., and Prentki, M. (2000). Glucose-regulated anaplerosis and cataplerosis in pancreatic β-cells. *Diabetes* **49,** 718–726.

Hammond, S. M., Caudy, A. A., and Hannon, G. J. (2001). Post-transcriptional gene silencing by double-stranded RNA. *Nat. Rev. Genet.* **2**(2), 110–119.

Ishihara, H., Wang, H., Drewes, L. R., and Wollheim, C. B. (1999). Overexpression of monocarboxylate transporter and lactate dehydrogenase alters insulin secretory responses to pyruvate and lactate in beta cells. *J. Clin. Invest.* **104,** 1621–1629.

Ivarsson, R., Quintens, R., Dejonghe, S., Tsukamoto, K., in't Veld, P., Rennstrom, E., and Schuit, F. C. (2005). Redox control of exocytosis: Regulatory role of NADPH, thioredoxin, and glutaredoxin. *Diabetes* **54,** 2132–2142.

Liu, Y.-Q., Jetton, T. L., and Leahy, J. L. (2002). β-cell adaptation to insulin resistance. Increased pyruvate carboxylase and malate-pyruvate shuttle activity in islets of nondiabetic Zucker fatty rats. *J. Biol. Chem.* **277,** 39163–39168.

Lu, D., Mulder, H., Zhao, P., Burgess, S. C., Jensen, M. V., Kamzplova, S., Newgard, C., and Sherry, A. D. (2002). ^{13}C NMR isotopomer analysis reveals a connection between pyruvate cycling and glucose-stimulated insulin secretion (GSIS). *Proc. Natl. Acad. Sci. USA* **99,** 2708–2713.

MacDonald, M. J. (1995). Feasibility of a mitochondrial pyruvate malate shuttle in pancreatic islets. Further implication of cytosolic NADPH in insulin secretion. *J. Biol. Chem.* **270,** 20051–20058.

MacDonald, M. J., Fahien, L. A., Brown, L. J., Hasan, N. M., Buss, J. D., and Kendrick, M. A. (2005). Perspective: Emerging evidence for signaling roles of mito-chondrial anaplerotic products in insulin secretion. *Am. J. Physiol. Metab.* **288,** E1–E15.

Malloy, C. R., Sherry, A. D., and Jeffrey, F. M. H. (1990). Analysis of tricarboxylic acid cycle of the heart using ^{13}C isotope isomers. *Am. J. Physiol.* **259,** H987–H995.

Mandella, R. D., and Sauer, L. A. (1975). The mitochondrial malic enzymes. I. Submito-chondrial localization and purification and properties of the $NAD(P)^+$-dependent enzyme from adrenal cortex. *J. Biol. Chem.* **250,** 5877–5884.

McManus, M. T., and Sharp, P. A. (2002). Gene silencing in mammals by small interfering RNAs. *Nat. Rev. Genet.* **3,** 737–747.

Pongratz, R. L., Kibbey, R. G., Shulman, G. I., and Cline, G. W. (2007). Cytosolic and mitochondrial malic enzyme isoforms differentially control insulin secretion. *J. Biol. Chem.* **282,** 200–207.

Teller, J. K., Fahien, L. A., and Davis, J. W. (1992). Kinetics and regulation of hepatoma mitochondrial NAD(P) malic enzyme. *J. Biol. Chem.* **267,** 10423–10432.

INSULIN SECRETION FROM β-CELLS IS AFFECTED BY DELETION OF NICOTINAMIDE NUCLEOTIDE TRANSHYDROGENASE

Kenju Shimomura,[*] Juris Galvanovskis,[†] Michelle Goldsworthy,[‡] Alison Hugill,[‡] Stephan Kaizak,[*] Angela Lee,[‡] Nicholas Meadows,[‡] Mohamed Mohideen Quwailid,[‡] Jan Rydström,[§] Lydia Teboul,[‡] Fran Ashcroft,[*] *and* Roger D. Cox[‡]

Contents

1. Introduction	452
2. Gene-Driven ENU Screens	453
2.1. Denaturing high-performance liquid chromatography (DHPLC)	453
2.2. High-resolution DNA melting analysis	454
2.3. Rederivation of live mice	456
3. Generating BAC Transgenics	457
3.1. Identification of a BAC clone	457
3.2. BAC transgenesis	458
4. Isolation of Islets of Langerhans from Mouse Pancreas	460
4.1. Isolation methods for obtaining islets	461
4.2. Insulin secretion experiment from static incubation	463
5. RNA Interference (RNAi) Methodologies on Insulinoma Cell Lines	464
5.1. Insulinoma cell lines	464
5.2. Gene silencing using siRNA	464
5.3. Confirmation of siRNA specific knockdown	465
6. Measuring Intracellular Calcium	467
7. Measuring Hydrogen Peroxide in Mitochondria	468

[*] Henry Wellcome Center for Gene Function, Department of Physiology, Anatomy, and Genetics, University of Oxford, Oxford, United Kingdom
[†] The Oxford Centre for Diabetes, Endocrinology, and Metabolism, Churchill Hospital, Oxford, United Kingdom
[‡] MRC Harwell, Diabetes Group, Harwell Science and Innovation Campus, Oxfordshire, United Kingdom
[§] Biochemistry and Biophysics, Department of Chemistry, Lundberg Laboratory, Göteborg University, Göteborg, Sweden

Methods in Enzymology, Volume 457
ISSN 0076-6879, DOI: 10.1016/S0076-6879(09)05025-3

8. Imaging of Mitochondrial Membrane Potential Changes by
   Confocal Microscopy   471
   8.1. Confocal imaging of mitochondrial potential in
        pancreatic β-cells   472
   8.2. Analysis of confocal images representing the mitochondrial
        membrane potential   474
9. Measuring NNT Activity   475
   9.1. Microassay of NNT   475
   9.2. Assay considerations   477
10. Conclusions   477
References   478

Abstract

Nicotinamide nucleotide transhydrogenase (NNT) is an inner mitochondrial membrane transmembrane protein involved in regenerating NADPH, coupled with proton translocation across the inner membrane. We have shown that a defect in Nnt function in the mouse, and specifically within the β-cell, leads to a reduction in insulin secretion. This chapter describes methods for examining Nnt function in the mouse. This includes generating *in vivo* models with point mutations and expression of Nnt by transgenesis, and making *in vitro* models, by silencing of gene expression. In addition, techniques are described to measure insulin secretion, calcium and hydrogen peroxide concentrations, membrane potential, and NNT activity. These approaches and techniques can also be applied to other genes of interest.

1. INTRODUCTION

We identified the nicotinamide nucleotide transhydrogenase (*Nnt*) gene as a possible candidate for the insulin secretory defect in C57BL/6J mice by quantitative trait loci mapping in an F2 mouse intercross between C57BL/6J and C3H mice (Toye *et al.*, 2005). In order to validate this gene, we silenced it in an insulin secretory cell line and showed impairment of insulin secretion. We also identified two ENU alleles from the Harwell ENU DNA and sperm archive and demonstrated a glucose intolerant phenotype with an insulin secretory defect (Freeman *et al.*, 2006a). We hypothesized that nicotinamide nucleotide transhydrogenase (NNT) is required to generate NADPH needed to produce reduced glutathione in the mitochondria of the β-cell and that a lack of reduced glutathione for use by glutathione peroxidase (GPX) leads to an increase in reactive oxygen species (ROS)/Hydrogen peroxide. We proposed that this causes uncoupling via activation of uncoupling protein 2 (UCP2) and a reduction in membrane potential and a loss of ATP synthesis and thus a reduced ability of glucose metabolism to increase ATP/ADP ratios required for K_{ATP} channel closure and insulin secretion (reviewed in Freeman and Cox (2006) and Freeman *et al.* (2006b)). Subsequent work in insulinoma cells using siRNAs has demonstrated the requirement for UCP2 for the effect on insulin secretion and membrane potential (manuscript in preparation). In C57BL/6J mice, we demonstrated that replacement of the *Nnt* gene using a bacterial artificial chromosome (BAC) transgene was able to complement the defect in glucose tolerance and insulin secretion (Freeman *et al.*, 2006c).

This chapter describes the assays and techniques that we used to achieve this work.

2. Gene-Driven ENU Screens

Gene-driven screens of mutagenized mice represent a powerful tool in the identification of allelic variants (Coghill *et al.*, 2002). N-ethyl-N-nitrosourea (ENU) is a germline mutagen that induces point mutations randomly and can be used to provide a mutant mouse resource (reviewed by Justice *et al.*, 1999). Male mice, exposed to ENU were used to create two parallel archives of frozen sperm and DNA. An archive was prepared from over 7000 inbred mice that were F1 hybrids between BALB/c mutagenized males crossed with C3H females (Quwailid *et al.*, 2004, note that later portions of the archive are C57BL/6J × C3H F1s). Screening of the DNA archive can be used to identify an allelic series for any given gene of interest.

At present the archive is screened using high throughput mutation scanning methods (see Sections 2.1 and 2.2). Inevitably, these miss a small proportion of mutations, but have the advantage of speed and cost efficiencies. When the cost of DNA sequencing becomes low enough we hope to sequence the DNA archive for all coding regions and thus generate a catalogue of mutations. Once a mutation is identified through DNA screening the mutant mice can be rederived easily by *in vitro* fertilization (IVF) for subsequent phenotyping of live mice (see e.g., http://www.har.mrc.ac.uk/services/fesa/for a service or training). As ENU induces point mutations, it can be expected that the full range of functional changes in any gene might be uncovered, including amorphs (null), hypomorphs, hypermorphs, and neomorphs. The DNA archive is available to academics as a community resource; alternatively, MRC Harwell is now offering full training to visiting scientists to perform gene-driven screens within our mutation detection facility (see: http://www.har.mrc.ac.uk/services/dna_archive/).

2.1. Denaturing high-performance liquid chromatography (DHPLC)

Initial screening of the Harwell ENU DNA archive was performed using DHPLC on the Transgenomic WAVE System (Coghill *et al.*, 2002; Quwailid *et al.*, 2004). Primers for polymerase chain reaction (PCR) amplification are designed to flank exons with at least a 20-bp gap to allow complete sequencing of the region of interest including splice sites. Optimal fragment size is 150–450 bp. To increase throughput, 12.5 ng of archive DNA from each of four individual mice is pooled together. Conditions for PCR are as follows: 200 μM dNTPs, 2.5 mM MgCl$_2$, 0.2 μM forward and reverse primers, 0.45 U Applied Biosystems AmpliTaq Gold, 0.025 U Stratagene Pfu Turbo polymerase and 1× AmpliTaq Gold buffer. A master mix of all PCR reagents is made and transferred to individual 96-well microtiter plates containing 50 ng of DNA pools (final PCR volume 25–50 μl). Touchdown PCR is performed on an MJ Research PTC 225 thermocycler using the following cycling conditions: 95 °C for 10 min, then 13 cycles of 94 °C for 20 s, primer annealing temperature plus 7 °C for 1 min with a decrease of 0.5 °C/cycle, and 72 °C for 1 min, followed by 19 cycles of 94 °C for 20 s, primer annealing temperature for 1 min, 72 °C for 1 min, and finally 72 °C for 5 min and a 15 °C hold. Following amplification, heteroduplexes are produced using the following conditions: 95 °C for 4 min, followed by 45 cycles of 93.5 °C for 1 min with a reduction of 1.5 °C/cycle, finally followed by holding at 25 °C. When partially denatured, PCR products containing heteroduplexes elute quicker from the WAVE column and are detected by a UV monitor as additional peaks appearing before the homoduplex wildtype peak (Fig. 25.1). Following identification of an interesting pool containing potential heteroduplexes, the individual DNAs making the pool are tested. The PCR amplification products containing heteroduplexes are purified using the QIAquick PCR purification kit (Qiagen) and then sent for sequence confirmation.

As DHPLC requires samples to be partially denatured to spot heteroduplexes, for some exons multiple temperatures are required to detect mutations across the entire PCR product. On a two-plate machine, two plates with two methods can be run in 24 h. Although a reliable method with ~95% detection efficiency (Dobson-Stone *et al.*, 2000) there are potentially higher throughput methods now available (see Section 2.2).

2.2. High-resolution DNA melting analysis

The primary mode of screening the Harwell DNA archive is high-resolution DNA melting analysis performed on the Idaho Technology LightScanner. High-resolution DNA melting analysis has been successfully used for several years to perform mutation scans and genotyping (Wittwer *et al.*, 2003). PCR is performed in the presence of the double-stranded DNA binding dye LCGreen. The LCGreen saturates heteroduplexes and they can be distinguished from the normal melt curve by an early decrease in fluorescence (Fig. 25.2).

Primer pairs for PCR amplification are designed using Idaho Technology's LightScanner Primer Design Software. Fragment size ranges between

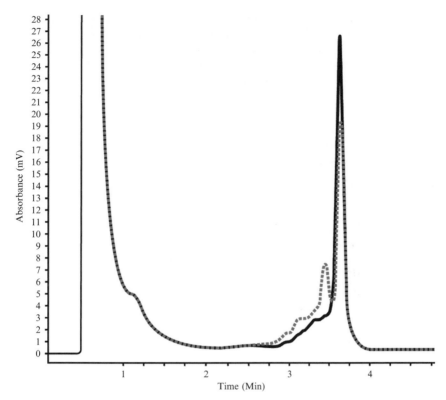

Figure 25.1   Absorbance trace of two pooled PCR products on the WAVE. Samples containing an ENU mutation cause the formation of heteroduplexs (dotted line), which are eluted from the WAVE column quicker than homoduplexes (solid line) and so appear as additional peaks appearing before the main sample peak. Natural polymorphisms between the inbred strains used to create the archive will also produce traces containing multiple peaks and may reduce the ability to detect an ENU induced mutation.

150 and 550 bp with primers flanking the region of interest by 20 bp to allow successful sequencing of the entire region. To increase throughput, 2.5 ng of archive DNA from four individual mice is pooled together. Conditions for PCR are as follows: 1×LCGreen plus dye and 1×HotShot Mastermix, 0.2 μM forward and reverse primers. Master mix of all PCR reagents are made, transferred to individual 96-well Framestar plates containing 10 ng of DNA pools (final PCR volume 10 μl) and overlaid with 20 μl of mineral oil. PCR is performed using a GRI G-Storm GS4 Thermal Cycler using the following cycling conditions: 95 °C for 2 min, then 44 cycles of 95 °C for 30 s, primer annealing temperature for 30 s, 72 °C for 30 s. Following initial PCR, heteroduplexes are produced by heating samples to 95 °C and then rapidly cooling to 15 °C. Samples are then centrifuged at 3000 rpm for 3 min before running on the LightScanner. Following identification of an interesting pool,

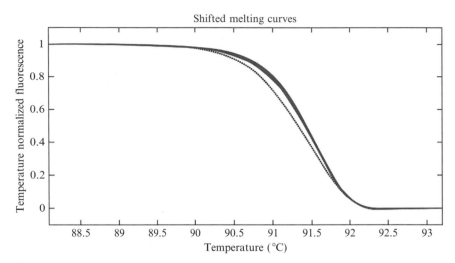

Figure 25.2   Melt profile of an ENU induced mutation on the LightScanner. LCGreen saturates heteroduplexes (dotted line, left) and causes an early decrease in fluorescence so the melt curve shifts to the left compared to homoduplexes (solid line, right). PCR must be optimal and the melt profile of the exon must be narrow or variants may not be detected.

the individual DNAs making the pool are then tested alongside at least 12 wildtype DNA controls. The PCR amplification products containing hetero-duplexes, are purified using QIAquick PCR purification kit (Qiagen) and sent for sequence confirmation.

This method is very rapid as once the PCR reactions are complete plates can be analyzed at the rate of 1 plate every 8 min per machine. Unfortunately, plate loading is not automated yet. We estimate that this method detects >90% of mutations.

2.3. Rederivation of live mice

Fifty nine new mutant lines have to date been successfully generated from the Harwell ENU archive. Embryos are either shipped at the two-cell stage to the user's facilities or can be rederived in-house for phenotyping. F1 archived mice carry approximately 30 potentially functional mutations (Keays *et al.*, 2006). The likelihood of more than one mutation confounding a given phenotype is very low; however, backcrossing of mice to C3H rapidly removes the unlinked mutations. After backcrossing for several generations homozygotes can be bred and if desired coding region surrounding the selected gene should be sequenced to confirm there are no mutations in closely linked genes although this is unnecessary particularly when several alleles are available to validate the phenotype (Keays *et al.*, 2007).

3. GENERATING BAC TRANSGENICS

We originally identified *Nnt* as a candidate gene in a Quantitative Trait Locus (QTL) mapping experiment (Toye *et al.*, 2005). Having identified it as a strong functional candidate, it was necessary to prove that it accounted for at least some of the glucose intolerance and insulin secretory phenotype in C57BL/6J mice. The effect was originally measured in an F2 C57BL/6J × C3H/HeH genetic environment. There are several ways to approach this problem and we decided to supplement the deleted copy of *Nnt* in C57BL/6J mice with a complete functional copy from the 129 inbred mouse. We achieved this by identifying a BAC containing the entire *Nnt* gene and injecting this into the pronucleus of a fertilized C57BL/6J egg (Freeman *et al.*, 2006c). The resulting offspring that carried transgenic integration events were examined for Nnt expression and phenotyped.

3.1. Identification of a BAC clone

To identify a BAC containing all 21 exons of *Nnt*, the RPCI-22 (129S6/SvEvTac) Mouse BAC Library (Children's Hospital Oakland Research Institute) was screened.

3.1.1. Generation of probe

Exons 4 and 17 of *Nnt* were PCR amplified to produce products of 443 and 462 bp, respectively. To make a probe, PCR products are purified using the QIAquick PCR purification kit (Qiagen). The purified PCR products are then radioactively labeled with [α-32P]dCTP at 3000 Ci/mmol using a Prime-It II Random primer labeling kit (Stratagene). Illustra NICK columns (GE Healthcare, Amersham, UK) are then used, to remove any unincorporated radiolabeled nucleotides from the labeling reaction. All kits are used according to manufacturer's instructions.

3.1.2. Hybridization

Library membranes (filters) are wet in a minimum volume of Church Buffer (10 g BSA, 2 ml 0.5 M EDTA, 500 ml 1 M NaHPO$_4$ (pH 7.2), 350 ml 20% SDS, up to 1 l with H$_2$O). Nylon mesh is placed between filters and filters placed in hybridization bottles with 25 ml Church Buffer. Filters are then prehybridized for 1 h at 65 °C after which one of the labeled probes is denatured at 100 °C for 5 min and then added. Filters are then left to hybridize overnight (16–18 h) at 65 °C.

After hybridization, filters are washed four times with wash solution 1 (5 g BSA, 1 ml 0.5 M EDTA, 40 ml 1 M NaHPO$_4$ (pH 7.2), 250 ml 20% SDS, up to 1 l with H$_2$O) at 65 °C for 20 min each wash. Filters are then washed with wash solution II (2 ml 0.5 M EDTA, 40 ml 1 M NaHPO4 (pH 7.2), 50 ml 20% SDS, up to 1 l H_2O) at 65 °C for 20 min until the level of radioactive counts per minute (cpm) removed remained constant.

Filters are then rinsed in water and wrapped individually in cling film and exposed to film for 2–72 h at −80 °C. Positive clones are then identified.

Filters are then stripped by incubating them at 65 °C in a shaking water bath prior to repeating the hybridization protocol again with the second labeled probe.

3.1.3. Identification of full length NNT BAC clone

Twenty two clones positive for both exons 4 and 17 were identified. To validate clones, each is streaked onto LB chloramphenicol (20 μg/ml) agar plates and incubated overnight at 37 °C. Individual colonies are then picked from each of these plates and PCR amplified directly for exons 1 and 21 of *Nnt*. One of the 22 clones identified, BAC RP22-455H18, contained both exons 1 and 21.

3.1.4. DNA extraction

A positive colony is inoculated into 10 ml of LB chloramphenicol (20 μg/ml) broth and left shaking overnight at 37 °C. The culture is then spun for 10 min at 2000 rpm and the supernatant poured off. The pellet is resuspended in 300 μl of 50 mM glucose, 25 mM Tris, 10 mM EDTA. Then 600 μl of 1% SDS, 0.2 M NaOH is added and the sample inverted several times (lysis for no longer than 5 min). Following this, 500 μl of 7.5 M Ammonium acetate is added and the sample inverted several times and then left on ice for 10 min, inverting at intervals. The sample is then spun at 13,000 rpm for 20 min and the supernatant poured into a fresh tube and spun for a further 10 min at 13,000 rpm. The supernatant is then poured into another fresh tube and 700 μl isopropanol added. The sample is inverted several times and spun for 20 min at 13,000 rpm. The supernatant is discarded and the pellet washed with 500 μl of 70% ethanol and then spun for 5 min at 13,000 rpm. The supernatant is discarded and the pellet left to air dry for 5–10 min after which it is resuspended in 100 μl 1× TE. RNase is added to a final concentration of 100 μg/ml.

All 21 exons of *Nnt* were then PCR amplified. BAC RP22-455H18 was shown to contain all 21 exons of the *Nnt* gene, as well as considerable 5′ and 3′ flanking intergenic DNA.

3.2. BAC transgenesis

A mouse transgenic model overexpressing Nnt was created to perform a complementation test of the defect in insulin secretion and glucose intolerance phenotypes observed in male C57BL/6J mice. The chosen strategy for transgenesis is the pronuclear injection of a BAC. This is because the large cargo capacity of such vectors is likely to include the sequences that support the regulation of the expression of the endogenous *Nnt* gene, as well as coding information.

BAC DNA is prepared either as described by these workers (Freeman *et al.*, 2006c) or currently with a method based on an earlier publication (Carvajal *et al.*, 2001). Great care is taken to optimize every step of the preparation: bacteria are isolated on freshly prepared plates. A colony is transferred into a 5-ml preculture, which is subsequently diluted into a 250-ml culture. The BAC DNA is extracted employing the HiSpeed Maxiprep kit (Qiagen) with the following modifications. The bacterial pellet is resuspended in 50 ml of ice-cold P1 buffer and incubated 10 min on ice. Then 50 ml of freshly prepared P2 buffer is added. The preparation is mixed by inversion and incubated 5–10 min at room temperature. This is followed by 50 ml of ice-cold N3 buffer and mixed by inversion. The tube is laid flat in a bed of ice with very gentle tilting for 20 min. After centrifugation, the supernatant is filtered on several layers of medical gauze and loaded on a column and washed as per manufacturer instructions. The DNA is eluted employing Elution Buffer prewarmed to 55 °C and precipitated as per manufacturer instructions. BAC DNA is resuspended overnight with very gentle agitation in BAC microinjection buffer (BAC MIB, 10 mM Tris–HCl at pH 7.5, 0.1 mM EDTA at pH 8.0, 100 mM NaCl, filtered through 0.2 μm). The DNA solution is dialyzed overnight against BAC MIB employing a Slide-A-Lyzer MINI dialysis unit (Pierce) and quantified by spectrophotometry or Nanodrop. Preparations yielding less than a total of 30 μg before dialysis are discarded.

Circular BAC DNA solution, no older than a month, is diluted with BAC MIB to a concentration of 1–2 ng/μl on the day of injection. Pronuclear injection of C57BL/6J embryos and rearing of potential founders are performed according to standards protocols (Nagy *et al.*, 2003).

Tail biopsies are taken from potential founders and processed employing a DNeasy extraction kit (Qiagen), according to the manufacturer. As the C57BL/6J strain was shown to lack exons 7–11 of *Nnt*, PCR assays amplifying each of these exons were used to screen for potential founders obtained by pronuclear microinjection. It is noteworthy that a great proportion of the transgenic insertions obtained with the microinjection of the *Nnt* BAC showed at least some degree of rearrangement so one or several of exons 7–11 of *Nnt* were missing from the transgene. In order to ascertain the integrity and functionality of the transgene in the positive founders, the expression of *Nnt* transcripts and protein was checked by RT-PCR and Western blot analysis, respectively (Freeman *et al.*, 2006c).

Once a transgenic line is established and validated, sperm or embryos are cryopreserved at an early generation. This is essential as loss of transgene activity through silencing or sequence rearrangements may occur in

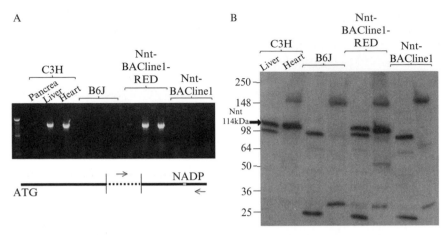

Figure 25.3 Expression of NNT in mice transgenic for the *Nnt* BAC transgene. (A) RT-PCR analysis of *Nnt* mRNA from C3H, C57Bl/6J, a transgenic line rederived from the original rescue line, and the original rescue line. One of the RT primers is located within the deleted region of the transcript (exons 7–11) and the other at the 3′ end including the NADP motif. These primers amplify a 1.3-kb product from C3H but not in Bl/6J. The BAC rescue line re-derived from frozen embryos stored shortly after the line was created (Nnt-BACline1-RED) also expresses *Nnt* at level similar to C3H. In contrast, the original line (Nnt-BACline1) that had been maintained by breeding over several years has lost *Nnt* expression, possibly due to genome rearrangement. (B) Proteins were separated by SDS-PAGE and the proteins transferred to a membrane for immunoblotting with a rabbit anti-NNT polypeptide antibody. C3H and "Nnt-BACline1-RED" mice express NNT as expected, whereas Bl/6J and "Nnt-BACline1" mice do not express NNT. Note that the polyclonal antibody detects some nonspecific bands in both C3H (intact) and Bl/6J (deleted) tissues. (See Color Insert.)

later generations. Figure 25.3 illustrates a *Nnt* BAC line rederived from frozen embryos after loss of expression was observed in the stock being maintained on the shelf over several years.

4. ISOLATION OF ISLETS OF LANGERHANS FROM MOUSE PANCREAS

To determine the phenotype of a mouse model, one can measure plasma glucose or insulin levels in response to a glucose challenge *in vivo* (for protocols, see EMPRESS: http://empress.har.mrc.ac.uk/browser/). However, a more sensitive description of insulin secretion can be obtained *in vitro* using isolated islets.

The cells of the endocrine pancreas are grouped into clusters known as islets of Langerhans. There are considered to be more than 10,000 islets per pancreas in rodents (Bonner-Weir and Orci, 1982). The size of an islet is 40–400 μm in diameter. Although islets are scattered throughout the pancreas, the majority are found in the body and tail of the organ, with only one-third being found in the head of the pancreas (Orci, 1982). There are four major endocrine cells in the islet: insulin-producing β-cells, glucagon-producing α-cells, somatostatin-producing δ-cells, and pancreatic polypeptide-producing PP-cells. In rodents, the insulin-producing β-cells constitute about 60–70% of islets cells. They form the central core of the islet with the other non-β-cells forming a mantle around them (Erlanden et al., 1976; Orci and Unger, 1975).

In pancreatic β-cells, insulin secretion is regulated by K_{ATP} channels. At substimulatory blood glucose levels, K_{ATP} channels in the β-cell membrane are open and efflux of K^+ ions through the channel generates a negative membrane potential. As a consequence, voltage-dependent Ca^{2+} channels (VDCC) remain closed (Ashcroft and Rorsmann, 1989). However, when the blood glucose concentration is increased by food intake, β-cell glucose uptake and metabolism is stimulated, producing changes in the cytosolic concentration of adenine nucleotides that lead to closure of the K_{ATP} channel. This induces a membrane depolarization which opens the VDCC, allowing Ca^{2+} influx into the cell. The resulting increase in the intracellular Ca^{2+} concentration ($[Ca^{2+}]_i$) induces exocytosis of insulin granules (Henquin, 2000).

Although single β-cells (or cultured β-cells such as HIT-T15 and β HC-9) are capable of secreting insulin in response to elevated level of glucose, it is known that the insulin release is enhanced when β-cells have contact with other cells in the islet (Bosco et al., 1989; Saito et al., 2003; Shimomura et al., 2004; Soria et al., 1991). Both electrical coupling between β-cells and the paracrine effects of hormones released from non-β-cells may contribute to the enhancement of insulin secretion (Kantengwa et al., 1997; Meda et al., 1983).

4.1. Isolation methods for obtaining islets

There are several ways to isolate islets from the pancreas. The simplest is the microdissection technique which involves physically cutting out islets with a sharp needle (30–31 G needle) under a binocular microscope (Henquin and Meissner, 1984). Although the microdissection technique may avoid damage caused by the use of enzymes, it only isolates a very limited number of islets (10–50) as well as the risk of contamination with surrounding tissues.

A better technique is to digest the pancreas using enzymes such as liberase. Liberase is a blend of highly purified type I and type II collagenases and proteases that is specifically designed for rodent islet isolation. Although enzymes can damage the islet, this can be minimized by using the proper enzyme concentration and by prompt and swift washing of digested tissues. This technique can provide larger numbers (150–300) of high quality islets.

In the liberase digestion technique, Hank's buffer with 0.2% BSA (Hank's BSA) is used as the media throughout the procedure.

In order to release the maximum number of high quality islets, it is necessary to disperse the liberase solution (at 0.5 mg/ml) throughout the pancreas. This can be achieved by injecting 2 ml of solution, using a syringe fitted with a 30½ G needle, into the bile duct after obstructing the exit to the duodenum by ligating the bile duct–duodenum junction with surgical string or by clamping the Vater papilla in the duodenum using a surgical clamp (McDaniel *et al.*, 1983; Wollheim *et al.*, 1990). The needle is inserted where the common hepatic duct and cystic duct join as illustrated in Fig. 25.4. When the injection is successful, the pancreas will inflate from head to tail. The inflation of the pancreas must be complete or the islet yield will be poor.

Once the injection is complete, the inflated pancreas should be carefully dissected away from other tissues, placed in 2 ml of liberase solution and kept at 37 °C for 20 min. After 20 min, the digested pancreas should be washed twice by adding 20 ml of ice cold Hank's solution and centrifuging at 2000 rpm for 2 min. After washing, the islets can be hand picked under

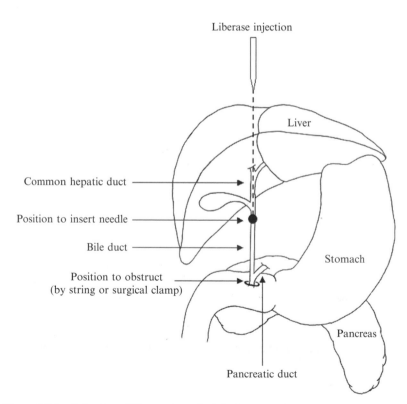

Figure 25.4   Injection of liberase into the bile duct in order to inflate the pancreas.

the microscope. The purpose of hand picking is to exclude non-islet tissues from the digested pancreas and obtain pure islets. It can be done in two ways. One way is to suck the islets up into a sterile pipette tip while visualizing them under a microscope. However, as this method often aspirates other tissues together with islets, the procedure needs to be repeated several times to obtain pure islets. Another way is to use a wire islet picker. This can be constructed by making a circle (with a diameter of ~1 mm) at one end of a length of wire and gluing the other end to a rod so that it can be easily manipulated. By using a wire islet picker like a spoon, it is possible to physically pick up one islet at a time.

4.2. Insulin secretion experiment from static incubation

The simplest and most effective way to measure glucose-stimulated insulin secretion (GSIS) from islets is by the static incubation technique (Freeman et al., 2006a; Tanaka et al., 2002). Various concentrations of glucose in Krebs-Ringer buffer (KRB) solution are used to stimulate insulin secretion. All solutions should contain 0.1% BSA (to which insulin will complex) to prevent insulin sticking to the incubation tubes. The islets should be incubated in a low-glucose solution (2 mM) for 30 min prior to the experiment. This will ensure that insulin secretion is not dictated by the glucose concentration to which the islets were previously exposed (either in the animal or the culture solution)

A 24-well plate can be used for the static incubation and each well should contain 2 ml of experimental solution containing either glucose at 2 mM, 10 mM, 20 mM, or (if desired) a sulphonylurea such as tolbutamide. Five islets are then placed in each well and incubated for 1 h at 37 °C.

Immediately after incubation, the supernatant (containing the secreted insulin) should be aspirated and centrifuged at 3000 rpm for 1 min to remove any cells. Insulin in the solution can then be measured immediately or the solution can be stored at -20 °C until desired.

The islets from each incubation tube are also collected under a microscope, transferred to an acidified-ethanol solution (95% ethanol and 5% acetic acid) and left overnight at -20 °C. This lyses the islets (and β-cells) completely and releases insulin into the solution, enabling the insulin content to be measured. Since the insulin content level is 20–200 times higher than the secreted insulin level, it is necessary to dilute these samples before measurement.

Insulin in the samples can be measured by an insulin enzyme linked immunosorbent-assay (ELISA assay) (Linco, Rat/mouse ELISA kit) or an RIA assay using the 5-μl protocol according to manufacturer's instructions. However, because the size of each islet varies, it is necessary to normalize the secreted insulin to insulin content in the same islets. Prior to the ELISA assay, supernatants are usually diluted 1 in 10–20, and content supernatants are diluted 1 in 200–400.

5. RNA INTERFERENCE (RNAi) METHODOLOGIES ON INSULINOMA CELL LINES

RNA interference (RNAi), a gene silencing technology extensively used in *Caenorhabditis elegans*, was utilized to study loss of function of Nnt in the insulin secreting cell line MIN6. Gene specific siRNA oligo duplexes are designed for the gene of interest and transfected into cells where they are incorporated into a ribonucleoprotein complex called RISC (RNA induced silencing complex). RISC mediates the unwinding of the siRNA duplex, by ATP-dependent RNA-helicase activity into a single stranded siRNA. The antisense strand of the siRNA duplex guides RISC to the homologous mRNA allowing it to bind in a sequence-specific manner, this mediates target mRNA endonucleolytic cleavage. This cleaved mRNA is then recognized by the cell as aberrant and destroyed, preventing translation resulting in the specific silencing of the target mRNA.

5.1. Insulinoma cell lines

We have used two insulin secreting cell lines (MIN6 and INS-1) for gene silencing in various projects. MIN6 is a mouse insulinoma cell line derived from a transgenic mouse expressing the large T-antigen of SV40 in pancreatic β-cells. Glucose stimulated insulin secretion occurs progressively in this cell line with the amount of insulin secreted at 25 mM glucose being 6–7-fold higher than that obtained at 5 mM glucose. This cell line was obtained at passage 40 and is maintained at high confluence passaged every 3–4 days in DMEM containing 25 mM glucose, supplemented with 15% heat inactivated fetal bovine serum in humidified 5% CO_2, 95% air at 37 °C. INS-1 is a rat cell line derived from an X-ray induced rat transplantable insulinoma (Asfari *et al.*, 1992), which retains partial glucose responsiveness, with an optimal secretion observed at 11.2 mM glucose which decreases slightly at higher levels (16.7 mM). We have used this cell line to silence *Sox4* gene expression using siRNA (Goldsworthy *et al.*, 2008). This cell line was obtained at passage 32 and is maintained in RPMI (11.11 mM glucose concentration) supplemented with 10 mM HEPES, 2 mM L-Glutamine, 1 mM sodium pyruvate, and 50 μM β-mercaptoethanol in humidified 5% CO_2, 95% air at 37 °C.

5.2. Gene silencing using siRNA

siRNA oligos were designed and synthesized by Eurogentec on their exclusive siRNA design software, with a total of four siRNA duplexes used. Each siRNA was 19-nucleotide long with two DNA overhangs at the 3' end and a Cy3 tag on the 5' end of the sense strand. The presence of a

Cy3 tag enabled transfected cells to be easily identified and transfection efficiency to be quantified. We used two independent *Nnt* siRNAs targeted to nonoverlapping regions of the gene, an siRNA duplex specific to glucokinase (*Gck*) as a positive control (*Gck* is a key glycolytic enzyme which when absent abolishes glucose stimulated insulin secretion) and a scrambled (nonsense) siRNA as a negative control. A number of companies now provide predesigned siRNAs that are ready to use.

Prior to transfection, cells are plated at a density of 1×10^5 and cultured on 24-well flat bottomed plates (Scientific Laboratory Supplies, Nottingham, UK) in their cell specific media and grown for 24–48 h in humidified 5% CO_2, 95% air at 37 °C. A cationic lipid mediated transfection (lipofectamine) is used to deliver the synthetic siRNA oligo duplex into the target cells. Incubation of lipofectamine and siRNA duplexes in serum-free media results in formation of unilamellar and multilamellar liposomes. The positively charged cationic lipids spontaneously interact with the negatively charged backbone of the siRNA to form complexes. These complexes then interact with the negatively charged cell membrane resulting in the delivery of the siRNAs into the cell. For each siRNA duplex, a lipofectamine complex is set up by the addition of 20 μl of lipofectamine to 2 ml of serum-free media, which is mixed gently by inversion and incubated at room temperature for 5 min to allow the liposomes to begin to form. After this incubation period, siRNA duplexes are added at a final concentration of 50 n*M*, mixed gently by inversion and incubated at room temperature for a further 20 min. Figure 25.5 illustrates the experiment schematically.

The 24-well plates of cells are prepared for transfection by aspiration of media and washing twice with PBS. Then, 200 μl of the previously prepared lipofectamine/siRNA complex is added to the appropriate wells. To avoid shearing of the complexes, the solution should be pipetted slowly and carefully down the side of each well with a 1-ml pipette and tip. Eighteen wells per siRNA should be transfected in order to perform an insulin secretion assay (in triplicate), extract RNA (to confirm silencing by real-time quantitative reverse transcription PCR) and protein (to confirm silencing by Western blotting). Transfected plates are incubated at 37 °C 5% CO_2, 95% air for 7–8 h after which time the wells should be topped up with an additional 200 μl of serum-containing media and incubated for a total of 24 h. After 24 h the cells from 6 wells for each siRNA duplex are trypsinized and the resulting cell pellet divided into two aliquots for RNA and Protein extraction and stored at -70 °C until use (Fig. 25.5).

5.3. Confirmation of siRNA specific knockdown

In order to confirm the observed loss of insulin secretion was due to loss of silencing of *Nnt* specifically both mRNA and protein levels were measured utilizing quantitative RT-PCR and Western blotting. Total RNA is

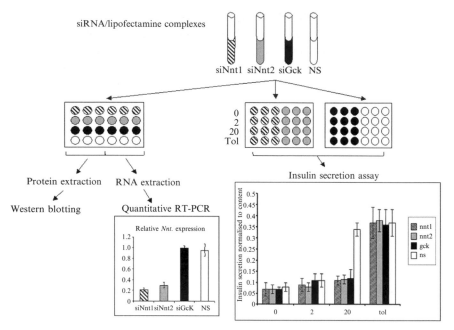

Figure 25.5 Schematic of the siRNA protocol. MIN6 cells were treated with four different silencing RNAs, *Nnt*1, *Nnt*2, *Gck*, and a nonsense siRNA.

extracted from silenced cells using a Qiagen RNeasy mini-kit (Qiagen, Crawley, UK) according to the manufacturers instructions; the optional on column DNase digestion is also performed. Concentration and quality of total RNA is measured on a spectrophotometer (Nanodrop) and cDNA is then generated by Superscript II enzyme (Invitrogen) and analyzed by quantitative RT-PCR using the taqman system based on real-time detection of accumulated fluorescence (ABI Prism 7700, Perkin-Elmer). Gene expression is normalized relative to the expression of glyceraldehyde-3-phosphate dehydrogenase (GAPDH). Inventoried Taqman probes with FAM tags can be purchased from Applied Biosystems. A student's *t* test was used to measure significance between nonsense siRNA transfected cells, *Gck* siRNA transfected cells, and *Nnt* siRNA transfected cells when amplifying for either *Gck* or *Nnt* gene expression.

Prior to Western blotting protein is extracted from stored siRNA transfected cells by the addition of 50 μl CelLytic^{TM} *M* Cell Lysis Reagent (Sigma) containing complete mini proteinase inhibitor cocktail (Roche), cells are incubated on ice for 60 min prior to centrifugation at 13,000 rpm for 15 min at 4 °C. The supernatant is removed and protein quantified and stored at -20 °C until use. For each well, 50 μg of total protein is heated in sample buffer (Laemmli Sample Buffer, BioRad) for 3 min at 100 °C prior to loading on 4–15% Tris–HCl Ready Gels (BioRad) and size separated by electrophoresis (100 V for 1 h). A SeeBlue (Invitrogen) protein ladder is included to allow molecular weights to be estimated. Proteins are transferred to Hybond-P membrane (Amersham, GE Healthcare) using a transblot system (Biorad) and hybridized with antibodies to GCK (Santa Cruz) and NNT (custom made). Bands are visualized on ECL film using ECL plus Western blotting detection system (Amersham, GE Healthcare) according to the manufacturers instructions. As expected Nnt protein is present in nonsense and *Gck* transfected MIN6 cells but not *Nnt* siRNA transfected cells. GCK protein was detected in nonsense and *Nnt* siRNA treated cells but not *Gck* siRNA transfected cells. Membranes were reprobed with a mouse actin monoclonal antibody (Chemicon International) as a protein loading control.

Once the knockdown of the target gene is confirmed (usually by qRT-PCR alone), cells may be tested for insulin secretion as described in Section 4.2. Cells used for confirming siRNA knockdown and cells for testing insulin secretion should be harvested at the same time to ensure that the knockdown reflects the change in insulin secretion.

6. MEASURING INTRACELLULAR CALCIUM

To measure intracellular calcium concentration in pancreatic islets the ratiometric Ca^{2+} probe fura-2 is used. The dye is introduced into the islets at room temperature by incubation in 3 μM of the lipophilic, cell-permeant acetoxymethyl ester (fura-2 AM), dissolved in EC Buffer (138 mM NaCl, 5.6 mM KCl, 1 mM MgCl$_2$, 2.6 mM CaCl$_2$, 5 mM HEPES, pH 7.4 with NaOH). The nonionic detergent pluronic acid is used at a dilution of 0.01% to facilitate uptake of fura-2 AM. After 20 min, an islet is transferred into a perfusion chamber mounted on the stage of an Axiovert 200 microscope (Carl Zeiss Jena, Germany) where it is held in place by a suction glass pipette. To allow for complete de-esterfication of intracellular AM ester and to remove external dye, the islet is perfused with warm EC buffer (33–35 °C) for 15 min. During imaging, the islet is continuously perfused with warm EC buffer (33–35 °C), containing different glucose concentrations or tolbutamide.

We used a fluorescence system (Photon Technology International) for ratiometric measurements with dual excitation at 340 and 380 nm. Emission is recorded at 510 nm. Pancreatic islets are imaged at two frames/s and fluorescence intensity (F) is detected with a photomultiplier tube. Ca^{2+} signals are presented as the ratio of F at different excitation wavelengths (F340/F380). Data are analyzed using FeliX32 Analysis software. Examples of calcium measurement on wild type cells are shown in Fig. 25.6.

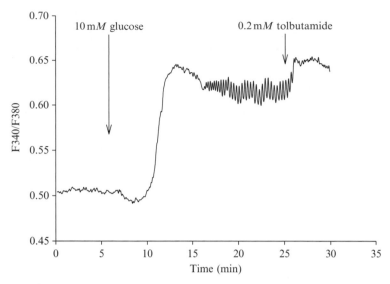

Figure 25.6 Typical calcium transient of a pancreatic islet stimulated with 10 mM glucose. Pancreatic islets exhibit a robust elevation of [Ca^{2+}]i when stimulated with glucose. This calcium peak occurs after a delay which can last for a couple of minutes. There is usually a decrease in [Ca^{2+}]i immediately before the calcium peak which is probably caused by Ca^{2+} sequestration in the ER driven by metabolism (Chow *et al.*, 1995; Tengholm *et al.*, 2001). Oscillations, which may follow the sharp rise in [Ca^{2+}]i, have typically three components: a faster component with a period of tens of seconds (as seen in the figure), a slower component with a period of 4–6 min and a combination of both, which shows faster oscillations superimposed on the slower plateaus (Bertram *et al.*, 2007). Application of tolbutamide, which blocks the K$_{ATP}$ channel, triggers a robust increase in [Ca^{2+}]i.

 ## 7. MEASURING HYDROGEN PEROXIDE IN MITOCHONDRIA

Dichlorodihydrofluorescein (DCF) and its derivatives have been used to study free radical production in a variety of systems. However, DCF has several significant disadvantages. Firstly, the dyes are not specific, that is, they are sensitive to multiple types of ROS. Secondly, they cannot be targeted to specific intracellular compartments. Thirdly, and most importantly, they can produce ROS themselves on photoexcitation. An alternative to DCF is HyPer (Evrogen, Russia), which does not suffer from these disadvantages.

HyPer is a genetically encoded fluorescent protein capable of specifically detecting intracellular hydrogen peroxide (H$_2$O$_2$) (Belousov *et al.*, 2006). It consists of a yellow-fluorescent protein (YFP) that has been inserted into the regulatory domain of the H$_2$O$_2$-sensing transcription factor OxyR, expressed by *Escherichia coli*. HyPer does not cause artificial ROS generation and can be used for detection of rapid changes of H_2O_2 concentrations within the physiological range. When the appropriate signal peptide is attached, HyPer can be expressed in different cell compartments. For example, addition of a mitochondrial targeting sequence (pHyPer-dMito) enables HyPer to be specifically expressed in mitochondria.

HyPer displays two excitation peaks (420 and 500 nm) and one emission peak (516 nm). Upon exposure to H_2O_2, the excitation peak at 420 nm decreases in proportion to the increase in the excitation peak at 500 nm. These spectral characteristics render it suitable for ratiometric measurements, thereby eliminating false results caused by photobleaching or differences in HyPer expression.

For detection of H_2O_2 in mitochondria, the insulinoma cell line MIN6 is transfected with a mammalian expression vector encoding mitochondria-targeted HyPer (Evrogen) using Lipofectamine (Invitrogen). The transfection protocol is carried out according to the manufacturer's instruction. HyPer fluorescence can be observed within 12 h of transfection and shows no significant decline for about 7 days.

Before imaging, cells are incubated for 1 h in EC buffer with 7% CO_2 and 93% air at 37 °C. During the experiment, cells are continuously perfused with EC Buffer at 37 °C and imaged on 35 mm Fluoro Dishes (World Precision Instruments) using an epifluorescence microscope (Nikon, Eclipse, TE 2000-U) with a CFI Super Fluor $40\times/1.3$ oil objective (Nikon). An IonOptix fluorescence system (Boston, Massachusetts) for ratiometric measurements with a dual excitation at 480/20 nm and 405/40 nm and emission at 530/40 nm filter (Chroma Technology) is used. Data are collected with a CCD camera (IonOptix). Cells are excited at one frame/3 s or one frame/10 s for experiments lasting for 35 min.

Using the technique described above, H_2O_2 levels in MIN6 cells can be measured in response to glucose stimulation. Within minutes of the addition of 20 mM glucose a steady increase in H_2O_2 is observed as a result of glucose metabolism (Fig. 25.7).

HyPer can also be used to measure H_2O_2 levels in response to glucose in MIN6 cells with altered levels of Nnt expression. Previous results suggest that Nnt is involved in detoxification of ROS (Freeman *et al.*, 2006a). MIN6 cells are cotransfected with pHyPer-dMito, nonsense and anti-*Nnt* siRNA duplexes (Section 5) in 24-well cell culture plates. Then, 24 h later ROS levels in response to glucose (0, 2, 20 mM) and menadione (50 μM) are measured. Fluorescence is measured for 30 min on a Fluostar OPTIMA (BMG Labtech), 485 nm excitation and 520 nm emission. Reduced expression of Nnt results in an increase in ROS detection in response to glucose but not with menadione (Fig. 25.8). Menadione preferentially generates free radicals inside the mitochondrion and the effects are too rapid for Nnt to assist in the detoxification of ROS accumulation.

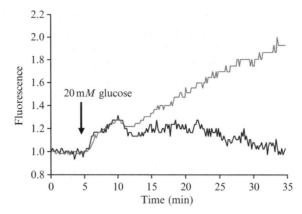

Figure 25.7 Changes in mitochondrial H_2O_2 concentration in si*Nnt* silenced MIN6 cells (70% knockdown efficiency) stimulated with 20 m*M* glucose added at 5 min. Representative traces of siRNA transfected MIN6 cells (grey) and nonsense siRNA transfected MIN6 cells (black) are depicted.

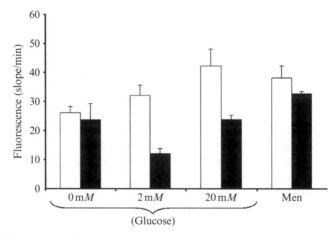

Figure 25.8 MIN6 cells cotransfected with anti-*Nnt* siRNA duplexes (white bars) or nonsense siRNA duplexes (black bars) and pHyPer-dMito were measured for ROS following treatment with glucose or menadione. Cells with reduced *Nnt* mRNA expression have higher levels of ROS after addition of 2 and 20 m*M* glucose. The experiment was performed in triplicate and the bars represent the mean and standard error for one experiment.

As reported for most sensors based on YFP, HyPer fluorescence depends on pH (Belousov *et al.*, 2006). The titration curve of HyPer shows that with increasing pH the fluorescence ratio of 480–405 nm increases concomitantly. Thus in order to ascribe changes in HyPer fluorescence to a change in H_2O_2 concentration, it is necessary to ascertain whether the pH stays unaltered throughout the experiment. The mitochondrial pH may be monitored by using a pH-sensitive probe such as mtAlpHi (Abad *et al.*, 2004) which is highly selective and can be targeted to the mitochondria. Using mtAlpHi, Wiederkehr and colleagues (2007) showed that glucose stimulation in both INS-1E cells and rat islets results in a gradual increase in matrix pH which reaches a steady state after 10–15 min.

8. IMAGING OF MITOCHONDRIAL MEMBRANE POTENTIAL CHANGES BY CONFOCAL MICROSCOPY

A negative potential difference $\Delta\psi_m$ of $\sim$-200 mV exists between the cell cytosol and the mitochondrial matrix. This potential difference falls across the inner membrane of the mitochondrion and is the major driving force for proton influx into the mitochondrion. Because proton influx is essential for ATP synthesis, any damage to proteins that ensure the presence of potential difference across the inner membrane will lead to decreased ATP levels in the cell cytosol. Therefore, monitoring of the mitochondrial potential provides means to estimate the efficiency of ATP production in cells, as well as to detect malfunctioning of proteins involved.

Direct measurements of mitochondrial membrane potential have been performed on individual mitochondria with the help of the patch clamp technique that is widely used to study electrical properties of cell membranes (Jonas *et al.*, 1999). This method, though being reliable and providing a detailed information about voltages across the inner mitochondrial membrane and conductances present in it, is technically difficult to perform as well as time and work consuming. Additionally, it only allows one to investigate the properties of a small sample of mitochondria present in cell cytoplasm.

Methods traditionally used to measure the mitochondrial membrane potential employ lipophilic cations, such as tetraphenylphosphonium, that can penetrate the lipid bilayer and accumulate into different cellular compartments according to their electrical potential. The redistribution of lipophilic ions between the bulk solution containing a preparation of mitochondria and the mitochondrial matrix is usually monitored electrically with the help of a specific ion–sensitive electrode.

Another method is to use of fluorescent lipophilic ions. These substances, besides being capable of penetrating lipid bilayers, fluoresce under irradiation with light of specific wavelengths. Examples of such fluorescent lipophilic cations or fluorescent probes are rhodamine 123, tetramethylrhodamine ethyl and methyl esters (usually referred to as TMRE and TMRM, respectively), 5,5′,6,6′-tetrachloro-1,1′,3,3′-tetraethylbenzamidazolocarbocyanine (known as JC-1), and others. These fluorescent substances, because of the electrical charge present on their molecules, are sensitive to the electrical potential and redistribute between various compartments of the system if electrical potentials within it have changed. Consequently, the application of these substances to cells and tissues allows one to visualize the changes in the electrical potential across the membranes separating cellular compartments. The physical visualization or imaging of these changes can be performed on any of the variety of fluorescence microscopes that are in use today. The confocal laser scanning microscopy with its highly and easily controlled conditions for sample irradiation and the collection of emitted light, the possibility to select the position and width of an optical slice within the sample from which the fluorescent light is collected, is a very convenient tool to follow and make visible the changes of mitochondrial membrane potential in various preparations such as isolated mitochondria, live cells, and tissues such as islets of Langerhans.

8.1. Confocal imaging of mitochondrial potential in pancreatic β-cells

Freshly prepared β-cells are allowed to attach to glass coverslips and then loaded with TMRM (50 nM) for 30 min at 37 °C in standard extracellular solution of the following composition: 138 mM NaCl, 5.6 mM KCl, 2.6 mM CaCl, 1.2 mM MgCl$_2$, 5 mM HEPES; pH is adjusted to 7.4 with NaOH.

Coverslips supporting TMRM-loaded cells are mounted on a custom-built microscope sample stage with an in-built perfusion bath. During the mounting procedure special care is taken to avoid the drying of cells. The sample stage is placed on a Zeiss Axiovert 200 M microscope and the cells are allowed to equilibrate for $\sim$10 min while being perfused with the standard extracellular solution containing 50 nM TMRM at room temperature of $\sim$22 °C. All subsequent experiments were also performed at this temperature.

The fluorescence of TMRM is monitored by Zeiss LSM 510 META confocal system attached to the Axiovert 200 M microscope. The fluorescence of the dye is excited with the 543 nm line of the 15 mW HeNe laser at 1.8% delivered beam energy. The emitted light is collected within the range of wavelengths of 554–597 nm. The peak of TMRM fluorescence is well within this spectral band, and the exclusion of longer wavelengths diminishes the background noise present in the signal. Time series of images are obtained for individual cells or clusters of cells with a sampling rate of 10 s.

Examples of confocal images of TMRM loaded β-cells are seen in Fig. 25.9.

It is important to realize that in mitochondrial potential measurements fluorescent voltage sensitive dyes can be used, broadly speaking, in two different ways which are sometimes called the redistribution and

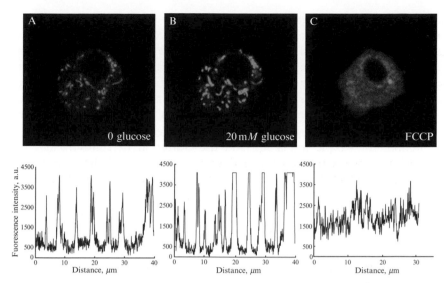

Figure 25.9 The visualization of the mitochondrial potential in MIN6 cells. Cells were loaded with TMRM at 50 nM concentration. Images were obtained on Zeiss confocal microscope LSM510 META. *Column A*: top panel—a confocal image of a loaded Min6 cell at 0 glucose; mitochondria are seen as elongated red dots; bottom panel—the intensity profile inside the image of the cell; sharp peaks represent mitochondria; the signal initiated in the cytosol is seen as low-level line between peaks. *Column B*: a confocal image of a loaded MIN6 cell at 20 mM glucose; the intensity of signal initiated in mitochondria is increased in comparison to image in column A; the cytosolic signal has decreased; this indicates the hyperpolarization of the mitochondrial membrane. *Column C*: a confocal image of a loaded MIN6 cell after application of FCCP; the peaks representing mitochondria are essentially diminished, the overall level of the signal is increased; this indicates depolarization of the mitochondrial membrane. The dark region in all three confocal images is the cell nucleus. (See Color Insert.)

quench/dequench methods (Duchen *et al.*, 2003). If the quench/dequench method is used, cells are loaded with above-mentioned dyes at comparatively high concentrations (1–20 μM) for time periods not longer than 15 min. In these conditions, dyes accumulate in mitochondria to levels that cause autoquenching of the dye. If mitochondria are subsequently transferred into a medium that causes their depolarization, the dye redistributes and partly leaves the mitochondrial matrix. This reduces autoquenching and the depolarization of mitochondria is optically detected as an increase in fluorescent signal. In investigations described here the redistribution method was used.

For cells that are loaded with TMRM at very low concentrations (50 nM), the possibility of autoquenching within the mitochondrial matrix is excluded. The changes in mitochondrial potential then lead to the redistribution of fluorescent signal between the matrix and cell cytosol.

A specific analysis of images has to be applied in order to extract information changes in mitochondrial potential. In this type of experiment, it is important to keep the energy delivered by the laser beam to cells low for two main reasons. One is the photobleaching of the dye which may become a major problem in prolonged dynamic measurements of the mitochondrial potential changes. Besides keeping the energy of the laser beam low, bleaching can be reduced by increasing the interval between sampling time. The possibility of increasing the concentration of the dye in cells in order to avoid effects of bleaching is severely limited by another reason why the beam energy has to be low. This is the phototoxicity of fluorescent dyes in general and TMRM in particular. Irradiation of fluorescent molecules leads to the increased production of free radicals within cells, and these may have adverse effects on various cellular processes. It is not so damaging in case of dyes that are evenly distributed within the whole cytosol. However, if dye is concentrated in the mitochondrial matrix, the effect of increased production of free radicals may be problematic. It was observed that mitochondria in MIN6 cells bathed in solution containing 50 nM TMRM depolarized quickly on irradiation with a 543 nm beam from 15 mW HeNe laser at 5% level. The effect was identical to the depolarization observed after application of a high concentration of FCCP. It was necessary to keep the beam energy below 2% to avoid phototoxic effects even at concentrations of TMRM as low as 50 nM.

8.2. Analysis of confocal images representing the mitochondrial membrane potential

Images seen in Fig. 25.9 contain information about the fluorescence changes caused by altered potential differences after application of different agents (high glucose, FCCP, menadione, etc.) on both the cell membrane and the inner membrane of mitochondria. If conclusions about the mitochondrial potential are to be made, the analysis of these images has to be performed in a way that targets properties of fluorescent signal characteristic or linked to the mitochondrial potential.

This can be achieved by measuring the standard deviation of the fluorescent signal intensity throughout the cell image (Duchen *et al.*, 2003; Goldstein *et al.*, 2000; Toescu and Verkhratsky, 2000). In basal conditions, the pancreatic insulin secreting cell has ~-70 mV across the cell membrane and an additional ~-200 mV across the inner mitochondrial membrane. At low TMRM concentrations (50 nM) in the extracellular solution this will lead to a slight increase in dye concentration inside cell cytosol and a much larger increase in dye concentration in the mitochondrial matrix (Fig. 25.9A–C). This distribution of the fluorescent probe results in a high–standard deviation of the signal intensity. Hyperpolarization of the mitochondrial membrane attracts additional amounts of TMRM into the matrix increasing the intensity of the signal from mitochondria and decreasing it in the cytosol (Fig. 25.9B). Such redistribution of fluorescent signal will be reflected in an increase of the overall standard deviation. In contrast, depolarization of the mitochondrial membrane results in the loss of dye from the mitochondria to the cell cytosol and, consequently, the standard deviation of the fluorescent signal decreases.

To account for different initial mean signals in cells, the standard deviation is usually scaled by the mean of the signal intensity.

9. Measuring NNT Activity

NNT is located in the mitochondrial inner membrane, with its active sites facing the matrix (Pederson $et\ al.$, 2008). The enzyme is thus not accessible to its hydrophilic substrates NAD(H) and NADP(H) (or the analogue 3-acetyl-pyridine-NAD$^+$, APAD) added externally. Normally, therefore, Nnt is assayed in submitochondrial particles (SMP) where the topology of the membrane is inverted. However, the sonication procedure involved in making these SMP requires relatively large amounts of mitochondria. A protocol describing the assay for NNT in large mitochondrial preparations from beef heart has been reported (Rydstrom, 1979). In the present procedure, which involves recovering much smaller amounts of mitochondria from mouse tissues (pancreas, heart, lungs, etc.), Nnt is assayed in isolated mitochondria that have been made permeable to its substrates by the presence of the detergents Brij 35 and lysolecithin. Lysolecithin has earlier been shown to work both as an effective detergent as well as maintaining Nnt in an active state, presumably through its property as a phospholipid (Rydstrom $et\ al.$, 1976). Brij 35 is a mild detergent, the weak denaturing effect of which is counteracted by lysolecithin. The composition of the detergent mixture was optimized regarding activity using purified $E.\ coli$ Nnt, and regarding permeability using intact mouse heart mitochondria (J. Rydström, unpublished data). Pure $E.\ coli$ Nnt is prepared as described (Meuller $et\ al.$, 1997).

9.1. Microassay of NNT

To prepare the mitochondria the procedure described in "Protocol: Isolation of Mitochondria from Tissues Using the Qproteome^{TM} Mitochondria Isolation Kit" (Qiagen) is followed in detail, until Step 11a (standard preparation of mitochondria), using 1.5–2-ml Eppendorf tubes, a cooled Eppendorf centrifuge, a TissueRuptor (Qiagen), and a 2-ml dounce homogenizer (teflon/glass). It does not appear to be necessary to continue with Step 12a, that is, a final wash. As applied to mouse heart and pancreas, acceptable modifications include doubling the amount of starting tissue material (100 mg instead of 60 mg wet weight). After the final centrifugation step, the mitochondrial pellet is about 2–3 mm in diameter (starting with a 3-week-old BALB/c mouse heart of about 100 mg wet weight). Following suspension of the pellet with 20 μl of mitochondria storage buffer (Qproteome^{TM} kit), the final volume of the mitochondrial suspension (using 100 mg of tissue) is 25–30 μl, giving an estimated protein concentration of 5–10 mg/ml, which is enough for at least four assays and a protein determination. The suspension should be kept on ice as freezing will result in a loss of activity.

The assay of Nnt in isolated mitochondria is carried out using a dual wavelength Shimadzu UV-3000 spectrophotometer, or an equivalent spectrophotometer with accurate signal to noise filters, with a minimal distance between cuvette and photomultiplier (in order to minimize noise due to the turbid sample). A normal split beam spectrophotometer is usually not sufficient, because it is too sensitive to the turbidity changes at the wavelengths used. At this wavelength, oxidation of NADPH in the reaction will contribute to an increased extinction coefficient. For the dual wavelength instrument, wavelengths are normally 375–400 nm (375 nm is the wavelength chosen for reduced APAD, and 400 nm is the reference, isosbestic, wavelength for APAD/APADH(NADP$^+$/NADPH) (under these conditions, the extinction coefficient is 7.63/mM/cm). A 1-ml cuvette with black sides is used to avoid stray light.

$$NADPH + (AcPy)NAD^+ \rightarrow NADP^+ + (AcPy)NADH$$

The assay medium is composed of 50 mM Tris–HCl (pH 8.0), 0.5% Brij 35, 1 mg/ml of lysolecithin, 300 μM each of APAD and NADPH. Stock solutions are 0.5 M Tris–HCl, 10% Brij-35, 10 mg/ml lysolecithin, and 20 mM each of APAD and NADPH. A drop of the Tris buffer (pH 8.0) is added to the stock NADPH (but not to the stock APAD) in order to maintain its preservation at 4 °C (NADPH is sensitive to low pH). Both substrate stock solutions are stable for at least a couple of days at 4 °C (but not in the freezer).

Full scale sensitivity of the spectrophotometer is set to at least 0.05 OD (can be lower, i.e., a higher sensitivity, but not higher). The assay is carried out with 1 ml of assay buffer and 10 μl of the mitochondrial suspension. A slight delay in the activity (maximum 30 s) has occasionally been noted, probably reflecting the fact that it may take time for the mitochondria to become permeable to the substrates. However, after this lag phase, the activity is essentially linear for several minutes. An example of the linear trace observed when measuring NNT activity in mouse pancreas mitochondria is shown in Fig. 25.10. The specificity of the assay is most apparent by the lack of activity observed in mitochondria from C57BL/6J tissue, which has a deletion in the *Nnt* gene resulting in a lack of function. The slopes from these traces can be used to generate values for the rate of activity per

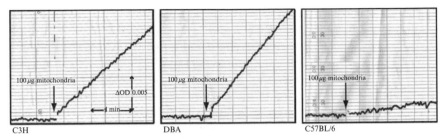

Figure 25.10 Raw traces of NNT activity in mitochondria from the pancreas of different mouse strains. C57BL6/J, which has a deletion in the NNT gene, shows minimal activity most of which is likely due to nonspecific reduction of (AcPy)NAD$^+$.

milligram of protein. By introducing a known amount of (AcPy)NADH at the end of the reaction the corresponding change in OD can be used as an indicator to calculate the change in (AcPy)NADH over time.

9.2. Assay considerations

The reason for choosing a higher assay pH is to minimize nonspecific interactions of NADH-linked reductases/dehydrogenases with NADPH. For the same reason it is best to avoid rotenone or other respiratory chain inhibitors, because these tend to have the same effect. The reduced APAD is not significantly oxidized due to its high redox potential.

It might be possible to also assay NNT by using a fluorometer, but again, the required high sensitivity combined with the turbidity will produce a noisy signal.

Whilst it is preferable to use a dual wavelength spectrophotometer, the availability of this equipment is increasingly scarce. We have successfully used a Varian Cary 4000 that can measure separate wavelengths almost simultaneously and therefore provides an accurate subtraction of the turbid background.

The concentrations of the substrates have been chosen based on the earlier estimated K_m-values for NAD$^+$ and NADPH, which are between 20 and 30 μM. Assay concentration should be at least 10 times the K_m. It should be noted though that permeabilized mitochondria are used, and traditional K_m values may therefore not apply. To be certain, one should double the substrate concentrations. Too high-substrate concentrations (and thus too high absorbance) tend to decrease the sensitivity of the spectrophotometer.

10. Conclusions

This chapter provides the background and techniques that were used in studying NNT and its role in insulin secretion. The importance of NNT was originally identified by its effect on glucose homeostasis in C57Bl/6J that carries a natural deletion of *Nnt*. To validate our findings, we have since combined a range of *in vitro* techniques including the use of ENU mutagenesis and BAC transgenesis to generate a number of *Nnt*-deficient alleles, as well as *in vivo* methodologies such as RNAi silencing in insulinoma cell lines which can be more readily performed. The mouse models that we have generated allow us to further investigate NNT function by using islets extracted from wild-type and mutant mice, and treating them with various reagents. To further dissect the mechanism of NNT function within the mitochondria, we have developed imagining techniques that quantify a range of intracellular and intramitochondrial variables in testing our hypothesis on NNT function. Finally, the Nnt activity assay on isolated mitochondria gives a more direct measure of Nnt function, and when combined with the other assays will provide us with a strong basis for validating drugs that target Nnt, which may have important medical potential in treating patients with type-2 diabetes.

REFERENCES

Abad, M. F. C., Di Benedetto, G., Magalhães, P. J., Filippin, L., and Pozzan, T. (2004). Mitochondrial pH monitored by a new engineered green fluorescent protein mutant. *J. Biol. Chem.* **279,** 11521–11529.

Asfari, M., Janjic, D., Meda, P., Li, G., Halban, P. A., and Wollheim, C. B. (1992). Establishment of 2-mercaptoethanol-dependent differentiated insulin-secreting cell lines. *Endocrinology* **130,** 167–178.

Ashcroft, F. M., and Rorsman, P. (1989). Electrophysiology of the pancreatic β-cell. *Prog. Biophys. Mol. Biol.* **54,** 87–143.

Belousov, V. V., Fradkov, A. F., Lukyanov, K. A., Staroverov, D. B., Shakhbazov, K. S., Terskikh, A. V., and Lukyanov, S. (2006). Genetically encoded fluorescent indicator for intracellular hydrogen peroxide. *Nat. Methods* **3,** 281–286.

Bertram, R., Sherman, A., and Satin, L. S. (2007). Metabolic and electrical oscillations: Partners in controlling pulsatile insulin secretion. *Am. J. Physiol. Endocrinol. Metab.* **293,** E890–E900.

Bonner-Wier, S., and Orci, L. (1982). New perspectives on the microvasculature of the islets of Langerhans in the rat. *Diabetes* **31,** 839–939.

Bosco, D., Orci, L., and Meda, P. (1989). Homologous but not heterologous contact increases the insulin secretion of individual pancreatic β-cells. *Exp. Cell Res.* **184,** 72–80.

Carvajal, J. J., Cox, D., Summerbell, D., and Rigby, P. W. (2001). A BAC transgenic analysis of the Mrf4/Myf5 locus reveals interdigitated elements that control activation and maintenance of gene expression during muscle development. *Development* **128,** 1857–1868.

Chow, R. H., Lund, P. E., Loser, S., Panten, U., and Gylfe, E. (1995). Coincidence of early glucose-induced depolarization with lowering of cytoplasmic Ca^{2+} in mouse pancreatic β-cells. *J. Physiol.* **485**(Pt. 3), 607–617.

Coghill, E. L., Hugill, A., Parkinson, N., Davison, C., Glenister, P., Clements, S., Hunter, J., Cox, R. D., and Brown, S. D. (2002). A gene-driven approach to the identification of ENU mutants in the mouse. *Nat. Genet.* **30,** 255–256.

Dobson-Stone, C., Cox, R. D., Lonie, L., Southam, L., Fraser, M., Wise, C., Bernier, F., Hodgson, S., Porter, D. E., Simpson, A. H. R. W., and Monaco, A. P. (2000).

Comparison of fluorescent single-strand conformation polymorphism analysis and denaturing high-performance liquid chromatography for detection of EXT1 and EXT2 mutations in hereditary multiple exostoses. *Eur. J. Hum. Genet.* **8,** 24–32.

Duchen, M. R., Surin, A., and Jacobson, J. (2003). Imaging mitochondrial function in intact cells. *Methods Enzymol.* **361,** 353–389.

Erlanded, S. L., Herge, O. D., Parsons, J. A., McEvoy, R. C., and Elde, R. P. (1976). Pancreatic islet cell hormones distribution of cell types in the islet and the evidence for presence of somatostatin and gastrin within the D cell. *J. Histochem. Cytochem.* **24,** 883–897.

Freeman, H. C., and Cox, R. D. (2006). Type-2 diabetes: A cocktail of genetic discovery. *Hum. Mol. Genet.* **15**(2), R202–R209; doi:10.1093/hmg/ddl191.

Freeman, H. C., Hugill, A., Dear, N. T., Ashcroft, F. M., and Cox, R. D. (2006c). Deletion of nicotinamide nucleotide transhydrogenase: A new quantitative trait locus accounting for glucose intolerance in C57BL/6J mice. *Diabetes* **55,** 2153–2156.

Freeman, H. C., Shimomura, K., Cox, R. D., and Ashcroft, F. M. (2006b). Nicotinamide nucleotide transhydrogenase: A link between insulin secretion, glucose metabolism and oxidative stress. *Biochem. Soc. Trans.* **34,** 806–810.

Freeman, H. C., Shimomura, K., Horner, E., Cox, R. D., and Ashcroft, F. M. (2006a). Nicotinamide nucleotranshydrogenase: A key role in insulin secretion. *Cell Metab.* **3,** 35–45.

Goldstein, J. C., Waterhouse, N. J., Juin, P., Evan, G. I., and Green, D. A. (2000). The coordinate release of cytochrome c during apoptosis is rapid, complete and kinetically invariant. *Nat. Cell Biol.* **2,** 156–162.

Goldsworthy, M., Hugill, A., Freeman, H., Horner, E., Shimomura, K., Bogani, D., Pieles, G., Mijat, V., Arkell, R., Bhattacharya, S., Ashcroft, F. M., and Cox, R. D. (2008). The role of the transcription factor *Sox4* in insulin secretion and impaired glucose tolerance. *Diabetes* **57,** 2234–2244.

Henquin, J. C. (2000). Triggering and amplifying pathways of regulation of insulin secretion by glucose. *Diabetes* **49,** 1751–1760.

Henquin, J. C., and Meissner, H. P. (1984). Significance of ionic fluxes and changes in membrane potential for stimulus-secretion coupling in pancreatic β-cells. *Experientia* **40,** 1043–1052.

Jonas, E. A., Buchanan, J., and Kaczmarek, L. K. (1999). Prolonged activation of mitochondrial conductances during synaptic transmission. *Science* **286,** 1347–1350.

Justice, M., Noveroske, K., Weber, J., Zheng, B., and Bradley, A. (1999). Mouse ENU mutagenesis. *Hum. Mol. Genet.* **8,** 1955–1963.

Kantengwa, S., Baetens, D., Sadoul, K., Buck, C. A., Halban, P. A., and Rouiller, D. G. (1997). Identification and characterization of integrin on primary and transformed rat islet cells. *Exp. Cell Res.* **237,** 394–402.

Keays, D. A., Clark, T. G., Campbell, T. G., Broxholme, J., and Valdar, W. (2007). Estimating the number of coding mutations in genotypic and phenotypic driven N-ethyl-N-nitrosourea (ENU) screens: Revisited. *Mamm. Genome* **18,** 123–124.

Keays, D. A., Clark, T. G., and Flint, J. (2006). Estimating the number of coding mutations in genotypic-and phenotypic-driven N-ethyl-N-nitrosourea (ENU) screens. *Mamm. Genome* **17,** 230–238.

McDaniel, M. L., Colca, J. R., Kotagal, N., and Lacy, P. E. (1983). A subcellular fractionation approach for studying insulin release mechanisms and calcium metabolism in islets of Langerhans. *Methods Enzymol.* **98,** 182–200.

Meda, P., Michaels, R. L., Halban, P. A., Orci, L., and Sheridan, J. D. (1983). *In vivo* modulation of gap junctions and dye coupling between β-cells of the intact pancreatic islet. *Diabetes* **32,** 858–868.

Meuller, J., Zhang, J., Hou, C., Bragg, P. D., and Rydström, J. (1997). Properties of a cysteine-free proton-pumping nicotinamide nucleotide transhydrogenase. *Biochem. J.* **324,** 681–687.

Nagy, A., Gertsenstein, M., Vintersten, K., Behringer, R. (eds.) (2003). Manipulating the Mouse Embryo: A Laboratory Manual. 3rd ed. Cold Spring Harbor Laboratory Press, Cold Spring Harbor, New York.

Orci, L. (1982). Macro- and micro-domains in the endocrine pancreas. *Diabetes* **31,** 538–565.

Orci, L., and Unger, R. H. (1975). Functional subdivision of islets of Langerhans and possible role of D-cells. *Lancet* **2,** 1243–1244.

Pedersen, A., Karlsson, B. G., and Rydström, J. (2008). Proton-translocating transhydrogenase: An update of unsolved and controversial issues. *J. Bioenerg. Biomembr.* **40,** 463–473.

Quwailid, M. M., Hugill, A., Dear, N., Vizor, L., Wells, S., Horner, E., Fuller, S., Weedon, J., McMath, H., Woodman, P., Edwards, D., Campbell, D., *et al.* (2004). A gene-driven ENU-based approach to generating an allelic series in any gene. *Mamm. Genome* **15,** 585–591.

Rydström, J. (1979). Assay of nicotinamide nucleotide transhydrogenases in mammalian, bacterial, and reconstituted systems. *Methods Enzymol.* **55,** 261–275.

Rydström, J., Hoek, J. B., Ericson, B. G., and Hundai, T. (1976). Evidence for a lipid dependence of mitochondrial nicotinamide nucleotide transhydrogenase. *Biochim. Biophys. Acta* **430,** 419–425.

Saito, T., Okada, S., Yamada, E., Ohshima, K., Shimizu, H., Shimomura, K., Sato, M., Pessin, J. E., and Mori, M. (2003). Syntaxin 4 and Synip (syntaxin 4 interacting protein) regulate insulin secretion in the pancreatic β HC-9 cell. *J. Biol. Chem.* **367,** 36718–36725.

Shimomura, K., Shimzu, H., Ikeda, M., Okada, S., Kakei, M., Matsumoto, S., and Mori, M. (2004). Fenofibrate, troglitazone and 15-deoxy-$\Delta^{12,14}$-prostaglandin J_2 close K_{ATP} channels and induce insulin secretion. *J. Pharmacol. Exp. Ther.* **310,** 1273–1280.

Soria, B., Chanson, M., Giordano, E., Bosco, D., and Meda, P. (1991). Ion channels of glucose-responsive and unresponsive β-cells. *Diabetes* **40,** 1069–1078.

Tanaka, Y., Tran, P. O., Harmon, J., and Robertson, R. P. (2002). A role of glutathione peroxidase in protecting pancreatic β-cells against oxidative stress in a model of glucose toxicity. *Proc. Natl. Acad. Sci. USA* **17,** 12363–12368.

Tengholm, A., Hellman, B., and Gylfe, E. (2001). The endoplasmic reticulum is a glucose-modulated high-affinity sink for Ca^{2+} in mouse pancreatic β-cells. *J. Physiol.* **530,** 533–540.

Toescu, E. C., and Verkhratsky, A. (2000). Assessment of mitochondrial polarization status in living cells based on analysis of the spatial heterogeneity of rhodamine 123 fluorescence staining. *Pflugers Arch.* **440,** 941–947.

Toye, A. A., Lippiat, J., Proks, P., Shimomura, K., Bentley, L., Hugill, A., Mijat, V., Goldsworthy, M., Moir, L., Haynes, A., Quarterman, J., Freeman, H., *et al.* (2005). A genetic and physiological study of impaired glucose homeostasis control in C57BL/6J mice. *Diabetologia* **48,** 675–686.

Wiederkehr, A., Park, K. S., Dupont, O., Demaurex, N., Pozzan, T., and Wollheim, C. B. (2007). ATP, mitochondrial calcium and mitochondrial pH responses during nutrient activation of the pancreatic β-cell. *Diabetologia* **50**(Suppl. 1), S74–S75(0167).

Wittwer, C. T., Reed, G. H., Gundry, C. N., Vandersteen, J. G., and Pryor, R. J. (2003). High-resolution genotyping by amplicon melting analysis using LCGreen. *Clin. Chem.* **49,** 853–860.

Wollheim, C. B., Meda, P., and Halban, P. A. (1990). Isolation of pancreatic islets and primary culture of the intact microorgans or of dispersed islet cells. *Methods Enzymol.* **192,** 188–223.

Author Index

A

Abad, M. F. C., 471
Abbott, W. G., 374
Abdollahi, M., 181
Abe, J., 118
Abou-Sleiman, P. M., 213, 276
Abramov, A. S., 222
Abrams, L., 217
Abrams-Carter, L., 217
Acín-Pérez, R., 338
Ackerley, S., 321
Adams, S. P., 152
Adams-Collier, C. J., 37
Ades, P. A., 352
Adey, D. B., 352
Adler, C., 6
Adolph, H., 38
Adornetto, A., 118
Aebersold, R., 98, 127
Affourtit, C., 396, 405, 406, 410, 411, 415, 417, 418, 419
Agami, R., 235
Aggeler, R., 64
Agnese, S., 118
Agostinelli, E., 118
Ahlman, B., 213, 215, 217
Ahmed, R., 257
Ahn, B. H., 138
Ahting, U., 4, 9, 13, 55, 82, 98
Akimoto, M., 336, 340, 344
Albanese, A., 213, 276
Alberici, L. C., 402
Albermann, K., 6
Albert, P. R., 171
Albert, S., 90
Aldred, S., 98
Aleshin, E. A., 134
Alessi, D. R., 115
Alexson, S. E., 186
Alger, J. R., 379
Al-Hakim, A., 171
Ali, Z., 213, 276
Allen, J. C., 344
Allen, J. F., 64
Alonso, A., 276
Alroy, J., 290, 295, 301, 303
Alshami, A., 352, 388

Al-Share, Q., 224
Alt, F. W., 138, 141, 142
Altermann, M., 113
Alto, N. M., 257
Altschul, S. F., 12
Amadio, S., 323
Amberger, J. S., 7
Ambrose, C. M., 160
Ames, B. N., 50
Amin, M. B., 336
Amo, T., 417, 419
An, Y., 52
Andersen, A. J., 355, 356, 357, 358
Andersen, J. S., 83, 138, 141, 143
Anderson, E., 219
Anderson, G. A., 53
Andersson, L., 84
Andreoli, C., 3, 4, 9, 11, 46, 51, 55, 82, 98
Anensen, N., 99
Angell, S., 433
Angelo, R. G., 172
Anna Maria, B., 90
Annen, A. S., 411
Annunziato, L., 118
Antonicka, H., 16
Antonio, T., 90
Antozzi, C., 344
Anvik, J., 38
Aoyama, T., 186
Aponte, A. M., 63, 64, 65, 195, 257
Aprille, J. R., 65
Arai, R., 9, 11
Arakane, F., 190
Arancia, G., 118
Araujo, E. P., 397, 400, 402
Arcelloni, C., 374
Ardail, D., 99
Argov, Z., 352
Ariyan, C., 352, 374, 386
Arkell, R., 464
Armistead, S., 419, 420
Arnold, S., 64, 195
Arnoult, D., 246, 292, 293, 296, 301, 303
Aronow, B. J., 186
Asano, M., 337
Ascoli, M., 172
Asfari, M., 411, 464
Ashburner, M., 15

Ashcroft, F., 451
Ashcroft, F. M., 452, 453, 457, 459, 461, 463, 464, 469
Ashcroft, S. J. H., 427
Asmann, Y. W., 352, 353, 360, 388
Atkins, J. F., 235
Attardi, G., 338, 339, 340
Augenlicht, L. H., 336
Aureliano, M., 209
Auricchio, A., 321
Avers, C. J., 290
Avvedimento, E. V., 118, 172
Ayliffe, M. A., 25
Azzu, V., 415

B

Baar, K., 217
Babcock, E. E., 374
Baddoo, A., 257
Badger, J. D. II, 321
Baetens, D., 461
Baffy, G., 397, 411, 418
Bagshaw, R. D., 98
Bailey, A. O., 152
Baines, C. P., 257
Baird, G. S., 313
Baker, D. P., 160
Balaban, R. S., 51, 63, 64, 65, 67, 195, 257
Baldwin, K. M., 215
Balgley, B. M., 52, 53, 54, 55, 56, 57, 59
Ball, C., 6
Ball, C. A., 15
Ball, H., 83, 110, 144
Ballabio, A., 321
Ballif, B. A., 64, 65
Baloh, R. H., 321
Banci, L., 257
Bank, W. J., 352
Banker, G., 322
Bann, J. G., 234
Bannai, H., 37
Barazzoni, R., 352
Barbera, M. J., 24, 25, 29, 36
Bardonova, L., 24
Bardwell, A. J., 181
Bardwell, L., 181
Barlow, D. P., 225
Barnouin, K., 101
Baron, R., 118
Barrett, J. C., 138
Bartlett, S., 388
Bartley, C., 411
Bartunik, H. D., 134
Bartz, S., 219
Baskerville, S., 219
Baskin, S., 257
Basu, R., 352, 353, 364

Basu, S., 82
Battezzati, A., 374
Baulcombe, D. C., 433
Baumann, A. K., 336
Baur, L. A., 374
Bayona-Bafaluy, M. P., 338
Beal, M. F., 213
Beausoleil, S. A., 64, 65, 125
Becker, C. F., 138
Beckman, K. B., 50
Beck-Nielsen, H., 83
Beechem, J. M., 64
Beer, M., 101
Beer, S. M., 83
Befroy, D., 215, 216, 352, 374, 386, 387
Befroy, D. E., 215, 352, 373, 382, 384, 386, 387, 388, 389
Behar, K. L., 382
Behringer, R., 459
Belenguer, P., 291
Bell, A. W., 98
Belle, A., 9
Belousov, V. V., 468, 470
Bemier, F., 454
Bemis, K. G., 51
Benaguid, A., 65
Bender, C., 98
Bender, E., 64
Bendtsen, J. D., 44
Bentivoglio, A. R., 213, 276
Bentley, L., 452, 457
Benz, R., 98
Berdal, A., 225
Bereiter-Hahn, J., 290
Berg, H. E., 215
Berger, F., 65
Berger, S. J., 53
Bergmann, U., 9, 11
Bergstrom, J., 214
Berkes, C. A., 217
Bernard, B. A., 9, 13
Bernardi, P., 98, 233, 307
Bernards, R., 235
Berner, D., 406
Bernstein, E., 433
Berria, R., 216, 375
Berthiaume, L. G., 149, 151, 152, 153, 155, 159, 163
Bertozzi, C. R., 151, 152, 153, 155, 159, 163
Bertram, R., 468
Bey, P., 187
Beyermann, M., 108
Bezzera, R. M., 402
Bhatnagar, R., 150
Bhattacharya, S., 464
Bianchi, K., 290
Bigelow, M. L., 352, 353, 360, 364, 388
Bijlmakers, M. J., 150

Bijur, G. N., 257
Bilz, S., 387
Bingley, M., 321
Bionda, C., 99
Biondi, R. M., 65
Birmingham, A., 219
Biro, M. C., 112
Bishop, T., 417
Bissar-Tadmouri, N., 290
Bixler, S. A., 160
Bizzozero, O. A., 161
Blackledge, M. J., 379, 388
Blackstock, W. P., 98
Blackstone, C., 233
Blagoev, B., 125
Blair, P. V., 51
Blake, J. A., 7, 15
Blakely, E. L., 14
Bland, J. L., 379
Blank, N. C., 85
Blass, J. P., 400
Blenis, J., 220, 222
Blinova, K., 63
Bloch, F., 387
Blom, J., 9, 11
Blomberg, A., 99
Blum, T., 38
Bocchini, C. A., 7
Bocek, P., 51
Boden, G., 374
Bodén, M., 35
Bodenmiller, B., 127
Bodmer, R., 213
Bodnar, A. G., 216
Boehm, A. M., 9, 11, 98
Boerner, J. L., 196
Boese, Q., 235
Bogani, D., 464
Bogardus, C., 374
Bogdanov, B., 53
Bogdanova, A., 51
Bogenhagen, D. F., 82
Bolender, N., 26, 27
Bolino, A., 323
Bolis, A., 323
Bolli, R., 257
Bolouri, M. S., 12, 13, 51, 98
Boneh, A., 12, 13, 25, 82
Bonekamp, N. A., 232
Bonini, N. M., 323
Bonner-Wier, S., 461
Boorsma, A., 9, 11
Booth, F. W., 354
Bordin, L., 118
Bork, P., 9, 13, 15
Börnsen, K. O., 112
Boschero, A. C., 397, 400, 402
Bosco, D., 461

Bose, S., 67
Boss, O., 397, 411, 418
Botstein, D., 6, 15
Boulianne, G. L., 328
Bourenkov, G. P., 134
Boushel, R., 368
Boutilier, R. G., 417
Bower, K., 9
Boxma, B., 25
Boyd, C. S., 257
Bradford, M. M., 398
Bradley, A., 453
Bradley, P. J., 24
Brady, S., 38
Brady, S. T., 321
Bragg, P. D., 475
Brand, A. H., 323
Brand, M. D., 138, 396, 405, 406, 407, 410, 411,
    415, 417, 418, 419, 421
Brandner, K., 45
Brandolin, G., 132
Brandt, D. R., 242
Braun, R. J., 101
Brdiczka, D., 99, 102
Breen, E., 415
Breger, R. K., 380
Bremer, E., 82
Brennan, R. G., 234
Brichese, L., 257
Brickey, D. A., 234
Bridges, B. J., 64, 75
Brill, L. M., 206, 207
Brindle, K. M., 379, 388
Britton, S. L., 224
Broccoli, V., 321
Brock, A., 206, 207
Brogan, R. J., 216, 218
Brookes, P. S., 118
Brossard, N., 25, 26
Brothers, G. M., 50
Brown, D. L., 154
Brown, G. C., 407, 410, 417
Brown, L. J., 426
Brown, M. D., 344
Brown, M. R., 52
Brown, P., 382
Brown, P. O., 9, 11
Brown, S. D., 453
Brown, S. E., 83
Brown, T. R., 379
Brownlees, J., 321
Broxholme, J., 456
Broyles, S. S., 276
Bruin, G. J. M., 112
Brummelkamp, T. R., 235
Brunak, S., 34, 44
Brunati, A. M., 117, 118, 124, 132, 133
Brusch, R. G., 323

Bruserud, O., 99
Bryant, M. L., 152
Bryant, S. H., 54, 55, 128
Buchanan, J., 471
Buck, C. A., 461
Buckingham, J. A., 407, 410, 417
Buckman, J. F., 326
Budd, S. L., 50
Bueno, M., 338
Büge, U., 199
Bugiani, M., 344
Buhler, S., 98, 102
Bui, E. T., 24
Bult, C. J., 7
Bunkenborg, J., 12, 13, 51, 83, 98, 138, 141, 142, 143
Buntinas, L., 50
Burchard, J., 219
Burger, G., 23, 25, 37, 39, 350
Burgess, S. C., 426, 448
Burke, D. J., 108, 125
Burneskis, J. M., 138
Burnett, P., 195
Burns, D. K., 374
Bursac, D., 24
Burum, D. P., 382
Busby, S. J., 375
Busciglio, J., 321
Bushby, K. M., 321
Buss, J. D., 426
Bussemaker, H. J., 9, 11
Bussini, S., 323
Buta, M. R., 336
Butler, H., 15
Butte, A. J., 216, 375
Buttgereit, F., 410, 417
Bylund, A. C., 214
Bylund-Fellenius, A. C., 354

C

Caballero, O. L., 336
Cacace, G., 85, 86
Cadenas, E., 257
Cadene, M., 112
Cadman, K. S., 388
Caenepeel, S., 256
Caesar, R., 99
Callahan, J. W., 98
Calvo, S., 12, 13, 51, 55
Calvo, S. E., 12, 13, 25, 82
Cameron, J. M., 276
Camp, D. G. II, 4, 9, 11, 46, 51, 98
Campanaro, S., 51
Campbell, D., 453
Campbell, T. G., 456
Campbell–Burk, S. L., 388
Cannon, B., 213, 216, 217, 218, 355, 357, 396

Cano, F., 171, 187
Cantley, L. C., 65, 125
Cao, J. H., 323
Cao, X., 388
Capaldi, R. A., 10, 13, 51, 64, 98
Caputo, V., 213, 276
Cardwell, E. M., 257
Carlsson, E., 216, 223, 375
Carlucci, A., 118
Carneiro, E. M., 397, 400, 402
Carr, B., 257
Carr, P. W., 86
Carr, S. A., 12, 13, 51, 55
Carrara, F., 344
Carreras, M. A. C., 174
Carroll, A. S., 9, 11, 14, 15
Carroll, S., 64, 65, 195, 257
Carvajal, J. J., 459
Carvalheira, J. B., 397, 400, 402
Casari, G., 225, 321
Casey, P. J., 118, 150, 151, 257, 276
Cassidy-Stone, A., 290
Castaldo, L., 118
Castilla, R., 171, 186, 187
Castillo, A. F., 187
Castillo, F., 171, 187
Castillo, R., 186
Caudy, A. A., 433
Cavallaro, G., 257
Cavet, G., 219
Cazettes, G., 257
Cedergren, R., 25, 26
Cereghetti, G. M., 233
Cerkasovová, A., 24
Cesaro, L., 117, 118, 132, 133
Chacinska, A., 9, 11, 25, 27, 45, 48, 51
Chada, S. R., 322
Chaffard, G., 406, 411
Chait, B. T., 112
Chakravarti, B., 83
Chakravarti, D. N., 83
Chalk, A. M., 218
Challiss, R. A., 379, 388
Chan, D. C., 246, 290, 292, 293, 296, 301, 303
Chance, B., 101, 352, 353, 362, 400
Chandrasekaran, K., 52, 53, 56, 57, 59
Chang, B., 12, 13, 25, 51, 55, 82
Chang, C. R., 233
Chang, D. T., 321, 322, 325
Chang, J., 51
Chang, P. V., 155
Chang, Y. I., 257
Chanson, M., 461
Chapman, A. L., 321
Charcinska, A., 98
Charydczak, G., 256
Chau, B., 219
Che, W., 118

Chen, A., 268, 312
Chen, C. H., 257
Chen, D., 10, 13, 51
Chen, H., 246, 290, 292, 293, 296, 301, 303
Chen, J., 53
Chen, J. N., 257, 258, 261, 263, 265, 267, 270
Chen, L., 257
Chen, L. B., 268, 312
Chen, S., 320, 322
Chen, W., 379, 388
Chen, W. K., 12, 13, 25, 82
Chen, Y., 83, 85, 93, 94, 110, 138, 141, 142, 144, 250
Chen, Y. C., 36
Cheng, C. Y., 323
Cheng, E. H.-Y., 99
Cheng, H. L., 138, 141, 142
Cheng, T., 83, 110, 144
Chernokalskaya, E., 51
Cherry, J. M., 6, 15
Chervitz, S. A., 6
Cheung, K. H., 9, 11
Chevet, E., 98
Chilibeck, P. D., 387, 388
Chin, L. S. L. L., 213
Chinnery, P. F., 14
Chipuk, J., 98
Chiu, C. P., 216
Choi, C. S., 388
Choi, J., 257, 258, 261, 263, 265, 267, 270
Choi, S. W., 417, 419
Chomicz, S., 419, 420
Chotai, D., 98
Choudhary, J. S., 98
Choudhury, A., 225
Chow, L. S., 352, 353, 388
Chow, R. H., 468
Christie, M. R., 427
Christin, L., 374
Christmas, R. H., 98
Chu, A. M., 9, 11
Chu, C. T., 257
Chung, C., 51, 55, 56
Chwalbinska-Moneta, J., 214
Ciaraldi, T. P., 217
Cifuentes-Diaz, C., 291
Clari, G., 118
Clark, B. J., 171, 186
Clark, C. G., 24, 25, 29, 36, 41
Clark, I. E., 323
Clark, J. I., 104
Clark, J. M., 104
Clark, M. G., 407, 410, 417
Clark, T. G., 456
Clark, W., 257
Cleary, M., 219
Clements, S., 453
Clerici, C., 225

Cline, G. W., 352, 374, 383, 384, 386, 388, 425, 426, 444
Clottes, E., 225
Cobb, M. H., 174
Codella, R., 388
Coenen-Schimke, J. M., 352, 353, 360, 364, 388
Coghill, E. L., 453
Cogswell, A. M., 354
Cohen, B. H., 359, 368
Cohen, M. V., 257
Cohen, P., 113, 119
Cohen, P. T. W., 256, 257, 258
Cohn, M. A., 125
Colberg, S. R., 374
Colca, J. R., 462
Cole, R. N., 10, 13, 51
Colonna, C., 171
Colwill, K., 325
Comb, M. J., 90, 115, 126
Confalonieri, P., 344
Constantini, P., 50
Constantin-Teodosiu, D., 214
Converso, D. P., 174
Cooper, J. W., 52
Cooper, R. H., 64, 73, 75
Corkey, B., 426
Cornejo Maciel, F., 169, 171, 186, 187, 188
Cornell, J., 51
Cornell, M., 9, 15
Cornwall, E. J., 417, 419
Corvi, M. M., 151, 153, 155, 159, 163
Costa, L. E., 400, 417
Costa, N. J., 83
Costello, M., 216, 375
Cotter, D., 82
Cottrell, J. S., 128
Coudert, A., 225
Couette, S., 225
Coumans, B., 224
Cousin, M. A., 232
Costello, M., 216, 375
Couture, P., 417
Cox, B., 51, 55, 56
Cox, D., 459
Cox, R. D., 451, 452, 453, 454, 457, 459, 463, 464, 469
Craigen, W. J., 99
Cravatt, B. F., 163
Creasy, D. M., 128
Cribbs, J. T., 231, 233, 243, 250
Cristianini, N., 28
Crossin, K. L., 320, 322
Crosthwaite, S., 225
Crouser, E. D., 98
Crunkhorn, S., 216, 375
Cuccurullo, M., 85, 86, 87
Cui, L., 215
Culter, C. L., 374
Cusi, K., 216, 375

Cwerkasòv, J., 24
Cymeryng, C., 186, 187, 188
Cymeryng, C. B., 186

D

Da Cruz, S., 10, 13, 98
Dada, L. A., 186, 187, 188
Dadali, E. L., 290
Dagda, R. K., 250, 257
Dahary, D., 225
Dahlmann, M., 65
Dahm, C. C., 83
Dahout-Gonzalez, C., 132
Dalal, N., 83
Dalmay, T., 433
Daly, M. J., 216, 223, 375
Dancis, A., 24
Dati, G., 323
Daum, G., 310
David, L., 4, 9, 11, 46, 51, 98
Davidson, M., 50
Davis, A. P., 15
Davis, J., 11, 13
Davis, J. W., 428
Davis, N. G., 152
Davis, R. W., 9, 11
Davis, S. C., 155
Davison, C., 453
Davoren, J. E., 320
Day, D. A., 25
De, S. D., 290
Dear, N. T., 453, 457, 459
Deas, E., 222
De Borman, B., 224
De Cobelli, F., 374
Deerinck, T. J., 257
De Feyter, H. M., 388
Degasperi, G. R., 395, 397, 400
de Graaf, R. A., 379
de Graaf, R. M., 25, 215, 384, 386, 387
Deichaite, I., 151
Dejonghe, S., 427
Dela, F., 368
Delage, E., 25, 26
Delbac, F., 25
Del Carro, U., 323
Delgadillo-Correa, M., 24
Del Maschio, A., 374
Del Turco, D., 213, 276
DeLuca, M., 360
Demaurex, N., 471
Demory, M. L., 196
Dencher, N., 114
Deng, C., 86, 87
Deng, C. X., 138
Deng, H., 220, 222
Deng, X., 257

den Hollander, J. A., 379
Denis, R. G., 397, 400
Denton, R. M., 64, 73, 74, 75, 195
DePaoli, A. M., 374
Dephoure, N., 9
DeProto, J., 324, 331
Dequiedt, F., 138
DeRisi, J. L., 9, 11
Deschenes, R. J., 150, 151
de Souza, C. F., 344
De Souza, C. T., 397, 400, 402
Detmer, S. A., 290, 301, 303
Deutsch, M., 291, 292, 294, 303
Deutschbauer, A. M., 9, 11
Devadas, B., 152, 154
Devenish, R. J., 305, 307, 309, 314, 315, 316
Devoe, D. L., 52, 53, 54
De Vos, K. J., 321
Devreese, B., 10, 13
Diaz-Triviño, S., 25, 29, 36
Di Benedetto, G., 471
Diemer, H., 10, 13
Diepstra, H., 16
Dietmann, S., 6
Diltz, C. D., 276
DiMarco, J. P., 100
DiMauro, S., 50, 51
Dimmer, K. S., 9, 11
Dina, G., 321, 323
Dinur-Mills, M., 16, 27
DiPietro, L., 352, 374, 386
Distefano, A., 388
Distler, A. M., 83, 97, 98, 99, 102, 103, 104
Diviani, D., 172
Dixit, V., 50
Dixon, J. E., 83, 118, 257, 275, 276, 277, 279, 283, 286
Dixon, J. R., 257, 276
Doan, J. W., 195, 208
Dobbins, R. L., 374, 375
Dobson-Stone, C., 454
Dodson, M. W., 323
Dolezal, P., 22, 24, 26
Dolfini, L., 262
Dolinski, K., 15
Dominko, T., 83
Domon, B., 127
Dong, J., 87, 88
Dong, Y., 257, 258, 261, 263, 265, 267, 270
Dönnes, P., 38
Doolittle, W. F., 25
Dorenbos, K. A., 52
Dorn, G. W. II, 257
Douglas, M. G., 314
Downey, J. M., 257
Downward, J., 222
Drawid, A., 11, 12
Dresner, A., 374

Drew, B., 355
Drewes, L. R., 427
Drisdel, R. C., 152
Drost, M. R., 375
Du, F., 388
Du, Y., 190
Duan, Y., 118
Duarte, A., 174, 186, 187
Dube, D. H., 152, 155
Duchen, M. R., 473, 474
Duckworth, A., 324, 331
Dudley, G. A., 215
Dufour, S., 215, 216, 352, 374, 383, 384, 386, 387, 388
Dujardin, G., 9, 13
Duncan, R. F., 257
Dunkley, P. R., 52
Dunlap, J., 225
Dunn, B., 6
Dunphy, J. T., 150, 151
Dupont, O., 471
Durand, A. G., 99
Durick, K., 257
Dwight, S. S., 6, 15
Dyal, P. L., 24
Dyer, R. B., 321
Dziura, J., 352, 386

E

Edelman, D. B., 320, 322
Edelman, G. M., 266
Edes, I., 65
Edwards, D., 453
Edwards, R., 355
Eguchi, Y., 99
Ehrat, M., 112
Eide, L., 321
Eisenberg, D., 27
Eisner, R., 38
Elander, A., 354
Elbashir, S., 218
Elde, R. P., 461
Elder, K., 325
Eleff, S. M., 336
Elias, J. E., 54, 55, 125
Ellingsen, O., 224
Ellisman, M. H., 257
Elorza, A., 290, 291, 292, 294, 295, 301, 303
Elpeleg, O., 16
Else, P. L., 417
Elstner, M., 3, 82
Emanuelsson, O., 34
Embley, T. M., 24, 25, 27
Eng, J. K., 98
Engberg, S. T., 186
Engelbrecht, J., 34
Ennion, S. J., 362

Enríquez, J. A., 338
Eppig, J. T., 7, 15
Ericson, B. G., 475
Eriksson, K. F., 216, 223, 375
Erlanded, S. L., 461
Ernberg, I., 216
Ernst, R. R., 382
Essen, B., 388
Esteves, T. C., 396
Evan, G. I., 474
Evans, F. J., 67
Evans, R. J., 362
Evans, W. J., 387
Evgrafov, O., 290
Ewald, A. J., 290, 303
Ewer, J., 321, 328
Eyrich, B., 83

F

Fagarasanu, A., 98
Faghihi, M. A., 216, 218, 222
Faherty, B. K., 54, 55
Fahien, L. A., 426, 427
Fahimi, F. M., 67, 68
Fahy, E., 10, 13, 30, 32, 51, 82, 98
Fairfull, L., 354
Falck, J. R., 151, 152, 153, 155, 159, 163
Falick, A. M., 114
Falk, P. K., 352
Falk Petersen, K., 374
Falvo, J. V., 9, 11, 14, 15
Fang, D., 219
Fang, X., 49, 52, 53, 54, 56, 57, 59
Fantz, D. A., 181
Farfari, S., 426
Farina, L., 344
Fathy, P., 83
Fazel, A., 98
Fedorov, Y., 219
Feliciello, A., 118, 172
Feltri, M. L., 323
Feng, S., 86, 87
Ferguson-Miller, S., 257
Fernández-Silva, P., 338
Fernstrom, M., 224
Ferreirinha, F., 321
Ferrell, R. E., 354, 374
Ferrick, D. A., 419, 420
Ficarro, S., 64, 118, 196, 197, 204, 207
Ficarro, S. B., 85, 108, 125, 206, 207
Fields, S., 9, 15
Filipovska, A., 83
Filippin, L., 471
Fink, M. P., 320
Finkel, T., 138
Finkielstein, C., 186, 187, 188
Finlay, B. J., 23

Finnigan, C., 98
Fischer, A., 24
Fischer, C. P., 216, 218
Fischer, E. H., 276
Fischer, H., 215, 217, 224
Fischle, W., 138
Fisher, J. K., 99
Fisher, W. H., 186, 188
Fiskum, G., 52, 53
Fiumera, H. L., 24
Flanders, W. D., 336
Flint, J., 456
Fliss, M. S., 336
Fong, Y. L., 234
Ford, G. C., 352
Foret, F., 51
Forner, F., 51
Forsén, S., 378
Foster, L. J., 51
Foster, P. G., 24
Fountoulakis, M., 10, 13
Fowler, S., 262, 286
Frachon, P., 303
Fradkov, A. F., 468, 470
Franchini, K. G., 397
Frank, G., 83
Fraser, M., 454
Fraser, S. E., 290, 303
Frattini, J., 387
Fredlund, K. M., 64
Fredriksson, K., 213, 215, 216, 217, 218, 222
Freeman, D., 388
Freeman, H., 452, 457, 464
Freeman, H. C., 452, 453, 457, 459, 463, 469
Freeman, H. N., 98
Freeman, S. K., 154
Freigang, J., 98
French, S., 51, 63, 64, 65, 67, 195, 257
Friauf, E., 128
Friedman, M., 388
Friedman, N. R., 359, 368
Friedrich, T., 25
Frisardi, M., 24
Frishman, D., 6
Fritz, S., 9, 11
Frolkis, M., 216
Fromm, H. J., 134
Frontera, W. R., 387
Froschauer, E., 305, 311
Frye, R., 138
Frye, R. A., 139, 140, 143
Fuchs, F., 9, 11
Fujiki, Y., 262, 286
Fujita, N., 37
Fukata, M., 151
Fukata, Y., 151
Fuller, S., 453
Futai, M., 314

G

Gabaldón, T., 4
Gabriel, J. L., 64
Gabriel, K., 26, 27
Gadaleta, M., 187
Gadian, D. G., 375, 379
Gaeta, L., 321
Gagneur, J., 9, 13
Galan, J., 89
Galeva, N., 113
Galia, H. C., 114
Galli, S., 174
Galvanovskis, J., 451
Gambaryan, S., 125
Gamper, N. L., 417
Gan, L., 138
Gandhi, S., 222
Gao, F., 257
Garber, E. A., 160
Garbi, C., 118
Garcia, R., 215, 216, 352, 386
Garg, A., 374
Garlid, K. D., 307
Gashue, J. N., 98
Gasparetti, A. L., 397, 400, 402
Gass, M. A. S., 112
Gassmann, M., 223
Gaston, D., 21, 25, 29, 36, 41
Gaucher, S. P., 10, 13, 51, 98
Gavigan, K. E., 352, 388
Gawryluk, R. M., 25, 26, 33, 36
Gebauer, P., 51
Geddes, J. W., 52
Geer, L. Y., 54, 55, 128
Geissler, A., 45
Gennis, R. B., 113
Gentry, M. S., 276
Gerber, L. K., 186
Gerber, S. A., 64, 65
Gerencser, A. A., 417, 419
Gerke, L. C., 9, 11, 14, 15
Germot, A., 24
Gerow, K., 374
Gerstein, M., 11, 12
Gertsenstein, M., 459
Gertz, M., 138
Getman, D. P., 154
Ghaemmaghami, S., 9
Ghaim, J., 113
Ghosh, S., 24, 374
Ghosh, S. S., 10, 13, 51, 98
Giaever, G., 9, 11
Gibson, B. W., 10, 13, 51, 98, 152
Gietz, R. D., 314
Gill, E. E., 25, 29, 36
Gillespie, D. A., 257
Gimeno, R. E., 217

Giordano, E., 461
Gispert, S., 213, 276
Giugliani, R., 344
Gjertsen, B. T., 99
Glaser, L., 156
Glenfield, P. J., 52
Glenister, P., 453
Glenn, G. M., 10, 13, 51, 98
Glocker, M. O., 98
Glossip, D., 181
Gnad, F., 86, 94, 125
Gnaiger, E., 353, 365, 366, 368
Godzik, A., 276
Goldberg, A. V., 25
Goldberger, O. A., 12, 13, 51, 55
Goldfine, A. B., 216, 375
Golding, G. B., 25, 26
Goldstein, J. C., 474
Goldstein, L. S., 323
Goldsworthy, M., 451, 452, 457, 464
Gomez, L., 243
Gonez, J., 102
Gong, B., 86, 87
Good, L., 216, 218, 222
Goodlett, D. R., 98
Goodman, T. N., 64
Goodpaster, B. H., 354, 374
Gorden, P., 374
Gordesky-Gold, B., 323
Gordon, J. A., 197
Gordon, J. I., 150, 154
Gorska-Andrzejak, J., 323
Goss, V. L., 90, 126
Goto, T., 374
Goto, Y., 338, 344
Gotoh, O., 338
Gottesman, M. E., 172
Goudreault, M., 325
Gould, K. L., 104
Gout, A. M., 25
Gouw, J. W., 85, 86, 91
Grabenbauer, M., 232
Graf, S. A., 291, 292, 294, 303
Gramolini, A. O., 51, 55, 56
Grant, A., 410, 417
Granzow, M., 223
Gray, M. W., 23, 25, 26, 33, 36, 350
Green, A., 374
Green, D. A., 474
Green, D. R., 98
Green, K., 396
Green, W. N., 152
Green-Church, K. B., 98
Greenhaff, P. L., 214, 215, 217, 224
Greenlund, L. J., 352
Gregory, R., 6
Greiner, R., 38
Grierson, A. J., 321

Griffin, E. E., 290, 301, 303
Griggs, R. C., 355
Grisar, T., 224
Grishin, N. V., 83, 110, 144
Gritsenko, M. A., 4, 9, 11, 46, 51, 98
Grivell, L., 9, 11
Gronendyk, J., 98
Gross, R. W., 152
Grossman, L. I., 64, 118, 196, 197, 204, 207, 208
Grujic, D., 397, 411, 418
Grunicke, H., 65
Grunwell, J. R., 155
Guan, K. L., 276
Guda, C., 30, 32
Guda, P., 30, 32, 82
Guerin, M., 307
Guiard, B., 9, 11, 25, 27, 51, 98
Guillery, O., 291
Guillou, E., 291
Güldener, U., 6
Gunawardena, S., 323
Gundry, C. N., 454
Gunter, K. K., 50
Gunter, T. E., 50
Guo, A., 90, 126
Guo, J., 219
Guo, M., 89, 323
Guo, T., 54
Guo, Y., 220, 222, 257
Guo, Z., 352, 353, 360
Guo, Z. K., 353, 388
Guo, Z.-M., 225
Gurney, K., 28
Gustafson, K., 328
Gustafsson, T., 214, 215, 217, 224
Gutierrez-Merino, C., 209
Gutmann, N. S., 171
Gygi, S. P., 54, 55, 64, 65, 125
Gylfe, E., 468

H

Haas, W., 54, 55
Haddad, F., 215
Haebel, S., 10, 13
Haffner, S. M., 374
Hagen, T., 397, 411, 418
Haggmark, T., 215
Hahn, H., 257
Haigh, S. E., 290, 291, 292, 294, 295, 301, 303
Håkansson, K., 85, 108
Hakkinen, A. M., 374
Halavaara, J., 374
Halban, P. A., 411, 461, 462, 464
Hall, J., 336
Haller, T., 353
Hallett, M. T., 51, 55, 56
Halliday, D., 355

Halse, R., 217
Hamamoto, M., 9, 11
Hamblin, K., 25, 29, 36, 41
Hamilton, A., 433
Hamilton, D. L., 217
Hammond, S. M., 432
Hamosh, A., 7
Hampl, V, 25
Hamuro, J., 186
Han, D., 51
Han, G., 87, 88
Han, H., 99
Han, J., 260, 261, 266
Han, X., 152
Han, X. J., 233, 234
Hand, S. C., 366
Handschin, C., 402
Hang, H. C., 155
Hangauer, M. J., 151, 153, 155, 159, 163
Hani, J., 6
Hannon, G. J., 433
Hanson, B. J., 51
Hao, P., 225
Harada, H., 37
Haram, P. M., 224
Harborth, J., 218
Harder, Z., 233
Harding, M., 24
Hardy, W. R., 325
Hargreaves, I., 222
Harley, C. B., 216
Harman, D., 352
Harman, J. M. P., 23
Harmon, J., 463
Harper, M. E., 410
Harris, M. A., 15
Harris, R. A., 51, 63, 64, 65, 195, 257
Harrison, R., 407
Harrison, S. M., 52
Harriss, J. V., 314
Hartmann, R., 205
Harvey, K., 213, 276
Hasan, N. M., 426
Haseloff, R. F., 108
Hashimoto, A., 9, 11
Hashimoto, T., 186
Hather, B., 215
Hatzinisiriou, I., 314, 316
Hawkins, J., 35
Hawkins, R. A., 384
Hay, B. A., 323
Hayashi, J.-I., 335, 336, 338, 340, 343, 344
Hayflick, L., 216
Haynes, A., 452, 457
Hays, L. G., 98
Hazelton, J. L., 52, 53
Hazuda, H. P., 374
He, J., 83, 354, 374, 375

He, L., 14
He, W.-Z., 225
Healy, D. G., 213, 276
Heath, J. W., 52
Heazlewood, J. L., 25, 82
Heck, A. J., 85, 86, 91, 125
Heegaard, N. H., 127
Heerdt, B. G., 336
Heikal, A. A., 313
Heinen, E., 224
Heiss, K., 4, 55
Heldin, C. H., 102
Hell, J. W., 250
Hellman, B., 468
Hellman, U., 102
Hendler, R. E., 374
Hennen, G., 224
Henquin, J. C., 461
Henriksson, J., 388
Henry, R. R., 217
Her, L. S., 323
Herge, O. D., 461
Herman, Z. S., 9, 11
Hernandez, H., 326
Hernandez, M., 25
Herschman, H. R., 260, 261, 266
Hersi, K., 325
Herzig, S., 290
Hess, S. T., 313
Hesselink, M. K., 388
Hester, E. T., 6
Hiardina, T., 99
Hibi, N., 337
Hickey, M. S., 352, 388
Hilhorst, M. J., 125
Hill, D. E., 12, 13, 25, 82
Hill, D. P., 15
Hillenkamp, F., 114
Hiltunen, J. K., 9, 11
Hirano, M., 50
Hiraoka, Y., 9, 11
Hirose, S., 233
Hirschey, M. D., 137
Hirt, R. P., 24, 25
Hixson, K. K., 4, 9, 11, 46, 53, 98
Hjerrild, M., 12, 13, 51, 98
Hodgson, S., 454
Hoek, J. B., 475
Hoffman, R. A., 378
Hoffmann, H. P., 290
Hofstadler, S. A., 54
Höglund, A., 38
Højlund, K., 83
Holfrich-Forster, C., 321, 328
Hollenbeck, P. J., 320, 321, 322, 324, 331
Holt, S. E., 216
Holton, F., 362
Home, W. C., 118

Honda, H. M., 256, 268, 270
Honda, S., 233
Honick, A. S., 321, 322, 325
Honjo, T., 99
Honma, Y., 336, 340, 344
Honzatko, R. B., 134
Hood, D. A., 352, 354
Hopkins, I. B., 52, 53
Hoppel, C. L., 67, 83, 97, 98, 99, 100, 101, 102, 103, 104, 359, 368
Hopper, R. K., 64, 65, 195, 257
Hoppins, S., 232
Horbinski, C., 257
Horikawa, I., 138
Horinouchi, S., 9, 11
Horiuchi, D., 324, 328, 330, 331
Horn, A., 10, 13
Hornbeck, P., 115
Horner, D. S., 24
Horner, E., 452, 453, 463, 464, 469
Hörtnagel, K., 4, 82
Horton, P., 11, 12, 13, 37
Horvath, C., 52
Hosaka, H., 343
Hosseini, S. H., 336
Hou, C., 475
Hoult, D. I., 375
Houmard, J. A., 352, 388
Houstis, N., 216, 223, 375
Houten, S. M., 388
Howe, C. J., 23
Howell, K. E., 98, 104, 109
Howson, R. W., 9, 11, 14, 15
Hoye, A. T., 320
Hrdy, I., 24
Hsiao, S. F., 257
Hu, P., 51, 55, 56
Hu, Y., 86
Hua, S., 11, 12, 13
Huang, C. Y., 25, 257
Huang, J.-Y., 137
Huang, L. J., 257
Hubbard, A. L., 262, 286
Hubbard, M., 113
Huber, M., 257
Huff, J. E., 98
Hug, L. A., 24
Hughes, V. A., 387
Hugill, A., 451, 452, 453, 457, 459, 464
Huh, J. R., 323
Huh, W. K., 9, 11, 14, 15
Hulbert, A. J., 417
Hullihen, J., 282
Hultman, E., 214, 353, 360
Hundai, T., 475
Hunt, D. F., 108, 125
Hunt, M. C., 186
Hunter, J., 453

Hunter, T., 125, 256
Hurd, D. D., 329
Hurd, T. R., 83
Hutchinson, D. S., 216, 218
Hüttemann, M., 64, 118, 193, 195, 196, 197, 204, 207, 208
Hwang, J. K., 36
Hwang, S.-I., 51
Hwang, Y., 388

I

Ichikawa, M., 336, 344
Ignatchenko, A., 51, 55, 56
Igout, A., 224
Ikeda, M., 461
Ilsley-Tyree, D., 219
Imami, K., 91
Imanishi, H., 336, 340, 344
Imhof, A., 98
Inada, N. M., 397, 400
Inoue, K., 338, 344
in't Veld, P., 427
Invernizzi, F., 344
Irth, H., 51
Irving, B. A., 353, 388
Ishida, Y., 99
Ishihama, Y., 91
Ishihara, H., 427
Ishihara, N., 233, 240
Ishikawa, K., 335, 336, 340, 344
Isobe, K., 343, 344
Issa, M. M., 336
Issaq, H., 101
Ito, S., 338, 343
Itoh, S., 118
Ivarsson, R., 427
Ivy, J. L., 374
Iwakura, Y., 337
Iwamura, Y., 343
Iyer, V. R., 9, 11

J

Jacintho, J. D., 50
Jackson, A., 219
Jackson, F. R., 321, 328
Jacob-Hirsch, J., 16
Jacobs, D., 181
Jacobson, J., 473, 474
Jacotot, E., 50
Jager, S., 215
Jahani-Asl, A., 233
Jain, M., 12, 13, 51, 55
Jakobsen, L. A., 127
Jaleel, A., 352
James, D. I., 303
Jan-Henrik, S., 85
Janjic, D., 411, 464

Jankovic, J., 220, 222
Jansson, E., 214, 215, 217, 224, 388
Jarausch, J., 199, 205
Jarosch, E., 262
Jarvie, P. E., 52
Jeannot, M. A., 113
Jedrychowski, M., 125
Jeffrey, F. M. H., 444, 446
Jekabsons, M. B., 417
Jen, J., 336
Jeneson, J. A., 375
Jenkins, A. B., 374
Jenkins, C. M., 152
Jensen, L. J., 9, 13
Jensen, M. V., 426, 448
Jensen, O. N., 88, 127
Jensen, S. S., 127
Jeong, S. Y., 290
Jetton, T. L., 426
Ji, J., 83
Ji, S., 83
Jia, Y., 6
Jiang, C., 323
Jiang, T., 83
Jiang, X., 86, 87, 88
Jin, J., 11, 13
Jofuku, A., 233, 240
Johnson, D. T., 51, 63, 64, 65, 195, 257
Johnson, G. A., 380
Johnson, J., 219
Johnson, P. J., 24, 25
Jonas, E. A., 471
Jones, E. W., 314
Jones, K. A., 388
Jones, T., 9, 11
Jope, R. S., 257
Joppich, C., 9, 11, 25, 27, 51, 98
Jorgensen, T. J. D., 88, 127
Joubert, F., 67
Joyner, M. J., 352, 353, 364
Jucker, B. M., 352, 383, 384, 386, 388
Judge, A. D., 219
Juin, P., 474
Julian, M. W., 98
Jun, A. S., 344, 355
Jung, S., 98
Junne, T., 262
Justice, M., 453
Juvik, G., 6

K

Kachholz, B., 138
Kaczmarek, L. K., 471
Kadenbach, B., 64, 195, 199, 205, 208
Kadin, J. A., 7
Kaech, S., 322
Kagan, J., 50, 83

Kagan, V. E., 50, 320
Kagawa, Y., 338, 343
Kah, R., 132
Kahl, J., 352, 353, 364
Kahn, C. R., 374
Kaijser, L., 214
Kaitsuka, T., 233, 234
Kaizak, S., 451
Kakei, M., 461
Kallen, C. B., 190
Kamal, M., 12, 13, 51, 98
Kamata, A., 9, 11
Kamzplova, S., 426, 448
Kanamori, T., 336
Kaneda, H., 336
Kang, B. P., 257
Kang, Y. J., 260, 261, 266
Kannan, A., 51, 55, 56
Kantengwa, S., 461
Karbowski, M., 246, 292, 293, 296, 301, 303
Karger, B. L., 51
Karlsson, G., 99
Karpilow, J., 219
Kashima, D. T., 11, 13
Kashyap, S., 216, 375
Kastaniotis, A. J., 9, 11
Katoh, A., 152
Katsonouri, A., 113
Katz, S., 290, 295, 301, 303
Kaufmann, N., 324, 331
Kay, L., 368
Keays, D. A., 456
Kecman, V., 28
Keen, G., 222
Keller, B. O., 98, 149, 151, 153, 155, 159, 163
Keller, H. J., 65
Keller, P., 213, 215, 217
Kelley, D. E., 354, 374, 375, 396
Kelly, P., 118, 257, 276
Kemp, G. J., 380, 388
Kendrick, M. A., 426
Kengo, N., 83
Kent-Braun, J. A., 352, 387, 388
Kerbey, A. L., 64, 74, 75
Kerner, J., 83, 97, 98, 99, 100, 101, 102, 103, 104
Kerr, D. S., 359, 368
Kho, Y., 83, 110, 144
Khvorova, A., 219, 235
Kibbey, R. G., 425, 426
Kientsch-Engel, R. I., 67, 68
Kiessling, K. H., 214
Kikuchi, A., 338
Kilvington, S., 24
Kim, H. S., 138
Kim, J., 138, 141, 142
Kim, S., 388
Kim, S. C., 83, 85, 93, 94, 110, 144
Kim, Y. B., 397, 411, 418

Kim, Y. L., 355
Kimple, M. E., 118, 257, 276
Kimura, Y., 233
King, H., 374
King, M. P., 338, 339, 340
King, S. R., 190
Kingsley, P. B., 379
Kinkl, N., 101
Kinoshita, S., 186
Kinzler, K. W., 336
Kirby, C., 134
Kishore, N. S., 152, 154
Kislinger, T., 51, 55, 56
Kiss, E., 65
Kitamura, T., 10, 11, 13
Klaus, K., 352
Klaus, K. A., 353, 388
Klebe, R. J., 314
Klemm, C., 108
Klingenberg, M., 65, 199
Klopstock, T., 3, 82
Klose, J., 10, 13
Klupsch, K., 222
Knapp, M., 171
Knebel, A., 119
Knoblach, B., 98
Ko, Y., 10, 13, 51
Kobayashi, M., 23
Kobayashi, S., 219
Kobayashi, Y., 9, 11
Koc, E. C., 85
Koc, H., 85
Koch, A., 232
Koch, L. G., 224
Koehler, C. M., 50, 255, 257, 258, 261, 262, 263, 265, 267, 270
Kohane, I., 216, 375
Kohlbacher, O., 38
Komatsuzaki, Y., 343
Kondo, H., 343
Kondoh, H., 160
Konstantinov, Y. M., 64
Koopman, W. J., 25
Korge, P., 255, 256, 257, 258, 261, 263, 265, 267, 268, 270
Kornbluth, S., 50
Kornfeld, K., 181
Korsmeyer, S. J., 99
Koshikawa, N., 336, 340, 344
Koster, J., 232
Kostiuk, M. A., 149, 151, 153, 155, 159, 163
Kotagal, N., 462
Kovacic, P., 50
Kowalak, J. A., 54, 55, 128
Koyama, A. H., 161
Kozany, C., 4, 9, 11, 46, 51, 98
Kozjak, V., 48
Krab, K., 199

Kramarova, T. V., 396
Kramer, C., 225
Kranias, E. G., 65
Kraunsoe, R., 368
Krause, E., 108
Krause, R., 9, 15
Krauss, S., 397, 411, 418
Kresbach, G. M., 112
Kress, G. J., 326
Krieger, D. A., 354
Kriketos, A. D., 374
Krishnan, K. J., 138
Kristal, B. S., 51
Kristian, T., 52, 53, 56, 57, 59
Kroemer, G., 98, 257
Krolewski, A. S., 374
Krssak, M., 374
Krüger, U., 114
Krushel, L. A., 266
Kuchar, J. A., 152
Kuhnke, G., 25
Kuhn-Nentwig, L., 199
Kulda, J., 24
Kulkarni, S., 325
Kumar, A., 9, 11
Kumar, C., 125
Kunji, E. R., 24, 25
Kuno, S. Y., 343
Kurata, T., 160
Kuwana, T., 50
Kuwazaki, S., 23
Kuznetsov, A. V., 366
Kweon, H. K., 85, 108
Kwong, J. Q., 336
Kyono, Y., 91

L

Lackner, L., 232
Lacy, P. E., 462
Laemmli, U. K., 398
Lai, J., 65
Lakaye, B., 224
Lakin-Thomas, P. L., 410
Lamantea, E., 344
Lambert, A. J., 138, 396
Lambo, R., 257
Lambrecht, B., 161
Landaker, E. J., 216, 375
Lander, E. S., 12, 13, 98
Lang, B. F., 23, 25, 26, 350
Langridge, J., 10, 13, 98
LaNoue, K. F., 65
Lanza, I. R., 349, 352, 353, 364, 388
Lapidot, M., 225
La Rocca, N., 118
Larsen, M. R., 88, 127
Larsen, R. G., 352

Larsson, O., 211, 215, 216, 217, 218, 224
Las, G., 290, 295, 301, 303
Lascaris, R., 9, 11
Lassmann, T., 216, 217, 218, 222
Lau, C., 65
Lau, K. F., 321
Laudeman, T., 54, 55
Lauquin, G. J. M., 132
Laurila, E., 216, 223, 375
Lavi, S., 257, 258
Lavorgna, G., 225
Laymon, R. A., 323
Lazarev, A., 51
Lazarow, P. B., 262, 286
Le, W., 220, 222
Leahy, J. L., 426
Leake, D., 219, 235
Lebon, V., 352, 383, 384, 386
Le Caer, J. P., 9, 13
Lee, A., 451
Lee, C., 52, 53
Lee, C. S., 49, 52, 53, 54, 55, 56, 57, 59
Lee, H., 388
Lee, I., 64, 193, 195, 196, 197, 204, 207, 208
Lee, I. H., 138
Lee, J., 85, 93, 94
Lee, K., 97
Lee, K. A., 90, 126
Lee, L., 118
Lee, S. J., 224
Leeuwenburgh, C., 355
Legeai, F., 11, 12, 13, 14, 15, 30, 32
Legendre-Guillemin, V., 321
Legros, F., 303
Le Guyader, H., 24
Lehar, J., 216, 223, 375
Lehmann, L. H., 151
Lehner, B., 225
Lehninger, A. L., 400, 417
Lehrke, M., 374
Lei, H., 379, 388
Leiber, M., 223
Leigh, J., 24
Leigh, P. N., 321
Leitner, A., 85
Lemaire, C., 9, 13
Lemay, S., 118
Lemieux, C., 25, 26
Lemire, B. D., 98
Lemire, K., 419, 420
Lenaers, E., 388
Lendeckel, W., 218
Lengauer, C., 336
Leo, S., 290
Leonard, J. V., 50, 232
Leon-Avila, G., 25
Leparc, G. G., 65
LePecq, J. B., 338

LePine, R., 444
Leroueil, M., 11, 13
Lescuyer, P., 10, 13
Levine, J. A., 352
Lewandrowski, U., 117, 118, 125, 128, 132, 133
Lewis, A. P., 98
Lezot, F., 225
Li, B., 219
Li, G., 406, 411, 464
Li, J., 54, 125
Li, L., 98, 113, 115
Li, Q., 152
Li, S. A., 233, 234
Li, Y., 86, 87, 292, 336
Li, Y.-X., 225
Li, Y.-Y., 225
Lian, W. N., 257
Lichtsteiner, S., 216
Liggett, S. B., 65
Lightfoot, S., 223
Likic, V., 22, 26
Lill, R., 25
Lillioja, S., 374
Lim, L., 219
Lim, S., 336
Lin, C. H., 257
Lin, D., 171
Lin, J., 215
Lin, M., 268, 312
Lin, W. J., 257
Lindahl, K. F., 336
Linden, M., 99
Lindenberg, K. S., 215
Linder, M. E., 150, 151
Lindgren, C. M., 216, 223, 375
Lindmark, D. G., 24
Lindner, L. W., 85
Lindquist, P. J., 186
Linn, T. C., 276
Linsley, P., 219
Lipman, D. J., 12
Lippiat, J., 452, 457
Lippincott-Schwartz, J., 292
Liron, T., 257
Lithgow, T., 22, 26, 50
Littlejohn, T. G., 25, 26
Liu, J., 138, 213
Liu, L., 225
Liu, S. S., 195
Liu, X., 134
Liu, Y., 9, 11, 83
Liu, Y.-Q., 426
Liu, Z., 388
Lively, C. M., 324, 328, 330, 331
Livigni, A., 118
Lloyd, D., 24
Loeffler, M., 50
Lombard, D. B., 138, 141, 142

Lombes, A., 291, 303
Lonie, L., 454
Lopez, M. F., 50, 51
López-Pérez, M. J., 338
Lorenzo, H. K., 50
Loros, J., 225
Lorusso, V., 257
Loser, S., 468
Lottspeich, F., 64, 118, 196, 197, 204, 207
Lowell, B. B., 388
Lozano, R. C., 171
Lu, C. H., 36
Lu, D., 426, 448
Lu, G., 255, 257, 258, 260, 261, 263, 265, 266, 267, 270
Lu, H. F., 154
Lu, P., 38
Lu, T., 152
Lu, Y. F., 233, 234
Lu, Z., 38
Luca, C., 90
Luchansky, S. J., 155
Luche, S., 10, 13
Lucocq, J. M., 25
Ludwig, B., 64
Lukasová, G., 24
Lukyanov, K. A., 468, 470
Lukyanov, S., 468, 470
Lunardi, J., 10, 13
Lund, P. E., 468
Lundgren, F., 354
Lundin, A., 360
Luo, W., 65
Lurin, C., 11, 12, 13, 14, 15, 30, 32
Lutz, K. J., 387
Luzi, L., 374

M

Ma, Y., 257
MacCoss, M. J., 98, 104, 109
Macdonald, I. A., 214
MacDonald, M. J., 426
Macdonell, C., 38
MacDougall, M., 225
Macek, B., 125, 237
MacFall, J. R., 380
Machicado, C., 338
Maciel, F. C., 174
MacKeigan, J. P., 220, 222
MacLachlan, I., 219
Madden, T. L., 12
Maechler, P., 406, 411
Maehama, T., 276
Magalhães, P. J., 471
Magee, A. I., 161
Mahaut-Smith, M. P., 362
Mahuran, D. J., 98

Mai, Z., 24
Maier, D., 6
Maj, M. C., 276
Maksimova, E., 219
Malfatti, E., 344
Malfer, C., 161
Malhotra, A., 257
Malka, F., 291
Mallik, B., 83
Malloy, C. R., 375, 444, 446
Maloberti, P., 169, 171, 174, 186, 187
Malorni, A., 85, 86, 87
Malorni, L., 85, 86
Mandavilli, B. S., 321
Mandelkow, E. M., 321
Mandella, R. D., 427
Mandich, D. V., 98
Manfredi, G., 336
Mangion, J., 50
Manieri, M., 215
Mann, M., 51, 86, 94, 125, 237
Mannhaupt, G., 6
Manning, G., 256
Manolis, S. C., 417
Manon, S., 307
Mans, A. M., 384
Manser, C., 321
Mansfield, L., 222
Mao, M., 219
Marande, W., 25
Marcotte, E. M., 27
Mardones, G., 98
Marelli, M., 98
Marin, R. M., 397
Maris, J., 352
Markey, S. P., 54, 55, 128
Markowitz, S. D., 336
Marsh, G. D., 387, 388
Marsh, M., 150
Marshall, B., 138
Marshall, F. F., 336
Marshall, W., 219
Marshall, W. F., 290
Marshall, W. S., 235
Martin, B. C., 374
Martin, D. D., 152
Martin, W., 23, 25, 27
Martinez, P., 10, 13
Martinou, J. C., 10, 13, 98, 290, 303
Martins de Brito, O., 233
Marto, J. A., 85
Maruo, F., 23
Maruyama, K., 186
Maruyama, O., 37
Marzo, I., 50
Masiar, P., 9, 11
Mason, D. E., 206, 207
Mason, G. F., 215, 352, 382, 384, 386, 387

Masselon, C. D., 53
Mastrangelo, A. J., 115
Mathes, I., 64, 118, 196, 197, 204, 207
Matlib, M. A., 186
Matschinsky, F. M., 406
Matsui, H., 233, 234
Matsumoto, S., 461
Matsuo, Y., 23
Matsuoka, Y., 99
Matsushita, M., 233, 234, 257
Matsuyama, A., 9, 11
Mattenberger, Y., 303
Matthay, M., 225
Matthews, P. M., 379
Mattson, M. P., 321
Matveeva, O., 235
Maurer, U., 98
Mauro, S., 90
May, G. H., 257
May, W. S., 257
May, W. S., Jr., 257
Mayer, K. F., 6
Maynard, D. M., 54, 55, 128
Mayya, V., 51
McBride, H., 233
McBride, H. M., 233
McCaffery, J. M., 301
McCleland, M. L., 108, 125
McClintock, K., 219
McConnell, H. M., 378
McConnell, J. P., 352, 353, 364
McCrady, S. K., 352
McCreary, C. R., 387, 388
McDaniel, M. L., 462
McDermott, C. J., 321
McDonald, T., 10, 13, 51
McDonald, W. H., 104
McElroy, W. D., 360
McEvoy, R. C., 461
McFarland, R., 14
McGarry, J. D., 374
McIntyre, E. A., 217
McKenna, M. C., 52, 53
McKusick, V. A., 7
McLoughlin, D. M., 321
McManus, M. T., 434
McMath, H., 453
McMillin-Wood, J., 354
McNiven, M. A., 232
McPherson, P. S., 321
McQuibban, A., 213
McStay, G., 98
McWherter, C. A., 154
Meadows, N., 451
Meda, P., 411, 461, 462, 464
Meeusen, S., 301
Meeusen, S. L., 290
Megeath, L. J., 323

Meggs, L. G., 257
Mehta, P. P., 152, 154
Meinertzhagen, I. A., 323
Meisinger, C., 4, 9, 11, 22, 25, 26, 27, 45, 48, 51, 83, 98
Meissner, H. P., 461
Meitinger, T., 3, 4, 55, 82, 98
Melanie, Y. W., 83
Mele, P., 186, 188
Mele, P. G., 171, 186, 187
Melov, S., 50, 51
Mendez, C. F., 169, 171, 186, 187, 188
Mendez, G., 366
Meng, L., 83
Menshikova, E. V., 354, 374, 375
Menzel, W., 223
Merchant, S., 262
Merglen, A., 406, 411
Merle, P., 205
Merrill, R. A., 250
Mersiyanova, I. V., 290
Messerschmitt, M., 9, 11
Meuller, J., 475
Mewes, H. W., 6
Meyer, H. E., 9, 11, 25, 27, 45, 51, 98
Meyerspeer, M., 380
Miatkowski, K., 160
Michaels, R. L., 461
Michalak, M., 98
Michan, S., 138
Michels, J., 98, 102
Michishita, E., 138
Miguel, J. S., 38
Mihara, K., 233, 240
Mijaljica, D., 314, 316
Mijat, V., 452, 457, 464
Mika, S., 85
Milani, A., 186
Milbrandt, J., 321
Miljan, E. A., 222
Millar, A. H., 25, 82
Miller, J. L., 85
Miller, K. E., 324, 331
Miller, P., 9, 11
Miller, W., 12
Miller, W. L., 171
Milner, M. R., 374
Minkiyu, P., 186, 188
Minkler, P., 102, 104
Mitchell, B. D., 374
Mitchell, D. A., 151
Mitchell, D. M., 113
Mitchell, P., 307
Miura, G. I., 151
Miwa, S., 396
Miyano, S., 37
Miyazaki, T., 118
Miyazaki, Y., 216, 375

Mizani, S. M., 65
Mizusawa, H., 343
Mochly-Rosen, D., 257
Modafferi, E. A., 52
Moebius, J., 125
Mogelsvang, S., 98
Mogensen, M., 83
Mohamed, H., 290, 291, 292, 294, 295, 301, 303
Mohammed, S., 85, 86, 91
Mohan, D., 53
Moir, L., 452, 457
Mokranjac, D., 9, 11
Mokrejs, M., 6
Molik, S., 25
Molina, A. J., 289, 290, 295, 301, 303
Moller, I. M., 64
Molloy, M., 101
Monaco, A. P., 454
Monahan, W. G., 379
Monroe, M. E., 4, 9, 11, 46, 51, 98
Montine, K. S., 11, 13
Montoya, J., 338
Moore, B., 235
Moore, R. J., 4, 9, 11, 46, 51, 98
Moorhead, P. S., 216
Mootha, V. K., 12, 13, 51, 55, 98, 215, 216, 223, 375
Mooyer, P. A., 232
Moraes, J. C., 397, 400
Moran, R., 419, 420
Moreno, R. D., 83
Morfini, G., 321
Mori, M., 461
Morin, G. B., 216, 325
Morino, K., 374, 387
Moritz, A., 90, 126
Moriyama, Y., 314
Morrice, N., 118, 119, 124
Morris, Q., 51, 55, 56
Morris, R. L., 324, 331
Morse, D., 352
Mortensen, P., 125
Mortimer, R. K., 6
Moser, E., 380
Moser, P. L., 368
Mostoslavsky, R., 138, 141, 142
Mottaz, H. M., 4, 9, 11, 46, 98
Mottola-Hartshorn, C., 268, 312
Mounier, R., 225
Moyle, J., 307
Mozdy, A. D., 290
Mruk, D. D., 323
Mucke, L., 138
Mudaliar, S., 217
Mueller, J. C., 4, 82
Mueller, L. N., 127
Mueller, M., 127
Mueller, O., 223

Muglia, M., 290
Mulder, H., 426, 448
Müller, M., 24, 25, 410, 417
Mumby, M., 257
Mumby, S. M., 151
Münsterkötter, M., 4, 6, 82
Muqit, M. M., 213, 276
Muramatsu, M., 99
Murphy, A., 138, 141, 142
Murphy, A. N., 10, 13, 51, 98, 277, 279, 283, 286
Murphy, L. O., 220, 222
Murphy, M. P., 83
Murphy, R. F., 314
Murray, A., 290
Mustelin, T., 276
Muzio, M., 50

N

Nagarajan, S. R., 154
Nagisa, S., 83
Nagy, A., 459
Nair, K. S., 349, 352, 353, 355, 360, 364, 388
Nairn, A. C., 233, 234
Najafi, H., 406
Najjar, S. M., 224
Nakada, K., 336, 340, 344
Nakai, K., 11, 12, 13, 37
Nakamura, N., 233
Nakano, K., 336
Nantel, A., 257
Nebesarova, J., 24
Nebreda, A. R., 65
Nechad, M., 218
Nechipurenko, Y., 235
Neckameyer, W. S., 213
Nedergaard, J., 213, 217, 218, 355, 357, 396
Neff, L., 118
Neher, R., 186
Neill, S., 340
Neilson, A., 419, 420
Neitzel, H., 10, 13
Nelis, E., 290
Neschen, S., 387
Nesvizhskii, A. I., 98
Neuman, I., 174, 186, 187, 188
Neumann, K., 25
Neupert, W., 9, 11, 98, 309
Neverova, M., 396
Newgard, C., 426, 448
Newman, J. M., 407, 410, 417
Newmeyer, D. D., 50, 257
Ng, A. V., 352, 387
Nicholls, D. G., 50, 417, 419, 422
Nichols, K., 219
Nicolay, K., 368, 375, 388
Nielsen, A. R., 216, 218
Nielsen, H., 34, 44

Niesman, I. R., 323
Niikura, M., 336, 344
Nikoulina, S., 217
Nikoulina, S. E., 217
Nishimaki, K., 343
Nixon, T. W., 382
Nobes, C. D., 407, 417
Noble, E. G., 387, 388
Nonaka, I., 336, 338, 344
Norbeck, J., 99
Norioka, S., 160
Norrbom, J., 215
Norris, K. L., 232, 243
North, B., 138
North, B. J., 139, 140, 143
Noveroske, K., 453
Nowikovsky, K., 305, 307, 309, 311, 314, 315, 316
Nozaki, M., 337
Nunnari, J., 232, 290, 301
Nussbaum, R., 213, 276

O

Obata, T., 65
Obayashi, T., 37
Ochiai, A., 337
Oda, H., 336
Oefner, P. J., 9, 11, 98
Offner, F. A., 368
Ogura, A., 344
Ogurtsov, A. Y., 235
Oh, R., 417, 419
Ohkuma, S., 314
Ohlmeier, S., 9, 11
Ohlsen, H., 215
Ohshima, K., 461
Ohta, S., 336, 338, 343
Oka, T., 233, 240
Okada, S., 461
O'Kane, C. J., 323
Oldham, S., 213
Olive, P. N., 407, 417
Oliveira, R. L., 402
Oliver, N. A., 338
Oliver, S. G., 9, 15
Olsen, J. V., 12, 13, 51, 86, 94, 98, 125, 237
Olveczky, B., 292
Olzmann, J. A., 213
Ong, S.-E., 12, 13, 25, 82
Oomens, C. C., 375
Ooms, B., 125
Opperdoes, F. R., 23
Oppliger, W., 262
Oral, E. A., 374
Orci, L., 461
Orrell, J., 419, 420
Ortiz de Montellano, P. R., 155

Ortner, M., 353
Osada, H., 257
Osawa, M., 118
Oseguera, M., 83
O'Shea, E. K., 9, 11, 14, 15
Ott, M., 139, 140, 143
Otto, S., 108
Ouellette, M., 216
Ouyang, Y., 213
Overly, C. C., 320, 321, 322
Owens, G. C., 320, 322
Ozawa, T., 10, 11, 13, 338

P

Pagliarini, D. J., 12, 13, 25, 82, 83, 118, 257, 276, 277, 279, 283, 286
Pagliato, E., 374
Paiement, J., 98
Pakay, J. L., 396
Pal, S., 326
Palcy, S., 98
Palmer, J. D., 24
Palmer, J. W., 67
Palsson, B. O., 4
Pan, C., 11, 13, 86, 94
Pan, D. A., 374
Panten, U., 468
Papa, S., 85, 86, 87, 257
Papas, K. K., 444
Papatheodorou, P., 138
Pappin, D. J., 128
Parikh, J. R., 85
Park, J. H., 321, 328
Park, J. Y., 138
Park, K. J., 37
Park, K. S., 217, 471
Parker, N., 396, 415, 417, 419
Parkinson, N., 453
Parland, W., 102, 104
Parman, Y., 290
Parone, P. A., 10, 13, 98
Parslow, D., 419, 420
Parsons, D. F., 101
Parsons, J. A., 461
Parsons, S. J., 196
Partikian, A., 292
Pasa-Tolic, L., 53, 54
Pask, H. T., 64, 74, 75
Patel, B. N., 324, 331
Paterson, D. H., 387, 388
Patitucci, A., 290
Patterson, G. H., 292
Patterson, J. K., 374
Patterson, N., 12, 13, 98
Patti, M. E., 216, 375
Patton, W. F., 51, 64
Pauli, J. R., 397

Paumen, M. B., 99
Pawson, T., 325
Paz, C., 169, 171, 174, 186, 187, 188
Pearlman, R. E., 25, 26, 33, 36
Pebay-Peyroula, E., 132
Pecina, P., 193, 195, 208
Pecinova, A., 193, 195, 208
Pedersen, P., 10, 13, 51
Pedersen, P. L., 282
Peeters, N., 11, 12, 13, 14, 15, 27, 30, 32
Pei, J., 83, 110, 144
Pelsman, A., 321
Peluffo, G., 83
Pepinsky, R. B., 160
Pereira-da-Silva, M., 397, 400, 402
Perez-Brocal, V., 25, 29, 36, 41
Pérez-Martos, A., 338
Perkins, D. N., 128
Perkins, G., 257
Perocchi, F., 9, 13
Peroni, O., 397, 411, 418
Perret, P., 397, 411, 418
Perrier, J., 99
Perrimon, N., 323
Perrino, L., 234
Perschil, I., 9, 11, 25, 27, 51, 98
Perseghin, G., 374
Peseckis, S. M., 151, 152
Pessin, J. E., 461
Pestronk, A., 321
Peterman, S. M., 99, 102, 103, 104
Peters, E. C., 206, 207
Petersen, K. F., 215, 216, 352, 373, 374, 382, 383,
    384, 386, 387, 389
Peterson, J. A., 155
Peterson, P., 352
Petroff, O. A., 382
Petros, J. A., 336
Petrovic, N., 213, 215, 216, 217, 218, 222, 396
Pette, D., 388
Pettit, F. H., 276
Peyrard-Janvid, M., 215, 217, 224
Pfanner, N., 4, 9, 11, 22, 25, 26, 27, 45, 48, 51,
    83, 98
Pflieger, D., 9, 13
Phillipe, H., 24
Phillips, D., 63
Phinney, B. S., 152
Pibouin, L., 225
Piehl, K., 214
Pieles, G., 464
Piemonte, F., 321
Pigino, G., 321
Pilling, A. D., 324, 328, 330, 331
Pilpel, Y., 225
Pilstrom, L., 214
Pine, P. S., 214
Pines, O., 16, 27

Ping, P., 256, 257, 270
Pinkse, M. W., 125
Pinkse, M. W. H., 85, 86, 91
Piper, M., 9, 11
Pirozzi, M., 321
Pizzo, P., 99
Plante, I., 25, 26
Plaut, G. W., 64
Plummer, G., 151, 153, 155, 159, 163
Pocsfalvi, G., 81, 85, 86, 87
Poderoso, C., 169, 174
Poderoso, J. J., 169, 174
Podesta, E. J., 169, 171, 186, 187
Polakiewicz, R. D., 90, 115, 126
Polevoda, B., 99
Pollheimer, J., 368
Polyak, K., 336
Pongratz, R. L., 425, 426
Poole, B., 314
Porath, J., 84
Porter, D. E., 454
Porter, R. K., 410, 417
Portoukalian, J., 99
Posestá, E. J., 174
Poucher, S. M., 214
Poulin, B., 38
Pozza, G., 374
Pozzan, T., 99, 471
Prada, P. O., 402
Prentki, M., 426
Prescher, J. A., 151, 152, 153, 155, 159, 163
Prescott, M., 314, 316
Preston, R. A., 314
Previs, S. F., 388
Previtali, S. C., 323
Price, T. B., 375
Pridgeon, J. W., 213
Privman, E., 257, 258
Proctor, D. N., 352
Prokisch, H., 3, 4, 9, 11, 13, 46, 51, 55, 82, 98
Proks, P., 452, 457
Prompers, J. J., 375, 388
Pryor, R. J., 454
Przybylski, M., 98, 102
Przyklenk, K., 195, 208
Pu, H., 83
Puchowicz, M. A., 359, 368
Puigserver, A., 99
Puigserver, P., 216, 223, 375
Pupko, T., 257, 258
Puschendorf, B., 65
Pypaert, M., 387

Q

Qi, D., 86, 87
Qian, L., 213
Qin, L., 225

Quarterman, J., 452, 457
Quattrini, A., 321, 323
Quintens, R., 427
Quwailid, M. M., 451, 453

R

Rabilloud, T., 10, 13, 101
Rabiner, L. R., 28
Rachman, J., 215, 217, 224
Rachubinski, R. A., 98
Rada, P., 24
Radda, G., 388
Radda, G. K., 375, 379, 388
Radi, R., 83
Rae, P. A., 171
Ragg, T., 223
Raghavakaimal, S., 352, 353, 364
Rajaiah, G., 151, 152, 153, 155, 159, 163
Ramalho-Santos, J., 83
Raman, M., 174
Randle, P. J., 64, 74, 75
Rapaport, D., 98
Rardin, M. J., 275, 277, 279, 283, 286
Rasis, M., 257, 258
Rasmussen, H. N., 355, 356, 357, 358
Rasmussen, U. F., 355, 356, 357, 358
Rattei, T., 6
Ravagnan, L., 257
Raval, A., 83
Ray, H. N., 12, 13, 51, 98
Rechavi, G., 16
Reed, G. H., 454
Reed, J. C., 98
Reed, L. J., 276
Reers, M., 268, 312
Reeve, A. K., 138
Rehling, P., 9, 11, 25, 27, 83, 98
Reid, E., 323
Reinders, J., 9, 11, 25, 27, 51, 83, 98
Reipert, S., 307, 309, 314, 315, 316
Ren, J., 352, 383, 384, 386, 388
Ren, S., 257, 258, 261, 263, 265, 267, 270
Renner, K., 368
Rennstrom, E., 427
Renolds, I. J., 321, 322, 325
Resh, M. D., 150, 151, 152
Retief, J., 214
Reyhorn, P., 160
Reynafarje, B., 400, 417
Reynolds, A., 219, 235
Reynolds, I. J., 326
Reynolds, N., 291, 292, 294, 303
Rezaei, K., 224
Rezaul, K., 51
Reznick, R. M., 215, 388
Rice-Evans, C., 257
Richards, R. E., 375

Richardson, A., 51
Richardson, J. E., 7
Richmond, G. S., 25, 29, 36, 41
Ridden, J., 215, 217, 224
Ridderstrale, M., 216, 223, 375
Rieff, H. I., 320, 321, 322
Rigby, P. W., 459
Riles, L., 6
Rioux, P., 25, 26
Ritov, V. B., 354, 374, 375
Rizza, R. A., 352
Rizzuto, R., 195, 290
Robb-Gaspers, L. D., 195
Robert, B., 225
Robertson, R. P., 463
Robinson, B. H., 276
Robinson, F. L., 174
Robinson, K., 219
Robinson, M., 51
Rocco, S. A., 397
Rocheford, M., 38
Rochelle, J., 290
Roden, M., 374
Rodriguez-Lafrasse, C., 99
Roe, T., 6
Roepstorff, P., 88, 127
Roger, A. J., 21, 24, 25, 29, 36, 41
Rogers, R., 24
Roglic, G., 374
Rogol, A., 171
Rojas, A., 276
Rojo, M., 291, 303
Rolf, M. G., 362
Rolfe, D. F., 407, 410, 417
Romanatto, T., 397, 400
Rooyackers, O., 213, 215, 216, 217, 218, 222
Rooyackers, O. E., 352
Ropelle, E., 397
Rorsman, P., 461
Rosado, C. J., 314, 316
Rosenbusch, J., 114
Rosenthal, B., 24
Rosenzweig, B. A., 214
Rosinke, B., 114
Rospert, S., 48
Ross, B., 388
Ross, E. M., 242
Ross, M. M., 108, 125
Rossi, L., 235
Rossier, J., 9, 13
Rossignol, F., 225
Rostas, J. A., 52
Roth, A. F., 152
Rothman, D. L., 215, 352, 373, 374, 375, 382, 383, 384, 386, 387, 389
Rotte, C., 23
Rouiller, D. G., 461
Roumier, T., 257

Rubi, B., 406, 411
Rubin, C. S., 172
Rück, A., 102
Rudd, S., 433
Rudnick, D. A., 152
Rudnick, P. A., 54
Ruepp, A., 4, 6, 55
Rui, X., 171
Rui, Y., 321
Ruiz-Pesini, E., 336
Ruiz-Trillo, I., 24
Rumi, I., 83
Rush, J., 90, 126
Rutter, G. A., 195, 406
Ruvolo, P., 257
Ruvolo, P. P., 257
Ryan, M. T., 48
Rydström, J., 451, 475
Ryningen, A., 99

S

Saad, M. J., 395, 397, 400, 402
Saada, A., 16
Saccone, R., 216, 375
Sadoul, K., 461
Sadygov, R., 104
Saenger, P., 171
Saetrom, P., 235
Sahlin, K., 83
Saito, T., 461
Sako, Y., 10, 11, 13
Saks, V., 368
Salcedo-Hernandez, R., 113
Saleem, R. A., 98
Saleh, M., 98
Salomon, A. R., 64, 118, 193, 196, 197, 204, 206, 207
Salowsky, R., 223
Saltin, B., 214, 388
Salvi, M., 117, 118, 124, 132, 133
Samavati, L., 193, 195, 196, 197, 204, 207
Samuel, V. T., 388
Samuelson, J., 24
Sanchez, L. B., 24, 25
Sancho, J., 338
Sandberg, R., 216
Sanderson, C. M., 225
Sankoff, D., 25, 26
Sano, Y., 186
Sanseverino, M. T., 344
Santoni, V., 101
Saraf, A., 104
Saraste, M., 199
Sardanelli, A., 257
Sardanelli, A. M., 257
Sasarman, F., 16
Satin, L. S., 468

Sato, M., 10, 11, 13, 461
Sato, W., 338
Sato, Y., 65, 233, 234
Satomi, Y., 160
Sauer, L. A., 427
Savage, D. B., 374
Savina, M. V., 417
Saxon, E., 152, 155
Saxton, M. W., 320, 321, 324, 328, 329, 330, 331
Scacco, S., 85, 86, 87, 257
Scaringe, S., 235
Scarpulla, R. C., 257
Schaefer, A. M., 14
Schäffer, A. A., 12
Schagger, H., 77, 185
Schapira, A. H., 50
Scharfe, C., 4, 9, 11, 46, 51, 55, 82, 98
Schatten, G., 83
Schatz, G., 262
Schechtman, D., 257
Scheele, C., 211, 213, 215, 216, 217, 218, 222
Schelter, J., 219
Schersten, T., 354
Schick, F., 374
Schimke, J. M., 353
Schimmer, B. P., 171
Schindler, J., 128
Schindler, P. A., 114
Schjerling, P., 368
Schlaeger, E. J., 10, 13
Schlesinger, M. J., 161
Schlicker, C., 138
Schlosser, G., 85, 86, 87
Schlunck, T., 98
Schmid, K., 262
Schmidt, M. F., 161
Schmidt, R., 6
Schmidt, R. E., 321
Schmitt, D., 99
Schmitt, S., 98
Schnaible, V., 98
Schon, E. A., 51
Schönfisch, B., 9, 11, 25, 27, 48, 51, 98
Schrader, M., 232
Schrauwen, P., 388
Schroeder, A., 223
Schroeder, A. J., 321, 328
Schroeder, M., 6
Schroeter, H., 257
Schuelke, M., 10, 13
Schuit, F. C., 427
Schulenberg, B., 51, 64
Schulz, V., 426
Schutgens, R. B., 407
Schutz, C., 125
Schwartz, D., 64, 65, 125
Schwarz, T. L., 319, 323, 326, 331
Schwer, B., 83, 138, 139, 140, 141, 143

Schweyen, R. J., 305, 307, 309, 311, 314, 315, 316
Scifo, P., 374
Scorrano, L., 233
Scorziello, A., 118
Scott, A. F., 7
Scott, J. D., 172, 257
Scott, M. S., 51, 55, 56
Scott, S. V., 290
Scovassi, A. I., 65
Secomb, T. W., 374
Sedat, J. W., 290
Seeley, P. J., 375
Segal–Bendirdjian, E., 338
Sekido, S., 9, 11
Sekiguchi, T., 338
Sekine, N., 411
Senderek, J., 290
Seol, D. W., 290
Seol, J. H., 323
Seppala-Lindroos, A., 374
Severson, D. L., 64, 75
Sewell, D. A., 216, 218
Shaag, A., 16
Shabalina, I. G., 396
Shabalina, S. A., 235
Shabanowitz, J., 108, 125
Shakhbazov, K. S., 468, 470
Sharp, P. A., 434
Sharp, Z. D., 314
Sharpe, P., 225
Shatkay, H., 38
Shatzkes, N., 387
Shaw, C. E., 321
Shaw, J. M., 290
Shaw, J. R., 219
Shaw, P. J., 321
Shaw, W. R., 323
Shawe–Taylor, J., 28
Shay, J. W., 216, 336
Sheiko, T., 99
Shen, J., 382
Shen, R. F., 64, 65, 195, 257
Shen, Y., 53
Shen, Y. Q., 37, 39
Sheng, S., 10, 13, 51
Shenolikar, S., 257
Sheridan, J. D., 461
Sherman, A., 468
Sherman, F., 99, 307
Sherry, A. D., 426, 444, 446, 448
Shestopalov, A. I., 51
Sheth, S. A., 12, 13, 25, 51, 55, 82
Sheu, S. S., 118
Shi, W., 54, 55, 128
Shi, Y., 225
Shidara, Y., 336
Shimazu, T., 137

Shimizu, H., 461
Shimizu, S., 99
Shimomura, K., 451, 452, 453, 457, 461, 463, 464, 469
Shinohara, Y., 99
Shirai, A., 9, 11
Shirihai, O. S., 289, 291, 292, 294, 303
Shisa, H., 338
Shitara, H., 336
Shore, G. C., 64
Short, D. K., 352, 353, 364
Short, K. R., 352, 353, 360, 364
Shoun, H., 23
Shulman, G. I., 215, 216, 352, 373, 374, 382, 383, 384, 386, 387, 388, 389, 426, 444
Shulman, R. G., 375, 379, 382, 388
Sickmann, A., 4, 9, 11, 25, 26, 27, 45, 51, 83, 98, 117, 118, 125, 128, 132, 133
Sicree, R., 374
Sidransky, D., 50, 83, 336
Siegel, E., 151
Siess, E. A., 67, 68
Signorile, A., 257
Sihag, S., 12, 13, 51, 98, 216, 223, 375
Sikorski, J. A., 154
Silberman, J. D., 24, 25, 29, 36
Silva, C., 196
Simerly, C., 83
Simizu, S., 257
Simoneau, J. A., 374, 375, 396
Simoni, A. M., 290
Simpson, A. H. R. W., 454
Sims, N. R., 398, 400
Sinclair, D., 138
Singh, R., 352, 353, 364
Sintasath, L., 323
Siu, K. W., 25, 26, 33, 36
Sjostrom, M., 354
Skezak, L. A., 383, 384, 386
Skovbro, M., 368
Slack, R. S., 233
Slater, E., 362
Sleutels, F., 225
Slezak, L. A., 352
Slotboom, D. J., 24
Small, I., 11, 12, 13, 14, 15, 27, 30, 32
Smeitink, J. A., 16
Smet, J., 10, 13
Smid, O., 24
Smiley, S. T., 268, 312
Smillie, K. J., 232
Smith, C. L., 246, 292, 293, 296, 301, 303
Smith, D. G., 25, 26, 33, 36
Smith, J. J., 50, 98
Smith, R. D., 53, 54, 115
Smith, S. J., 320, 322
Smits, P., 16
Smotrys, J. E., 151

Snel, B., 9, 15
Snider, J., 152
Snow, B. E., 50
Snyder, M., 9, 11
Soares, S. S., 209
Soderling, T. R., 234
Soderling, Y., 257
Soejima, A., 338
Soeldner, J. S., 374
Solano, A. R., 186, 187, 188
Soltys, C. L., 163
Soma, R., 343
Song, C. X., 257
Song, S., 138
Song, T., 52, 53, 54, 55
Sonnhammer, E. L., 218
Sono, S., 387
Sood, V., 219
Sorek, R., 225
Soria, B., 461
Southam, L., 454
Souza, J. M., 83
Sovijarvi, A., 374
Spamer, C., 388
Span, A. S. W., 23
Spannagl, M., 6
Sparagna, G. C., 50
Speers, A. E., 163
Spek, E. J., 90, 126
Spencer, D. E., 25, 26, 33, 36
Spencer, J. P., 257
Speranza, F., 257
Spiegelman, B. M., 402
Spierings, D., 98
Spinazzola, A., 12, 13, 51, 55
Sprung, R., 83, 85, 93, 94, 110, 144
Sreekumar, R., 353, 388
Srere, P. A., 358
Srivastava, S., 50, 83
Stadlmann, S., 368
Stahl, E., 12, 13, 51, 98
Stangherlin, A., 233
Stanley, B., 10, 13, 51
Stansbie, D., 64, 75
Stanyer, L., 222
Stapnes, C., 99
Staroverov, D. B., 468, 470
Stechmann, A., 25, 29, 36, 41
Steegborn, C., 138
Steenaart, N. A., 64
Stefani, E., 260, 261, 266
Stegehuis, D. S., 51
Stein, D. T., 374
Steinmetz, L. M., 9, 11, 13, 115
Stern, A., 257, 258
Stern, M. P., 374
Stettler-Gill, M., 206, 207
Stevens, R. J., 354

Steward, O., 52
Stiles, L., 290, 295, 301, 303
Stocco, D. M., 171, 186, 190
Stocker, S., 223
Stojanovski, D., 83
Stoppiglia, L. F., 397
Storlien, L. H., 374
Stowers, R. S., 323
St-Pierre, J., 215, 417
Strack, S., 231, 233, 243, 250, 257
Straight, A., 290
Strauss, J. F. III, 171, 190
Streeper, R. S., 138, 141, 142
Strijkers, G. J., 375
Stringaro, A., 118
Stroh, A., 199
Strub, J. M., 10, 13
Struglics, A., 64
Strupat, K., 114
Stucki, J. W., 151
Stukenberg, P. T., 108, 125
Stump, C. S., 352, 353, 360
Stümpflen, V., 6
Sturm, M., 85
Su, J., 225
Subramaniam, S., 30, 32, 82
Subramanian, A., 216, 223, 375
Sudarsanam, S., 256
Suen, D. F., 232, 243
Sugawara, T., 171
Sugiana, C., 12, 13, 25, 82
Sugiyama, N., 91
Sui, Z., 155
Sullivan, P. G., 52
Summerbell, D., 459
Sun, B., 152
Sun, C. Q., 336
Sun, H., 255
Sun, Z., 11, 12, 13
Sundberg, C. J., 214, 215, 217, 224
Surin, A., 473, 474
Susin, S. A., 50
Sutak, R., 24, 25
Sutovsky, P., 83
Suzuki, S., 23
Svensson, L. T., 186
Swaminathan, R., 292
Sweet, I. R., 406
Swift, A. L., 419, 420
Swoap, S., 224
Szabadkai, G., 290
Szafron, D., 38
Szczepaniak, L. S., 374
Szoko, E., 51

T

Tachezy, J., 22, 24, 25, 26
Tagashira, Y., 337

Tager, J. M., 407
Taguchi, N., 233, 240
Tahara, E., 337
Takada, R., 160
Takada, S., 160
Takahama, S., 336
Takai, D., 338
Takakura, H., 99
Takao, T., 160
Takaya, N., 23
Takei, K., 232, 233, 234
Takemitsu, M., 336, 338
Takenaga, K., 336, 340, 344
Takeo, K., 23
Takimoto, A., 23
Tal, M., 16, 27
Tam, Y. Y., 98
Tamada, Y., 37
Tamagnine, J., 419, 420
Tamas, I., 25, 29, 36
Tamura, Y., 257
Tan, Y., 115
Tanaka, M., 338
Tanaka, S., 118
Tanaka, Y., 463
Tandler, B., 67, 100, 101, 102, 103
Tao, W. A., 89
Tapscott, S. J., 217
Tarr, P. T., 215
Tashiro, Y., 314
Tasto, J. J., 104
Tate, C. A., 354
Taya, C., 336
Taylor, A. W., 387, 388
Taylor, F. R., 160
Taylor, G. S., 275
Taylor, R. S., 98
Taylor, R. W., 14
Taylor, S. I., 374
Taylor, S. S., 257
Taylor, S. W., 10, 13, 51, 98
Teboul, L., 451
Technikova-Dobrova, Z., 257
Techritz, S., 10, 13
Teich, J., 419, 420
Teller, J. K., 427
Tengholm, A., 468
Tennant, M. E., 321
Terskikh, A. V., 468, 470
Tesch, P. A., 215, 216, 218
Testolin, G., 374
Tetko, I., 6, 82
Thaete, F. L., 374
Theander, S., 406, 411
Thellin, O., 224
Thiemann, M., 232
Thies, E., 321
Thingholm, T. E., 88, 127

Thomas, A. P., 195
Thomas, B., 225
Thomas, D. Y., 257
Thompson, K. L., 214
Thompson, R. T., 387, 388
Thonberg, H., 215
Thore, A., 360
Tibaldi, E., 117
Tielens, A. G., 23, 25
Timmis, J. N., 25
Timmons, J. A., 211, 213, 214, 215, 216, 217, 218, 222, 224
Tiwari, P., 321
Tjaden, U. R., 51
Tjader, I., 213, 215, 217
Toescu, E. C., 474
Tokatlidis, K., 262
Tokuda, M., 233
Tokumitsu, Y., 65
Toledo, F. G., 354
Tolic, N., 53
Tomita, M., 91
Tomizawa, K., 233, 234
Tompkins, A., 118
Toninello, A., 117, 118, 124, 132, 133
Tonks, N. K., 276
Tonti-Fillippini, J. S., 25
Tosches, N., 9, 11
Toshihiko, U., 83
Tosu, M., 337, 338
Tovar, J., 24, 25
Towbin, H., 186, 188
Towler, D., 156
Toye, A. A., 452, 457
Tozzi, G., 321
Tran, D. D., 321
Tran, P. O., 463
Treisman, J. E., 151
Trezeguet, V., 132
Tronstad, K. J., 99
Trounce, I., 338, 340
Trounce, I. A., 344, 355
Truscott, K. N., 45
Trush, M. A., 336
Trushina, E., 321
Tsang, H. T., 323
Tsao, J., 171
Tsaousis, A. D., 21, 25
Tsatsos, P. H., 113
Tsien, R. Y., 313
Tsujimoto, Y., 99
Tsukada, S., 257
Tsukada, Y., 337
Tsukamoto, K., 427
Tsutsumi, R., 151
Tudor, E. L., 321
Turcotte, L. P., 374, 375
Turkaly, J., 98, 100

Turkaly, P., 100, 101, 102, 103
Turmel, M., 25, 26
Turnbull, D. M., 14, 138
Turner, N., 417
Turpaz, Y., 214
Tuschl, T., 218
Twig, G., 290, 291, 292, 294, 295, 301, 303
Tyndall, G. L., 352, 388
Tyurina, Y. Y., 50

U

Uchida, T., 225
Ueffing, M., 101
Ueno, M., 402
Ueno, N., 160
Ugurbil, K., 379, 388
Ui, M., 65
Uitto, P. M., 125
Umezawa, Y., 10, 11, 13
Unger, R. H., 461
Ungibauer, M., 199
Urbonas, A., 257
Ure, D., 321
Urs, L., 90
Usachev, Y. M., 250
Usadel, H., 336
Usuda, N., 186
Uziel, G., 344

V

Vafai, S. B., 12, 13, 25, 82
Valdar, W., 456
Valente, E. M., 213, 276
Valenzuela, D., 138, 141, 142
Valette, A., 257
Valle, G., 51
Valsecchi, V., 321
van Alen, T. A., 25
Van Beeumen, J., 10, 13
van Breukelen, B., 85, 86, 91
Van Coster, R., 10, 13
van den Bosch, H., 407
van Der Bliek, A. M., 27
van der Giezen, M., 24, 25, 29, 36, 41
Van der Greef, J., 51
van der Laan, M., 83
van der Spek, H., 9, 11
van der Staay, G. W., 25
Vandersteen, J. G., 454
Van Dorsselaer, A., 10, 13, 114
Van Eyk, J. E., 10, 13, 51
van Hellemond, J. J., 23, 25
Van Houten, B., 321
Van Remmen, H., 51
Vanrobaeys, F., 10, 13
van Roermund, C. W., 232
van Schilfgaarde, R., 397

van Suylichem, P. T., 397
Van Vactor, D., 324, 331
Vanzulli, A., 374
Varnes, M. E., 359, 368
Vassilopoulos, A., 138
Vasudevan, A., 151
Veenstra, T., 101
Veerkamp, J. H., 374, 375
Vehkavaara, S., 374
Velloso, L. A., 395, 397, 400, 402
Vercesi, A. E., 395, 397, 400, 402
Verdin, E., 83, 137, 138, 139, 140, 141, 143
Verdin, R. O., 83, 138, 141, 143
Vergari, R., 257
Verkhratsky, A., 474
Verkman, A. S., 292
Verma, M., 50, 83
Vianna, C. R., 215
Vidal, M., 12, 13, 25, 82
Vidal-Puig, A. J., 388, 397, 411, 418
Vi-Fane, B., 225
Vilas, G. L., 152
Vilbois, F., 10, 13, 98
Villen, J., 64, 65, 125
Vintersten, K., 459
Vittone, J. L., 352
Vivares, C. P., 25
Vizor, L., 453
Vo, T. D., 4
Vogel, H., 213
Vogel, S. M., 374
Vogelstein, B., 336
Volinia, S., 65
Vomel, M., 328, 331
von Adrichem, J. H. M., 112
von Heijne, G., 34, 44
von Jagow, G., 77, 185, 199
von Mering, C., 9, 13, 15
Vos, H. R., 85, 86, 91
Voth, M., 290
Votyakova, T. V., 326
Vriend, G., 16
Vrieze, B. T., 321
Vyssokikh, M., 99

W

Wade, A. C., 152
Wadzinski, B. E., 257
Wagner, K., 83
Wagner, L., 54, 55, 128
Wagner, R., 26, 27
Wagner, Y., 9, 11, 25, 27, 51, 98
Wahlestedt, C., 215, 216, 217, 218, 222, 224
Waizenegger, T., 98
Walden, T. B., 216, 217, 218
Walford, G. A., 12, 13, 25, 82
Walker, J. E., 64

Walker, M., 217
Wallace, D. C., 336, 338, 340, 344, 355
Wallimann, T., 368
Walsh, L. P., 190
Walter, P., 290
Walter, U., 125
Walzer, G., 290, 295, 301, 303
Wan, J., 152
Wanders, R. J., 232, 388, 407
Wang, D., 213
Wang, E. A., 160
Wang, G. W., 257
Wang, H., 427
Wang, J., 213
Wang, L., 257
Wang, M., 51
Wang, O. L., 257
Wang, Q., 83
Wang, W., 52, 53, 54, 56, 57, 59
Wang, W. L., 257
Wang, X., 319, 323, 326, 331
Wang, Y., 52, 53, 255, 257, 258, 260, 261, 263,
　265, 266, 267, 270
Ward, M. A., 98
Ward, W. F., 51
Warden, C. H., 396
Warnock, D. E., 10, 13, 51, 98
Warram, J. H., 374
Warringer, J., 99
Washburn, M., 104
Washburn, M. P., 109
Wasiak, S., 233
Watari, H., 190
Waterham, H. R., 232
Waterhouse, N. J., 474
Waters, J., 320, 322
Watkins, S., 374
Webb, W. W., 313
Weber, J., 453
Weber, K., 218
Weedon, J., 453
Wegener, C., 328, 331
Weh, S. F., 257
Wehrli, F. W., 380
Weidner, M. L., 352, 388
Weinbach, N., 9, 11
Weiss, A., 104
Weiss, J. N., 255, 256, 257, 258, 261, 263, 265,
　267, 268, 270
Weissman, J. S., 9, 11, 14, 15
Welling, B., 16
Wells, C. D., 325
Wells, S., 453
Welte, W., 98, 102
Wen, D., 160, 161
Wendt, S., 102
Weng, S., 6
Wennmalm, K., 216, 217, 218

Werbin, H., 336
Wernerman, J., 213, 215, 217
Wernstedt, C., 102
Westerbacka, J., 374
Westermann, B., 9, 11, 309
Wharton, S. B., 321
Wheeler, M. B., 397, 411, 418
Whelan, J., 25
White, F. M., 108, 125
White, M., 83, 110, 144
White, M. F., 387
Whitehouse, S., 64, 74, 75
Whitehurst, A. W., 174
Whittaker, R. G., 14
Wibom, R., 353, 360
Wicks, K. L., 352
Wieckowski, M. R., 290
Wiedemann, N., 22, 26, 45, 48
Wiederkehr, A., 471
Wieland, O. H., 67, 68
Wieringa, B., 368
Wiesner, J., 125
Wikstrom, J. D., 290, 291, 292, 294, 295, 301, 303
Wikström, M., 199
Wild, S., 374
Wiles, M., 355
Wiley, S., 10, 13, 51, 98
Wiley, S. E., 118, 257, 276, 277, 279, 283, 286
Wilkins, J. A., 52
Williams, B. A., 25
Williams, G., 353, 362
Williams, G. R., 101, 400
Williams, K. P., 160
Willson, J. K., 336
Wiltshire, C., 257
Wing, R. R., 374
Winkler, C., 9, 11, 98
Wipf, P., 320
Wise, C., 454
Wishart, D. S., 38
Wisloff, U., 224
Wisniewski, J. R., 12, 13, 51, 98
Withers, K. W., 417
Wittwer, C. T., 454
Witzmann, F. A., 64, 65, 195, 257
Wolf, C., 108
Wollheim, C. B., 406, 411, 427, 462, 464, 471
Wolters, D., 104, 109
Wolters, G. H., 397
Wood, J. D., 321
Wood-Kaczmar, A., 222
Woodman, P., 453
Woods, R. A., 314
Worby, C. A., 118, 257, 276
Wortelkamp, S., 125
Wright, W. E., 216
Wu, C. C., 98, 104, 109
Wu, G., 257

Wu, L., 51, 336
Wu, M., 290, 295, 301, 303, 419, 420
Wu, P. H., 215
Wu, R., 87, 88

X

Xenarios, I., 10, 13, 27, 98
Xia, K., 213
Xiang, R., 52
Xiao, H., 83, 110, 144
Xiao, L., 83, 110, 144
Xiao, W., 4, 9, 11, 46, 51, 98
Xie, J., 10, 13
Xie, S., 85, 93, 94
Xie, W., 220, 222
Xie, X., 12, 13, 51, 55
Xie, Z., 321
Xiong, H., 213
Xu, H., 225
Xu, M., 54, 55, 128
Xu, X., 86, 87
Xu, Y., 83, 85, 93, 94, 110, 144
Xue, G. P., 24

Y

Yadava, N., 417, 419
Yaffe, M. B., 65
Yakin, A., 218
Yamada, E., 461
Yamada, H., 232
Yamada, J., 186
Yamagata, K., 336
Yamaguchi, A., 336, 340, 344
Yamaguchi, M., 23
Yamamoto, A., 314
Yancopoulos, G., 138, 141, 142
Yang, J., 152
Yang, L., 52, 53, 54, 55, 56, 57, 59, 213
Yang, P., 86, 87
Yang, X., 54, 55, 83, 86, 128
Yang, X. J., 99
Yang, Y., 138, 141, 142, 213
Yao, Z., 222
Yashiroda, Y., 9, 11
Yates, J. R. III, 98, 104, 109, 152
Ye, M., 86, 87, 88
Yeaman, S. J., 217
Yeh, E., 328
Yi, E., 98
Yki-Jarvinen, H., 374
Yoneda, Y., 99
Yonekawa, H., 336, 338, 344
Yoo, S. J., 323
Yoon, Y., 232
Yoshida, M., 9, 11
Yoshida, M. C., 337, 338
Yoshida, Y., 232
Yoshimori, T., 314

You, J., 51
Youle, R. J., 232, 243, 246, 292, 293, 296, 301, 303
Young, A., 355
Young, A. A., 374
Yu, C., 155, 374
Yu, C. S., 36
Yu, H., 195
Yu, K., 193

Z

Zahedi, R. P., 9, 11, 25, 27, 51, 83, 98, 125
Zamzami, N., 50
Zappia, M., 290
Zaucha, J. A., 257
Zaviani, M., 51, 55
Zawadzki, J. K., 374
Zeimet, A. G., 368
Zeitlin, S., 321
Zeng, C., 160
Zervos, P. R., 64
Zeth, K., 98
Zeviani, M., 12, 13
Zha, X., 115
Zha, X.-M., 90, 126
Zhang, B., 10, 13, 51, 98
Zhang, C. Y., 215, 397, 411, 418
Zhang, D., 388
Zhang, H., 90, 115, 126
Zhang, J., 11, 12, 13, 257, 475
Zhang, N., 115, 388
Zhang, X., 86, 87
Zhang, Y., 388
Zhang, Z., 12, 213
Zhao, P., 426, 448
Zhao, Y., 85, 93, 94
Zheng, B., 453
Zheng, J., 98, 113
Zheng, J. Q., 321
Zheng, W., 54
Zheng, X. X., 397, 411, 418
Zheng, Y. T., 257
Zhou, C., 82
Zhou, H., 86, 87, 88
Zhu, D., 11, 13
Zhu, H., 336
Zhu, X. H., 388
Zhuang, Z., 54
Ziegler, M., 65
Zinser, E., 310
Zollner, A., 6
Zorzi, W., 224
Zou, H., 86, 87, 88
Zubacova, Z., 24
Zuchner, S., 290
Zufall, N., 9, 11, 98
Zunino, R., 233
Zupec, M. E., 154
Zurgil, N., 291, 292, 294, 303
Zwart, R., 225

Subject Index

A

Acetylated mitochondrial proteins
  detection
    heavy mitochondrial fraction preparation
      from mouse liver, 144–145
    immunoprecipitation, 145–146
  SIRT3
    deacetylation assay *in vivo*
      cell culture, transfection, and RNA
        interference, 143
      immunoprecipitation and Western blot,
        143–144
      plasmids and RNA interference
        oligonucleotides, 143
    peptide substrate assay *in vitro*, 141
    protein substrate assay *in vitro*
      incubation conditions, 142
      substrate preparation, 141–142
    purification
      bacterial expression systems, 140–141
      mammalian cell culture, 139–140
    sirtuin types in mitochondria, 138
Acot2
  acyl-CoA thioesterase activity assay, 188
  functional overview, 186–187
  phosphoprotein characterization, 188–190
  phosphorylation sites, 187
  purification from rat adrenal glands, 187–188
Anaplerosis, *see* Malic enzyme
ANN, *see* Artificial neural network
Antisense oligonucleotides,
  see Uncoupling protein 2
Apoptosis
  assay in primary astrocytes, 244–245
  PINK1 role, 220–222
Artificial neural network, mitochondrial
  protein localization prediction, 27–28
ATP production, mitochondria
  bioluminescence assay
    calculations, 363–364
    incubation conditions, 362–363
    limitations, 369
    principles, 360–361
    stock solution preparation, 361–362
  enzyme markers, 352–353
  mitochondria isolation from skeletal
    muscle for assay
    differential centrifugation, 357–358

homogenization, 355–357
    overview, 354–355
    yield and integrity assessment, 358–360
  oxidative phosphorylation overview,
    350–352
  phosphorous-31 saturation-transfer magnetic
    resonance spectroscopy, 379–380
Axonal transport, mitochondria
  Drosophila larva axon transport
    fluorescence microscopy, 329
    fly stocks and culture, 328–329
    image analysis, 329–331
    larva staging and dissection, 329
    principles, 322–324
  hippocampal neuron transport in rat
    confocal microscopy, 325
    embryonic neural dissection, 324
    fluorescent proteins
      constructs, 325
      transient transfection and dye
        application, 325
    image analysis, 325–328
    primary neuronal culture, 324
    principles, 322
  overview, 320–321
12-Azidododecanoic acid, synthesis, 154
14-Azidotetradecanoic acid
  protein labeling, *see* Palmitoylated
    mitochondrial proteins
  synthesis, 155

B

Bacterial artificial chromosome transgenesis,
  see Nicotinamide nucleotide
  transhydrogenase
Beta cell
  bioenergetics
    coupling efficiency of oxidative
      phosphorylation, 408–410
    membrane potential, 407, 421–422
    noninvasive measurements, 418–422
  glucose-stimulated insulin secretion
    regulation, 406, 426
  INS-1E cell line model
    coupling efficiency measurements
      trypsinized cells, 416–418
      uncoupling protein 2 knockdown
        effects in attached cells, 419–420

Beta cell (*cont.*)
    modular-kinetic measurement of
        bioenergetics, 420–422
    origins, 411
    tissue culture, 411, 413
    uncoupling protein 2
        RNA interference, 415–416
        Western blot from mitochondria and
            whole-cell lysates, 413–415
    malic enzyme expression, *see* Malic enzyme
    nicotinamide nucleotide transhydrogenase,
        see Nicotinamide nucleotide
        transhydrogenase
    uncoupling protein 2 function, 406
Blue native gel electrophoresis, mitochondrial
    proteins labeled with phosphate-32, 77–78

C

Calcium flux, measurement in islets of
    Langerhans, 467–468
Capillary isotachophoresis
    brain mitochondria protein analysis in mouse
    data analysis and proteome
        characterization, 54–59
        mass spectrometry coupling, 53–54
        mitochondria isolation, 52–53
        sample preparation, 53
    mitochondrial proteome enrichment
        principles, 51–52
Carnitine palmitoyltransferase-I
    functional overview, 98
    palmitoylation, 151
    posttranslational modification analysis
        denaturing gel electrophoresis
            electroelution, 102
            mass spectrometry, 104
            reduction, alkylation, and enzymatic
                digestion, 103–104
            running conditions, 102
        immunological analysis
            antibody preparation and
                immobilization, 105
            immunocapture, 106
            immunoprecipitation, 106
            Western blot, 103, 106–107
        mass spectrometry of isolated protein
            digestion, 107–108
            phosphopeptide isolation and analysis,
                108–109
        mitochondria isolation, 100
        outer membrane preparation, 100–101
        shotgun approach
            digestion, 110
            mass spectrometry, 110–111
            overview, 109–110
CBOrg, mitochondrial protein prediction in large
    datasets, 29–31

CELLO, sequence feature-based prediction of
    mitochondrial localization, 36
Cholesterol transport
    mitochondrial proteins, 171
    steroidogenic acute regulatory protein
        phosphorylative regulation,
        see Steroidogenic acute
        regulatory protein
CITP, *see* Capillary isotachophoresis
comparison of programs, 40–43
Confocal microscopy
    axonal transport of mitochondria, 325–328
    extracellular signal-regulated kinase in
        mitochondrial kinase complex,
        176–177
    mitochondria morphology in Mdm38 null
        mutant, 308–309
    mitochondrial membrane potential imaging
        beta cell imaging, 472–474
        image analysis, 474–475
        overview, 471–472
    photoactivateable green florescent protein in
        mitochondria, 296, 298
Coupling efficiency, oxidative phosphorylation
    INS-1E cells
        trypsinized cells, 416–418
        uncoupling protein 2 knockdown effects in
            attached cells, 419–420
    overview of measurement, 408–411
CPT-I, *see* Carnitine palmitoyltransferase-I
Cybrids, *see* Mitochondrial DNA-exchanged
    cybrids
Cytochrome *c* oxidase
    electron transport chain, 194–195
    mitochondria purification
        maintaining protein phosphorylation
        animal tissue handling and
            pretreatment, 197
        depolymerized vanadate preparation, 197
        general considerations, 196–197
        ground tissue studies of signaling
            pathways, 197–198
        homogenization and centrifugation,
            198–199
        vanadate treatment, 198, 209
    phosphorylation analysis
        phosphopeptides
            immobilized metal affinity
                chromatography, 206–207
            mass spectrometry, 206–208
        Western blot, 205–206
    purification
        ammonium sulfate precipitation, 203–204
        anion-exchange chromatography,
            202–203
        homogenization of mitochondria, 201–202
        purity assessment, 199–200

D

Denaturing high-performance liquid chromatography, *see* Nicotinamide nucleotide transhydrogenase

Diabetes, *see* Beta cell; Magnetic resonance spectroscopy; Malic enzyme; Nicotinamide nucleotide transhydrogenase; Uncoupling protein 2

DNA microarray, PINK1 expression analysis, 223–224

Drp1, *see* Dynamin-related protein 1

Dual specific phosphatase 18
  activity assay, 279
  functional overview, 276
  mitochondria localization studies
    crude mitochondria isolation by differential centrifugation, 280–281
    histodenz gradient purification of mitochondria, 281
    inner membrane association studies
      integral versus peripheral association, 285–286
      trypsin digestion, 283–285
    subfractionation of mitochondria, 282–283
  purification of recombinant protein
    bacteria culture, 278
    expression constructs, 277–278
    nickel affinity chromatography, 278–279
    site-directed mutagenesis, 277–278

Dual specific phosphatase 21
  functional overview, 276
  mitochondria localization studies
    crude mitochondria isolation by differential centrifugation, 280–281
    histodenz gradient purification of mitochondria, 281
    inner membrane association studies
      integral versus peripheral association, 285–286
      trypsin digestion, 283–285
    subfractionation of mitochondria, 282–283

Dynamin-related protein 1
  functional overview, 232
  phosphorylation
    characterization of mutants
      apoptosis/necrosis assays in primary astrocytes, 244–245
      GTPase activity assays, 242–243
      viability assays in PC12 cells, 243–244
    enzymes, 233–234
    mitochondrial shape analysis with ImageJ
      fluorescence microscopy, 246
      image enhancement, 246–249
      macro for morphometry, 250–251
      morphometry, 249–250

phosphorous-32 metabolic labeling and immunoprecipitation, 238–239
    phosphorylation-specific antibody generation, 239–240
    protein kinase A reaction, 242
    site-directed mutagenesis
      of modification site, 234, 236
      sites, 233
    transfection and stable cell line generation, 237
    Western blot, 239
  purification of recombinant protein, 240–242
  RNA interference, 235–236
  structure, 232–233

E

Electron microscopy, mitochondria morphology in Mdm38 null mutant, 308, 310

ERK1/2, *see* Extracellular signal-regulated kinase

Extracellular signal-regulated kinase, mitochondrial kinase complex in steroidogenesis
  confocal microscopy, 176–177
  expression constructs, 173–174
  subcellular fractionation, 174–175
  transient transfection, 174
  Western blot analysis, 175–177

F

Fgr
  assay, 121–122
  mitochondrial protein modification
    extraction and chromatography, 122–123
    kinase incubation and tyrosine phosphorylation, 123–124
    mitochondria isolation from rat brain, 122
  purification, 119–121
  substrate identification with mass spectrometry
    immobilized metal affinity chromatography of phosphopeptides, 127–128
    immunoaffinity chromatography of phosphopeptides, 125–127
    multiple reaction monitoring, 130–131
    overview, 125–126
    precursor ion scanning, 129–130
    tandem mass spectrometry, 128–129

Functional genomics, *see* PINK1

G

GFP, *see* Green fluorescent protein

Green fluorescent protein, *see also* Photoactivateable green florescent protein

Green fluorescent protein, *see also*
    Photoactivateable green florescent protein
    (*cont.*)
    HyPer and hydrogen peroxide assay in
        mitochondria, 468–471
    Rosella
        construct, 313–314
        mitochondria targeting, 314
        mitophagy assay
            fluorescence microscopy, 315–316
            transformation of yeast, 314–315
            yeast culture, 315
GTPase, dynamin-related protein 1 activity assays,
    242–243

H

Hidden Markov Model, mitochondrial protein
    localization prediction, 27–28
High-performance liquid chromatography
    denaturing high-performance liquid
        chromatography, 453–454
    phosphopeptide enrichment with
        two-dimensional liquid
        chromatography, 90–92
HMM, *see* Hidden Markov Model
HPLC, *see* High-performance liquid
    chromatography
Hydrogen peroxide, assay in
    mitochondria, 468–471
Hydrogenosome
    evolution and diversity, 24
    protein targeting and import, 27
3-Hydroxy-3-methylglutaryl-CoA synthase,
    palmitoylation, 163
HyPer, hydrogen peroxide assay in
    mitochondria, 468–471

I

ImageJ, mitochondrial shape analysis
    fluorescence microscopy, 246
    image enhancement, 246–249
    macro for morphometry, 250–251
    morphometry, 249–250
IMAC, *see* Immobilized metal affinity
    chromatography
Immobilized metal affinity chromatography
    cytochrome *c* oxidase phosphopeptides, 206–207
    phosphopeptide enrichment from
        mitochondria
        alkaline-induced β-elimination
            and Michael addition
            for increased sensitivity, 89
        media preparation, 84–85
        metal oxide single-tube and microtip
            techniques, 85–88
        methyl esterification for increased
            selectivity, 88–89

spin columns, 85
    phosphopeptides for mitochondrial tyrosine
        kinase substrate identification, 127–128
Immunoaffinity chromatography
    phosphopeptide enrichment from
        mitochondria, 89–90
    phosphopeptides for mitochondrial tyrosine
        kinase substrate identification, 125–127
Immunoprecipitation
    acetylated mitochondrial proteins, 145–146
    carnitine palmitoyltransferase-I, 106
    dynamin-related protein 1, 238–239
    SIRT3 deacetylation assay, 143–144
INS-1E cell, *see* Beta cell
Insulinoma, *see* Beta cell; Malic enzyme;
    Nicotinamide nucleotide transhydrogenase
Isobaric tagging for relative and absolute
    quantitation, mitochondrial quantitative
    phosphoproteomics, 94
iTRAQ, *see* Isobaric tagging for relative and
    absolute quantitation

L

LETM1, *see* Potassium/proton
    exchanger, mitochondria
Lyn
    assay, 121–122
    mitochondrial protein modification
        extraction and chromatography, 122–123
        kinase incubation and tyrosine
            phosphorylation, 123–124
        mitochondria isolation from rat brain, 122
    purification, 119–121
    substrate identification with mass
        spectrometry
        immobilized metal affinity
            chromatography of
            phosphopeptides, 127–128
        immunoaffinity chromatography of
            phosphopeptides, 125–127
        multiple reaction monitoring, 130–131
        overview, 125–126
        precursor ion scanning, 129–130
        tandem mass spectrometry, 128–129

M

Magnetic resonance spectroscopy
    mitochondrial metabolism studies in diabetes
        ATP synthesis measurement with
            phosphorous-31 saturation-transfer
            spectroscopy, 379–380
        exchange reactions and magnetization
            transfer spectroscopy, 377–379
        insulin resistance and resting mitochondrial
            metabolism, 386–389
        metabolic flux measurement with
            carbon-13 spectroscopy, 380–384

overview, 374–375
prospects for study, 389
tricarboxylic acid cycle metabolic flux
    modeling
    isotope mass balance equations, 385
    muscle metabolite concentration, 386
    overview, 384
principles and advantages for *in vivo*
    studies, 375–377
Malic enzyme
anaplerosis
    induction of enzyme, 426
    relative rates of pathways from carbon-13
        glutamate isoprotomer distribution
        carbon-13 labeling of cells, 444
        nuclear magnetic resonance, 444–446
        overview, 442–444
assays
    cytosolic enzyme
        lysate preparation, 436
        ME1 mixture components, 436–437
        protein assay, 437
    mitochondrial enzyme
        cytochrome *c* assay, 439–440
        glycerol-3-phosphate dehydrogenase
            assay, 440
        kinetic analysis, 440–441
        mass spectrometry for determining
            carbon-13 enrichment, 441–442
        mitochondria isolation, 438–439
        protein assay, 439
    principles, 435
expression analysis
    real-time polymerase chain reaction,
        430–432
    reverse transcription, 429–430
    RNA isolation, 428–429
functional overview, 426–428
malate–pyruvate cycles, 427–428
metabolic flux calculation with tcacalc,
    446–448
RNA interference in insulinoma cell line
    incubation conditions, 434–435
    small interfering RNA design, 433
    transfection optimization, 434
Mass spectrometry
brain mitochondria protein analysis in mouse
    capillary isotachophoresis coupling, 53–54
    data analysis and proteome
        characterization, 54–59
carnitine palmitoyltransferase-I
    posttranslational modification analysis
    phosphorylation
        digestion, 107–108
        phosphopeptide isolation
            and analysis, 108–109
        shotgun approach
            digestion, 110

mass spectrometry, 110–111
overview, 109–110
cytochrome *c* oxidase phosphopeptides,
    206–208
malic enzyme flux analysis, 440
mitochondrial proteins labeled with
    phosphate-32 from gels, 71–72
palmitoylated mitochondrial proteins labeled
    with azido-palmitate analogs
    database searching, 159–160
    gel electrophoresis, digestion, and peptide
        extraction, 159
phosphopeptides from mitochondria
    electrospray ionization, 93
    matrix-assisted laser
        desorption ionization, 92–93
    tandem mass spectrometry, 93–94
Src tyrosine kinase substrate identification
    immobilized metal affinity chromatography
        of phosphopeptides, 127–128
    immunoaffinity chromatography of
        phosphopeptides, 125–127
    multiple reaction monitoring, 130–131
    overview, 125–126
    precursor ion scanning, 129–130
    tandem mass spectrometry, 128–129
Mdm38, *see* Potassium/proton exchanger,
    mitochondria
MEK1/2, mitochondrial kinase complex in
    steroidogenesis
    function, 172
    subcellular fractionation, 174–175
    Western blot analysis, 175–176
Membrane potential
beta cell bioenergetics, 407, 421–422
confocal microscopy imaging of
    mitochondrial membrane potential
    beta cell imaging, 472–474
    image analysis, 474–475
    overview, 471–472
Mdm38 null mutant studies, 312–313
Methyl 12-azidododecanoate, synthesis, 154
Methyl 12-hydroxydodecanoate, synthesis, 153
Methyl 12-hydroxytetradecanoate, synthesis,
    145–155
Methylmalonyl semialdehyde dehydrogenase,
    palmitoylation, 151
Methyl 12-methanesulfonyloxydodecanoate,
    synthesis, 154
Mitochondria
ATP production, *see* ATP production,
    mitochondria
axonal transport, *see* Axonal transport,
    mitochondria
diabetes studies, *see* Beta cell; Magnetic
    resonance spectroscopy; Malic enzyme;
    Nicotinamide nucleotide
    transhydrogenase; Uncoupling protein 2

Mitochondria (*cont.*)
  evolution and diversity, 23–25, 50
  fusion/fission dynamics monitoring with
    photoactivateable green
    florescent protein
    artifacts and controls
      photobleaching and saturation, 301–302
      phototoxicity, 301
      thresholding, 302
    co-labeling with mitochondria-specific
      fluorescent dyes, 294–295
    confocal microscopy, 296
    mitochondrial matrix targeting, 292–293
    overview of dynamics, 290
    photoconversion, 292, 295–296
    polyethylene glycol fusion assay, 302–303
    principles, 291
    viral vectors for delivery, 293
    whole-cell assay
      data analysis, 299
      dilution monitoring, 299–300
      overview, 296–297
      photoconversion of mitochondrial
        subpopulations, 297–298
      spread monitoring, 300
      time-lapse confocal imaging, 298
  PINK1 function, *see* PINK1
  potassium/proton exchange, *see* Potassium/
    proton exchanger, mitochondria
  proteome, *see* Proteome, mitochondria
  respiration, *see* Respiration, mitochondria
  shape analysis, *see* ImageJ
Mitochondrial DNA-exchanged cybrids
  cell fusion between $\rho0$ cells and cytoplasts
    cell suspension preparation, 342
    fusion, 342
    selection, 342–343
  donor cells
    enucleation, 340–341
    preparation, 340
  isolation scheme, 337
  prospects for study, 343–344
  $\rho0$ cell establishment from mice
    colony picking, 338
    confirmation of mitochondrial
      DNA loss, 338–340
    ditercalinium treatment, 338
    overview, 337–338
  rationale for generation, 336
Mitogen-activated protein kinase cascade,
  see MEK1/2
MitoP2
  applications, 5
  candidate gene identification, 15–16
  creation rationale, 4
  disease protein searches, 13–14
  dual-targeted proteins, 16–17
  functional category searches, 12–13

  mitochondrial protein localization with
    in silico predictions, 12
  output lists, 14–15
  proteome data sources, 6–8
  proteome, transcriptome, and
    sublocalization experiments
    human, 10–12
    mouse, 10–12
    yeast, 9–10
  reference set, 8
  search mask, 6
  support vector machine prediction score, 8
MITOPRED, mitochondrial protein prediction
  in large datasets, 30, 32–33
MitoProt, mitochondrial protein
  prediction with amino-terminal
  targeting sequence, 33–34
Mitosome
  evolution and diversity, 24–25
  protein targeting and import, 27
MMSDH, *see* Methylmalonyl semialdehyde
  dehydrogenase
MRS, *see* Magnetic resonance spectroscopy
MS, *see* Mass spectrometry
Multiple reaction monitoring, *see* Mass
  spectrometry

N

Nicotinamide nucleotide transhydrogenase
  assays
    general considerations, 477
    microassay, 475–477
  N-ethyl-*N*-nitrosourea mutagenesis screens
    denaturing high-performance liquid
      chromatography, 453–454
    high-resolution DNA melting analysis,
      454–456
    overview, 453
    rederivation of live mice, 456
  hydrogen peroxide assay in mitochondria,
    468–471
  insulin secretion role, 452–453
  islets of Langerhans
    calcium flux measurement, 467–468
    insulin secretion assay, 463
    isolation from mouse pancreas, 460–463
    mitochondrial membrane potential imaging
      with confocal microscopy
      beta cell imaging, 472–474
      image analysis, 474–475
      overview, 471–472
  RNA interference
    confirmation of knockdown, 465–467
    insulinoma cell lines, 464
    knockdown, 464–465
  transgenic mouse overexpression
    bacterial artificial chromosome clone
      identification

DNA extraction, 458
  full-length clones, 458
  hybridization, 457–458
  probe generation, 457
  bacterial artificial chromosome
    transgenesis, 458–460
NMR, see Nuclear magnetic resonance
NNT, see Nicotinamide nucleotide
  transhydrogenase
Nuclear magnetic resonance, carbon-13
  glutamate isoprotomer distribution studies
    of anaplerotic pathway rates, 444–446

O

Oxygraph, see Respiration, mitochondria

P

PAGFP, see Photoactivateable green florescent
  protein
Palmitoylated mitochondrial proteins
  catalysis, 151
  cysteine residue thioester bond analysis
    alkali treatment of nylon membrane, 162
    alkylation with N-ethylmaleimide, 161
    neutral hydroxylamine treatment of gels,
      161–162
    principles, 160–161
    treatment of proteins in solution, 162
  detection with azido-palmitate analog
    anion-exchange chromatography of soluble
      proteins, 158
    applications, 162–163
    azido-fatty acid preparation, 155
    14-azidotetradecanoyl-CoA derivatives
      production, 156
    protein labeling in vitro, 156–159
    mass spectrometry of labeled proteins
      database searching, 159–160
      gel electrophoresis, digestion, and
        peptide extraction, 159
    mitochondria isolation, 157
    principles, 152–153
    synthesis
      12-azidododecanoic acid, 154
      4-azidotetradecanoic acid, 155
      methyl 12-azidododecanoate, 154
      methyl 12-hydroxydodecanoate, 153
      methyl 12-hydroxytetradecanoate,
        145–155
      methyl 12-
        methanesulfonyloxydodecanoate,
        154
      tagged phosphine probes, 155
  enzyme targets, 151–152
  functional overview, 150–151
PA-SUB, mitochondrial protein prediction, 38
PCR, see Polymerase chain reaction

PGC-1a, see Proliferator-activated receptor-γ
  cofactor-1α control of uncoupling protein
  2 expression, 402–403
Phosphorylated mitochondrial proteins, see also
  Acot2; Cytochrome c oxidase; Dynamin-
  related protein 1; PINK1
  bioinformatic analysis of tyrosine
    phosphorylation functional effects,
    131–134
  carnitine palmitoyltransferase-I, see Carnitine
    palmitoyltransferase-I
  dynamics and cell function regulation, 257
  phosphatases, see Dual specific phosphatase 18;
    Dual specific phosphatase 21; Protein
    phosphatase 2Cm
  phosphopeptide enrichment for protein
    identification
    immobilized metal affinity chromatography
      alkaline-induced β-elimination and
        Michael addition for increased
        sensitivity, 89
      media preparation, 84–85
      metal oxide single-tube and microtip
        techniques, 85–88
      methyl esterification for increased
        selectivity, 88–89
      spin columns, 85
    immunoaffinity enrichment with
      phosphotyrosine antibody, 89–90
    mass spectrometry
      electrospray ionization, 93
      matrix-assisted laser desorption
        ionization, 92–93
      tandem mass spectrometry, 93–94
    overview and rationale, 82–84
    quantitative proteomics, 894
    two-dimensional liquid chromatography,
      90–92
  phosphorous-32 labeling studies
    blue native gel electrophoresis, 77–78
    chase experiments with cold phosphate, 74
    dose and time course for labeling,
      72–73
    incubation conditions for radiolabeling, 68
    mitochondrial isolation, 66–67
    phosphate loading, 67–68
    principles, 65–66
    purification of labeled proteins, 76
    pyruvate dehydrogenase phosphorylation
      suppression, 75–76
    rationale, 65
    two-dimensional gel electrophoresis
      assignments, 73
      image acquisition and analysis, 71
        mass spectrometry analysis of spots,
        71–72
    sample preparation, 68–71
    specificity of phosphorylations, 64–65

Phosphorylated mitochondrial proteins, *see also*
    Acot2; Cytochrome *c* oxidase; Dynamin-
    related protein 1; PINK1 (*cont.*)
steroidogenic mitochondrial kinase complex,
    see Steroidogenesis
tyrosine kinases
    Fgr
        assay, 121–122
        purification, 119–121
    Lyn
        assay, 121–122
        purification, 119–121
    protein substrates, 118
    Src tyrosine kinase modification
        extraction and chromatography, 122–123
        kinase incubation and tyrosine
            phosphorylation, 123–124
        mitochondria isolation from rat brain, 122
    substrate identification with mass
        spectrometry
        immobilized metal affinity
            chromatography of
            phosphopeptides, 127–128
        immunoaffinity chromatography of
            phosphopeptides, 125–127
        multiple reaction monitoring, 130–131
        overview, 125–126
        precursor ion scanning, 129–130
        tandem mass spectrometry, 128–129
    types, 118
Photoactivateable green florescent protein,
    mitochondria fusion/fission dynamics
    monitoring
    artifacts and controls
        photobleaching and saturation, 301–302
        phototoxicity, 301
        thresholding, 302
    colabeling with mitochondria-specific
        fluorescent dyes, 294–295
    confocal microscopy, 296
    mitochondrial matrix targeting, 292–293
    overview of dynamics, 290
    photoconversion, 292, 295–296
    polyethylene glycol fusion assay, 302–303
    principles, 291
    viral vectors for delivery, 293
    whole-cell assay
        data analysis, 299
        dilution monitoring, 299–300
        overview, 296–297
        photoconversion of mitochondrial
            subpopulations, 297–298
        spread monitoring, 300
        time-lapse confocal imaging, 298
PINK1
    functional genomics
        models, 213
        overview, 212–213

gene expression analysis
    cell line expression studies
        brown adipose tissue, 217–218
        muscle primary myotubules,
            216–217
    DNA microarray analysis, 223–224
    quantitative real-time polymerase chain
        reaction, 224
    RNA isolation, 222–223
    skeletal muscle expression studies
        in humans
        exercise effects, 214–215
        inactivity effects, 215
        mitochondrial dysfunction in disease
            effects, 215–216
    natural antisense analysis, 225
    RNA interference studies
        apoptosis effects, 220–222
        mechanisms of small interfering
            RNAs, 218
        off-target effects, 219–220
    splice variant analysis and verification,
        225–226
PKA, *see* Protein kinase A
Polyacrylamide gel electrophoresis,
    see Blue native gel electrophoresis;
        Two-dimensional gel electrophoresis
Polymerase chain reaction
    malic enzyme expression analysis with real-
        time polymerase chain
        reaction, 430–432
    mitochondrial DNA quantification, 364
    mitochondrial respiration assay, 400
    PINK1 quantitative real-time polymerase
        chain reaction, 224
    uncoupling protein 2 expression analysis with
        real-time polymerase chain reaction
        amplification reaction, 399–400
        complementary DNA preparation, 399
Potassium/proton exchanger, mitochondria
    assays
        fluorescent dyes entrapped in
            submitochondrial particles, 311
        membrane potential, 312–313
        mitophagy assay with Rosella, 313–316
        potassium acetate-induced swelling,
            310–311
    Mdm38 null mutant studies
        mitochondria morphology analysis,
            308–310
        mutant generation, 307
PP2Cm, *see* Protein phosphatase 2Cm
PProwler, mitochondrial protein prediction with
    amino-terminal targeting sequence, 35
Predotar, mitochondrial protein prediction in
    large datasets, 32
Progesterone, cell-free assay of production,
    179–180

Proliferator-activated receptor-γ cofactor-1α,
    control of expression, 402–403
Protein kinase A
    dynamin-related protein 1 phosphorylation
        in vitro, 242
    mitochondrial kinase complex
        regulation, 172
    RNA interference studies, 173
Protein phosphatase 2Cm
    activity assays of recombinant enzyme
        colorimetric assay, 264–265
        radioactive assay, 264
    adenovirus-mediated gene delivery, 266–267
    deficiency effect assays
        cell viability, 267–268, 270–271
        mitochondrial permeability transition
            pore regulation, 270
        oxidative phosphorylation and
            respiration, 268–269
    gene cloning, 258, 260
    mitochondria localization
        immunofluorescence microscopy, 260–261
        matrix localization, 262
        osmotic shock assay, 262–263
        solubility assay, 262
        subcellular fractionation, 261–262
        targeting prediction, 258–259
    phosphatases
        database mining in mitochondria, 258
        family members, 256–257
    RNA interference, 265–266
    site-directed mutagenesis, 265
Proteome, mitochondria
    amino-terminal targeting sequence
            prediction programs
        MitoProt, 33–34
        PProwler, 35
        TargetP, 34–35
    capillary isotachophoresis separations, *see*
            Capillary isotachophoresis
    database, *see* MitoP2
    meta-analysis with YimLOC, 39–40
    mitochondrial localization
        computational prediction
            comparison of programs, 40–43
            data preparation, 28–29
            overview, 27–28
            PA-SUB, 38
            PSORT, 37–38
            SherLoc, 38–39
        targeting and import, 26–27
        overview of studies, 25–26
    posttranslational modifications,
        see Acetylated mitochondrial proteins;
            Carnitine palmitoyltransferase-I;
            Palmitoylated mitochondrial proteins;
            Phosphorylated mitochondrial proteins

    protein prediction in large datasets
        CBOrg, –31
        MITOPRED, 30, 32–33
        Predotar, 32
    sequence feature-based prediction of
            mitochondrial localization
        CELLO, 36
        overview, 35–36
    two-dimensional polyacrylamide gel
            electrophoresis, 51
PSORT, mitochondrial protein prediction,
        37–38
Pyruvate dehydrogenase, phosphorylation
    suppression in labeling studies, 75–76

R

Respiration, mitochondria
    assay for uncoupling protein 2 studies, 400
    enzyme markers, 352–353
    mitochondria isolation from skeletal
            muscle for assay
        differential centrifugation, 357–358
        homogenization, 355–357
        overview, 354–355
        yield and integrity assessment, 358–360
    oxidative phosphorylation
        overview, 350–352
        oxygraph assay, 365–369
RNA interference
    dynamin-related protein 1, 235–236
    malic enzyme in insulinoma cell line
        incubation conditions, 434–435
        small interfering RNA design, 432–433
        transfection optimization, 434
    nicotinamide nucleotide transhydrogenase
        confirmation of knockdown, 465–467
        insulinoma cell lines, 464
        knockdown, 464–465
    PINK1 studies
        apoptosis effects, 220–222
        mechanisms of small interfering RNAs, 218
        off-target effects, 219–220
    protein kinase A knockdown and
            mitochondrial kinase complex
            regulation, 173
    protein phosphatase 2Cm, 265–266
    SIRT3 deacetylation assay, 143
    uncoupling protein 2 from INS-1E cells,
            415–416, 419–420
Rosella
    construct, 313–314
    mitochondria targeting, 314
    mitophagy assay
        fluorescence microscopy, 315–316
        transformation of yeast, 314–315
        yeast culture, 315

S

SherLoc, mitochondrial protein
    prediction, 38–39
SILAC, *see* Stable-isotope labeling by
    amino acids in cell culture
SIRT3
    deacetylation assays
        assay *in vivo*
            cell culture, transfection, and RNA
                interference, 143
            immunoprecipitation and
                Western blot, 143–144
            plasmids and RNA interference
                oligonucleotides, 143
        peptide substrate assay *in vitro*, 141
        protein substrate assay *in vitro*
            incubation conditions, 142
            substrate preparation, 141–142
    purification
        bacterial expression systems, 140–141
        mammalian cell culture, 139–140
Site-directed mutagenesis
    dual specific phosphatase 18, 277–278
    dynamin-related protein 1 phosphorylation
        site, 234, 236
    protein phosphatase 2Cm, 265
    steroidogenic acute regulatory protein
        phosphorylation site, 182–183
Src tyrosine kinases, *see* Fgr; Lyn
Stable-isotope labeling by amino acids in cell
    culture, mitochondrial quantitative
    phosphoproteomics, 94
StAR protein, *see* Steroidogenic
    acute regulatory protein
Steroidogenesis
    Acot2
        acyl-CoA thioesterase activity assay, 188
        functional overview, 186–187
        phosphoprotein characterization, 188–190
        phosphorylation sites, 187
        purification from rat adrenal glands, 187–188
    cell lines for mitochondria
        studies, 171–172
    mitochondria abundance in
        steroidogenic cells, 170–171
    mitochondrial kinase complex
        extracellular signal-regulated kinase
            confocal microscopy, 176–177
            expression constructs, 173–174
            subcellular fractionation, 174–175
            transient transfection, 174Western blot
                analysis, 175–177
        function, 172–173
        MEK1/2
            function, 172
            subcellular fractionation, 174–175
            Western blot analysis, 175–176

progesterone production in
    cell-free assay, 179–180
protein kinase A
    regulation, 172
    RNA interference studies, 173
steroidogenic acute regulatory protein
    mutagenesis of phosphorylation site,
        182–183
    phosphopeptide analysis, 184–185
    phosphorylation *in vitro* by ERK1,
        183–184
    pull-down assay with kinase complex,
        181–182
    sequence analysis, 181
    serine-232 phosphorylation effects on
        function, 185–186
    target of kinases, 171
    transfection, 178
phosphorylative regulation, 171
Steroidogenic acute regulatory protein,
    mitochondrial kinase complex
    phosphorylation
    mutagenesis of phosphorylation site, 182–183
    overview, 171
    phosphopeptide analysis, 184–185
    phosphorylation *in vitro* by ERK1, 183–184
    pull-down assay with kinase
        complex, 181–182
    sequence analysis, 181
    serine-232 phosphorylation effects on
        function, 185–186
    steroidogenic acute regulatory protein
        phosphorylation *in vitro* by ERK1,
        183–184
    transfection, 178
Support vector machine
    mitochondrial protein localization
        prediction, 27–28
    prediction score in MitoP2, 8
SVM, *see* Support vector machine

T

Tandem mass spectrometry,
    see Mass spectrometry
TargetP, mitochondrial protein prediction with
    amino-terminal targeting sequence, 34–35
tcacalc, metabolic flux calculation, 443–446
TIM, *see* Translocase of inner membrane
Translocase of inner membrane,
    mitochondrial protein import, 26
Tricarboxylic acid cycle, metabolic
    flux modeling with magnetic
    resonance spectroscopy
    isotope mass balance equations, 385
    muscle metabolite concentration, 386
    overview, 384
Two-dimensional polyacrylamide gel
    electrophoresis

mitochondrial proteins labeled with
    phosphate-32
  assignments, 73
  image acquisition and analysis, 71
  mass spectrometry analysis of spots, 71–72
  sample preparation, 68–71
  mitochondrial proteomics, 51
Tyrosine kinases, *see* Fgr; Lyn; Phosphorylated
  mitochondrial proteins

U

UCP2, *see* Uncoupling protein 2
Uncoupling protein 2
  antisense oligonucleotide knockdown, 401
  beta cells
    function, 406
    INS-1E cell line model
      RNA interference, 415–416
      Western blot from mitochondria and
        whole-cell lysates, 413–415
    knockdown effects on coupling
      efficiency in attached
      cells, 419–420
  cold exposure-induced expression,
    400–401
  expression analysis with real-time
    polymerase chain reaction
    amplification reaction, 399–400
    complementary DNA preparation, 399
  functional overview, 396
  insulin secretion relationship, 396–397
  mitochondrial respiration assay, 400

proliferator-activated receptor-γ cofactor-1α
  control of expression, 402–403
  tissue distribution, 396
  Western blot
    gel electrophoresis and blotting, 398–399
    mitochondria isolation, 398
    pancreatic islet isolation, 397–398

W

Western blot
  carnitine palmitoyltransferase-I
    posttranslational modifications, 103,
    106–107
  cytochrome *c* oxidase phosphorylation analysis,
    205–206
  dynamin-related protein 1, 239
  mitochondrial kinase complex in
    steroidogenesis
      extracellular signal-regulated kinase,
      175–177
      MEK1/2, 175–176
  SIRT3 deacetylation assay, 143–144
  uncoupling protein 2
    gel electrophoresis and blotting, 398–399
    INS-1E cell protein, 413–415
    mitochondria isolation, 398
    pancreatic islet isolation,
    397–398

Y

YimLOC, mitochondrial protein
  prediction with meta-analysis, 39–40

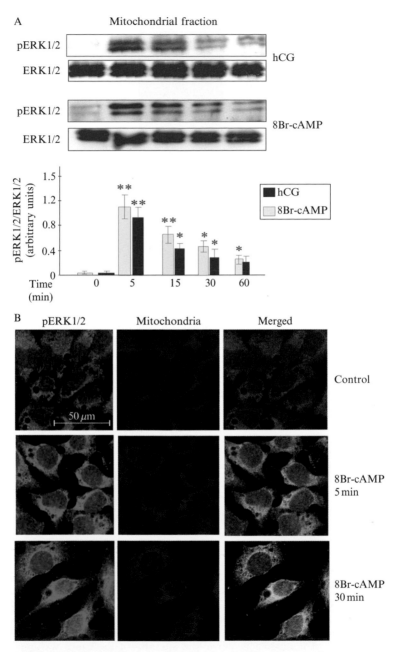

Cristina Paz *et al*., Figure 10.1   Subcellular distribution of hCG/8Br-cAMP-stimulated pERK1/2 activation. MA-10 cells were incubated with or without 20 ng/ml of hCG or 0.5 m*M* 8Br-cAMP (A) for the indicated times. Next, subcellular fractions were obtained and 40 µg of the total protein of each fraction were subjected to SDS–PAGE and Western

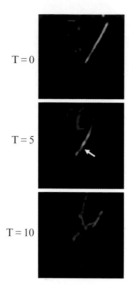

Anthony J. A. Molina and Orian S. Shirihai, Figure 16.1 Mitochondrial fusion results in the transfer of PAGFPmt. At time zero, a photolabeled PAGFPmt positive mitochondrion is depicted as an isolated unit within an INS1 cell. Five min later, this mitochondrion has moved and undergone fusion with an adjacent mitochondrion. The arrow points to the presumptive fusion site. PAGFPmt has been transferred to another mitochondrion which now appears yellow due to the mixing of the green and red fluorophore. At $T = 10$ min, more fusion events have occurred and the average PAGFPmt fluorescence intensity has gone down due to fusion dilution photoconverted molecules.

---

blot to detect phospho-ERK1/2 (indicated with pERK1/2). After stripping, total ERK1/2 (indicated with ERK1/2) was detected in the same membrane. The Western blots show the results of a representative experiment performed three times. Graphic bar below shows the quantification of immunoblots above; the intensity of the bands was quantitated using total ERK1/2 as loading control. Bars denote relative levels of pERK1/2 presence in arbitrary units. $\star\star p = 0.01$, $\star p = 0.05$ vs. time 0. (B) Immunofluorescent staining for pERK1/2 (green) and mitochondria (red) in MA-10 cells after treatment with or without 0.5 mM of 8Br-cAMP for the indicated times. Merged images are shown in the right panel.

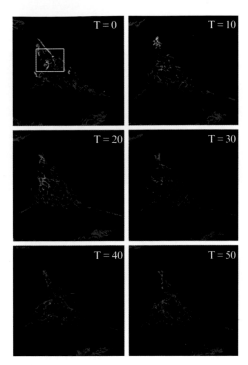

Anthony J. A. Molina and Orian S. Shirihai, Figure 16.2 Monitoring mitochondrial fusion within the entire population. At time zero, the white box designates the area targeted with 2-photon laser for photoconversion of PAGFPmt to an active form. Notice that by the time the imaging acquisition has started, noticeable movement of mitochondria has already occurred. Subsequent images are obtained every 10 min for 50 min. Images are a single slice through the middle of an INS1 cell. The PAGFPmt signal is diluted over time leading to the appearance of yellow mitochondria. The yellow mitochondria appear more red over time due to further dilution of the green fluorophore.

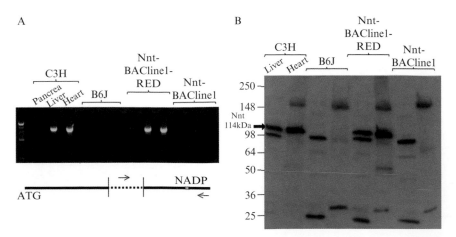

Kenju Shimomura *et al.*, Figure 25.3 Expression of NNT in mice transgenic for the *Nnt* BAC transgene. (A) RT-PCR analysis of *Nnt* mRNA from C3H, C57Bl/6J, a transgenic line rederived from the original rescue line, and the original rescue line. One of the RT primers is located within the deleted region of the transcript (exons 7–11)

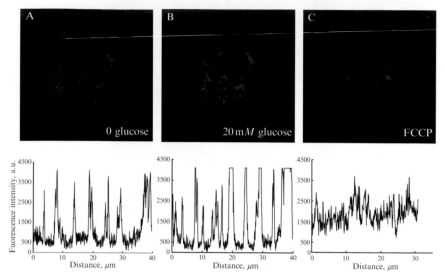

Kenju Shimomura *et al.*, Figure 25.9 The visualization of the mitochondrial potential in MIN6 cells. Cells were loaded with TMRM at 50 nM concentration. Images were obtained on Zeiss confocal microscope LSM510 META. *Column A*: top panel— a confocal image of a loaded Min6 cell at 0 glucose; mitochondria are seen as elongated red dots; bottom panel—the intensity profile inside the image of the cell; sharp peaks represent mitochondria; the signal initiated in the cytosol is seen as low-level line between peaks. *Column B*: a confocal image of a loaded MIN6 cell at 20 mM glucose; the intensity of signal initiated in mitochondria is increased in comparison to image in column A; the cytosolic signal has decreased; this indicates the hyperpolarization of the mitochondrial membrane. *Column C*: a confocal image of a loaded MIN6 cell after application of FCCP; the peaks representing mitochondria are essentially diminished, the overall level of the signal is increased; this indicates depolarization of the mitochondrial membrane. The dark region in all three confocal images is the cell nucleus.

and the other at the 3′ end including the NADP motif. These primers amplify a 1.3-kb product from C3H but not in Bl/6J. The BAC rescue line re-derived from frozen embryos stored shortly after the line was created (Nnt-BACline1-RED) also expresses *Nnt* at level similar to C3H. In contrast, the original line (Nnt-BACline1) that had been maintained by breeding over several years has lost *Nnt* expression, possibly due to genome rearrangement. (B) Proteins were separated by SDS-PAGE and the proteins transferred to a membrane for immunoblotting with a rabbit anti-NNT polypeptide antibody. C3H and "Nnt-BACline1-RED" mice express NNT as expected, whereas Bl/6J and "Nnt-BACline1" mice do not express NNT. Note that the polyclonal antibody detects some nonspecific bands in both C3H (intact) and Bl/6J (deleted) tissues.